Basic Research and Clinical Aspects of Adamantinomatous Craniopharyngioma

Juan Pedro Martinez-Barbera
Cynthia Lilian Andoniadou
Editors

Basic Research and Clinical Aspects of Adamantinomatous Craniopharyngioma

Springer

Editors
Juan Pedro Martinez-Barbera
Developmental Biology and Cancer Programme
Birth Defects Research Centre
Institute of Child Health
University College London
London, UK

Cynthia Lilian Andoniadou
Craniofacial Development and Stem Cell Biology
King's College London
London, UK

ISBN 978-3-319-51888-6 ISBN 978-3-319-51890-9 (eBook)
DOI 10.1007/978-3-319-51890-9

Library of Congress Control Number: 2017933569

Printed on acid-free paper

This Springer imprint is published by Springer Nature
The registered company is Springer International Publishing AG
The registered company address is: Gewerbestrasse 11, 6330 Cham, Switzerland

Preface

This edited book, *Basic Research and Clinical Aspects of Adamantinomatous Craniopharyngioma*, aims to present an up-to-date picture of the current efforts for the effective treatment and management of the paediatric tumour adamantinomatous craniopharyngioma (ACP) and how patient management can be tailored in the future.

The book was triggered by a lack of comprehensive publications bringing together a global view of all relevant aspects of these tumours. Here we bring a multidisciplinary group of international experts that present the current views on ACP diagnosis and management, as well as recent research generating mouse models for these tumours and yielding findings that have the potential to impact on future patient treatments.

The first two chapters of this book focus on the pathogenesis of human ACP and the molecular analyses that have been carried out in the tumours to date. The next chapter provides an overview of the available models for human ACP and the insights gained from animal research. Chapters focus on the diagnosis and management of ACP patients as well as the clinical sequelae associated with the disease. Finally, the last chapter provides an overview of the current clinical trials and the long-term management of ACP patients.

The intended audience of this book will consist of clinical students and practitioners as well as basic researchers studying pituitary tumourigenesis.

We would like to convey our appreciation to all authors who have contributed chapters and to the team at Springer for their expert support throughout this process.

London, UK

Juan Pedro Martinez-Barbera
Cynthia Lilian Andoniadou

Contents

Contents

About the Editors

Dr. Cynthia Lilian Andoniadou is a stem cell researcher and lecturer at the Department of Craniofacial Development and Stem Cell Biology of King's College London focusing on the biology of normal and pathogenic pituitary stem cells.

Professor Juan Pedro Martinez-Barbera is a leading expert in pituitary development and associated congenital disorders. He is a researcher at the Birth Defects Research Centre, Developmental Biology and Cancer Programme of the UCL Great Ormond Street Institute of Child Health.

Contributors

Cynthia Lilian Andoniadou Craniofacial Development and Stem Cell Biology, King's College London, London, UK

Kristian Aquilina Department of Neurosurgery, Great Ormond Street Hospital for Children, London, UK

Ute Bartels Department of Oncology, The Hospital for Sick Children, Toronto, ON, Canada

Julie A. Bradley University of Florida Health Proton Therapy Institute, Jacksonville, FL, USA

Michael Buchfelder Department of Neurosurgery, University Hospital Erlangen, Erlangen, Germany

Mehul T. Dattani Genetics and Genomic Medicine Programme, UCL Great Ormond Street Institute of Child Health, London, UK

The London Centre for Paediatric Endocrinology and Diabetes, Great Ormond Street Hospital for Children NHS Foundation Trust, London, UK

Eva Marie Erfurth Department of Endocrinology, Skånes University Hospital, Lund, Sweden

Department of Clinical Sciences, Medical Faculty of Lund University, Lund, Sweden

Hoong-Wei Gan Genetics and Genomic Medicine Programme, UCL Great Ormond Street Institute of Child Health, London, UK

The London Centre for Paediatric Endocrinology and Diabetes, Great Ormond Street Hospital for Children NHS Foundation Trust, London, UK

Todd C. Hankinson Children's Hospital Colorado, Pediatric Neurosurgery, Aurora, CO, USA

Annett Hölsken University Hospital, Friedrich—Alexander University Erlangen-Nürnberg (FAU), Erlangen, Germany

Daniel J. Indelicato University of Florida Health Proton Therapy Institute, Jacksonville, FL, USA

Babita Jyoti University of Florida Health Proton Therapy Institute, Jacksonville, FL, USA

Niki Karavitaki Institute of Metabolism and Systems Research (IMSR), College of Medical and Dental Sciences, University of Birmingham and Centre for Endocrinology, Diabetes and Metabolism, Birmingham Health Partners, Birmingham, UK

John-Paul Kilday Department of Oncology, Royal Manchester Children's Hospital, Manchester, UK

Juan Pedro Martinez-Barbera Developmental Biology and Cancer Programme, Birth Defects Research Centre, Institute of Child Health, University College London, London, UK

Hermann L. Müller Department of Pediatrics and Pediatric Hematology and Oncology, Klinikum Oldenburg AöR, Medical Campus University Oldenburg, Oldenburg, Germany

Leslie Robinson Children's Hospital Colorado, Pediatric Neurosurgery, Aurora, CO, USA

Ronny L. Rotondo University of Florida Health Proton Therapy Institute, Jacksonville, FL, USA

Sandro Santagata Brigham and Women's Hospital, Department of Pathology, Harvard Institute of Medicine, Boston, MA, USA

Helen A. Spoudeas The London Centre for Paediatric Endocrinology and Diabetes, Great Ormond Street Hospital for Children NHS Foundation Trust, London, UK

Ashley J. Vardon Institute of Metabolism and Systems Research (IMSR), College of Medical and Dental Sciences, University of Birmingham and Centre for Endocrinology, Diabetes and Metabolism, Birmingham Health Partners, Birmingham, UK

Pathogenesis of Human ACP

Annett Hölsken

Abstract In 2002 J.T. Rutka stated in an editorial commentary that: "There is perhaps no other primary brain tumour that evokes more passion, emotion, and, as a result, controversy than does the craniopharyngioma". This statement clarifies the situation of medical specialists such as endocrinologists, neurosurgeons, neuropathologists, neurooncologists, paediatric oncologists and neuro-radiologists involved in treatment of craniopharyngioma (CP) patients. Although CPs are classified by the WHO as benign tumours, their size and relationship to multiple critical structures can incur massive clinical complications. Here we elucidate the histological and genomic hallmarks, setting adamantinomatous (ACP) apart from papillary craniopharyngiomas (PCP). In addition, molecular mechanisms that we already know to be involved in the pathogenesis of ACP, e.g. Wnt, SHH and EGFR signalling influencing proliferation, morphology, migration and recurrence by the establishment of a presumably tumour stem cell niche, are discussed. Our understanding of the driving force of ACP tumourigenesis opens new avenues for targeted treatment options, e.g. the inhibition of the EGFR.

Keywords Wnt • EGFR • SHH • CD44 • CD133 • Cytokeratin • p21 • BMP4 • Ki-67 • β-catenin mutation

Definition and Historical Notes

As defined in the World Health Organisation (WHO) classification of Tumours of the Central Nervous System (CNS) "craniopharyngiomas (CPs) are benign, partly cystic epithelial tumours of the sellar region presumably derived from Rathke's pouch epithelium" [1, 2].

A. Hölsken (✉)
University Hospital, Friedrich—Alexander University Erlangen-Nürnberg (FAU), Erlangen, Germany
e-mail: annett.hoelsken@uk-erlangen.de

J.P. Martinez-Barbera, C.L. Andoniadou (eds.), *Basic Research and Clinical Aspects of Adamantinomatous Craniopharyngioma*,
DOI 10.1007/978-3-319-51890-9_1

The name craniopharyngioma was coined by Harvey Cushing (1869–1937) in 1932. Hitherto prevalent synonyms such as Rathke's pouch tumour, craniopharyngeal duct tumour, adamantinomatous tumour, pituitary adamantinoma or dysodontogenic epithelial tumour [3] had been used. The first comprehensive publication on CPs called "Hypophysengang-geschwülste" ("craniopharyngeal duct tumours") was published in 1904 by Jakob Erdheim (1874–1937), a pathologist at the Allgemeines Krankenhaus der Stadt Wien (Vienna General Hospital) [3–5]. In his honour, these tumours are also sometimes called Erdheim tumours. Nevertheless, prior to his publication several pathologists including J. Abercromie (1828), F.A. Zenker (1857) as well as F.W. Mott and J.O.W. Baratt (1899) had made similar observations, describing lesions near the pituitary gland thought to be craniopharyngiomas [5].

Variants

CPs are categorised into adamantinomatous (ACP) and papillary (PCP) subtypes by their different histological, genetic and clinical behaviour [2, 6]. Morphological and histological as well as genetic distinctions of CP subtypes will be discussed in more detail in a separate paragraph. ACP and PCP are completely different entities as they were shown to have a distinctive gene expression and methylation profile [7]. From an anatomical point of view, CPs can be classified according to the relationship of the tumour with the optic chiasm or the diaphragma sellae [8].

Epidemiology

The incidence of CPs varies between 0.5 and 2.5 cases in 1 million people per year irrespective of gender [2]. There is a bimodal distribution pattern of ACPs with peaks occurring during childhood between the ages of 5–15 years and later in adulthood between the ages of 45–60 years. PCPs appear exclusively in adulthood peaking between the ages of 40–55 years [2]. In paediatric patients nearly all CPs were from the adamantinomatous subtype. In adults ACP are more common than PCP [9]. Due to the fact the overall incidence of brain tumours in adults is much higher than in children, the majority of CPs occur in adults (1.2–4.6% of all intracranial tumours) [9–11]. Nevertheless, at a rate of 5–10%, ACP is the most common non-neuroepithelial intracranial tumour in children [2] and accounts for 56% of paediatric sellar and suprasellar tumours [11]. The rate of CPs in children was observed to be more frequent in Nigerian (18% of all CNS tumours) [2, 12] and Japanese children with an annual incidence of 5.25 cases per million in the paediatric population [2, 13].

Morphological and Histological Characteristics of CP Subtypes

Macroscopically, PCPs appear predominantly solid and well-circumscribed in contrast to ACPs which are often lobulated and cystic. The cysts are sometimes filled with a machine oil-like, thick brownish-yellow fluid that contains cholesterol and brittle debris.

Histologically ACPs are characterised by epithelial cells with picket-fence-like palisaded nuclei resting on a basal membrane. These cells surround a polymorphous mass of squamous epithelium. The internal epithelium reveals heterogeneous structures (Fig. 1), such as compact cells resembling whorl like arrays, loosely packed stellate cells, also called stellate reticulum, and nodules of wet keratin (Fig. 1). Regressive changes such as cholesterol deposits and calcifications are frequently noted in ACPs but rarely observed in PCP variants [2, 14]. PCPs consist of circumscribed predominantly solid and well-differentiated squamous epithelium forming pseudopapillae. The absence of nuclear palisading, wet keratin, calcification and cholesterol deposits serve as distinguishing features to ACP subtype. It is worth noting that subtype classification of some cases based solely on the histology is not clearly possible, which in the past may have led to the diagnosis of a mixed subtype. However, recent studies have shown that a clear distinction between ACP and PCP could be achieved by the subtype specific molecular appearance [7]. Histologically, an additional important characteristic attribute of ACPs are finger-like outgrowing tumour protrusions that invade into brain parenchyma (Fig. 2) [2, 3, 14, 15]. As a reaction of tumour infiltration the adjacent brain tissue is characterised by a dense fibrous network of astroglia that replaces the area of a previous degenerative lesion. In this process, also called reactive gliosis, some astrocytes form Rosenthal fibres (Fig. 1), which are modified processes of astrocytes exhibited as oval or elongated eosinophilic masses. The occurrence of Rosenthal fibres should not be misinterpreted as an astrocytic tumour.

ACPs exhibit striking morphological and histological similarities to odontogenic tumours of the jaw. Most analogies were found to calcifying odontogenic cysts featuring proliferating ameloblastic epithelium with calcifications, wet keratin, ghost cells and cyst formations [16–19]. ACPs can also resemble ameloblastoma or adenomatoid odontogenic tumours [17, 20]. Moreover, odontogenic epithelial differentiations of ACPs are demonstrated by the detection of enamel proteins as amelogenin, enamelysin and enamelin which are exclusively expressed in teeth. This observation underscores the assumption that the ACP variant arises from embryonic remnants of the stomodeum representing Rathke's pouch and teeth forming progenitor tissue [21, 22]. This is substantiated by the finding of a rudimentary tooth in an ACP of a 4-year-old boy [20]. It is worth noting that no similarity to odontogenic tumours could be established in PCPs [16, 17].

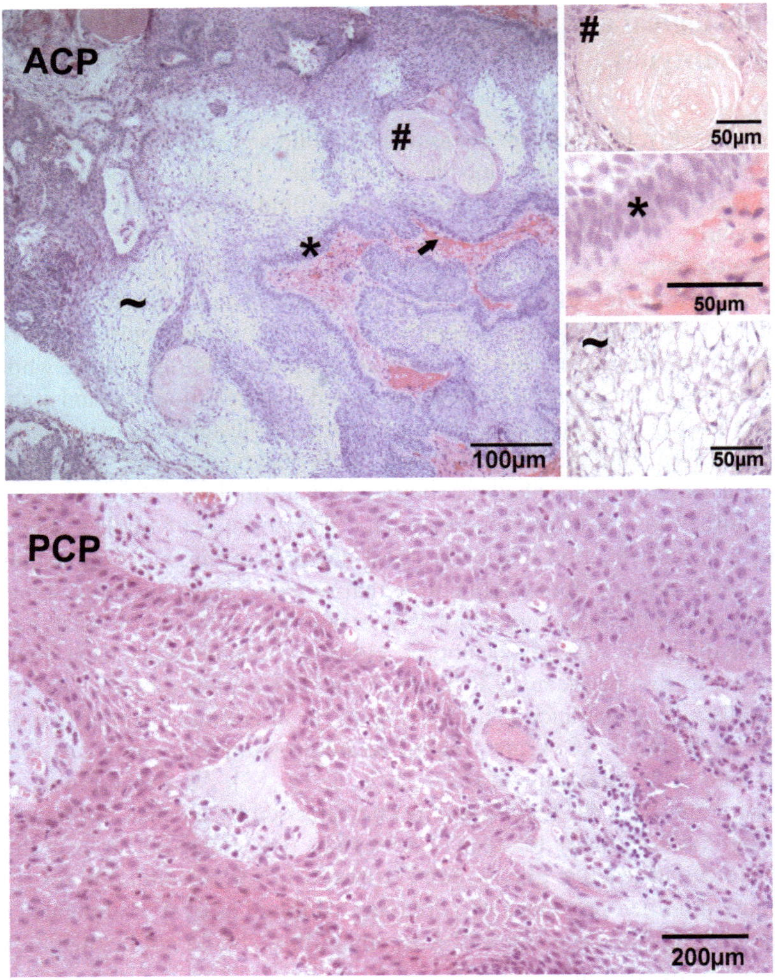

Fig. 1 Histological characteristics of CP subtypes HE stained section of ACP shows wet keratin (*hash symbol*), the palisaded cells (*asterisk*) circumscribing the tumour and the reticulo stellate cells (*tilde symbol*). These histological characteristics are magnified on the right side. Rosenthal fibres in the adjacent brain tissue are labelled by an *arrow*. PCPs consist of compact squamous epithelium

Genetic Distinctions of CP Subtypes

Genetic aberrations in PCP were first identified in 2014 by Brastianos et al. using exome sequencing [23]. The BRAF V600E mutation could be found in up to 100% of PCP analyses [7, 23, 24].

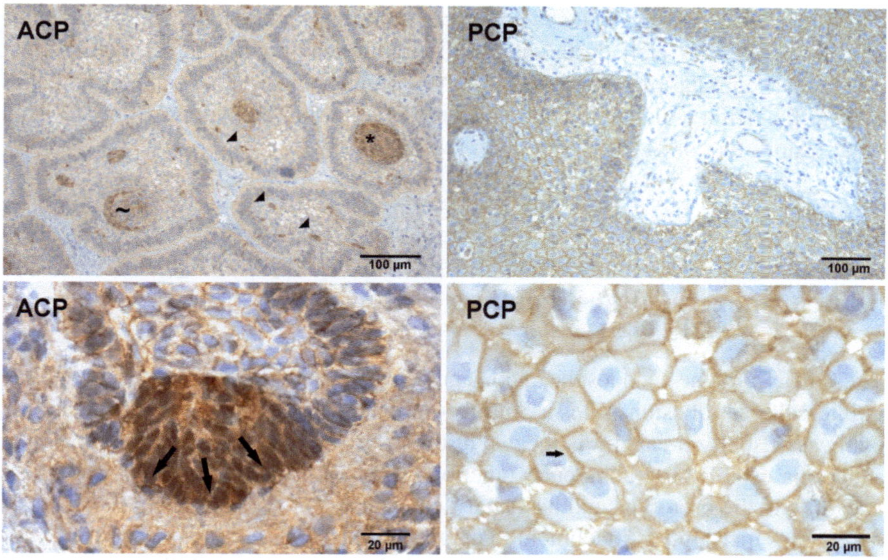

Fig. 2 Immunohistochemistry of β-catenin in CP. Nuclear β-catenin accumulation was found in most of the ACPs, in whorl like structures (*asterisk*) or in single cells (*filled triangle*). Note that not all cells are configured in whorls always show nuclear β-catenin enrichment (*tilde*). The lower figure illustrates the detection of β-catenin in whorl like structures forming a tip of outgrowing tumour cells (*arrows*). The nuclear accumulation is absent and β-catenin occurrence is restricted to the cell membrane (*up arrow*) in PCP

Sekine et al. first described activating mutations of the β-catenin encoding gene CTNNB1 in ACPs as early as 2002 [6]. However, until now, there was no CTNNB1 mutation described in PCP [6, 7, 23, 25].

Nearly all studies identified a distinctive occurrence of CTNNB1 mutations in ACP and BRAF mutations in PCP. However, one study showed the coexistence of both mutations in ACP [26]. It is worth noting that negative mutational screening is also sufficient to differentiate Rathke's Cleft cyst with epithelial metaplasia from PCP which sometimes bear a histologically striking resemblance [24, 27, 28].

In ACP it was assumed that CTNNB1 mutations lead to nuclear accumulation of β-catenin all over the tumour. Interestingly, only distinctive cells were identified in ACPs revealing β-catenin accumulations [6]. This was observed to occur in up to 94% of this CP subtype [25]. Neoplastic cells with nuclear accumulation of β-catenin seemed not to be equally distributed throughout the tumour (Fig. 2). They were prevalently identified either in the centre of epithelial whorls or adjacent to ghost cells. However, whorl like configured cells do not always consistently express nuclear β-catenin (Fig. 2). There are also single cells, dispersed throughout the tumour exhibiting nuclear β-catenin. In contrast, nuclear accumulations within palisaded cells are rare and sometimes seen in the outer epithelial margin bordering either the brain parenchyma or mesenchymal scarring (Fig. 2).

Further lesions of the sellar region such as pituitary adenomas, PCPs, arachnoidal cysts, Rathke's cleft cysts and xanthogranulomas do not show such distinct β-catenin accumulation as illustrated for ACP (Figs. 2, 5 and 7). This finding makes nuclear β-catenin a diagnostic hallmark of ACPs, setting them apart from PCP and other lesions that originate in the sellar region [25, 29].

β-catenin: A Multifunctional Protein and Key Component of the Canonical Wnt Signalling Pathway

The CTNNB1 gene encodes the 88 kDa protein β-catenin, which plays a crucial role in intercellular adhesion and is a central component of the canonical Wnt signalling pathway. Usually β-catenin is located at the cell membrane where it interacts with cadherins establishing adherence junctions (Fig. 3) [30]. Epithelial cells express E-cadherin, the prototype and best analysed cadherin of this transmembranous glycoprotein superfamily. Attachment to neighbouring cells is maintained by the extracellular domain of E-cadherin in a calcium dependent homophilic manner. It is anchored intracellularly to the cytoskeleton via β–catenin and α-catenin, which link the whole complex to actin (Fig. 3). Maintenance of cellular adhesion is important to keep cellular polarity, differentiation and tissue integrity [31]. However, distribution of β-catenin within the cell is adjustable and can vary rapidly from membranous anchorage to cytoplasmic occurrence impairing adherence junctions [32]. Dissociation of β-catenin from the adherence complex could be provoked by the epidermal growth factor receptor (EGFR) or src driven phosphorylation at Tyrosine of codon 654 (Y654) which reduces its binding affinity to E-cadherin [33, 34]. Fyn, Fer or c-Met promote phosphorylation of β-catenin tyrosine residue 142, which results in dissociation of α-catenin from β-catenin (Figs. 3 and 4) with concomitant loss of cadherin adhesion [33]. In the cytoplasm, β-catenin is tagged for proteosomal degradation, when Wnt signalling is inactive (Fig. 3) [35]. Its cytoplasmatic amount is thus strictly regulated by a multiprotein-complex containing the tumour suppressor adenomatous polyposis coli (APC), the scaffold proteins Axin and Axin2, the phosphokinases GSK-3β and casein kinase 1 (CK1). Following binding to this destruction-complex, β-catenin becomes phosphorylated within its N-terminal region encoded by exon 3 of the β-catenin gene (CTNNB1) (Figs. 3 and 4) [36–38]. The phosphorylation of Threonine and Serine residues in this region occurs in a well-regulated manner. Initially CK1 phosphorylates β-catenin at a Serine (S) encoded by codon 45 (S45) which provides the GSK-3β priming site [39]. Following GSK-3β binding Threonine 41 (T41) and S37, S33 and S29 are subsequently modified by phosphorylation (Fig. 4). The phosphorylation of S45 and T41 can also be maintained by I kappa B kinase alpha (IKK-α) [40] independent of CK1 priming (Fig. 4). Phosphorylation of S33 and S37 residues is essential for the recognition of β-catenin by the β-TrCP complex (Fig. 3) [41, 42]. Multiple ubiquitin molecules are added covalently to Lysine residues of β-catenin, which is recognised by the proteasome leading to its subsequent degradation [41].

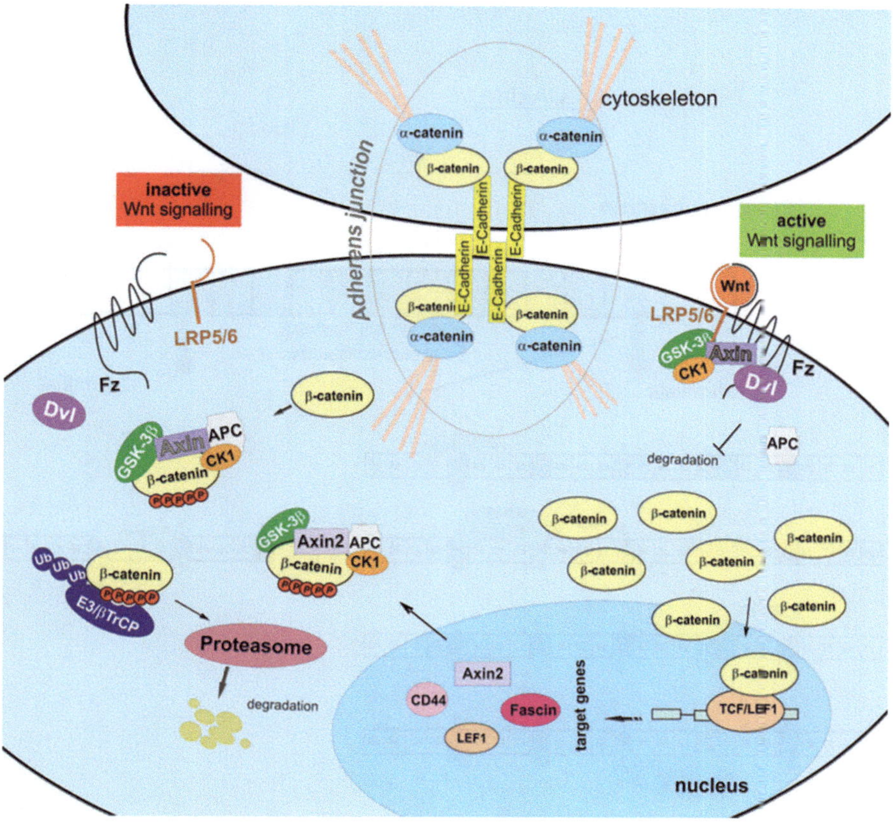

Fig. 3 Cellular β-catenin distribution and its regulation by Wnt signalling. The β-catenin protein is located at the cell membrane integrated in adherens junction complexes. The cytoplasmic pool is controlled by the destruction complex composed of APC, the scaffold proteins Axin/Axin2, as well as protein kinases CK1 and GSK-3β. The complex binds β-catenin and marks it by phosphorylation (*red circles*) for βTrCP driven ubiquitination leading to its subsequent proteosomal degradation. This process is inhibited if Wnt signalling is active. Wnt molecules bind to the frizzled (Fz) receptor and co-receptor LRP5/6. Dishevelled (Dvl) inhibits GSK-3β and therefore the designation of β-catenin proteosomal degradation. The protein accumulates within the cytoplasm and translocates to the nucleus. Here it possesses the function of a transcription co-factor by interacting with transcription factors of the TCF family [T-cell factor (TCF)/lymphocyte enhancer factor 1 (LEF1)] initiating target gene expression, e.g. Axin2, Fascin, CD44 and EGFR. Axin2 acts as a negative feedback regulator because it terminates Wnt signalling via acceleration of the β-catenin degradation

The activation of Wnt signalling (Fig. 3) leads via Dishevelled (Dvl) to membranous translocation of GSK3β, and the scaffold protein Axin [43], which in turn results in the decreased targeting of β-catenin for proteasomal degradation, accompanying with its cytoplasmic stabilisation and subsequent nuclear translocation [44]. Within the nucleus, β-catenin initiates gene expression via interaction with transcription factors of the LEF1/TCF family [31]. The target genes of β-catenin are promoters of cell proliferation, differentiation and also cell migration [32, 45].

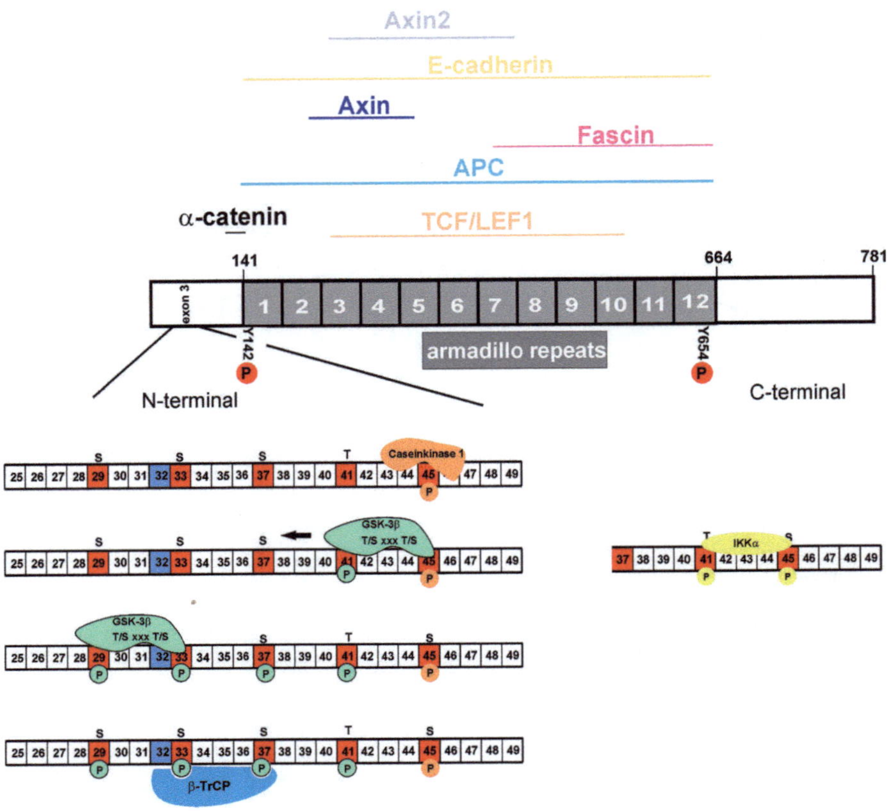

Fig. 4 β-catenin structure, its interaction sites with other proteins and its modifications. The central region of β-catenin consists of a 12 times repeated segment of 42 amino acids. This central part, also called armadillo repeats, is highly conserved and forms a positively charged groove which binds negatively charged proteins. A subset of proteins and their β-catenin interaction sites, which are depicted above, marked by a line. The functionality of β-catenin depends on its capacity to interact with different proteins, which is influenced by posttranslational modifications as phosphorylation (P). The N-terminal region in particular the proportion encoded by exon 3 is essential for β-catenin endurance. Protein kinases (casein kinase 1, IKKα and GSK-3β) phosphorylate Serine (S) and Threonine (T) residues establishing a β-TrCP binding site crucial for following ubiquitinylation targeting β-catenin for proteosomal degradation. Phosphorylation of Tyrosine (Y) codon 142 (Y142) and Y654 downregulates β-catenin interaction with α-catenin and E-cadherin, respectively. Modified from Xu, W. et al. (2007)

Mutations in exon 3 (CTNNB1) are not Sufficient to Induce Nuclear β-Catenin Accumulation in ACP Per se

Molecular genetic studies of various ACP demonstrated a broad range of missense mutations within CTNNB1 exon 3, compromising residues essential for proteasomal degradation labelling, e.g. codons 32, 33, 34, 35, 37, 41, 43 and 45 (Fig. 4) [6, 7, 25, 26, 46]. In addition to point mutations, deletions in exon 3 were also observed [7, 23, 25].

The underlying cause leading to β-catenin accumulation only in distinct cells (Fig. 2) was evaluated in more detail. Activating mutations within exon 3 of the β-catenin gene (CTNNB1) were assumed to cause an increase of β-catenin levels within the cytoplasm and nucleus as a consequence of impaired degradation [6, 25]. If mutations within CTNNB1 promote cellular stabilisation of β-catenin, then it suggests that only those cells harbouring elevated β-catenin levels that also exhibit a particular genetic profile in relation to surrounding cells without β-catenin accumulations. Unexpectedly, as selective mutational analyses of β-catenin accumulating vs. non-accumulating cells revealed, both cell populations harbour activating mutations in exon 3 of the CTNNB1, without identifying a cell-specific mutation pattern [47]. These results imply that alterations in β-catenin gene regions, encoding phosphorylation sites and binding sites of β-TrCP-complex, which are crucial for protein degradation, are not sufficient to cause nuclear accumulation per se. In contrastto Sekine and colleagues, Kato et al. were not able to verify mutations beyond the epithelial tumour cell component [6, 46]. The identification of multiple mutations in a single CP sample [23, 47] suggests an increased genetic instability. However, chromosomal instability appears unlikely in CP [48]. As described in the following, the impact of β-catenin mutations on cellular transformation is associated with its degree of impaired degradation, depending on the affected codon of exon3. The majority of ACP mutations occur in codons 33 and 37, which were mostly found in benign entities, compared to codons 41 and 45 which are associated with more malignant transformed phenotypes [40]. Mutations in codons 32 and 34 do not display changes in GSK-3β phosphorylation, but decrease ubiquitination as they represent the ubiquitination recognition motive (Fig. 4). Cells with affected codon D32 were shown to be transformed to a higher degree in relation to S33 or G34 mutations, whereas resulting β-catenin accumulation does not increase target gene expression [42, 49]. It was shown that target gene expression occurs when β-catenin is not phosphorylated on Threonine 41 and Serine 37 [50]. Missense mutations of CTNNB1 codon 35 are seldom, and were only detected in hepatocellular carcinomas [51, 52]. This implies that identified mutations within the β-catenin gene exhibit various impacts on ACP tumourigenesis. Since these results revealed that β−catenin accumulations are not caused by mutations per se in exon 3 of the CTNNB1 gene, additional mechanisms have to be involved to achieve its cytoplasmic and nuclear stabilisation.

Many other mechanisms are likely to play a role, e.g. upregulation of β-catenin binding partners competing with multi-destruction complex component binding sites, thus preventing degradation (Fig. 4). Identified binding partners include TCF-family transcription factors, Axin2 and APC [53, 54]. In addition, armadillo repeat units 10–12 are essential for the passive nuclear-pore-complex translocation of β-catenin [55]. Although compromised nuclear shuttling of β-catenin may lead to nuclear accumulation, but no mutations within the CTNNB1 region encoding for the armadillo repeats (exon 8–13) were identified in ACP [47], thus indicating an intact passive nuclear transport machinery. Furthermore, a genetically impaired membranous anchorage of β-catenin facilitated by exon 4 [33] was also excluded

[47]. This notion indicates that interaction of β-catenin with other proteins is not compromised by genetic alterations.

A large body of β-catenin binding partners, such as E-cadherin, LEF1, Axin/Axin2, APC and Fascin, [53, 54, 56–58] were identified as being expressed in ACPs. Importantly, proteins that share the same β-catenin binding sites (Fig. 4) compete with each other [59]. If binding of the destruction complex is permitted, β-catenin is stabilised [56]. In addition, binding partners regulate the subcellular localisation of β-catenin [60]. This finding supports the cellular reallocation of β-catenin in ACPs. Stabilisation of β-catenin, as a result of activated SHH or EGFR signalling, could provide an additional mechanism [61–64]. In summary, activating alterations in exon 3 of CTNNB1 are not restricted to nuclear β-catenin accumulating tumour cells.

It is important to note that the impact of compromised β-catenin in the genesis of ACP could be proven in a genetically modified mouse model [65].

Systematic analyses of the β-catenin associated Wnt signal transduction pathway offer the opportunity to unravel new insights into the pathogenesis of ACP. The current knowledge about the impact of β-catenin accumulating cells on cellular proliferation, differentiation and migration in ACP cells will be illuminated in the following.

ACPs Show Activation of Wnt Signalling Target Genes in Cell Clusters with Nuclear β-Catenin Accumulation

Cellular localisation of β-catenin is not per se sufficient to promote target gene activation [66]. However, it has been shown that nuclear β-catenin accumulations, as a hallmark of ACP, are sufficient to induce Wnt signalling target gene expression, e.g. Axin2, BMP4 and CD44 (Fig. 5) [7, 47, 67]. This was identified on mRNA as well as protein level in tumour cell clusters exhibiting nuclear β-catenin accumulations. Comparative gene expression analyses of both CP variants revealed that Axin2 and BMP4 are increased in ACP. Furthermore, the target gene Fascin1 [68] and EGFR (unpublished data) was identified as being regulated by β-catenin, using knockdown experiments in primary ACP cell cultures.

Axin2

Axin2 also known as conductin (mice) and Axil (rats) is a homolog of Axin. Axin2 is only detected in cells with activated Wnt signalling in contrast to Axin, which is expressed ubiquitously during development. Further studies revealed that Axin2 is a direct target of the Wnt pathway [69]. Its cellular function is comparable with that of Axin because it is also able to bind APC, GSK3β and β-catenin [36] (Fig. 3). Axin2 facilitates as a component of the destruction complex GSK 3β–driven phosphorylation and degradation of β-catenin and acts, therefore, as a negative feedback regulator of the Wnt signalling pathway. The expression of Axin2 is therefore an

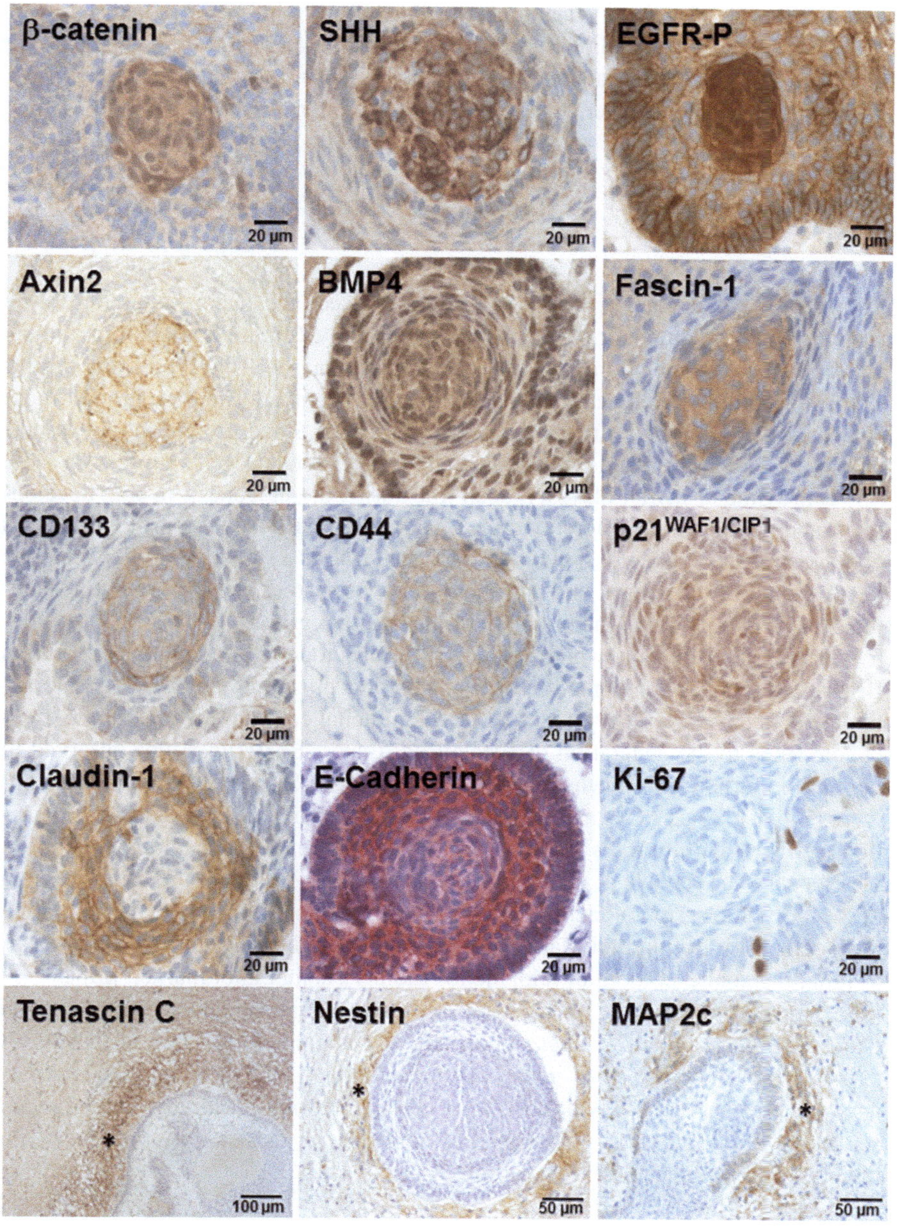

Fig. 5 Expression of proteins in relation to β-catenin accumulating whorls. Cells with nuclear β-catenin accumulation are often configured in a whorl shape. The cells with an activated Wnt signalling pathway also exhibit activated SHH and EGFR signal transduction. It was shown in various studies that the target genes (Axin2, BMP4 and Fascin-1), stem cell markers (CD133 and CD44) and the cell cycle inhibitor p21$^{Waf1/CIP1}$ co-localise in cells with nuclear β-catenin. Proliferating Ki-67 positive cells were mostly identified in the palisading cell layer. Cell adhesion component expression, e.g. claudin-1 (Tight junctions) and E-cadherin (adherens junctions) were decreased in whorl like structures. The brain-tumour boundary (*asterisk*) is characterised by increased tenascin-C, Nestin and MAP2c levels. The letter could also be found in the picket fence-like cells

important mechanism in terminating the action of β-catenin [69, 70]. Interruption of Axin2 by mutations was shown to be associated with an enhanced Wnt signalling that causes tooth agenesis and tumour genesis [71]. The latter has been recognised, for example, in colon cancer, medulloblastomas and hepatoblastomas [72].

BMP4

The bone morphogenetic protein 4 (BMP4) is a secreted factor belonging to the transforming growth factor beta (TGF-β) superfamily. Mature BMP4s form dimers which bind to transmembranous serine/threonine kinase of the type I and type II TGF-β receptors, thereby composing multimeric receptors. The formation of these heteromeric complexes initiates a cascade of several Smad protein modifications. Finally, Smads translocate into the nucleus and co-activate gene expression [73]. The expression of BMP4 plays a crucial role during organogenesis in both development and cellular transdifferentiation. BMP4 knockout leads to early lethality as observed in mice embryos. It plays an important role as a signalling molecule in the development of the central and peripheral nervous system, musculature and skeleton. BMP4 is essential in the initiation of the pituitary gland [73–76]. The presumptive ventral diencephalon, which contacts the oral ectoderm (stomodeum), produces BMP4 and induces the formation of Rathke's pouch primordium [75]. An interaction mediated by several cytokines between epithelial cells and surrounding mesenchyme leads to the generation of specialised tissues. BMP4 was also observed to play an important role downstream of Wnt signalling during early tooth development [77–79]. Since ACPs are presumed to emerge from the stomodeum, the detection of BMP4 within ACPs supports this hypothesis. In tooth development, BMP4 was shown to be capable of maintaining Wnt signalling by activating the expression of the LEF1 transcription factor [80], which is expressed in β-catenin accumulating ACP cells [19]. Both proteins are expressed in the enamel knot, which functions as a signalling centre in tooth development. Another enamel knot marker, $p21^{WAF1/CIP1}$ [79], was also shown to be highly expressed in β-catenin accumulating tumour cell clusters (Fig. 5) [81, 82]. These observations substantiate the notion that ACPs express various tooth specific genes, exhibit an odontogenic morphological appearance [17, 19] and are likely to arise from oral ectodermal precursors. In colorectal cancer BMP4 expression was explicitly described in association with β-catenin mutations and enhanced Wnt signalling [83].

Fascin

Fascin was identified as a target gene of β-catenin in colon cancer [84]. Accordingly, the Fascin promoter and the first intron have consensus sequences matching with β-catenin interacting transcription factors of the TCF/LEF family [85]. Fascin, also known as Fascin-1, is a globular protein of 55 kDa that consists of 493 amino acids and acts as an actin bundling protein. It is expressed in vascular endothelial cells, fibroblasts, neuronal tissue and mature antigen presenting cells, but is predominantly absent in epithelial cells [86]. In humans, two additional paralogs of Fascin, Fascin-2 and Fascin-3, are expressed exclusively in the retina and the testis, respectively [86]. Fascin is involved in the migration machinery and reorganisation of the actin cytoskeleton. It cross-links filamentous actin (F-actin), which is formed by actin monomers that polymerise to fibres and coil with a second fibre. These structures are characterised by enhanced cytoskeleton rigidity and serve as scaffolds in cellular protrusions, the so-called filopodia. Filopodia formation plays a crucial role in cell migration, wound healing, neurite outgrowth, and serves as a precursor of dendritic spines in neurons [87]. The initiation and maintenance of filopodia formation is strongly regulated via extra- and intracellular signals. However, Fascin recently became a focus of interest, as aberrant expression was detected in various carcinomas, in particular at the invasion front of colorectal cancer and gastric cancer, where it has been identified as a target gene of β-catenin TCF signalling [84, 88]. Tumours with an enhanced Fascin expression were shown to be associated with increased invasive growth behaviour which resulted in an unfavourable prognosis for patients [89, 90].

Increased Fascin expression could be involved in sequestration and therefore restricted accumulation of β-catenin in ACP. The β-catenin protein binds Fascin by its armadillo repeat sequence which competes with E-cadherin and APC interaction (Fig. 4) [58, 59]. This suggests a cytoplasmic reallocation of β-catenin, since changes in cell adhesion, as well as a reduction of proteosomal destruction, occur [56].

Activated EGFR

Several studies showed that activated canonical Wnt signalling can be associated with an activated epidermal growth factor receptor (EGFR) cascade [91, 92] and vice versa [93], which is mediated by direct interactions between EGFR and β-catenin [94]. EGFR expression was shown to be increased in ACP and activated EGFR (phosphorylation of T1068) is detectable in the cytoplasm, as well as the nucleus of β-catenin accumulating cells (Fig. 5) [95]. The EGFR is a 170 kDa tyrosine kinase, also referred to as HER1, ErbB1, mENA and PIG61, which belongs to the human epidermal growth factor receptor (HER) family, comprising four members (HER1-4). As a transmembrane glycoprotein, it consists of an extracellular domain facilitating ligand binding, a transmembrane region and an intracellular region that comprises the tyrosine kinase domain and autophosphorylation sites (Fig. 6). Numerous growth factors bind to and activate the EGFR, including the epidermal growth factor (EGF), transforming growth factor alpha (TGF-alpha), amphiregulin, epiregulin, heparin-binding EGF (HB-EGF) and betacellulin [96, 97]. Ligand binding initiates a conformational change that causes receptor dimerisation and possesses protein-tyrosine kinase activity.

The EGFR protein-tyrosine kinase domain is composed of amino acid 686 to 960 which is encoded by exon 18 to 21 of the EGFR gene [96, 98]. The receptor dimer undergoes transphosphorylation on multiple tyrosine (Y) residues, which leads to the recruitment of a plethora of enzymes and adapter proteins acting as signal transducers and activators of at least five different intracellular cascades [99].

The major downstream signal transduction pathways activated by receptors of the HER family are the Ras-Raf mitogen-activated protein kinase (MAPK) pathway (MAPK/ERK cascade) and the phosphatidylinositol 3-kinase (PI3K)/Akt pathway (Fig. 6). Phospholipase Cγ (PLCγ), the signal transducer and activator of transcription (STAT) and src/FAK cascades are also stimulated [97, 99]. This signalling network possesses various cellular processes, e.g. gene expression, proliferation, inhibition of apoptosis, angiogenesis and cell migration [97].

The activation of the EGFR in an autonomous fashion through gene amplification or mutational events paves the way for human epithelial cancer development. The EGFR is overexpressed in the majority of solid tumours, including breast cancer, head and neck cancer, non-small cell lung cancer (NSCLC), renal cancer, ovarian cancer, and colon cancer and in a smaller percentage, also in bladder cancer, pancreatic cancer and gliomas [97]. EGFR overexpression produces an intense signal generation and activation of downstream signalling pathways (Fig. 6), resulting in a high rate of tumour growth, poor tumour differentiation and elevated invasive characteristics. In several lesions, mutations of the EGFR have been described in the tyrosine kinase domain, or mutations leading to truncation as a result of exons 2–7 deletion comprising amino acids 6–273 of the extracellular ligand domain [99], resulting in a ligand independent activation. This truncated EGFR variant, designated as EGFRvIII, is found in gliomas (38%) and NSCLC (5%) [97, 99–101]. Mutations within the tyrosine kinase domain are observed in 26% of NSCLC. Nearly 90% of alterations affect hotspots within exon 18–21 that code for tyrosine kinase. In frame deletions in exon 19, codons 746–750 account for 45–50% and substitution of leucine 858 for at least 35–45%. Insertion in exon 20 was found in 5% and substitutions spanning exon 18–21 occur rarely. Mutations exhibit oncogenic activities by selective activation of anti-apoptotic pathways through increased PI3K/AKT signalling [99].

In ACPs, DNA screening revealed that no activating mutations of the cytoplasmic EGFR autophosphorylation site encoded by exons 17–28 were detectable. The collection of normal controls showed a similar distribution of SNPs, suggesting that they are irrelevant in craniopharyngioma pathogenesis. Further examination of extracellular receptor regions excludes the occurrence of the EGFRvIII variant responsible for autonomous activation. In addition, no EGFR amplification was identified in ACP [95]. As non-genetic alterations are found, activation of the EGFR seems to occur in an autocrine fashion, by secreted soluble ligands or by components of its environment.

Indeed, EGF was found to be upregulated in β-catenin accumulating cell clusters in the embryonic mouse model of ACP and microarray gene expression studies of ACP identified increased amphiregulin levels, implying EGFR activation in an autocrine manner [102, 103]. Recent studies revealed that tenascin-C, a component of the extracellular matrix, is also capable of acting as an EGFR ligand, inducing its dimerisation and subsequent activation [104, 105]. It is worth nothing that ACPs

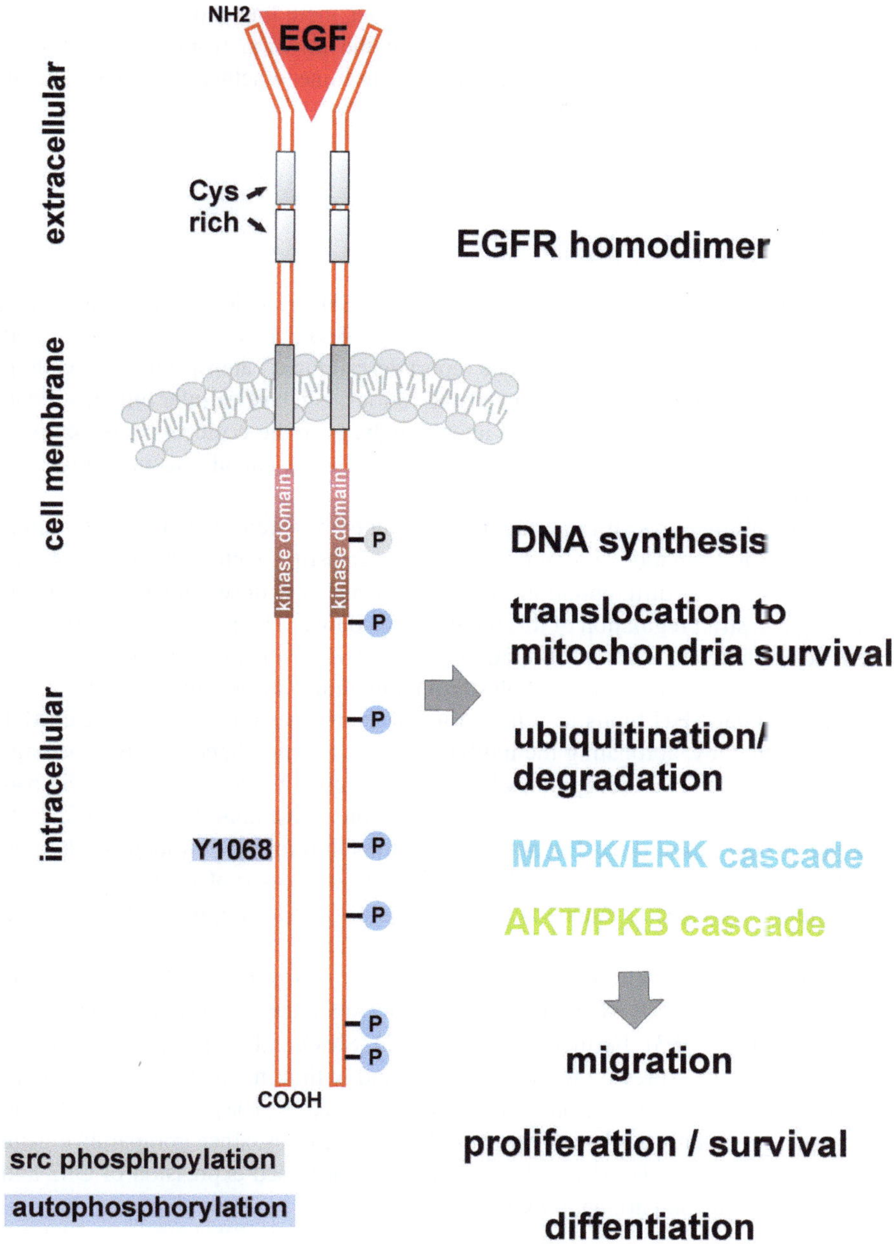

Fig. 6 Signal transduction of an activated EGFR. This illustration shows a subset of the most important tyrosine residues involved in EGFR signalling. Ligand binding to the extracellular region induces EGFR dimerisation. Intracytoplasmic tyrosine residues are autophosphorylated (*blue*) by its intrinsic tyrosine kinase, or phosphorylated by src kinase (*grey*). Adapter proteins bind to these phosphorylated tyrosine residues, thereby initiating the activation of diverse downstream cellular cascades, associated with cellular migration, proliferation, survival and differentiation. Adapted from www.cell-signalling.com

establish a circumscribed tumour surrounding area, including explicit expression of the extracellular matrix protein tenascin-C at the invasion front (Fig. 5) [106]. However, additional studies are needed to determine whether this extracellular matrix protein corresponds to EGFR activation in ACP.

SHH

Several reports support a role of Sonic Hedgehog (SHH) signalling in the pathogenesis of ACP. No mutations in components of the SHH pathway could be identified in ACP [107]. However, there was one report describing a Patch 1 mutation leading to multiple Gorlin syndrome associated aberrations, e.g. basocellular tumours, bone and craniofacial malformations, and interestingly, even the occurrence of a craniopharyngioma. This mutation was observed to induce β-catenin accumulations in craniopharyngioma tissue [108].

HH signalling plays an important role during embryogenesis in cell differentiation and organ formation, e.g. in the brain, skin, intestine, teeth and pituitary gland [109–114]. Proper differentiation processes occur in a dose-dependent manner, pointing to a strict regulation of this pathway. In adults, SHH pathway is involved in stem cell maintenance and proliferation [115]. SHH as a paracrine acting factor is posttranscriptional modified by cholesterol maintaining a lipid anchor [113].

Extracellular SHH binds as a ligand to the transmembrane receptor Patched 1 (PTCH1), thereby, abrogating the inhibitory effect to smoothened (SMO), belonging to the frizzled class receptors, which subsequently induces downstream activation of the zinc finger proteins glioma-associated oncogene homologue 1–3 (Gli1–3) via the protein Fused (Fu). The activation of the Gli transcription factors could be inhibited by the proteinkinase A (PKA) and the Suppressor of fused (SuFu). The SHH inhibitory receptor Patch1 is a target gene of Gli transcription factors, and its upregulation represents a negative feedback loop [115].

A role of SHH pathway in ACP was first identified in the ACP mouse models [103]. In humans, ACP whorl like structures were shown to express SHH (Fig. 5) and Patch1 [102, 103]. Furthermore, Gli1, Gli3, SMO and SUFU were publicised not only as being expressed in the cytoplasm and within the nucleus in whorl configured cells, but also in the palisaded cell layer of ACP [108]. The SHH pathway seems to play mainly a relevant role in ACP pathogeneses, since comparative gene expression and methylation profiling revealed an increased expression of Gli2 and Ptch1 in ACP in comparison to PCP [7].

Impact of Activated Signalling Pathways on Proliferation and Differentiation in ACPs

Although the β-catenin-TCF pathway has a well-established role in activating cell proliferation processes [52], a decreased proliferation activity, as indicated by negative Ki67 staining [46, 81] and increased staining levels of the cell cycle inhibitor

$p21^{WAF1/CIP1}$, was shown [81, 116]. The proliferation activity correlating with the Ki-67 staining was shown to be low (<1% in high power microscopic inspection) or absent in epithelial whorls with β-catenin accumulations (Fig. 5). Furthermore, the distribution of $p21^{WAF1/CIP1}$ is prominent in β-catenin accumulating whorls (Fig. 5). The $p21^{WAF1/CIP1}$ immunoreactivity is also detectable beyond these cell clusters, e.g. in surrounding nests of ghost cells, although to a variable extent. In contrast, proliferating tumour cells are confined to the outer squamous epithelium, especially the basal cell layer [81]. During squamous differentiation of endometrial carcinoma cells, β-catenin is known to induce $p21^{WAF1/CIP1}$ expression, thereby inhibiting cell proliferation [117]. A similar mechanism may play a role in ACP.

While nuclear β-catenin expression seems to have no impact on the proliferation rate of the affected cells, a systematic examination of epithelial differentiation revealed a reliable distinction in cytokeratin expression of nuclear β-catenin accumulating versus non-accumulating cell clusters [46, 81]. Analyses of different cytokeratins (CK5/6, CK8 and CK18) reveal varying epithelial differentiation in aberrant nuclear β-catenin expressing cells, compared to non-accumulating cells (Fig. 7). All ACPs showed expression of high molecular weight cytokeratins 5/6/7and 14 [81, 118].

Mainly the internal cords and strands of stellate epithelial cells as well as the superficial cell layer were immunoreactive, whereas the expression in the outer palisaded cell layer was variable.

Intriguingly, nuclear β-catenin expressing cell clusters of ACPs were always negative for CK5/6 (Fig. 7). In contrast, CK5/6 negative β-catenin accumulating cell clusters revealed an increased expression of CK8 (Fig. 7). The same was observed for CK18, another marker for simple epithelium known to co-localise with CK8 (Fig. 7) [81].

Human cytokeratins represent a major part of epithelial cytoskeleton and belong to the family of water insoluble proteins ranging in size from 40 to 70 kDa [119]. Various subsets of keratin proteins occur in a given epithelium, depending on cell type, stage of differentiation and embryonic development, cellular growth environment and type of malignancy [120].

Cytokeratins 5/6 are known as high molecular weight cytokeratins (58 and 56 kDa, respectively) and are part of the basic (type II) subfamily of cytokeratins [119]. They are closely related and found in non-keratinising stratified squamous epithelia, complex stratified epithelia, epidermis, hair follicles as well as mammary glands and serve as differentiation markers [120]. Low molecular keratins, type I CK18 (45 kDa) and type II CK 8 (52 kDa), are largely expressed as a pair in all simple epithelia, ductal epithelia and pseudostratified epithelia, and are useful for the histopathological differentiation of carcinomas. Basal cell carcinoma, myoepitheliomas and squamous cell carcinomas are negative for CK8 and CK18 [119, 120]. Both cytokeratin are thought to have an ancient, common origin and may represent ancestral precursors of other keratin genes [121]. Their expression is required during early embryogenesis, detectable as early as the 4-cell stage in the mouse embryo, and precedes that of other cytoplasmic intermediate filaments [122]. They can be expressed ectopically in squamous cell carcinomas arising from stratified epithelia, especially in less differentiated cancers with a more invasive phenotype and reduced expression of differentiation-specific keratins. Aberrant expression

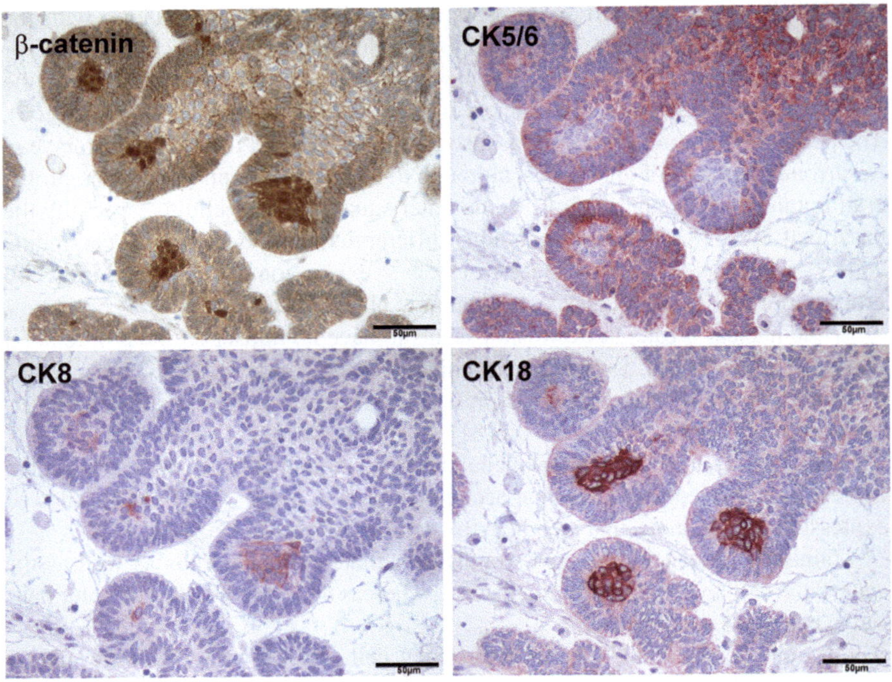

Fig. 7 Simple epithelial differentiation in nuclear β-catenin expressing tumour cell clusters. β-catenin staining revealed neoplastic whorls with distinct nuclear β-catenin expression. The palisading cell layer as well as the vast majority of compact epithelial tumour cells revealed membranous staining only. Serial sections confirmed lack of CK5/6 expression in nuclear β-catenin accumulating cell clusters. However, these cells not only express CK8 but also CK18, another marker for simple epithelia. The scale bar corresponds in all pictures to 50 µm

in squamous carcinomas is thought, therefore, to represent either a passive indicator for loss of differentiation, or alternatively, a positively selected change in differentiation that favours tumour progression [120, 123].

Various lines of evidence indicate, however, that increased activity of β-catenin correlates with loss of epithelial differentiation in colon cancer and was regarded as an initial signal for transdifferentiation mechanisms [117, 124, 125] which was demonstrated to be associated with loss of E-cadherin expression [126–128]. Therefore, E-cadherin is used as a marker of epithelial differentiation in cancers [129–131]. In ACP controversial observation has been made. E-cadherin expression in whirl like structures (Fig. 5) seems to be less intense when compared to the surrounding area, thereby suggesting impaired adherence junction formation. One study identified no changes of E-cadherin expression in β-catenin accumulating cells of ACPs [132] although, another study demonstrated that low E-cadherin expression associates with ACP recurrence [133]. Even if adherens junctions seem not to be compromised, claudin-1, a component of tight junctions, was identified to be downregulated in β-catenin accumulating cell clusters (Fig. 5) [116]. In relation to PCP less overall expression of claudin-1 in ACP is documented. Moreover, reduction of claudin-1 protein levels point towards an invasive growth pattern in ACP [116].

In conclusion, β-catenin is likely to play a pivotal role in tumour morphogenesis and epithelial differentiation of ACPs. Aberrant nuclear β-catenin accumulation is associated with a specific morphology or phenotype rather than just proliferation.

Occurrence of Tumour Stem Cells in ACP

In ACP distinct β-catenin accumulating cells show an activation of the Wnt, SHH and EGFR pathway, which are reported to crosstalk with each other and to be involved in tumour stem cell maintenance [134–136].

Moreover, these clusters express the stem cell marker CD133 and CD44 [67] and are quiescent as lack of the proliferation marker Ki-67 [46, 81] and, as an enhancement of the cell cycle inhibitor $p21^{WAF1/CIP1}$ indicates (Fig. 5) [81, 116].

The expression of additional stem cell markers SOX9, KLF4 and Oct4 was also identified by immunohistochemistry not only in ACP but also in PCP [137].

Transgenic ACP mouse models confirmed that overexpression of β-catenin in pituitary precursor cells are sufficient to induce the development of tumours showing characteristics comparable with the human counterpart [65]. This model implies that β-catenin accumulating cells function as a kind of tumour signalling centre.

Wnt and EGFR Pathways were Identified as Being Involved in the Cellular Motility of ACP

ACP tumour cells with nuclear β-catenin accumulations were identified to exhibit changes in tight junction mediated cell adhesion, e.g. by reduced claudin-1 levels [116] and areas with finger-like tumour outgrowth into the adjacent brain parenchyma (Figs. 2, 5 and 7) [68, 82], thereby inducing morphological changes of the glial cells (Rosenthal fibres). Tenascin-C, Nestin and Map2c (Fig. 5) were shown to be expressed at the tumour-brain boundary as a gradient [106]. Human ACP tumour growth examined with an orthotropic xenotransplant mouse model clearly revealed that those cells with β-catenin accumulations extend into the surroundings [82]. This is in accordance with the observation that recurrence of CPs, which depend on invasion, was found to associate significantly with cells forming whorl like arrays [138]. The impact of β-catenin accumulations on epithelial cell migration as described for other tumours [45, 139] was analysed functionally in ACPs using primary cell cultures. It was shown that activated Wnt signalling has an impact on CP tumour cell migration and that inhibition of β-catenin expression in primary ACP cell cultures using siRNA leads to diminished tumour cell migration [68]. Notably, β-catenin promotes migration indirectly because it regulates the expression of target genes necessary for invasion and dissociation [140]. One target gene known to influence epithelial migration is the aforementioned protein Fascin. In ACPs, Fascin was detected by immunohistochemistry to be increased mainly in β-accumulating cells (Fig. 5). Therefore, the suggestion was put forward that it works as a facilitator of ACP cell migration. This notion was confirmed by the observation that decreased migratory potential after β-catenin silencing is also associated with decreased expression levels of Fascin [68]. The direct impact of Fascin expression on migration, however, was proven by a selective knockdown experiment showing that its expression is in accordance with the migratory capacity [68].

EGFR activation in ACP (Fig. 5) associates with an increased tumour cell motility, as primary cell cultures experiments revealed [95]. The inhibition of EGFR signalling using the small molecule Gefitinib, also known as Iressa which belongs to anilinoquinazolines and is approved for therapeutic treatment of NSCLC, was shown to be effective in ACP [95]. It inhibits EGFR autophosphorylation in a highly selective and reversible manner by competing with ATP [141]. In addition, Gefitinib exhibits a good bioavailability and high EGFR selectivity [141]. Treatment of Gefitinib in a concentration of 500 nM was sufficient to inhibit EGF induced EGFR activation almost completely in ACP cell cultures as immunoblotting disclosed [95]. In general, responsiveness of ACP to Gefitinib is an important finding, as inhibition of the EGFR affords a therapeutic opportunity also affecting Wnt signalling. The prevention of EGFR signalling was shown to prohibit expression of the β-catenin target gene Fascin that accompanies tumour cell motility. The therapeutic application could occur directly and intracranially into the tumour by an Ommaya reservoir in high doses in order to bypass a systemic exposition. This kind of treatment is currently practised for interferon alpha treatment of CPs [142].

From a clinical point of view, it is a very important issue to inhibit ACP invasion into surrounding brain structures, as it counteracts tumour relapse and continuing deterioration of vital structures.

Furthermore, inhibition of EGFR signalling provides an interesting opportunity to affect also those cells, believed to expose tumour stem cell characteristics.

Further research is warranted to identify and prove pharmaceutical targets for successful adjuvant chemotherapy in patients with ACPs.

References

1. Rutka JT (2002) Craniopharyngioma. J Neurosurg 97(1):1–2 discussion 2
2. Louis D, Ohgaki H, Wiestler O, Cavenee W, Ellison D, Figarella-Branger D, et al (2016) WHO classification of tumours of the central nervous system, Revised 4th Edition. IARC, Lyon, pp 324–328. ISBN 978-92-832-4492-9
3. Garnett MR et al (2007) Craniopharyngioma. Orphanet J Rare Dis 2(1):1
4. Erdheim J (1904) Über Hypophysenganggeschwülste und Hirncholesteatome. Sitzber. d. k. Akad. Wiss. Wien 113:537–726
5. Lindholm J, Nielsen EH (2009) Craniopharyngioma: historical notes. Pituitary 12(4):352–359

6. Sekine S et al (2002) Craniopharyngiomas of adamantinomatous type harbor beta-catenin gene mutations. Am J Pathol 161(6):1997–2001
7. Hölsken A et al (2016) Adamantinomatous and papillary craniopharyngiomas are characterized by distinct epigenomic as well as mutational and transcriptomic profiles. Acta Neuropathol Commun 4(1):1–13
8. Hoffman H (1994) Surgical management of craniopharyngioma. Pediatr Neurosurg 21(Suppl. 1):44–49
9. Weiner HL et al (1994) Craniopharyngiomas: a clinicopathological analysis of factors predictive of recurrence and functional outcome. Neurosurgery 35(6):1001–1011
10. Gupta N, Banerjee A, Haas-Kogan D (2010) Pediatric CNS tumors. Springer, Berlin
11. Miller D (1994) Pathology of craniopharyngiomas: clinical import of pathological findings. Pediatr Neurosurg 21(Suppl. 1):11–17
12. Izuora G et al (1988) Childhood intracranial neoplasms Enugu, Nigeria. West Afr J Med 8(3):171–174
13. Kuratsu J-I, Ushio Y (1996) Epidemiological study of primary intracranial tumors in childhood. Pediatr Neurosurg 25(5):240–247
14. Karavitaki N et al (2006) Craniopharyngiomas. Endocr Rev 27(4):371–397
15. Karavitaki N et al (2005) Craniopharyngiomas in children and adults: systematic analysis of 121 cases with long-term follow-up. Clin Endocrinol 62(4):397–409
16. Bernstein ML, Buchino JJ (1983) The histologic similarity between craniopharyngioma and odontogenic lesions: a reappraisal. Oral Surg Oral Med Oral Pathol 56(5):502–511
17. Paulus W et al (1997) Odontogenic classification of craniopharyngiomas: a clinicopathological study of 54 cases. Histopathology 30(2):172–176
18. Sekine S et al (2003) Beta-catenin mutations are frequent in calcifying odontogenic cysts, but rare in ameloblastomas. Am J Pathol 163(5):1707–1712
19. Sekine S et al (2004) Expression of enamel proteins and LEF1 in adamantinomatous craniopharyngioma: evidence for its odontogenic epithelial differentiation. Histopathology 45(6):573–579
20. Seemayer T, Blundell J, Wiglesworth F (1972) Pituitary craniopharyngioma with tooth formation. Cancer 29(2):423–430
21. Soukup V et al (2008) Dual epithelial origin of vertebrate oral teeth. Nature 455(7214):795–798
22. Sartoretti-Schefer S et al (1997) MR differentiation of adamantinous and squamous-papillary craniopharyngiomas. AJNR Am J Neuroradiol 18(1):77–87
23. Brastianos PK et al (2014) Exome sequencing identifies BRAF mutations in papillary craniopharyngiomas. Nat Genet 46(2):161–165
24. Schweizer L et al (2015) BRAF V600E analysis for the differentiation of papillary craniopharyngiomas and Rathke's cleft cysts. Neuropathol Appl Neurobiol 41(6):733–742
25. Buslei R et al (2005) Common mutations of beta-catenin in adamantinomatous craniopharyngiomas but not in other tumours originating from the sellar region. Acta Neuropathol 109(6):589–597
26. Larkin SJ et al (2014) BRAF V600E mutations are characteristic for papillary craniopharyngioma and may coexist with CTNNB1-mutated adamantinomatous craniopharyngioma. Acta Neuropathol 127(6):927–929
27. Kim JH, Paulus W, Heim S (2015) BRAF V600E mutation is a useful marker for differentiating Rathke's cleft cyst with squamous metaplasia from papillary craniopharyngioma. J Neuro-Oncol 123:181–191
28. Alomari AK et al (2015) Craniopharyngioma arising in a Rathke's cleft cyst: case report. J Neurosurg Pediatr 15(3):250–254
29. Hofmann BM et al (2006) Nuclear beta-catenin accumulation as reliable marker for the differentiation between cystic craniopharyngiomas and rathke cleft cysts: a clinico-pathologic approach. Am J Surg Pathol 30(12):1595–1603

30. Xu W, Kimelman D (2007) Mechanistic insights from structural studies of beta-catenin and its binding partners. J Cell Sci 120(Pt 19):3337–3344
31. Behrens J et al (1996) Functional interaction of beta-catenin with the transcription factor LEF-1. Nature 382(6592):638–642
32. Nelson WJ, Nusse R (2004) Convergence of Wnt, beta-catenin, and cadherin pathways. Science 303(5663):1483–1487
33. Lilien J, Balsamo J (2005) The regulation of cadherin-mediated adhesion by tyrosine phosphorylation/dephosphorylation of beta-catenin. Curr Opin Cell Biol 17(5):459–465
34. Roura S et al (1999) Regulation of E-cadherin/Catenin association by tyrosine phosphorylation. J Biol Chem 274(51):36734–36740
35. Kemler R (1993) From cadherins to catenins: cytoplasmic protein interactions and regulation of cell adhesion. Trends Genet 9(9):317–321
36. Behrens J et al (1998) Functional interaction of an axin homolog, conductin, with beta-catenin, APC, and GSK3beta. Science 280(5363):596–599
37. Rubinfeld B et al (1996) Binding of GSK3beta to the APC-beta-catenin complex and regulation of complex assembly. Science 272(5264):1023–1026
38. Rubinfeld B et al (1993) Association of the APC gene product with beta-catenin. Science 262(5140):1731–1734
39. Amit S et al (2002) Axin-mediated CKI phosphorylation of β-catenin at Ser 45: a molecular switch for the Wnt pathway. Genes Dev 16(9):1066–1076
40. Provost E et al (2003) Functional correlates of mutations in beta-catenin exon 3 phosphorylation sites. J Biol Chem 278(34):31781–31789
41. Aberle H et al (1997) beta-catenin is a target for the ubiquitin-proteasome pathway. EMBO J 16(13):3797–3804
42. Al-Fageeh M et al (2004) Phosphorylation and ubiquitination of oncogenic mutants of beta-catenin containing substitutions at Asp32. Oncogene 23(28):4839–4846
43. Gao C, Chen Y-G (2010) Dishevelled: The hub of Wnt signaling. Cell Signal 22(5):717–727
44. Liu X, Rubin JS, Kimmel AR (2005) Rapid, Wnt-induced changes in GSK3beta associations that regulate beta-catenin stabilization are mediated by Galpha proteins. Curr Biol 15(22):1989–1997
45. Müller T et al (2002) Regulation of epithelial cell migration and tumor formation by β-catenin signaling. Exp Cell Res 280(1):119–133
46. Kato K et al (2004) Possible linkage between specific histological structures and aberrant reactivation of the Wnt pathway in adamantinomatous craniopharyngioma. J Pathol 203(3):814–821
47. Holsken A et al (2009) Target gene activation of the Wnt signaling pathway in nuclear beta-catenin accumulating cells of adamantinomatous craniopharyngiomas. Brain Pathol 19(3):357–364
48. Yoshimoto M et al (2004) Comparative genomic hybridization analysis of pediatric adamantinomatous craniopharyngiomas and a review of the literature. J Neurosurg 101(1 Suppl):85–90
49. Provost E et al (2005) Functional correlates of mutation of the Asp32 and Gly34 residues of beta-catenin. Oncogene 24(16):2667–2676
50. Staal FJ et al (2002) Wnt signals are transmitted through N-terminally dephosphorylated β-catenin. EMBO Rep 3(1):63–68
51. Legoix P et al (1999) Beta-catenin mutations in hepatocellular carcinoma correlate with a low rate of loss of heterozygosity. Oncogene 18(27):4044–4046
52. Van Nhieu JT et al (1999) Nuclear accumulation of mutated β-catenin in hepatocellular carcinoma is associated with increased cell proliferation. Am J Pathol 155(3):703–710
53. Huber AH, Nelson WJ, Weis WI (1997) Three-dimensional structure of the armadillo repeat region of beta-catenin. Cell 90(5):871–882
54. von Kries JP et al (2000) Hot spots in beta-catenin for interactions with LEF-1, conductin and APC. Nat Struct Biol 7(9):800–807

55. Koike M et al (2004) beta-Catenin shows an overlapping sequence requirement but distinct molecular interactions for its bidirectional passage through nuclear pores. J Biol Chem 279(32):34038–34047
56. Henderson BR et al (2002) Lymphoid enhancer factor-1 blocks adenomatous polyposis coli-mediated nuclear export and degradation of beta-catenin. Regulation by histone deacetylase 1. J Biol Chem 277(27):24258–24264
57. Wong NACS, Pignatelli M (2002) β-catenin—a linchpin in colorectal carcinogenesis? Am J Pathol 160(2):389–401
58. Tao YS et al (1996) beta-Catenin associates with the actin-bundling protein fascin in a non-cadherin complex. J Cell Biol 134(5):1271–1281
59. Hülsken J, Birchmeier W, Behrens J (1994) E-cadherin and APC compete for the interaction with beta-catenin and the cytoskeleton. J Cell Biol 127(6):2061–2069
60. Krieghoff E, Behrens J, Mayr B (2006) Nucleo-cytoplasmic distribution of beta-catenin is regulated by retention. J Cell Sci 119(Pt 7):1453–1463
61. Hu T, Li C (2010) Convergence between Wnt-β-catenin and EGFR signaling in cancer. Mol Cancer 9(1):1
62. Sharma M, Chuang WW, Sun Z (2002) Phosphatidylinositol 3-kinase/Akt stimulates androgen pathway through GSK3β inhibition and nuclear β-catenin accumulation. J Biol Chem 277(34):30935–30941
63. Ulloa F, Itasaki N, Briscoe J (2007) Inhibitory Gli3 activity negatively regulates Wnt/β--catenin signaling. Curr Biol 17(6):545–550
64. Noubissi FK et al (2009) Wnt signaling stimulates transcriptional outcome of the Hedgehog pathway by stabilizing GLI1 mRNA. Cancer Res 69(22):8572–8578
65. Gaston-Massuet C et al (2011) Increased Wingless (Wnt) signaling in pituitary progenitor/stem cells gives rise to pituitary tumors in mice and humans. Proc Natl Acad Sci U S A 108(28):11482–11487
66. Prieve MG, Waterman ML (1999) Nuclear localization and formation of β-catenin–lymphoid enhancer factor 1 complexes are not sufficient for activation of gene expression. Mol Cell Biol 19(6):4503–4515
67. Holsken A et al (2013) Adamantinomatous craniopharyngiomas express tumor stem cell markers in cells with activated Wnt signaling: further evidence for the existence of a tumor stem cell niche? Pituitary 17:546–556
68. Holsken A et al (2010) Tumour cell migration in adamantinomatous craniopharyngiomas is promoted by activated Wnt-signalling. Acta Neuropathol 119:631–639
69. Jho EH et al (2002) Wnt/beta-catenin/Tcf signaling induces the transcription of Axin2, a negative regulator of the signaling pathway. Mol Cell Biol 22(4):1172–1183
70. Lustig B et al (2002) Negative feedback loop of Wnt signaling through upregulation of conductin/axin2 in colorectal and liver tumors. Mol Cell Biol 22(4):1184–1193
71. Lammi L et al (2004) Mutations in AXIN2 cause familial tooth agenesis and predispose to colorectal cancer. Am J Hum Genet 74(5):1043–1050
72. Taniguchi K et al (2002) Mutational spectrum of beta-catenin, AXIN1, and AXIN2 in hepatocellular carcinomas and hepatoblastomas. Oncogene 21(31):4863–4871
73. Liu A, Niswander LA (2005) Bone morphogenetic protein signalling and vertebrate nervous system development. Nat Rev Neurosci 6(12):945–954
74. Rahman MS et al (2015) TGF-β/BMP signaling and other molecular events: regulation of osteoblastogenesis and bone formation. Bone Res 3:15005
75. Takuma N et al (1998) Formation of Rathke's pouch requires dual induction from the diencephalon. Development 125(23):4835–4840
76. Treier M et al (1998) Multistep signaling requirements for pituitary organogenesis in vivo. Genes Dev 12(11):1691–1704
77. Ohazama A, Tucker A, Sharpe PT (2005) Organized tooth-specific cellular differentiation stimulated by BMP4. J Dent Res 84(7):603–606

78. Liu F et al (2008) Wnt/β-catenin signaling directs multiple stages of tooth morphogenesis. Dev Biol 313(1):210–224
79. Thesleff I (2003) Epithelial-mesenchymal signalling regulating tooth morphogenesis. J Cell Sci 116(Pt 9):1647–1648
80. Kratochwil K et al (1996) Lef1 expression is activated by BMP-4 and regulates inductive tissue interactions in tooth and hair development. Genes Dev 10(11):1382–1394
81. Buslei R et al (2007) Nuclear beta-catenin accumulation associates with epithelial morphogenesis in craniopharyngiomas. Acta Neuropathol 113(5):585–590
82. Stache C et al (2015) Insights into the infiltrative behavior of adamantinomatous craniopharyngioma in a new xenotransplant mouse model. Brain Pathol 25(1):1–10
83. Kim JS et al (2002) Oncogenic beta-catenin is required for bone morphogenetic protein 4 expression in human cancer cells. Cancer Res 62(10):2744–2748
84. Vignjevic D et al (2007) Fascin, a novel target of β-catenin-TCF signaling, is expressed at the invasive front of human colon cancer. Cancer Res 67(14):6844–6853
85. Grothey A et al (2000) C-erbB-2HER-2 upregulates fascin, an actin-bundling protein associated with cell motility, in human breast cancer cell lines. Oncogene 19(42):4864–4875
86. Hashimoto Y, Skacel M, Adams JC (2005) Roles of fascin in human carcinoma motility and signaling: prospects for a novel biomarker? Int J Biochem Cell Biol 37(9):1787–1804
87. Mattila PK, Lappalainen P (2008) Filopodia: molecular architecture and cellular functions. Nat Rev Mol Cell Biol 9(6):446–454
88. Kim SJ et al (2010) Galectin-3 increases gastric cancer cell motility by up-regulating fascin-1 expression. Gastroenterology 138(3):1035–1045 e2
89. Hashimoto Y et al (2006) Prognostic significance of fascin expression in advanced colorectal cancer: an immunohistochemical study of colorectal adenomas and adenocarcinomas. BMC Cancer 6(1):1
90. Yoder BJ et al (2005) The expression of fascin, an actin-bundling motility protein, correlates with hormone receptor–negative breast cancer and a more aggressive clinical course. Clin Cancer Res 11(1):186–192
91. Civenni G, Holbro T, Hynes NE (2003) Wnt1 and Wnt5a induce cyclin D1 expression through ErbB1 transactivation in HC11 mammary epithelial cells. EMBO Rep 4(2):166–171
92. Yu W-H et al (2002) CD44 anchors the assembly of matrilysin/MMP-7 with heparin-binding epidermal growth factor precursor and ErbB4 and regulates female reproductive organ remodeling. Genes Dev 16(3):307–323
93. Lu Z et al (2003) Downregulation of caveolin-1 function by EGF leads to the loss of E-cadherin, increased transcriptional activity of β-catenin, and enhanced tumor cell invasion. Cancer Cell 4(6):499–515
94. Schroeder JA et al (2002) ErbB-β-catenin complexes are associated with human infiltrating ductal breast and murine mammary tumor virus (MMTV)-Wnt-1 and MMTV-c-Neu transgenic carcinomas. J Biol Chem 277(25):22692–22698
95. Holsken A et al (2011) EGFR signaling regulates tumor cell migration in craniopharyngiomas. Clin Cancer Res 17:4367–4377
96. Roskoski R (2004) The ErbB/HER receptor protein-tyrosine kinases and cancer. Biochem Biophys Res Commun 319(1):1–11
97. Herbst RS (2004) Review of epidermal growth factor receptor biology. Int J Radiat Oncol Biol Phys 59(2):S21–S26
98. Zhang X et al (2006) An allosteric mechanism for activation of the kinase domain of epidermal growth factor receptor. Cell 125(6):1137–1149
99. Laurent-Puig P, Lievre A, Blons H (2009) Mutations and response to epidermal growth factor receptor inhibitors. Clin Cancer Res 15(4):1133–1139
100. Gan HK, Kaye AH, Luwor RB (2009) The EGFRvIII variant in glioblastoma multiforme. J Clin Neurosci 16(6):748–754
101. Willmore-Payne C, Holden JA, Layfield LJ (2006) Detection of EGFR-and HER2-activating mutations in squamous cell carcinoma involving the head and neck. Mod Pathol 19(5):634–640

102. Gump JM et al (2015) Identification of targets for rational pharmacological therapy in childhood craniopharyngioma. Acta Neuropathol Commun 3:30
103. Andoniadou CL et al (2012) Identification of novel pathways involved in the pathogenesis of human adamantinomatous craniopharyngioma. Acta Neuropathol 124(2):259–271
104. Iyer AKV et al (2008) Cell surface restriction of EGFR by a tenascin cytotactin-encoded EGF-like repeat is preferential for motility-related signaling. J Cell Physiol 214(2):504–512
105. Swindle CS et al (2001) Epidermal growth factor (EGF)-like repeats of human tenascin-C as ligands for EGF receptor. J Cell Biol 154(2):459–468
106. Burghaus S et al (2010) A tumor-specific cellular environment at the brain invasion border of adamantinomatous craniopharyngiomas. Virchows Arch 456(3):287–300
107. Sarubi J et al (2001) Clonal composition of human adamantinomatous craniopharyngiomas and somatic mutation analyses of the patched (PTCH), Gsα and Gi2α genes. Neurosci Lett 310(1):5–8
108. Gomes DC et al (2015) Sonic Hedgehog pathway is upregulated in adamantinomatous craniopharyngiomas. Eur J Endocrinol 172(5):603–608
109. Treier M et al (2001) Hedgehog signaling is required for pituitary gland development. Development 128(3):377–386
110. Palma V et al (2005) Sonic hedgehog controls stem cell behavior in the postnatal and adult brain. Development 132(2):335–344
111. Athar M et al (2006) Hedgehog signalling in skin development and cancer. Exp Dermatol 15(9):667–677
112. Dassule HR et al (2000) Sonic hedgehog regulates growth and morphogenesis of the tooth. Development 127(22):4775–4785
113. Taipale J, Beachy PA (2001) The Hedgehog and Wnt signalling pathways in cancer. Nature 411(6835):349–354
114. Gulino A, Ferretti E, De Smaele E (2009) Hedgehog signalling in colon cancer and stem cells. EMBO Mol Med 1(6–7):300–302
115. di Magliano MP, Hebrok M (2003) Hedgehog signalling in cancer formation and maintenance. Nat Rev Cancer 3(12):903–911
116. Stache C et al (2014) Tight junction protein claudin-1 is differentially expressed in craniopharyngioma subtypes and indicates invasive tumor growth. Neuro-Oncology 16(2):256–264
117. Saegusa M et al (2004) β-catenin simultaneously induces activation of the p53-p21WAF1 pathway and overexpression of cyclin D1 during squamous differentiation of endometrial carcinoma cells. Am J Pathol 164(5):1739–1749
118. Tateyama H et al (2001) Different Keratin Profilesin Craniopharyngioma Subtypes and Ameloblastomas. Pathol Res Pract 197(11):735–742
119. Moll R (1993) Cytokeratins as markers of differentiation. Expression profiles in epithelia and epithelial tumors. Veroff Pathol 142:1–197
120. Chu P, Weiss L (2002) Keratin expression in human tissues and neoplasms. Histopathology 40(5):403–439
121. Blumenberg M (1988) Concerted gene duplications in the two keratin gene families. J Mol Evol 27(3):203–211
122. JACKSON BW et al (1980) Formation of cytoskeletal elements during mouse embryogenesis. Differentiation 17(1–3):161–179
123. Markey AC et al (1991) Expression of simple epithelial keratins 8 and 18 in epidermal neoplasia. J Invest Dermatol 97(5):763–770
124. Mariadason JM et al (2001) Down-regulation of β-catenin TCF signaling is linked to colonic epithelial cell differentiation. Cancer Res 61(8):3465–3471
125. Naishiro Y et al (2001) Restoration of epithelial cell polarity in a colorectal cancer cell line by suppression of β-catenin/T-cell factor 4-mediated gene transactivation. Cancer Res 61(6):2751–2758

126. Behrens J et al (1993) Loss of epithelial differentiation and gain of invasiveness correlates with tyrosine phosphorylation of the E-cadherin/beta-catenin complex in cells transformed with a temperature-sensitive v-SRC gene. J Cell Biol 120(3):757–766
127. Brabletz T et al (2001) Variable beta-catenin expression in colorectal cancers indicates tumor progression driven by the tumor environment. Proc Natl Acad Sci U S A 98(18): 10356–10361
128. Jamora C et al (2003) Links between signal transduction, transcription and adhesion in epithelial bud development. Nature 422(6929):317–322
129. Liu D et al (2001) E-cadherin expression associated with differentiation and prognosis in patients with non–small cell lung cancer. Ann Thorac Surg 71(3):949–954
130. Owens DW et al (2000) E-cadherin at the cell periphery is a determinant of keratinocyte differentiation in vitro. Biochem Biophys Res Commun 269(2):369–376
131. Schipper J, Unger A, Jahnke K (1994) E-cadherin as a functional marker of the differentiation and invasiveness of squamous cell carcinoma of the head and neck. Clin Otolaryngol Allied Sci 19(5):381–384
132. Preda V et al (2015) The Wnt signalling cascade and the adherens junction complex in craniopharyngioma tumorigenesis. Endocr Pathol 26(1):1–8
133. Qi ST et al (2012) Epithelial–mesenchymal transition and clinicopathological correlation in craniopharyngioma. Histopathology 61(4):711–725
134. Mimeault M, Batra SK (2010) Frequent deregulations in the hedgehog signaling network and cross-talks with the epidermal growth factor receptor pathway involved in cancer progression and targeted therapies. Pharmacol Rev 62(3):497–524
135. Reya T, Clevers H (2005) Wnt signalling in stem cells and cancer. Nature 434(7035): 843–850
136. Reya T et al (2001) Stem cells, cancer, and cancer stem cells. Nature 414(6859):105–111
137. Garcia-Lavandeira M et al (2012) Craniopharyngiomas express embryonic stem cell markers (SOX2, OCT4, KLF4, and SOX9) as pituitary stem cells but do not coexpress RET/GFRA3 receptors. J Clin Endocrinol Metab 97(1):E80–E87
138. Tena-Suck ML et al (2009) Expression of epithelial cell adhesion molecule and pituitary tumor transforming gene in adamantinomatous craniopharyngioma and its correlation with recurrence of the tumor. Ann Diagn Pathol 13(2):82–88
139. Demir R et al (2009) Malignant progression of invasive tumour cells seen in hypoxia present an accumulation of β-catenin in the nucleus at the tumour front. Exp Mol Pathol 87(2): 109–116
140. Brabletz T et al (2001) Variable β-catenin expression in colorectal cancers indicates tumor progression driven by the tumor environment. Proc Natl Acad Sci 98(18):10356–10361
141. Herbst RS, Fukuoka M, Baselga J (2004) Gefitinib—a novel targeted approach to treating cancer. Nat Rev Cancer 4(12):979–987
142. Bartels U (2012) Intracystic therapies for cystic craniopharyngioma in childhood. Front Endocrinol (Lausanne) 3:39

Transcriptomic and Genomic Analyses of Human Craniopharyngioma

Leslie Robinson, Sandro Santagata, and Todd C. Hankinson

Abstract Recent studies of the genome and transcriptome of human craniopharyngioma have contributed significant insights into our understanding of these tumors. Genomic studies of adamantinomatous craniopharyngioma (ACP) have revealed mutations in the *CTNNB1* gene, resulting in nuclear and cytoplasmic accumulation of β-catenin protein. Papillary craniopharyngioma (PCP), however, is characterised by the BRAF V600E mutation, which suggests novel therapeutic strategies. These mutations appear to be mutually exclusive, implying that ACP and PCP are distinct tumor types. Studies of the craniopharyngioma transcriptome, while few, have raised additional therapeutic possibilities, such as potentially targeting matrix metallopeptidases. While a phase II clinical trial for targeted therapy in PCP is forthcoming, additional studies and confirmation of the current preliminary findings are required for advancing novel therapies for ACP.

Keywords Craniopharyngioma genomics • Craniopharyngioma transcriptomics • CTNNB1 • β-catenin • WNT pathway • BRAF

Introduction

Historically, therapy for human craniopharyngioma (CP) has been limited by an incomplete understanding of the biology of these highly morbid lesions. As described in previous chapters, there have been considerable advances in our knowledge over the last decade and a half, based on improved laboratory techniques and studies of novel animal models. While progress has been somewhat slower than that

L. Robinson • T.C. Hankinson (✉)
Children's Hospital Colorado, Pediatric Neurosurgery,
13123 E. 16th Ave, Box 330, Aurora, CO 80045, USA
e-mail: Leslie.Robinson@ucdenver.edu; Todd.Hankinson@childrenscolorado.org

S. Santagata
Brigham and Women's Hospital, Department of Pathology, Harvard Institute of Medicine,
HIM-921, 77 Avenue Louis Pasteur, Boston, MA 02115, USA
e-mail: SSantagata@bics.bwh.harvard.edu

J.P. Martinez-Barbera, C.L. Andoniadou (eds.), *Basic Research and Clinical Aspects of Adamantinomatous Craniopharyngioma*,
DOI 10.1007/978-3-319-51890-9_2

seen with some other pediatric and adult brain tumors, such as medulloblastoma [1], ependymoma [2], diffuse lower grade gliomas [3], and glioblastoma [4] analyses of the CP genome and transcriptome have provided considerable insight into the origins of this tumor and the drivers of growth. Further development of these lines of research will guide the development of novel therapies against CP.

This chapter summarises the current state of knowledge regarding the genome and transcriptome of both adamantinomatous (ACP) and papillary (PCP) craniopharyngioma. As the data suggest that these tumors are genetically distinct entities, they are presented as such. With regard to each tumor type, we briefly discuss the most well-established pathways that contribute to tumorigenesis and growth.

Adamantinomatous Craniopharyngioma

β-Catenin and the WNT/Wingless Pathway

The WNT/wingless pathway is well-established in oncogenesis and harbors the only known recurrent genome level aberration in ACP.

In the normal physiological state, Wnt/wingless signaling is involved in organogenesis and adult stem cell maintenance [5, 6]. Among the key members of the canonical WNT signaling pathway is β-catenin [7, 8]. This protein plays an integral role in development, cellular proliferation, differentiation, and cell migration [7, 9–11]. The WNT pathway is maintained in its default off state by a β-catenin destruction complex that restrains β-catenin within the cytosol and facilitates its ubiquitination and proteasomal degradation [12]. Exogenous Wnt signaling transitions this pathway away from the off state [12]. Nuclear β-catenin, therefore, is a hallmark characteristic of active canonical Wnt signaling [5].

In the inactive state, β-catenin is a submembranous component of the cadherin complex [13, 14]. It forms adherens junctions at the cell membrane, where it interacts with E-cadherin and A-cadherin to tether the actin cytoskeleton to the membrane. There it helps preserve cytoskeletal architecture and is involved in polarity and intercellular connectivity [6, 12, 13, 15]. Typically, β-catenin in the cytoplasm is bound and inactivated by a destruction complex [13]. This destruction complex is likely controlled by Axin, a tumor suppressor protein that serves as a multimeric scaffolding protein [12]. Axin recruits and binds to adenomatosis polyposis coli (APC), a tumor suppressor protein, as well as GSK3beta, casein kinase 1α, and protein phosphatase 2A (PP2A) [9, 12, 13]. APC is phosphorylated by these kinases, resulting in recruitment of β-catenin to the complex [12, 13]. These kinases tag β-catenin for degradation through phosphorylation of the serine (S33, S37, S45) or threonine (T41) residues of exon 3 on the *CTNNB1* gene [12, 13]. This phosphorylation pattern is recognised by a component of E3 ubiquitin ligase, b-TrCP, which ubiquitinates β-catenin, thereby targeting it for proteasome-mediated degradation [9, 12]. When β-catenin is phosphorylated within the multiprotein degradation complex, Wnt signaling is inactive [9].

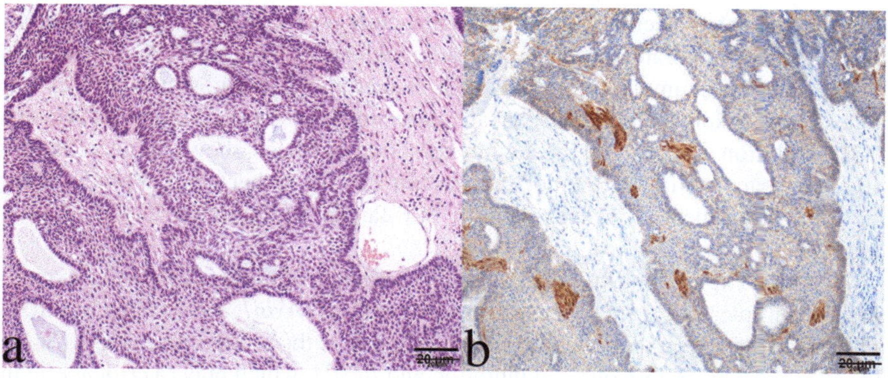

Fig. 1 Histological specimens of adamantinomatous craniopharyngioma demonstrating epithelial regions admixed with microcysts and a reactive glial surround: (**a**) Haematoxylin and eosin stain; (**b**) Immunohistochemistry for β-catenin demonstrates nuclear and cytoplasmic staining in the epithelial cells without uptake in the glial or cystic regions of the tissue

The Wnt protein family includes approximately 20 different proteins that bind to the Frizzled (Fz) family of receptors. When these Wnt proteins bind the highly conserved, cysteine rich extracellular domains of Fz receptors, the Wnt pathway is activated [12] resulting in an intracellular signaling cascade that promotes dimerisation of Fz with LRP5/6 [10, 12]. The Fz/LRP5/6 dimer can bind to Axin, GSK3β, and CK1, which facilitates their accumulation at the cell membrane, preventing them from joining the β-catenin destruction complex [9, 10, 12, 13, 16]. In the absence of the destruction complex, β-catenin protein accumulates and ultimately translocates to the nucleus, where it activates tcf/let transcription factors. This allows transcription of β-catenin target genes and stimulation of cellular proliferation and other Wnt regulated cellular processes [9, 10, 12, 13, 16, 17].

Neoplastic cells with elevated β-catenin levels and activated target genes demonstrate increased migratory capacity and invasiveness [13]. In ACP, this is demonstrated by β-catenin accumulation within cell clusters that are localised at invasive protrusions of tumor into normal brain [13] [Fig. 1]. These processes may be facilitated by *fascin-1*(fascin), a β-catenin tcf signaling target gene, which is involved in the migration mechanism and reorganisation of the actin cytoskeleton, as needed for cellular motility [13, 18]. The promoter region for fascin contains a TCF-binding region, which allows β-catenin to regulate fascin expression [12]. Fascin, an actin-bundling protein, functions by cross-linking F-actin, which is critical for filopodia formation and rigidity, and is frequently seen in migrating cells [12, 13]. Many transformed cell types demonstrate increased fascin expression. This is particularly notable in colorectal cancer, where fascin is found at the invasive edges [13]. Fascin is also overexpressed in ACP cell clusters that have nuclear accumulation of β-catenin [18]. Fascin, in association with β-catenin, is felt to promote migration and invasion of ACP cells. β-catenin can bind fascin by its armadillo repeat sequence which promotes reallocation of β-catenin, changes cell adhesion properties, and reduces β-catenin destruction [13]. Armadillo repeat units 10–12 are integral to translocation of β-catenin via the passive nuclear-pore-complex [15]. This armadillo sequence of β-catenin is also involved in the E-cadherin and APC interactions [13]. β-catenin accumulation can reduce E-cadherin expression, which reduces cell adhesion and likely leads to cells that are more motile and, as a result, more invasive [13]. Inhibition of fascin or β-catenin expression has been shown to decrease the migratory capacity of ACP tumor cells in culture [13]. Other binding products for the armadillo repeat sequences include TCP-family transcription factors, Axin2, and APC [15].

Increased expression and/or activation of β-catenin likely leads to active expression of multiple target genes. These genes include *Lef1*, *Axin1*, *c-myc*, and CyclinD1, which is a major regulator of cell-cycle progression to the proliferative stages [5, 10, 11, 16]. *Axin2*, a well-recognised inhibitor, and *BMP4*, which plays a critical role in tooth development and increases cell proliferation in conjunction with β-catenin accumulation, are also target genes with increased expression as a result of increased Wnt signaling [9, 12, 16]. Both Axin2 and BMP4 were shown to co-localise to the nucleus with β-catenin with increased mRNA levels [9, 12, 13, 15]. Lef1, an enamel protein, expression is activated by BMP4 [15, 16].

ACP Genome Analyses

CTNNB1 Mutation

Exon 3 of the *CTNNB1* gene (β-catenin) encodes a degradation targeting motif. Mutation may confer resistance of the protein to the usual destruction complex [11, 16] and manifest through nuclear accumulation of β-catenin [5, 7, 10, 11, 19]. The first descriptions of this genome level aberration in ACP demonstrated *CTNNB1* mutations in 69–100% [6, 16, 20] of ACP specimens and the absence of such mutations in PCP and other parasellar tumors [6, 16]. Using massively parallel sequencing of 12 specimens of ACP, followed by targeted genotyping in an additional 53 specimens *CTNNB1* mutation, were identified in 95% (11/12 and 51/53) of specimens [21]. *CTNNB1* mutations were absent in PCP. Other groups have identified *CTNNB1* mutations in approximately 70–80% of ACPs [12, 16, 19, 22]. Although the source of thus variation remains unknown, potential explanations could include intratumoral heterogeneity, sampling error, or even the existence of multiple subtypes of ACP.

As described above, the majority of *CTNNB1* mutations in ACP are upregulating and involve phosphorylation of serine or threonine sites encoded by exon 3, which ultimately encode the degradation targeting box of β-catenin, which corresponds to the GSK3β binding domain [5, 7, 9, 11–13, 19, 20]. Mutations of *CTNNB1* have been identified at a number of different codons [9, 15, 16]. These include codon 32, which is adjacent to the phosphorylation site of GSK3β, as well as codons 33, 37, and 41, which affect the serine and threonine residues that are targeted by GSK3β [6, 9, 14, 15, 19].

The presence of the *CTNNB1* mutation in ACP may offer insight into the etiology of this tumor. Mutations of the GSK3β binding domain of *CTNNB1* have been identified in other lesions, including calcifying odontogenic cysts and pilomatricoma [7, 16, 19], a benign skin tumor that is histologically characterised by anucleate squamous cells, similar to ACP. As such, ACP shares genetic and histological characteristics with tumors of odontogenic origin. However, it has been postulated that ACP arises from undifferentiated anterior pituitary epithelial stem cells and differentiates toward an oral epithelial phenotype [18, 20, 23, 24]. One model proposes that there is a small population of mutated cells, which express nucleocytoplasmic β-catenin, which are histologically identified within well-demarcated clusters [17]. This theory is supported by recent in vitro and animal models that are discussed in Chap. 2 and 3. Briefly, a leading hypothesis regarding ACP tumorigenesis is that para–/autocrine signaling emanating from this small population of *CTNNB1* mutated cells (i.e. cancer stem cells) and that this drives tumor growth and invasion. It has also been postulated that microRNA derangements also in part help drive tumor behaviour [7]. It may therefore be that, while the tissue of origin differs, the phenotypic maturation of ACP tissue converges with that of odontogenic lesions [25, 26].

Epidermal Growth Factor Receptor

Although genome level dysregulation of the epidermal growth factor pathway does not appear to exist in ACP, there is evidence that upregulation of this pathway, potentially by epigenetic mechanisms, contributes to ACP cell growth and infiltration.

The epidermal growth factor receptor (EGFR) is a 170 kDa receptor tyrosine kinase (RTK) that is also known as HER1, ErbB1, mENA, and PIG61 [10, 18]. It is a transmembrane glycoprotein composed of an extracellular domain that is involved in ligand binding, a transmembrane portion, and an intracellular portion that makes up the tyrosine kinase domain and is the site for autophosphorylation [18]. EGFR can be activated by multiple growth factors, including epidermal growth factor (EGF), TGF-α, amphiregulin, epiregulin, heparin-binding EGF (HB-EGF), and betacellulin [18]. When the ligands bind the extracellular domain, the receptor undergoes dimerisation, which activates the intracellular tyrosine kinase domain. This results in phosphorylation of multiple tyrosine residues (e.g. Y992, Y1045, Y1068, Y1086, Y1148, and Y1173) [11, 18]. Binding of the ligand to the extracellular domain also promotes recruitment of other enzymes and proteins, which act as signal transducers and can lead to initiation of multiple intracellular cascades involved in regulation of cell proliferation, differentiation, apoptosis, and motility [18]. Alterations in EGFR signaling have been noted in multiple human tumor types [20] resulting from various mechanisms including EGFR gene amplification, activating mutations, or overexpression of ligands and receptors [18]. In some tumor types, a common location for mutation is the area that encodes the tyrosine kinase domain of EGFR, between exons 18 and 21 [18].

The mechanism of EGFR pathway contribution to ACP behaviour may lie in the interaction between EGFR and β-catenin [12, 18, 27]. When activated and phosphorylated, one interesting proposed mechanism is that EGFR co-localises with nuclear β-catenin and fascin, particularly at the infiltration border of ACP, within whorl-like cell clusters [12, 18]. As described above, in ACP, these clusters have been associated with activation of the canonical Wnt signaling pathway. This suggests a convergent role for both the EFGR and Wnt pathways in the pathogenesis of ACP [18]. This concept is supported by laboratory data [10, 20] and culture models that correlate EGFR pathway activation with ACP growth and migration of ACP cells [18]. Additionally, EGFR appears to be involved in the regulation of stem cell like properties in ACP through the regulation of the expression of stem cell markers and the activation of the EGFR pathway in β-catenin accumulating cells [5]. Lastly, the ACP transcriptome demonstrates high levels of EGF pathway genes, including *AREG*, *EGFR*, and *ERBB3* [20].

SHH

Sonic hedgehog (SHH) plays an integral role in the maintenance of stem cell milieus in adult tissues [28] and in the normal development of several organs, including Rathke's pouch [12]. The SHH pathway is involved in pituitary formation during early embryogenesis. The transmembrane protein Patched 1 (PTCH1) receptor complex binds the hedgehog ligand, which activates the pathway [22]. Once the ligand binds to the receptor complex, the frizzled class receptor, smoothened (SMO), is freed from PTCH1 inhibition. It then activates glioma-associated oncogene family zinc fingers—the GLI1, GLI2, and GLI3 transcription factors [22]. Pathological upregulation of SHH signaling contributes to mitogenic behaviour in numerous neoplasms, including glioma, basal cell carcinoma, lung cancer, and breast cancer [28, 29]. Among pediatric brain tumors, the most widely known example of hedgehog dysregulation is in the SHH subgroup of medulloblastoma [1].

As described in Chap. 2, expression of SHH has been observed in a murine model of ACP and in human ACP. In both mouse and human tissues, expression of SHH has been shown to co-localise in cells that have distinct nuclear localisation of β-catenin [28]. The pattern of expression, combined with our understanding of the SHH pathway in normal human pituitary development, supports the hypothesis that both autocrine and paracrine SHH signaling may contribute to ACP tumorigenesis. The importance of SHH signaling in the pathogenesis of ACP has been supported by two studies of the human ACP transcriptome, both of which have demonstrated significant upregulation of SHH pathway genes [20, 22] [Fig. 2]. In addition, a very recent study comparing the transcriptomes of ACP with PCP showed significantly higher expression of SHH pathway genes *Axin2*, *Gli2*, and *PTCH1* in ACP [25]. This important finding further supports the possibility of WNT and SHH signaling pathways converging to support the growth of ACP. This study also showed that methylation profiling patterns are distinct between ACP and PCP, further establishing that ACP

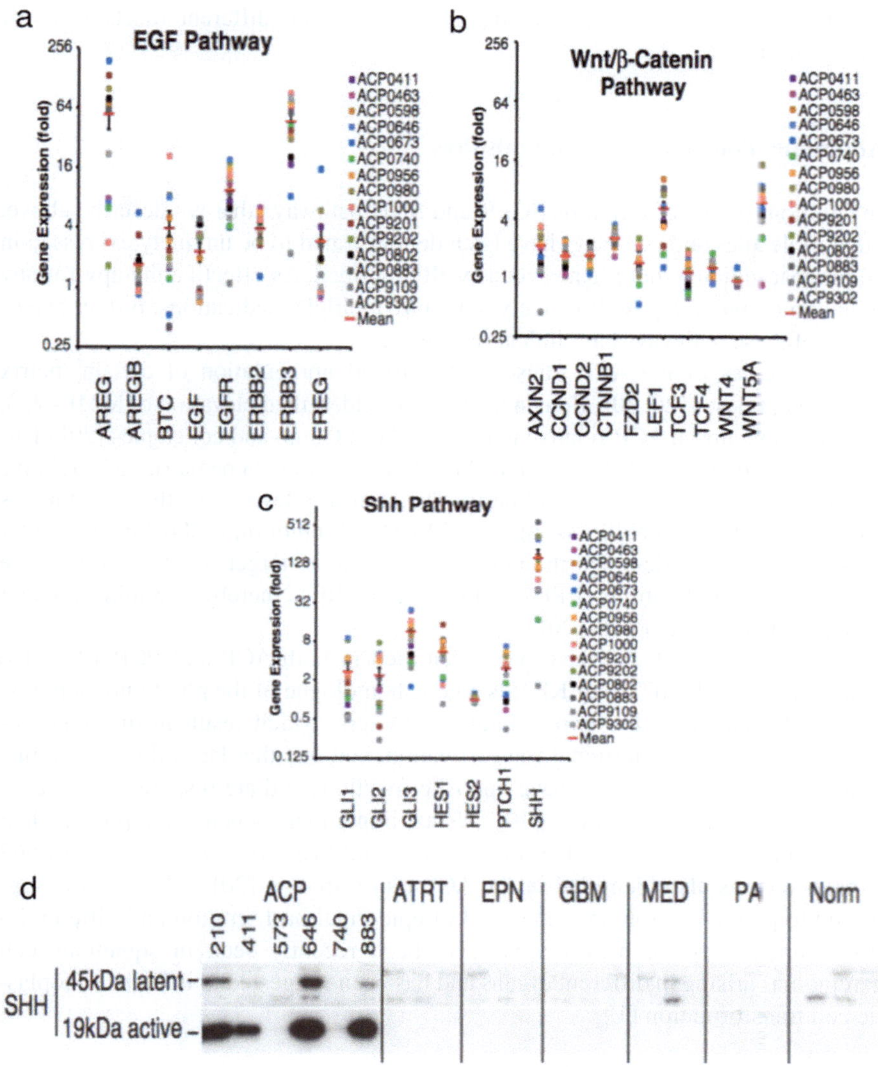

Fig. 2 Expression of the indicated developmental and cancer-related genes in individual ACP samples: (**a**) Epidermal growth factor genes; (**b**) WNT pathway; (**c**) Sonic hedgehog pathway; (**d**) Western blot analysis demonstrating overexpression of the latent pre-forms and active cleaved protein isoforms of SHH in ACP relative to other common pediatric brain tumors and normal brain. *AT/RT* atypical teratoid/rhabdoid tumor, *EPN* ependymoma, *GBM* glioblastoma, *MED* medulloblastoma, *PA* pilocytic astrocytoma, *Norm*, normal brain). Adapted from Gump et al [20]

and PCP are indeed biologically distinct lesions with different mechanisms of pathogenesis.

Additional Potential Relevant Pathways

In addition to the activation of EGFR and SHH pathways that we describe above, other molecules and pathways have been demonstrated to be uniquely expressed in craniopharyngioma, but are not yet as well described. As effective therapy for any brain tumor, including ACP, is likely to require multiple medications, further exploration of these pathways has clinical relevance.

Two studies of the ACP transcriptome found upregulation of certain matrix metallopeptidases [20, 30]. Gong and colleagues identified elevation of MMPs 2, 3, 7, and 9 in recurrent ACP, relative to primary ACP. Gump and colleagues [20] identified elevated levels of MMPs 9 and 12 mRNA transcripts in pediatric ACP, relative to other brain tumors and normal brain. The clinical relevance of these findings is increased by the availability of the oral MMP9/12 inhibitor, AZD1236. The latter group additionally identified the upregulation of three targets of the oral tyrosine kinase inhibitor dasatinib (EPHA2, LCK, and SRC), thereby revealing another potential therapeutic option [20].

Elevated levels of p63 have been identified in both ACP and PCP. P63, also known as p51, p40, p73L, or KET, is a gene homologue of the p53 tumor suppressor family. The gene has two separate promoters, which result in different isoforms, one with an N-terminal transactivation, and one that lacks this N-terminal domain. These two isoforms have opposite functions and are responsible for cell-cycle arrest and proliferation [19]. Using immunohistochemistry, p63 nuclear immunopositivity was identified in both ACP and PCP [10, 19]. Evidence of p63 expression was also identified in the ACP transcriptome [20]. While p63 is suggested to play a key role in the control of epidermal proliferation and differentiation, impaired p63 expression has also been recently seen in squamous cell carcinomas arising in different organs and has been suggested to influence neoplastic cell transformation [19].

Papillary Craniopharyngioma

Distinction from ACP

As described above, the distinct mutations identified in ACP and PCP, the different transcriptome profiles, and the differences in genome wide methylation patterns support the view that ACP and PCP are biologically distinct entities with different mechanisms of pathogenesis. These genomic studies support the earlier findings from histological studies that demonstrated nuclear and cytoplasmic accumulation of β-catenin in ACP whereas β-catenin localised exclusively to the cell membranes of tumor cells in PCP, similar to the pattern of localisation in other *CTNNB1* wild type tumors of the sellar region [10, 12, 16, 19].

BRAF

Mutation of the *BRAF* gene results in oncogenic potential through constitutive activation of a serine-threonine kinase that regulates MAP kinase signaling and affects cell division and differentiation. A multitude of human tumors harbor *BRAF* anomalies, most commonly *BRAF* V600E mutation. Successful therapy has been reported in cutaneous melanoma [31, 32] in addition to less common lesions, including ameloblastoma [33], hairy cell leukemia [34], and pleomorphic xanthoastrocytoma [35]. As described above, a recent genomic analysis of PCP [21] used a combination of whole exome sequencing and targeted genotyping. In this study, *BRAF* V600E mutations were identified in 92.8% of PCP specimens, but none of the ACP specimens examined. Notably, only three PCP specimens were analysed with whole exome sequencing of both normal DNA and PCP tumor DNA, thus the presence of other lower frequency recurrent genomic events in PCP cannot be entirely excluded. Additional studies have validated that nearly all PCP harbor *BRAF* V600E mutations [36–38]. One report suggests that a subset of craniopharyngioma may harbor coexisting mutations in both *CTNNB1* and *BRAF* genes [8]. Such a co-occurrence of mutations would be of great clinical interest and warrants further exploration and validation.

The potential clinical significance of *BRAF* V600E mutation in PCP has been described with regard to both diagnosis and therapy. A mutation-specific antibody (clone VE1) which recognises the BRAF V600E mutant protein but not the wild type protein [39] is now routinely used in clinical practice to support the diagnosis PCP and to help distinguish it from other masses of the sellar lesion including pituitary adenomas, ACP, and Rathke's cleft cysts [36, 38, 40]. Despite cross-reactivity with normal pituicytes and with ciliated cells, this antibody generally allows pathologists to rapidly and inexpensively confirm the diagnosis of PCP. Difficult cases can be assessed using targeted genotyping of the BRAF V600E mutant allele.

From a therapeutic standpoint, the combination of the RAF inhibitor dabrafenib and MEK inhibitor trametinib was recently used to successfully treat an adult patient with a multiply recurrent craniopharyngioma [41]. This patient had an exceptional response to treatment [Fig. 3]. Despite the frequent development of resistance to BRAF and MEK inhibitors that emerges in melanoma patients, whole exome sequencing failed to find evidence of such genetic resistance mechanisms in the PCP tissue that remained post-treatment. The rationale for targeting both BRAF and MEK is supported by the recent example of a patient with a BRAF V600E mutant PCP that was treated with single agent vemurafenib [42]. That tumor was

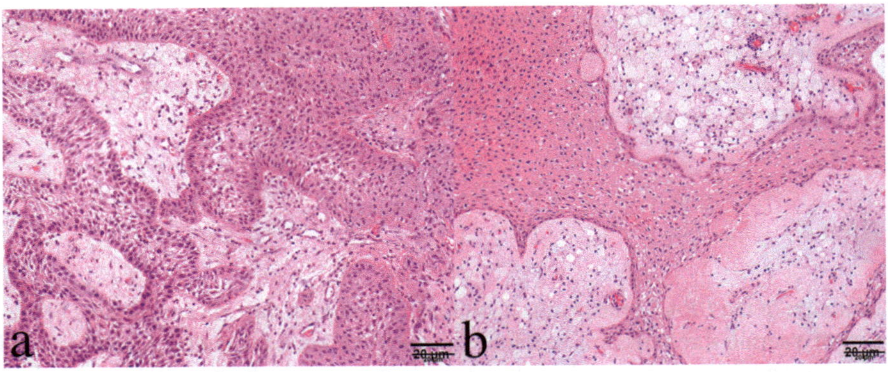

Fig. 3 Effect of dual therapy using dabrafenib and trametinib on multiply recurrent papillary craniopharyngioma. Haematoxylin and eosin stained specimens prior to (**a**) and following (**b**) treatment. Note the absence of epithelial tumor components and the engorgement of some fibrovascular cores with foamy macrophages

also exceptionally responsive to targeted treatment, however, when vemurafenib treatment was stopped the tumor regrew and subsequently became refractory to vemurafenib. This example suggests that combining BRAF and MEK inhibition may be needed for durable control of tumor growth.

A multi-institutional phase II study of combined BRAF and MEK inhibition for patients with PCP is forthcoming [43]. This trial will allow for correlative studies to assess if mechanisms of resistance emerge post-treatment as well as whether mutant DNA can be reliably detected in the patients' blood as a non-invasive means for diagnosis and monitoring therapy.

Summary

Over the last decade, substantial progress has been achieved in our understanding of the processes underlying craniopharyngioma pathogenesis. Genomic studies have identified distinct epigenomic and mutational profiles in the vast majority of ACP and PCP tumors. Thus, ACP and PCP are distinct entities and will require distinct therapeutic approaches. In the case of PCP, advances made in the context of other tumors that harbor *BRAF* mutations have facilitated the initiation of clinical trials of novel systemic therapies for this tumor. In the case of ACP, therapy directed at inhibiting β-catenin is not yet available. However, evidence from both genome and transcriptome studies has implicated additional opportunities such as the SHH and EGFR signaling pathways and matrix metalloproteinases that offer more readily available therapeutic options. Further validation of these findings may soon advance biomarker driven and molecularly guided systemic therapies for ACP as well.

References

1. Taylor MD, Northcott PA, Korshunov A, Remke M, Cho Y-J, Clifford SC et al (2012) Molecular subgroups of medulloblastoma: the current consensus. Acta Neuropathol 123(4):465–472
2. Witt H, Mack SC, Ryzhova M, Bender S, Sill M, Isserlin R et al (2011) Delineation of two clinically and molecularly distinct subgroups of posterior fossa ependymoma. Cancer Cell 20(2):143–157
3. Cancer Genome Atlas Research Network, Brat DJ, Verhaak RG, Aldape KD, Yung WK, Salama SR, Cooper LA et al (2015) Comprehensive, integrative genomic analysis of diffuse lower-grade gliomas. N Engl J Med 372(26):2481–2498. doi:10.1056/NEJMoa1402121
4. Brennan CW, Verhaak RG, McKenna A, Campos B, Noushmehr H, Salama SR et al (2013) The somatic genomic landscape of glioblastoma. Cell 155(2):462–477. doi:10.1016/j.cell.2013.09.034
5. Hölsken A, Stache C, Schlaffer SM, Flitsch J, Fahlbusch R, Buchfelder M et al (2014) Adamantinomatous craniopharyngiomas express tumor stem cell markers in cells with activated Wnt signaling: further evidence for the existence of a tumor stem cell niche? Pituitary 17(6):546–556
6. Kato K, Nakatani Y, Kanno H, Inayama Y, Ijiri R, Nagahara N et al (2004 Jul) Possible linkage between specific histological structures and aberrant reactivation of the Wnt pathway in adamantinomatous craniopharyngioma. J Pathol 203(3):814–821
7. Campanini ML, Colli LM, Carvalho Paixao BM, Freitas Cabral TP, Amaral FC, Machado HR et al (2010) *CTNNB1* gene mutations, pituitary transcription factors, and microRNA expression involvement in the pathogenesis of adamantinomatous craniopharyngiomas. Horm Canc 1(4):187–196
8. Larkin SJ, Preda V, Karavitaki N, Grossman A, Ansorge O (2014) BRAF V600E mutations are characteristic for papillary carniopharyngioma and may coexist with CTNNB1-mutated adamantinomatous craniopharyngioma. Acta Neuropathol 127(6):927–929
9. Cani CMG, Matushita H, Carvalho LRS, Soares IC, Brito LP, Almeida MQ et al (2011) *PROP1* and *CTNNB1* expression in adamantinoumatous craniopharyngiomas with or without b-catenin mutations. Clinics (Sao Paulo) 66(11):1849–1854
10. Esheba GE, Hassan AA (2015) Comparative immunohistochemical expression of b-catenin, EGFR, ErbB2, and p63 in adamantinomatous and papillary craniopharyngiomas. J Egypt Natl Canc Inst 27(3):139–145
11. Zuhur SS, Tanik C, Erol RS, Musluman AM, Kabukcuoglu F, Altuntas Y (2013) Immunohistochemical expression of ErbB2 in adamantinomatous craniopharyngiomas: a possible target for immunotherapy. Turk Neurosurg 23(1):55–60
12. Larkin SJ, Ansorge O (2013) Pathology and pathogenesis of craniopharyngiomas. Pituitary 16(1):9–17
13. Hölsken A, Buchfelder M, Fahlbusch R, Blumcke I, Buslei R (2010) Tumour cell migration in adamantinomatous craniopharyngiomas is promoted by activated Wnt-signalling. Acta Neuropathol 119(5):631–639
14. Sekine S, Shibata T, Kokubu A, Morishita Y, Moguchi M, Nakanishi Y et al (2002) Craniopharyngiomas of adamantinoumatous type harbor b-catenin gene mutations. Am J Pathol 161(6):1997–2001
15. Hölsken A, Kreutzer J, Hofmann BM, Hans V, Oppel F, Buchfelder M et al (2009) Target gene activation of the Wnt signaling pathway in nuclear b-catenin accumulating cells of adamantinomatous craniopharyngiomas. Brain Pathol 19(3):357–364
16. Buslei R, Nolde M, Hofmann B, Meissner S, Eyupoglu IY, Siebzehnrubl F et al (2005) Common mutations of b-catenin in adamantinomatous craniopharyngiomas but not in other tumors originating from the sellar region. Acta Neuropathol 109(6):589–597
17. Buseli R, Hölsken A, Hofmann B, Kreutzer J, Siebzehnrubl F, Hans V et al (2007) Nuclear b-catenin accumulation associates with epithelial morphogenesis in craniopharyngiomas. Acta Neuropathol 113(5):585–590

18. Hölsken A, Gebhardt M, Buchfelder M, Fahlbusch R, Blumeke I, Buslei R (2011) EGFR signaling regulates tumor cell migration in craniopharyngiomas. Clin Cancer Res 17(13):4367–4377
19. Cao J, Lin JP, Yang LX, Chen K, Huang ZS (2010) Expression of aberrant beta-catenin and impaired p63 in craniopharyngiomas. Br J Neurosurg 24(3):249–256
20. Gump JM, Donson AM, Birks DK, Amani VM, Rao KK, Griesinger AM et al (2015) Identification of targets for rational pharmacological therapy in childhood craniopharyngioma. Acta Neuropathol Commun 3:30
21. Brastianos PK, Taylor-Weiner A, Manley PE, Jones RT, Dias-Santagata D, Thorner AR et al (2014) Exome sequencing identifies BRAF mutations in papiilary craniopharyngiomas. Nat Genet 46(2):161–165
22. Gomes DC, Jamra SA, Leal LF, Colli LM, Campanini ML, Oliveria RS et al (2015) Sonic Hedehog pathway is upregulated in adamantinomatous craniopharyngiomas. Eur J Endocrinol 172(5):603–608
23. Gaston-Massuet C, Andoniadou CL, Signore M, Jayakody SA, Charolidi N, Kyeyune R et al (2011) Increased Wingless (Wnt) signaling in pituitary progenitor/stem cells gives rise to pituitary tumors in mice and humans. Proc Natl Acad Sci U S A 108(28):11482–11487
24. Sekine S, Takata T, Shibata T, Mori M, Morishita Y, Noguchi M et al (2004) Expression of enamel proteins and LEF1 in adamantinomatous craniopharyngioma: evidence for its odontogenic epithelial differentiation. Histopathology 45(6):573–579
25. Hölsken A, Sill M, Merkle J, Schweizer L, Buchfelder M, Flitsch J et al (2016) Adamantinomatous and papillary craniopharyngiomas are characterized by distinct epigenomic as well as mutational and transcriptomic profiles. Acta Neuropathol Commun. 4(1):20
26. Martinez-Barbera JP, Andoniadou CL (2016) Concise review: paracrine role of stem cells in pituitary tumors: a focus on adamantinomatous craniopharyngioma. Stem Cells 34(2):268–276
27. Lee CH, Hung HW, Hung PH, Shich YS (2010) Epidermal growth factor receptor regulates beta-catenin location, stability, and transcriptional activity in oral cancer. Mol Cancer 9:64
28. Andoniadou CL, Gaston-Massuet C, Reddy R, Schneider RP, Blasco MA, Le Tissier P et al (2012) Identification of novel pathways involved in the pathogenesis of human adamantinoumatous cranipharyngioma. Acta Neuropathol 124(2):259–271
29. Katoh Y, Katoh M (2009) Hedgehog target genes: mechanisms of carcinogenesis induced by aberrant hedgehog signaling activation. Curr Mol Med 9(7):873–886
30. Gong J, Zhang H, Xing S, Li C, Ma Z, Jia G et al (2014) High expression levels of CXCL12 and CXCR4 predict recurrence of adamantinomatous craniopharyngiomas in children. Cancer Biomark 14(4):241–251
31. Flaherty KT, Infante JR, Daud A, Gonzalez R, Kefford RF, Sosman J et al (2012) Combined BRAF and MEK inhibition in melanoma with BRAF V600 mutations. N Engl J Med 367(18):1694–1703
32. Flaherty KT, Puzanov I, Kim KB, Ribas A, McArthur GA, Sosman JA et al (2010) Inhibition of mutated, activated BRAF in metastatic melanoma. N Engl J Med 363(9):809–819
33. Kaye FJ, Ivey AM, Drane WE, Mendenhall WM, Allan RW (2014) Clinical and radiographic response with combined BRAF-targeted therapy in stage 4 ameloblastoma. J Natl Cancer Inst 107(1):378. doi:10.1093/jnci/dju378
34. Dietrich S, Pircher A, Endris V, Peyrade F, Wendtner CM, Follows GA, Hüllein J, Jethwa A, Ellert E, Walther T, Liu X, Dyer MJ, Elter T, Brummer T, Zeiser R, Hermann M, Herold M, Weichert W, Dearden C, Haferlach T, Seiffert M, Hallek M, von Kalle C, Ho AD, Gaehler A, Andrulis M, Steurer M, Zenz T (2016) BRAF inhibition in hairy cell leukemia with low dose vemurafenib. Blood 127(23):2847–2855
35. Hyman DM, Puzanov I, Subbiah V, Faris JE, Chau I, Blay JY et al (2015) Vemurafenib in multiple nonmelanoma cancers with BRAF V600 mutations. N Engl J Med 373(8):726–736

36. Kim JH, Paulus W, Heim S (2015) BRAF V600E mutation is a useful marker for differentiating Rathke's cleft cyst with squamous metaplasia from papillary craniopharyngioma. J Neuro-Oncol 123(1):189–191. doi:10.1007/s11060-015-1757-6
37. Marucci G, de Biase D, Zoli M, Faustini-Fustini M, Bacci A, Pasquini E, Visani M, Mazzatenta D, Frank G, Tallini G (2015) Targeted BRAF and CTNNB1 next-generation sequencing allows proper classification of nonadenomatous lesions of the sellar region in samples with limiting amounts of lesional cells. Pituitary 18(6):905–911. doi:10.1007/s11102-015-0569-y Erratum in: Pituitary. 2016 Feb;19(1):113
38. Schweizer L, Capper D, Hölsken A, Fahlbusch R, Flitsch J, Buchfelder M, Herold-Mende C, von Deimling A, Buslei R (2015 Oct) BRAF V600E analysis for the differentiation of papillary craniopharyngiomas and Rathke's cleft cysts. Neuropathol Appl Neurobiol 41(6):733–742. doi:10.1111/nan.12201
39. Capper D, Preusser M, Habel A, Sahm F, Ackermann U, Schindler G et al (2011) Assessment of BRAF V600E mutation status by immunohistochemistry with a mutation-specific monoclonal antibody. Acta Neuropathol 122(1):11–19
40. Jones RT, Abedalthagafi MS, Brahmandam M, Greenfield EA, Hoang MP, Louis DN, Hornick JL, Santagata S (2015 Apr) Cross-reactivity of the BRAF VE1 antibody with epitopes in axonemal dyneins leads to staining of cilia. Mod Pathol 28(4):596–606. doi:10.1038/modpathol.2014.150
41. Brastianos PK, Shankar GM, Gill CM, Taylor-Weiner A, Nayyar N, Panka DJ et al (2015) Dramatic response of BRAF V600E mutant papillary craniopharyngioma to targeted therapy. J Natl Cancer Inst 108(2):djv310
42. Aylwin SJ, Bodi I, Beaney R (2015) Pronounced response of papillary craniopharyngioma to treatment with vemurafenib, a BRAF inhibitor. Pituitary 19(5):544–546
43. Brastianos PK, Santagata S (2016) Endocrine tumors: BRAF V600E mutations in papillary craniopharyngioma. Eur J Endocrinol 174(4):R139–R144

Genetically Modified Mouse Models of Adamantinomatous Craniopharyngioma

Cynthia Lilian Andoniadou and Juan Pedro Martinez-Barbera

Abstract Over the last few years, there have been significant advances in the understanding of the pathogenesis of human adamantinomatous craniopharyngioma. Global and gene-specific expression studies of human biopsies have helped to delineate the network of pathways involved in the initiation and establishment of these tumours. Interestingly, genetically modified and patient-derived xenograft mouse models have revealed the function of some of the dysregulated pathways as well as identified a role for specific tumour cells in the formation of these neoplasias. It has emerged that undifferentiated embryonic precursors and adult pituitary stem cells are critical in the pathogenesis of murine tumours resembling human ACP. In this chapter, we aim to provide a general overview of pituitary development, since there is strong case for a developmental origin for human ACP. We also discuss the in vitro and in vivo evidence demonstrating the presence of stem cells in the adult pituitary as well as their role of the pathogenesis of craniopharyngioma.

Keywords Pituitary • Stem cells • Sox2 • Craniopharyngioma • WNT pathway • Mouse models

Normal Development of the Pituitary Gland

The pituitary gland is a master endocrine organ that controls critical physiological processes including growth, reproduction, metabolism and the stress response. The mature gland, consisting of the anterior and posterior pituitary (AP and PP, respectively), starts to develop around 9.0 days post coitum (dpc) in mice and around

C.L. Andoniadou (✉)
Craniofacial Development and Stem Cell Biology, King's College London, Tower Wing Floor 27, Guy's Campus, London SE1 9RT, UK
e-mail: cynthia.andoniadou@kcl.ac.uk

J.P. Martinez-Barbera
Developmental Biology and Cancer Programme, Birth Defects Research Centre, Institute of Child Health, University College London, 30 Guilford Street, WC1N 1EH London, UK
e-mail: j.martinez-barbera@ucl.ac.uk

J.P. Martinez-Barbera, C.L. Andoniadou (eds.), *Basic Research and Clinical Aspects of Adamantinomatous Craniopharyngioma*,
DOI 10.1007/978-3-319-51890-9_3

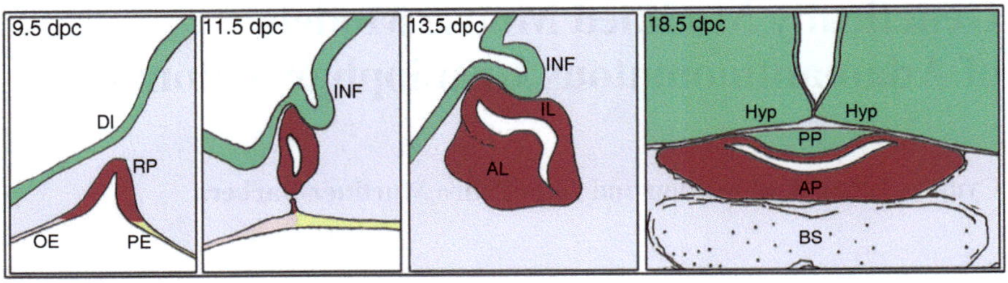

Fig. 1 Cartoon outlining key steps during mouse pituitary development. Rathke's pouch (RP) forms at 9.5 dpc (days post coitum) as a thickening of the oral ectoderm (OE) at the boundary between OE and pharyngeal endoderm (PE) that invaginates toward the ventral diencephalon (DI). The epithelium of RP contains the undifferentiated progenitors that give rise to all hormone-producing cells by the end of gestation. By 11.5 dpc, the invaginating RP pinches off from the oral ectoderm and establishes contact with the infundibulum (INF), a recess of the floor of the DI. RP progenitors proliferate rapidly from 9.5 dpc and the progeny migrate ventrally and initiate differentiation to populate the primordium of the anterior lobe (AL) of the pituitary. The presumptive intermediate lobe (IL) is indicated. At 18.5 dpc, just prior to birth, the pituitary comprises the posterior pituitary (PP), of neural origin, and the anterior pituitary (AP), derived from the anterior and intermediate lobes. The pituitary is located between the hypothalamus (Hyp) and the basisphenoid cartilage (BS). Reproduced with permission from Wiley (Martinez-Barbera and Andoniadou (2016) *Stem Cells*)

week 4 in humans from two distinct germ layers (Fig. 1). The AP derives from Rathke's pouch, a region of the oral epithelium of the early embryo, which contains the undifferentiated precursors of all of the hormone-producing endocrine cells of the mature pituitary as well as tissue-specific pituitary stem cells of the adult organ. The endocrine cell types are somatotrophs, lactotrophs, thyrotrophs, gonadotrophs, corticotrophs and melanotrophs, secreting growth hormone (GH), prolactin (PRL), thyroid stimulating hormone (TSH), the gonadotrophins follicle stimulating hormone and luteinising hormone (FSH and LH), adrenocorticotropic hormone (ACTH) and melanocyte stimulating hormone (MSH), respectively [1]. The PP is of neural origin and originates from a diverticulum at the floor of the prospective hypothalamus called the infundibulum. The PP is richly vascularised, hypocellular and contains the axons of the oxytocin and arginine vasopressin neurons of the hypothalamus. It is connected to and under the control of the hypothalamus through the pituitary stalk, which comprises the portal vessels and hypothalamic axons. Together, these structures form a functional unit, the hypothalamo-pituitary (HP) axis.

Normal morphogenesis and differentiation of the pituitary gland is regulated by the activity of several transcription factors and signalling pathways [2–4]. One of these pathways is WNT/β-catenin, a major regulator of several cellular processes during organogenesis and adulthood [5, 6]. The activity of the WNT/β-catenin pathway is controlled by the stability of β-catenin, a transcriptional activator lacking a

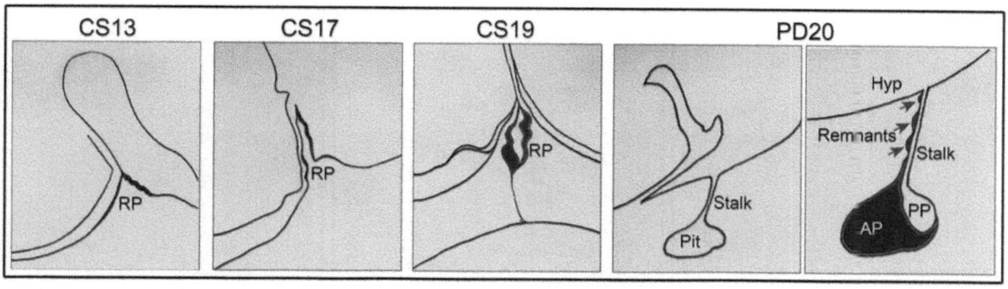

Fig. 2 Cartoon outlining key steps during human pituitary development. By Carnegie Stage (CS) 13, invagination of Rathke's pouch (RP) has initiated. This makes contact with the ventral diencephalon from CS17. By CS19 the rudimentary pouch has formed and is detaching from the oral epithelium. The postnatal pituitary gland (Pit), presented here at postnatal day (PD) 20, is connected to the hypothalamus (Hyp) via the pituitary stalk (Stalk), with axons terminating in the posterior pituitary (PP). A distinct feature of the human pituitary compared to mouse is the longer pituitary stalk, on which are found remnants of anterior pituitary (AP) tissue

DNA-binding domain. In the absence of WNT ligands, β-catenin is phosphorylated by a destruction complex and degraded rapidly via the ubiquitination pathway, preventing its stabilisation and the activation of the WNT cascade. Binding of WNT ligands to their receptors disrupts the destruction complex preventing the phosphorylation of β-catenin and its subsequent degradation. Stabilised β-catenin can translocate into the nucleus and interact with DNA-binding transcription factors of the TCF/LEF family to activate the expression of target genes (Fig. 2).

In humans, the morphogenesis of the HP axis shares a common embryonic origin with mice and the same genes and pathways are deployed during its development. However, there are some species-specific differences that are important to keep in mind. In both human and mouse embryos, the precursors of the anterior pituitary in Rathke's pouch are initially in contact with the overlying hypothalamic primordium and their descendants must migrate into the sellar region to form the anterior pituitary. The primordium of the posterior pituitary (i.e. infundibulum) is initially embedded within the hypothalamus and moves downwards to its final position adjacent to the anterior pituitary within the sella turcica (Fig. 3). However, the pituitary stalk is markedly longer in human embryos and pituitary cells must migrate further during morphogenesis [7]. Developmental defects in these morphogenetic movements lead to Pituitary Stalk Interruption Syndrome (PSIS) in humans, which is characterised by the presence of a thin or absent pituitary stalk, commonly associated with a hypoplastic or aplastic anterior pituitary or an ectopic posterior pituitary (e.g. attached to the floor of the 3rd ventricle, optic chiasm or median eminence) [8, 9]. Another peculiarity of the human HP axis, a feature not reported in mice, is the presence of pituitary tissue along the dorso-ventral axis of the stalk. These pituitary 'nests' were initially observed by Jakob Erdheim in 1904 [10], suggesting that remnants of RP could be trapped during normal development of the HP axis, a finding that could have implications in the location of pituitary tumours within the sella region, such as whether supra- or intra-cellar.

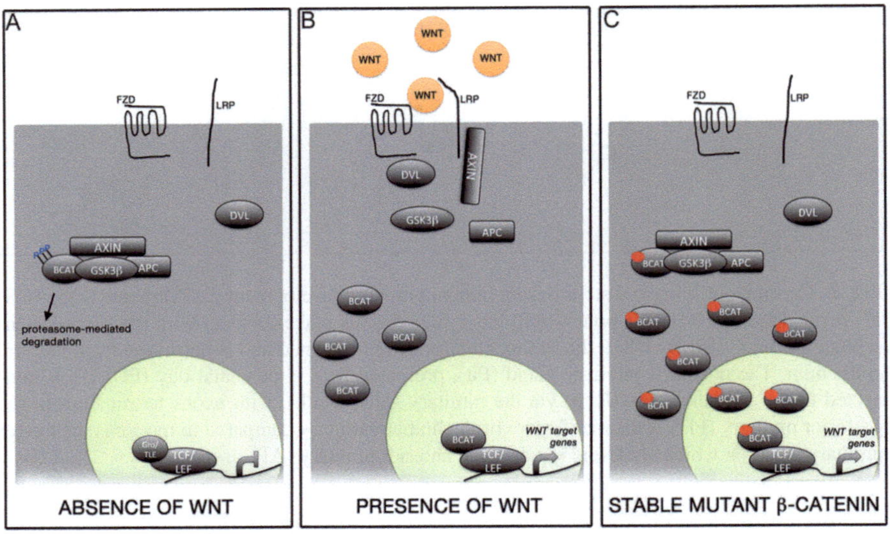

Fig. 3 WNT/β-catenin signalling pathway. (**a**) In the absence of WNT ligands, β-catenin is in a destruction complex formed by AXIN, APC and GSK3β and phosphorylated in specific residues encoded by exon 3 of *CTNNB1*. This results in rapid degradation of β-catenin following ubiquitination, and consequently, the repression of target genes brought about by the interaction of TCF/LEF1 with co-repressors (e.g. groucho (Gro)/TLE family members). (**b**) Binding of WNT ligands to Frizzled (FZD) receptors and co-receptors LRP leads to the recruitment of dishevelled (DVL) and AXIN to the cytoplasmic membrane. This causes the disruption of the destruction complex, releasing β-catenin, which cannot be efficiently phosphorylated and degraded. The stabilisation of β-catenin results in increased cytoplasmic protein levels and eventually nuclear translocation, where its interaction with TCF/LEF1 factors mediates the expression of pathway target genes. (**c**) Expression of a stable form of β-catenin lacking the amino acids encoded by exon 3 of *CTNNB1* (*small red circle* in BCAT) results in protein stabilization, since mutant β-catenin cannot be targeted for degradation by phosphorylation at these residues. This causes the activation of the WNT/β-catenin pathway in a cell-autonomous manner, even in the complete absence of WNT ligands. Reproduced with permission from Wiley (Martinez-Barbera (2015) *Neuropathol Appl Neurobiol*)

Pathology of Human ACP

Human ACPs are cystic tumours that develop within the sellar region, frequently compromising the hypothalamus, pituitary gland, pituitary stalk and optic chiasm. The histopathology of these tumours is characterized by the formation of a peripheral basal cell layer of palisading epithelium, loose aggregates of stellate cells, nodules containing anucleated 'ghost cells' and 'wet keratin', large areas of regressive changes, i.e. hemosiderin deposits, cholesterol clefts, multinucleated foreign-body giant cells, inflammation and calcifications (Fig. 4a) [11–14]. Human ACPs are usually benign tumours but there have been rare cases of malignant ACP, perhaps linked to the use of radiotherapy [15–17]. In these malignant neoplasms, P53 is overexpressed [18], alluding to the possibility that other mutations may have been induced as a result of the irradiation thus generating more aggressive tumours.

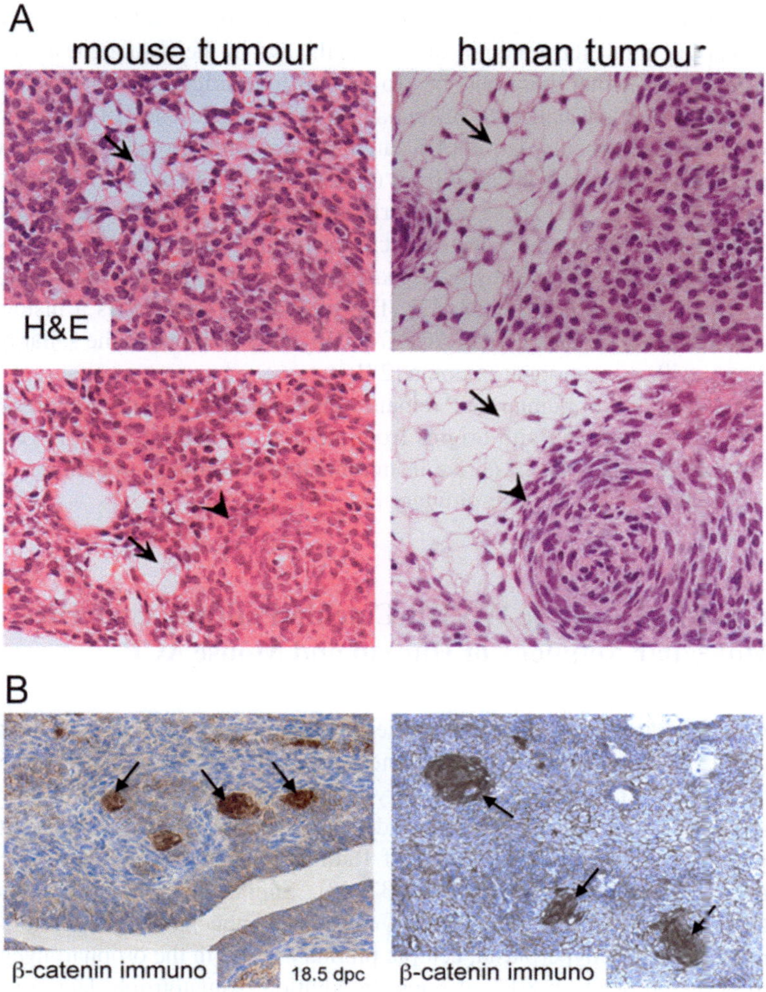

Fig. 4 Histomorphological features of mouse and human ACP. (**a**) Haematoxylin and eosin staining of mouse and human tumours showing the presence of microcystic changes (stellate reticulum; *arrows*) and whorl-like nodular structures (cell clusters; *arrowheads*). (**b**) Immunohistochemistry with a specific anti-β-catenin antibody showing the presence of cell clusters with nucleo-cytoplasmic accumulation of β-catenin. Reproduced with permission from PNAS (Gaston-Massuet et al. (2011) *Proc Natl Acad Sci USA*)

Mutations in *CTNNB1* have been identified in the majority of the human ACP samples analysed [19–24]. These are mostly restricted to exon 3 and result in the substitution or deletion of the critical regulatory amino acids that control stabilisation of the protein, predicted to lead to its nucleo-cytoplasmic accumulation and over-activation of the WNT/β-catenin pathway [25–29]. Global expression analyses have identified several genes/pathways that are dysregulated in human ACP, some of which could be targeted using drugs that are currently approved [30, 31]. However, little is known about the function of such pathways in the pathogenesis of these neoplasias, e.g. whether these may be pro- or anti-tumourigenic.

A histopathological hallmark of human ACP is the presence of cells showing nucleo-cytoplasmic accumulation of β-catenin, commonly grouped forming whirl-like structures close to the invasive front (hereby referred to as 'cell clusters') [20, 21, 24, 32, 33] (Fig. 4b). These are clearly a minority of cells and are distributed evenly throughout the tumour. Clusters accumulating β-catenin activate the WNT/β--catenin pathway, as evidenced by the elevated expression of pathway targets such as *AXIN2*, *LEF1* and *BMP4* [26, 28, 32]. The presence of clusters is a histological hallmark that differentiates human ACP from any other tumour of the sellar region, including papillary craniopharyngioma [27].

The cell of origin of human ACP remains unknown. The pattern of expression of cytokeratins is consistent with origins from the oral epithelium and it is generally believed that these neoplasms derive from remnants of Rathke's pouch, the embryonic primordium of the anterior pituitary [34, 35].

Over-Activation of the WNT/β-Catenin Pathway Underlies the Molecular Aetiology of Human and Mouse ACP

Genetic studies in mice have demonstrated the requirement of the WNT/β-catenin pathway for normal pituitary morphogenesis and differentiation in both mice and humans [36]. WNT/β-catenin is necessary for terminal differentiation of hormone-producing cells at late stages of pituitary development, whilst at early stages it must be regulated for normal proliferation of Rathke's pouch progenitors, e.g. if not inhibited results in overproliferation [37–42]. Supporting this notion, the up-regulation of the WNT/β-catenin pathway in Rathke's pouch (RP) progenitors leads to dramatic pituitary hyperplasia, ectopic pituitary tissue in the oropharyngeal cavity and perinatal death in 80% of pups. In addition, somatotroph, lactotroph and thyrotroph differentiation is impaired [26]. This genetically modified mouse model ($Hesx1^{Cre/+};Ctnnb1^{lox(ex3)/+}$) harbours a mutated form of β-catenin that is transcriptionally active, but lacks the amino acids encoded by exon 3 of *Ctnnb1* (the gene encoding β-catenin), preventing its phosphorylation and subsequent degradation, leading to the over-activation of the WNT/β-catenin pathway, akin to human mutations in exon 3 [43].

An interesting observation of the previously described embryonic mouse line is that pups which survive weaning (around 20%) show variable growth defects, consistent with an impairment in somatotroph differentiation, and fully penetrant pituitary tumours [26]. Histopathological analysis revealed characteristic hallmarks of human ACP, such as large cysts, cell clusters with nucleo-cytoplasmic accumulation of β-catenin and expression of *Axin2*, *Lef1* and *Bmp4*, indicating the activation of the WNT/β-catenin pathway [25–27]. These neoplastic cells did not express markers of proliferation or pituitary hormones and were negative for synaptophysin, a marker of neurons and endocrine cells. Therefore, cluster cells in mouse and human ACP have exited the cell cycle, are undifferentiated and, possibly not of either neural or endocrine origin. However, contrary to the human histopathology, mouse ACP did not display wet keratin or calcification, suggesting species-specific differences between mouse and human ACP. Critically, human ACP develop over years, but mouse ACP develop over a period of weeks. Despite these differences, the mouse tumours are more similar to human ACP than to any other malformation or tumour of the sellar region including pituitary adenoma, Rathke's cleft cysts and even the papillary type of craniopharyngioma. Together, the human and mouse studies have provided robust evidence for a role of *CTNNB1* mutations and the over-activation of the WNT/β-catenin pathway in the molecular aetiology of ACP.

Childhood-Onset ACP is Likely to be a Tumour of Embryonic Origin

The molecular and genetic studies using the embryonic ACP mouse model revealed two key findings, of relevance to the pathogenesis of human ACP, and in particular childhood-onset ACP. Analyses of mouse embryos at different stages of development established these form β-catenin-accumulating cell clusters with a similar gene expression profile to the human clusters and that these are established by the end of gestation [26]. Second, that in these mice, tumours only form when oncogenic β-catenin is expressed in Rathke's pouch undifferentiated precursors, but not when committed progenitors or fully differentiated hormone-producing cells are targeted [26].

Are these findings applicable to the human tumours? The similarities between the mouse and human clusters are remarkable, not only morphologically but also at the molecular level, strongly suggesting that they represent equivalent structures in both species. In fact, whole transcriptome gene expression profiling of the mouse cluster cells revealed numerous genes/pathways that were dysregulated in mouse and also in human tumours [30, 31, 44–46]. The possibility that ACP patients may have cell clusters in the pituitary gland is supported by clinical evidence. In a retrospective analysis of 90 children diagnosed with ACP (median age at diagnosis 8.3 years), Müller and colleagues reported that a significant decrease in height score was noted at 10–12 months of age and persisted until the diagnosis of ACP [47]. Taken into consideration that somatotrophs are clearly defective in the embryonic ACP model by the end of gestation and postnatally result in dwarfism [26], it is prudent to suggest that clusters may develop during gestation of human patients. In fact, ACP was recently diagnosed in a human foetus at 18 weeks [48].

Taking into account the previously described anatomical differences in the length of the pituitary stalk as well as the more pronounced cephalic flexure of the developing human embryo [49], it is possible that tumour epithelial cells (i.e. destined to be cluster cells) could be trapped in the suprasellar region near the hypothalamus. These could eventually give rise to a tumour with little or no sellar component at all. This is compatible with the remnants of pituitary tissue identified along the dorso-ventral axis of the human pituitary stalk. It could be speculated that the position of the tumour-initiating cell targeted with the *CTNNB1* mutation could determine the location of the ACP tumour in relation to the sella. Therefore, it is possible that the distinct behaviour of human ACP, for instance regarding its infiltrative behaviour, could be related to the position of the tumour initiating cells rather than to specific differences in its genetic/epigenetic landscape. Further molecular analyses will clarify this hypothesis.

Pituitary Stem Cells are Present in Mouse and Human Pituitaries

In recent years, solid evidence has gathered for the existence of a tissue-specific population of undifferentiated progenitors or stem cells within the hormone-secreting anterior pituitary gland of the mouse (pituitary stem cells, PSCs). These long-lived, undifferentiated cells have the capacity to directly commit or generate daughter cells of all cell lineages.

The transcription factor SOX2 is highly associated with pluripotency in the early embryo, in embryonic stem (ES) cells and induced pluripotent stem (iPS) cells, as well as the stem cell state in multiple embryonic and adult tissues. In the pituitary it was shown to define a population, which exclusively contains cells capable of both adherent and non-adherent clonogenic colony formation [44, 50]. In rodents at post-natal stages, SOX2-positive cells show a high degree of overlap with other proposed stem cell markers in vivo, such as PROP1 [51, 52], PRX1/2 [53] and SOX9 [54]. SOX2-expressing cells with similar clonogenic potential have been identified in human pituitaries [55].

Although SOX2-positive cells do not express hormones and SOX2 expression does not generally overlap with expression of the committed progenitor markers, the population has a heterogeneous nature in terms of properties, where only a low proportion (2.5–5%) are shown to be capable of in vitro self-renewal. Purifying pituitary cells based on S100β expression using S100β-GFP transgenic animals enriches in vitro colony formation, a property lying exclusively within the SOX2-positive compartment [56], hence suggesting that the overlapping population is likely to be enriched in stem cells and the strong association of S100β with multiple proposed pituitary stem cell populations makes this protein a good candidate marker. Indeed, there is significant overlap between SOX2 and S100β positivity [50, 56]. On its own, S100β marks multiple populations, displaying expression in the SOX2-positive marginal zone region, in the morphologically distinct folliculostellate cells and reported to be expressed in pituicytes. Multiple functions have been put forward for S100β-positive folliculostellate cells, reviewed comprehensively elsewhere [57], including providing structural support and enabling communication within the pituitary thus facilitating pulsatile hormone secretion [58, 59]. A further attribute of these cells is the production of paracrine signals that contribute to the regulation of surrounding endocrine cells. For example, synthesis and secretion of FSH in gonadotrophs is mediated by activin, and regulated by the activin-binding protein follistatin, both of which are produced by folliculostellate cells [60]. Numerous other paracrine factors secreted by these cells have been identified, including interleukin-6 (IL-6), leukemia inhibitory factor (LIF), vascular endothelial growth factor (VEGF) and basic fibroblastic growth factor (bFGF), all of which contribute to pituitary function [61–63].

Contribution of Stem Cells in the Long-Term Maintenance of the Anterior Pituitary

Despite a plethora of identified markers, until recently there was no evidence to support that pituitary stem cells function as such in vivo. This changed with the generation of two similar inducible mouse strains expressing CreERT2 under the regulation of the *Sox2* promoter, generated by the Hochedlinger and Martinez-Barbera/Pevny labs [56, 64]. In these, Cre recombinase is expressed by cells expressing *Sox2*, but will not be active in the nucleus until the administration of tamoxifen, allowing temporal control of recombinase action. Although the expression construct of these two strains is not identical, the expression of CreERT2 faithfully reproduces that of *Sox2* in both cases. *Sox2*-expressing cells were found to give rise to all committed progenitor cell types (PIT1 (POU1F1), TPIT (TBX19), SF1 (NR5A1)), and thus all hormone-secreting cells types of the anterior lobe (GH, PRL, TSH, ACTH, LH/FSH). In addition, $SOX2^+$ stem cells were shown to compose a long-lived population, either persisting long-term or maintained as a self-renewing pool of stem cells. Similar results were observed using a mouse expressing CreERT2 under the control of the *Sox9* gene (*Sox9-CreERT2* mouse line) [54]. This is not surprising as SOX2- and SOX9-expressing cells represent the same population. The above experiments demonstrate the presence of a long-lived population that retains pituitary stem cell properties throughout normal life.

Paracrine Contribution of Stem Cells to Tumourigenesis

As previously noted, mouse ACP form only when embryonic progenitors are targeted. Confirming this notion, expression of oncogenic β-catenin specifically in $SOX2^+$ stem cells of the embryonic or adult pituitary ($Sox2^{CreERT2/+};Ctnnb1^{lox(ex3)/+}$ mouse model) also result in tumours that are similar to human ACP [56]. These mouse tumours contain cell clusters showing nucleo-cytoplasmic β-catenin and are undifferentiated (i.e. do not express any hormone), as the human tumours.

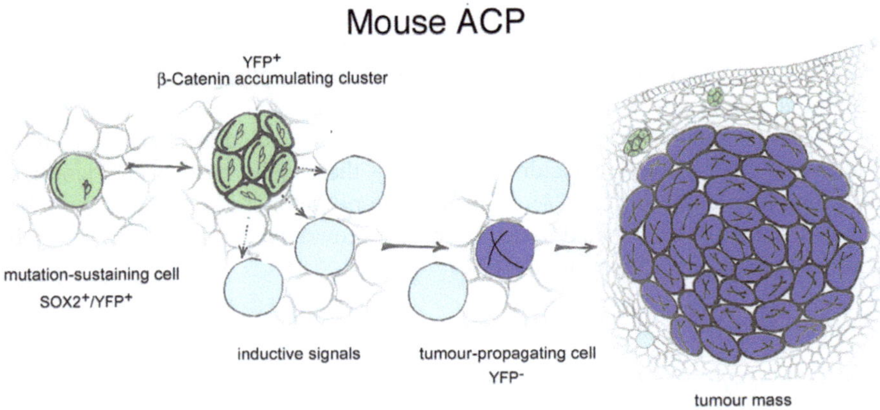

Fig. 5 Paracrine model of involvement of pituitary stem cells in tumourigenesis. When targeted to express oncogenic β-catenin, SOX2⁺ cells (*green*) accumulate nucleo-cytoplasmic β-catenin, proliferate transiently and stop dividing to form distinct clusters. Clusters secrete signals to the surrounding cells that may induce transformation and proliferation, resulting in tumour growth originating from a cell that is not derived from the targeted SOX2⁺ stem cell. Reproduced with permission from Elsevier (Andoniadou et al. (2013) *Cell Stem Cell*)

The postnatal mouse model has been used to trace the descendant cells of pituitary stem cells (PSCs) which are mutated, a technique referred to as 'genetic tracing' [65]. $SOX2^+$ cells are targeted to simultaneously express oncogenic β-catenin and Yellow Fluorescent Protein (YFP); the former is the oncogenic stimulus and the latter allows identification of all the daughter cells derived from the mutated $SOX2^+$ PSCs. These experiments have revealed that the typical cell clusters derive from $SOX2^+$ cells, but intriguingly, the bulk of the tumour mass does not. In other words, mutated $SOX2^+$ cells are not transformed into tumour-propagating cancer stem cells (CSCs) by oncogenic β-catenin; instead, mutated $SOX2^+$ PSCs generate cell clusters that have the capacity to induce tumours in a paracrine fashion and the tumour cells are not derived from these $SOX2^+$ cells [56] (Fig. 5). Strengthening the notion of this paracrine action, cluster cells express a vast array of growth factors, chemokines and cytokines and act as signalling centres, possibly changing the tumour microenvironment and facilitating tumourigenesis [44, 56]. Of relevance, expression of several of these cytokines and growth factors have been shown to play a role in normal pituitary physiology as well as in pituitary adenoma, a more prevalent pituitary tumour in humans [61, 66, 67].

Conclusions

Aspects of the pathogenesis of human ACP have been revealed and several dysregulated genes and pathways have been identified. These findings may impact the diagnosis, prognosis and treatment of patients in the years to come. It is established that mutations in *CTNNB1* resulting in the over-activation of the WNT/β-catenin pathway underlie the aetiology of ACP. The development of mouse models for human ACP not only has contributed to understanding the pathogenesis of these aggressive human tumours, but also importantly, they represent a suitable platform to investigate the function of the dysregulated pathways. In the long term, pre-clinical studies using these murine models will reveal potential therapeutical targets and speed up their use in the clinic.

Acknowledgements We would like to apologise to authors whose work was not included in the interest of space. This work was supported by grants 15/190 from Children with Cancer UK (CWCUK) and Great Ormond Street Hospital Children's Charity (GOSHCC) and grants MR/M000125/1 from the Medical Research Council (MRC) to JPMB and MR/L016729/1 to CLA.

References

1. Kelberman D, Rizzoti K, Lovell-Badge R, Robinson IC, Dattani MT (2009) Genetic regulation of pituitary gland development in human and mouse. Endocr Rev 30(7):790–829. doi:10.1210/er.2009-0008
2. Davis SW, Castinetti F, Carvalho LR, Ellsworth BS, Potok MA, Lyons RH, Brinkmeier ML, Raetzman LT, Carninci P, Mortensen AH, Hayashizaki Y, Arnhold IJ, Mendonca BB, Brue T, Camper SA (2010) Molecular mechanisms of pituitary organogenesis: In search of novel regulatory genes. Mol Cell Endocrinol 323(1):4–19. doi:10.1016/j.mce.2009.12.012
3. Mollard P, Hodson DJ, Lafont C, Rizzoti K, Drouin J (2012) A tridimensional view of pituitary development and function. Trends Endocrinol Metab 23(6):261–269. doi:10.1016/j.tem.2012.02.004
4. Watkins-Chow DE, Camper SA (1998) How many homeobox genes does it take to make a pituitary gland? Trends Genet 14(7):284–290
5. Clevers H, Nusse R (2012) Wnt/beta-catenin signaling and disease. Cell 149(6):1192–1205. doi:10.1016/j.cell.2012.05.012
6. Huelsken J, Birchmeier W (2001) New aspects of Wnt signaling pathways in higher vertebrates. Curr Opin Genet Dev 11(5):547–553
7. Larsen W (2001) Human embryology, 3rd edn. Elsevier, Philadelphia, PA, USA
8. Abernethy LJ (1998) Imaging of the pituitary in children with growth disorders. Eur J Radiol 26(2):102–108
9. Wang Q, Hu Y, Li G, Sun X (2014) Pituitary stalk interruption syndrome in 59 children: the value of MRI in assessment of pituitary functions. Eur J Pediatr 173(5):589–595. doi:10.1007/s00431-013-2214-1
10. Erdheim J (1904) Über Hypophysenganggeschwulste und Hirmcholesteatome. Sitzungsb Kais Akad Wissen Math Naturw Klin 113:537–726
11. Desiderio C, Martelli C, Rossetti DV, Di Rocco C, D'Angelo L, Caldarelli M, Tamburrini G, Iavarone F, Castagnola M, Messana I, Cabras T, Faa G (2013) Identification of thymosins beta4 and beta 10 in paediatric craniopharyngioma cystic fluid. Childs Nerv Syst 29(6):951–960. doi:10.1007/s00381-013-2069-9
12. Martelli C, Iavarone F, Vincenzoni F, Rossetti DV, D'Angelo L, Tamburrini G, Caldarelli M, Di Rocco C, Messana I, Castagnola M, Desiderio C (2014) Proteomic characterization of pediatric craniopharyngioma intracystic fluid by LC-MS top-down/bottom-up integrated approaches. Electrophoresis 35(15):2172–2183. doi:10.1002/elps.201300578
13. Pettorini BL, Inzitari R, Massimi L, Tamburrini G, Caldarelli M, Fanali C, Cabras T, Messana I, Castagnola M, Di Rocco C (2010) The role of inflammation in the genesis of the cystic

component of craniopharyngiomas. Childs Nerv Syst 26(12):1779–1784. doi:10.1007/s00381-010-1245-4
14. Young SC, Zimmerman RA, Nowell MA, Bilaniuk LT, Hackney DB, Grossman RI, Goldberg HI (1987) Giant cystic craniopharyngiomas. Neuroradiology 29(5):468–473
15. Gao S, Shi X, Wang Y, Qian H, Liu C (2011) Malignant transformation of craniopharyngioma: case report and review of the literature. J Neuro-Oncol 103(3):719–725. doi:10.1007/s11060-010-0407-2
16. Negoto T, Sakata K, Aoki T, Orito K, Nakashima S, Hirohata M, Sugita Y, Morioka M (2015) Sequential pathological changes during malignant transformation of a craniopharyngioma: a case report and review of the literature. Surg Neurol Int 6:50. doi:10.4103/2152-7806.154274
17. Sofela AA, Hettige S, Curran O, Bassi S (2014) Malignant transformation in craniopharyngiomas. Neurosurgery 75(3):306–314 . doi:10.1227/neu.0000000000000380discussion 314
18. Ishida M, Hotta M, Tsukamura A, Taga T, Kato H, Ohta S, Takeuchi Y, Nakasu S, Okabe H (2010) Malignant transformation in craniopharyngioma after radiation therapy: a case report and review of the literature. Clin Neuropathol 29(1):2–8
19. Brastianos PK, Taylor-Weiner A, Manley PE, Jones RT, Dias-Santagata D, Thorner AR, Lawrence MS, Rodriguez FJ, Bernardo LA, Schubert L, Sunkavalli A, Shillingford N, Calicchio ML, Lidov HG, Taha H, Martinez-Lage M, Santi M, Storm PB, Lee JY, Palmer JN, Adappa ND, Scott RM, Dunn IF, Laws ER Jr, Stewart C, Ligon KL, Hoang MP, Van Hummelen P, Hahn WC, Louis DN, Resnick AC, Kieran MW, Getz G, Santagata S (2014) Exome sequencing identifies BRAF mutations in papillary craniopharyngiomas. Nat Genet 46(2):161–165. doi:10.1038/ng.2868
20. Buslei R, Nolde M, Hofmann B, Meissner S, Eyupoglu IY, Siebzehnrubl F, Hahnen E, Kreutzer J, Fahlbusch R (2005) Common mutations of beta-catenin in adamantinomatous craniopharyngiomas but not in other tumours originating from the sellar region. Acta Neuropathol 109(6):589–597. doi:10.1007/s00401-005-1004-x
21. Kato K, Nakatani Y, Kanno H, Inayama Y, Ijiri R, Nagahara N, Miyake T, Tanaka M, Ito Y, Aida N, Tachibana K, Sekido K, Tanaka Y (2004) Possible linkage between specific histological structures and aberrant reactivation of the Wnt pathway in adamantinomatous craniopharyngioma. J Pathol 203(3):814–821. doi:10.1002/path.1562
22. Larkin SJ, Preda V, Karavitaki N, Grossman A, Ansorge O (2014) BRAF V600E mutations are characteristic for papillary craniopharyngioma and may coexist with CTNNB1-mutated adamantinomatous craniopharyngioma. Acta Neuropathol 127:927. doi:10.1007/s00401-014-1270-6
23. Oikonomou E, Barreto DC, Soares B, De Marco L, Buchfelder M, Adams EF (2005) Beta-catenin mutations in craniopharyngiomas and pituitary adenomas. J Neuro-Oncol 73(3):205–209. doi:10.1007/s11060-004-5232-z
24. Sekine S, Shibata T, Kokubu A, Morishita Y, Noguchi M, Nakanishi Y, Sakamoto M, Hirohashi S (2002) Craniopharyngiomas of adamantinomatous type harbor beta-catenin gene mutations. Am J Pathol 161(6):1997–2001
25. Buslei R, Holsken A, Hofmann B, Kreutzer J, Siebzehnrubl F, Hans V, Oppel F, Buchfelder M, Fahlbusch R, Blumcke I (2007) Nuclear beta-catenin accumulation associates with epithelial morphogenesis in craniopharyngiomas. Acta Neuropathol 113(5):585–590. doi:10.1007/s00401-006-0184-3
26. Gaston-Massuet C, Andoniadou CL, Signore M, Jayakody SA, Charolidi N, Kyeyune R, Vernay B, Jacques TS, Taketo MM, Le Tissier P, Dattani MT, Martinez-Barbera JP (2011) Increased Wingless (Wnt) signaling in pituitary progenitor/stem cells gives rise to pituitary tumors in mice and humans. Proc Natl Acad Sci U S A 108(28):11482–11487. doi:10.1073/pnas.1101553108
27. Hofmann BM, Kreutzer J, Saeger W, Buchfelder M, Blumcke I, Fahlbusch R, Buslei R (2006) Nuclear beta-catenin accumulation as reliable marker for the differentiation between cystic craniopharyngiomas and rathke cleft cysts: a clinico-pathologic approach. Am J Surg Pathol 30(12):1595–1603. doi:10.1097/01.pas.0000213328.64121.12

28. Holsken A, Kreutzer J, Hofmann BM, Hans V, Oppel F, Buchfelder M, Fahlbusch R, Blumcke I, Buslei R (2009) Target gene activation of the Wnt signaling pathway in nuclear beta-catenin accumulating cells of adamantinomatous craniopharyngiomas. Brain Pathol 19(3):357–364. doi:10.1111/j.1750-3639.2008.00180.x
29. Sekine S, Takata T, Shibata T, Mori M, Morishita Y, Noguchi M, Uchida T, Kanai Y, Hirohashi S (2004) Expression of enamel proteins and LEF1 in adamantinomatous craniopharyngioma: evidence for its odontogenic epithelial differentiation. Histopathology 45(6):573–579. doi:10.1111/j.1365-2559.2004.02029.x
30. Gump JM, Donson AM, Birks DK, Amani VM, Rao KK, Griesinger AM, Kleinschmidt-DeMasters BK, Johnston JM, Anderson RC, Rosenfeld A, Handler M, Gore L, Foreman N, Hankinson TC (2015) Identification of targets for rational pharmacological therapy in childhood craniopharyngioma. Acta Neuropathol Commun 3:30. doi:10.1186/s40478-015-0211-5
31. Holsken A, Sill M, Merkle J, Schweizer L, Buchfelder M, Flitsch J, Fahlbusch R, Metzler M, Kool M, Pfister SM, von Deimling A, Capper D, Jones DT, Buslei R (2016) Adamantinomatous and papillary craniopharyngiomas are characterized by distinct epigenomic as well as mutational and transcriptomic profiles. Acta Neuropathol Commun 4:20. doi:10.1186/s40478-016-0287-6
32. Holsken A, Buchfelder M, Fahlbusch R, Blumcke I, Buslei R (2010) Tumour cell migration in adamantinomatous craniopharyngiomas is promoted by activated Wnt-signalling. Acta Neuropathol 119(5):631–639. doi:10.1007/s00401-010-0642-9
33. Larkin SJ, Ansorge O (2013) Pathology and pathogenesis of craniopharyngiomas. Pituitary 16(1):9–17. doi:10.1007/s11102-012-0418-4
34. Bernstein ML, Buchino JJ (1983) The histologic similarity between craniopharyngioma and odontogenic lesions: a reappraisal. Oral Surg Oral Med Oral Pathol 56(5):502–511
35. Xin W, Rubin MA, McKeever PE (2002) Differential expression of cytokeratins 8 and 20 distinguishes craniopharyngioma from rathke cleft cyst. Arch Pathol Lab Med 126(10):1174–1178. doi:10.1043/0003-9985(2002)126<1174:deocad>2.0.co;2
36. Gaston-Massuet C, McCabe MJ, Scagliotti V, Young RM, Carreno G, Gregory LC, Jayakody SA, Pozzi S, Gualtieri A, Basu B, Koniordou M, Wu CI, Bancalari RE, Rahikkala E, Veijola R, Lopponen T, Graziola F, Turton J, Signore M, Mousavy Gharavy SN, Charolidi N, Sokol SY, Andoniadou CL, Wilson SW, Merrill BJ, Dattani MT, Martinez-Barbera JP (2016) Transcription factor 7-like 1 is involved in hypothalamo-pituitary axis development in mice and humans. Proc Natl Acad Sci U S A. doi:10.1073/pnas.1503346113
37. Brinkmeier ML, Potok MA, Cha KB, Gridley T, Stifani S, Meeldijk J, Clevers H, Camper SA (2003) TCF and Groucho-related genes influence pituitary growth and development. Mol Endocrinol 17(11):2152–2161. doi:10.1210/me.2003-0225
38. Brinkmeier ML, Potok MA, Davis SW, Camper SA (2007) TCF4 deficiency expands ventral diencephalon signaling and increases induction of pituitary progenitors. Dev Biol 311(2):396–407. doi:10.1016/j.ydbio.2007.08.046
39. Cha KB, Douglas KR, Potok MA, Liang H, Jones SN, Camper SA (2004) WNT5A signaling affects pituitary gland shape. Mech Dev 121(2):183–194. doi:10.1016/j.mod.2003.12.002
40. Kioussi C, Briata P, Baek SH, Rose DW, Hamblet NS, Herman T, Ohgi KA, Lin C, Gleiberman A, Wang J, Brault V, Ruiz-Lozano P, Nguyen HD, Kemler R, Glass CK, Wynshaw-Boris A, Rosenfeld MG (2002) Identification of a Wnt/Dvl/beta-Catenin -- > Pitx2 pathway mediating cell-type-specific proliferation during development. Cell 111(5):673–685
41. Olson LE, Tollkuhn J, Scafoglio C, Krones A, Zhang J, Ohgi KA, Wu W, Taketo MM, Kemler R, Grosschedl R, Rose D, Li X, Rosenfeld MG (2006) Homeodomain-mediated beta-catenin-dependent switching events dictate cell-lineage determination. Cell 125(3):593–605. doi:10.1016/j.cell.2006.02.046
42. Zhang Z, Florez S, Gutierrez-Hartmann A, Martin JF, Amendt BA (2010) MicroRNAs regulate pituitary development, and microRNA 26b specifically targets lymphoid enhancer factor 1 (Lef-1), which modulates pituitary transcription factor 1 (Pit-1) expression. J Biol Chem 285(45):34718–34728. doi:10.1074/jbc.M110.126441

43. Harada N, Tamai Y, Ishikawa T, Sauer B, Takaku K, Oshima M, Taketo MM, Miyoshi H, Murai N, Oshima H (1999) Intestinal polyposis in mice with a dominant stable mutation of the beta-catenin gene. EMBO J 18(21):5931–5942
44. Andoniadou CL, Gaston-Massuet C, Reddy R, Schneider RP, Blasco MA, Le Tissier P, Jacques TS, Pevny LH, Dattani MT, Martinez-Barbera JP (2012) Identification of novel pathways involved in the pathogenesis of human adamantinomatous craniopharyngioma. Acta Neuropathol 124(2):259–271. doi:10.1007/s00401-012-0957-9
45. Gomes DC, Jamra SA, Leal LF, Colli LM, Campanini ML, Oliveira RS, Martinelli CE Jr, Elias PC, Moreira AC, Machado HR, Saggioro F, Neder L, Castro M, Antonini SR (2015) Sonic Hedgehog pathway is upregulated in adamantinomatous craniopharyngiomas. Eur J Endocrinol 172(5):603–608. doi:10.1530/eje-14-0934
46. Gong J, Zhang H, Xing S, Li C, Ma Z, Jia G, Hu W (2014) High expression levels of CXCL12 and CXCR4 predict recurrence of adamanti-nomatous craniopharyngiomas in children. Cancer Biomark 14(4):241–251. doi:10.3233/cbm-140397
47. Muller HL, Emser A, Faldum A, Bruhnken G, Etavard-Gorris N, Gebhardt U, Oeverink R, Kolb R, Sorensen N (2004) Longitudinal study on growth and body mass index before and after diagnosis of childhood craniopharyngioma. J Clin Endocrinol Metab 89(7):3298–3305. doi:10.1210/jc.2003-031751
48. Kostadinov S, Hanley CL, Lertsburapa T, O'Brien B, He M (2014) Fetal Craniopharyngioma: Management, Postmortem diagnosis and literature review of an intracranial tumor detected in utero. Pediatr Dev Pathol 17(5):409–412. doi:10.2350/14-06-1506-cr.1
49. Xue L, Cai JY, Ma J, Huang Z, Guo MX, Fu LZ, Shi YB, Li WX (2013) Global expression profiling reveals genetic programs underlying the developmental divergence between mouse and human embryogenesis. BMC Genomics 14:568. doi:10.1186/1471-2164-14-568
50. Fauquier T, Rizzoti K, Dattani M, Lovell-Badge R, Robinson IC (2008) SOX2-expressing progenitor cells generate all of the major cell types in the adult mouse pituitary gland. Proc Natl Acad Sci U S A 105(8):2907–2912. doi:10.1073/pnas.0707886105
51. Yoshida S, Kato T, Susa T, Cai LY, Nakayama M, Kato Y (2009) PROP1 coexists with SOX2 and induces PIT1-commitment cells. Biochem Biophys Res Commun 385(1):11–15. doi:10.1016/j.bbrc.2009.05.027
52. Yoshida S, Kato T, Yako H, Susa T, Cai LY, Osuna M, Inoue K, Kato Y (2011) Significant quantitative and qualitative transition in pituitary stem/progenitor cells occurs during the postnatal development of the rat anterior pituitary. J Neuroendocrinol 23(10):933–943. doi:10.1111/j.1365-2826.2011.02198.x
53. Higuchi M, Yoshida S, Ueharu H, Chen M, Kato T, Kato Y (2014) PRRX1 and PRRX2 distinctively participate in pituitary organogenesis and a cell-supply system. Cell Tissue Res 357(1):323–335. doi:10.1007/s00441-014-1861-5
54. Rizzoti K, Akiyama H, Lovell-Badge R (2013) Mobilized adult pituitary stem cells contribute to endocrine regeneration in response to physiological demand. Cell Stem Cell 13(4):419–432. doi:10.1016/j.stem.2013.07.006
55. Garcia-Lavandeira M, Quereda V, Flores I, Saez C, Diaz-Rodriguez E, Japon MA, Ryan AK, Blasco MA, Dieguez C, Malumbres M, Alvarez CV (2009) A GRFa2/Prop1/stem (GPS) cell niche in the pituitary. PLoS One 4(3):e4815. doi:10.1371/journal.pone.0004815
56. Andoniadou CL, Matsushima D, Mousavy Gharavy SN, Signore M, Mackintosh AI, Schaeffer M, Gaston-Massuet C, Mollard P, Jacques TS, Le Tissier P, Dattani MT, Pevny LH, Martinez-Barbera JP (2013) Sox2(+) stem/progenitor cells in the adult mouse pituitary support organ homeostasis and have tumor-inducing potential. Cell Stem Cell 13(4):433–445. doi:10.1016/j.stem.2013.07.004
57. Devnath S, Inoue K (2008) An insight to pituitary folliculo-stellate cells. J Neuroendocrinol 20(6):687–691. doi:10.1111/j.1365-2826.2008.01716.x
58. Asa SL, Tannenbaum GS (2006) Cell-cell communication in the pituitary: orchestrator of pulsatile growth hormone secretion? Trends Endocrinol Metab 17(8):299–300. doi:10.1016/j.tem.2006.08.004

59. Fauquier T, Guerineau NC, McKinney RA, Bauer K, Mollard P (2001) Folliculostellate cell network: a route for long-distance communication in the anterior pituitary. Proc Natl Acad Sci U S A 98(15):8891–8896. doi:10.1073/pnas.151339598
60. Bilezikjian LM, Blount AL, Leal AM, Donaldson CJ, Fischer WH, Vale WW (2004) Autocrine/paracrine regulation of pituitary function by activin, inhibin and follistatin. Mol Cell Endocrinol 225(1-2):29–36. doi:10.1016/j.mce.2004.02.010
61. Arzt E, Pereda MP, Castro CP, Pagotto U, Renner U, Stalla GK (1999) Pathophysiological role of the cytokine network in the anterior pituitary gland. Front Neuroendocrinol 20(1):71–95. doi:10.1006/frne.1998.0176
62. Denef C (2008) Paracrinicity: the story of 30 years of cellular pituitary crosstalk. J Neuroendocrinol 20(1):1–70. doi:10.1111/j.1365-2826.2007.01616.x
63. Renner U, Pagotto U, Arzt E, Stalla GK (1996) Autocrine and paracrine roles of polypeptide growth factors, cytokines and vasogenic substances in normal and tumorous pituitary function and growth: a review. Eur J Endocrinol 135(5):515–532
64. Arnold K, Sarkar A, Yram MA, Polo JM, Bronson R, Sengupta S, Seandel M, Geijsen N, Hochedlinger K (2011) Sox2(+) adult stem and progenitor cells are important for tissue regeneration and survival of mice. Cell Stem Cell 9(4):317–329. doi:10.1016/j.stem.2011.09.001
65. Barker N, van Oudenaarden A, Clevers H (2012) Identifying the stem cell of the intestinal crypt: strategies and pitfalls. Cell Stem Cell 11(4):452–460. doi:10.1016/j.stem.2012.09.009
66. Arzt E, Chesnokova V, Stalla GK, Melmed S (2009) Pituitary adenoma growth: a model for cellular senescence and cytokine action. Cell Cycle 8(5):677–678
67. Graciarena M, Carbia-Nagashima A, Onofri C, Perez-Castro C, Giacomini D, Renner U, Stalla GK, Arzt E (2004) Involvement of the gp130 cytokine transducer in MtT/S pituitary somatotroph tumour development in an autocrine-paracrine model. Eur J Endocrinol 151(5): 595–604

Clinical Diagnosis of Human ACP

Ashley J. Vardon and Niki Karavitaki

Abstract Craniopharyngiomas are epithelial tumours in the sellar/parasellar area accounting for 2–5% of all primary intracranial neoplasms and for up to 15% of intracranial tumours in children. The adamantinomatous histological subtype is the most common and the majority of them (94–95%) have a suprasellar component. Their potential proximity to, and the subsequent pressure effects on vital structures of the brain (visual pathways, brain parenchyma, ventricular system, major blood vessels and hypothalamo-pituitary system) can result to multiple clinical manifestations, the severity of which depends on the location, size and growth potential of the tumour. The main manifestations at diagnosis are neurological, visual and hypothalamo-pituitary with headaches, nausea/vomiting, visual disturbances, growth failure (in children) and hypogonadism (in adults) being the most frequently reported. One or more pituitary hormone deficits may be already present at the time of craniopharyngioma detection in a significant number of patients necessitating early diagnosis and appropriate treatment.

Keywords Craniopharyngioma • Adamantinomatous • Hypopituitarism • Hypothalamic • Suprasellar tumour • Pituitary hormone deficits

Introduction

Craniopharyngiomas are rare epithelial tumours arising along the path of the craniopharyngeal duct, the canal connecting the stomodeal ectoderm with the evaginated Rathke's pouch. They may be diagnosed during childhood or adult life and are often associated with an unpredictable growth pattern. Despite their benign histological appearance, their often infiltrative tendency into critical parasellar structures and their aggressive behaviour may result in significant morbidities and mortality (even

A.J. Vardon • N. Karavitaki (✉)
Institute of Metabolism and Systems Research (IMSR), College of Medical and Dental Sciences, University of Birmingham and Centre for Endocrinology, Diabetes and Metabolism, Birmingham Health Partners, Birmingham, UK
e-mail: n.karavitaki@bham.ac.uk

J.P. Martinez-Barbera, C.L. Andoniadou (eds.), *Basic Research and Clinical Aspects of Adamantinomatous Craniopharyngioma*,
DOI 10.1007/978-3-319-51890-9_4

after apparently successful therapy) making them a challenging disease entity for many specialists.

In 1857, Zenker was the first to identify masses of cells resembling squamous epithelium along the pars distalis and pars tuberalis of the pituitary [1] and in 1902, Fr. Saxer reported a tumour consisting of these cells [2]. The first attempt for surgical removal of such a lesion ("from a patient presenting the symptoms associated with hypophyseal growths but without acromegaly") by Dr. Halstead in St. Luke's Hospital (Chicago) was reported in 1910 [3]. During the following years, different terminologies were used for them (including hypophyseal duct or craniopharyngeal duct or Rathke's pouch tumours, interpeduncular or dysontogenetic or suprasellar or craniobuccal cysts, suprasellar epitheliomas and adamantinomas), until 1932, when the name "craniopharyngioma" was introduced by H. Cushing [2].

Craniopharyngiomas account for 2–5% of all the primary intracranial neoplasms and for up to 15% of the intracranial tumours in children [2]. Their incidence is reported as 0.13 per 100,000 person-years [4] and may be detected at any age, even in the prenatal and neonatal periods [5, 6]. A bimodal age distribution with peak incidence rates of ages 5–14 and 50–74 years has been described [4]. In population-based studies from the USA and Finland, no gender differences have been found [4, 7]. Two main histological subtypes have been reported: the adamantinomatous and the papillary. The adamantinomatous is the most common and it may occur at all ages predominantly affecting young people during their first two decades of life [8–10].

Location

Craniopharyngiomas may arise anywhere along the craniopharyngeal canal, but most of them are located in the sellar/parasellar region. They show a variable growth pattern which is closely related to their clinical presentation and outcomes. The majority (94–95%) has a suprasellar component (purely suprasellar 20–41%/both supra- and intrasellar 53–75%), whereas the purely intrasellar ones represent the least common variety (5–6%) [11, 12]. Suprasellar tumours may extend into the anterior, middle or posterior fossa. Other rare described locations include the nasopharynx, the paranasal area, the sphenoid bone, the ethmoid sinus, the intrachiasmatic area, the temporal lobe, the pineal gland, the posterior cranial fossa, the cerebellopontine angle, the midportion of the midbrain or completely within the third ventricle [2].

Presenting Clinical Manifestations

The presenting clinical manifestations of craniopharyngiomas show a wide variety and are not specific for this tumour or for the adamantinomatous subtype. They are attributed to pressure effects the mass exerts on vital structures of the brain (visual pathways, brain parenchyma, ventricular system, major blood vessels and hypothalamo-pituitary system). Their severity depends on the location, size and growth potential of the craniopharyngioma [10–17] and their duration until diagnosis has been reported between 1 week and 372 months [2]. In a series of childhood craniopharyngioma, tumour size and hypothalamic involvement were not related with the duration of history before the diagnosis [18].

The presenting signs and symptoms can be classified into three main groups: neurological, visual and hypothalamo-pituitary necessitating thorough neurological, ophthalmological and endocrine assessment. A summary of the main clinical manifestations and their frequency based on published series of children and adults is shown in Table 1. Headaches (in 7–81%), nausea/vomiting (in 21–54%), visual disturbances (in 39–80%), growth failure (in children) (in 8–93%) and hypogonadism (in adults) (in 10–85%) are the most frequently reported. Headaches and nausea/vomiting result either from mass effect from the tumour itself or from secondary hydrocephalus caused by obstruction of the foramen of Monro, the third ventricle or the aqueduct of Sylvius. Notably, hydrocephalus has been described more commonly in children [11, 17, 20]. Visual disturbances include bitemporal hemianopia from inferior chiasmatic compression, homonymous hemianopia, scotoma and optic atrophy with papilloedema. Temporal alterations of the pattern for the visual field defects due to intermittent emptying of the cyst fluid into the ventricular system may occur [21]. It has also been suggested that in childhood craniopharyngioma, compromised growth rate is already evident in early infancy, whereas an increase in the weight tends to present later and is a predictor of obesity [22]. In a large series of patients comparing the presenting manifestations between childhood and adult diagnosed disease, headaches, nausea/vomiting, papilloedema and cranial nerve palsies were more frequent in children (probably associated with the high rates of hydrocephalus in this age group) [11].

Other less common or rare features at presentation include motor disorders, as hemi- or monoparesis [11, 12, 17], seizures [12], psychiatric symptoms, as emotional lability, hallucinations, paranoic delusions [11, 12, 14, 23], autonomic disturbances [12], precocious puberty [23, 24], syndrome of inappropriate secretion of antidiuretic hormone [25], chemical meningitis due to spontaneous cyst rupture [26], hearing loss [27], anosmia [12], nasal obstruction [28], epistaxis [29], photophobia [30], emaciation [31], Weber's syndrome [32] and Wallenberg's syndrome (signs due to occlusion of the posterior inferior cerebellar artery) [33].

The pituitary function may be severely affected already at the time of presentation contributing to the reported signs and symptoms. Based on a number of series in which, however, different diagnostic tests and criteria have been adopted, GH deficiency is present in 35–100% of the evaluated patients, FSH/LH deficiency in 50–91%, ACTH deficiency in 21–68%, TSH deficiency in 20–37% and diabetes insipidus (DI) in 6–38% (Table 2). The rates of GH deficiency reported in some studies may overestimate the true incidence of this disorder, due to selection bias of the patients (particularly adults) tested, because the symptomatic ones are more likely to have been assessed [2].

Table 1 Main clinical manifestations of craniopharyngiomas in children and adults at presentation

Reference, number and age of patients	Headache	Nausea/ vomiting	Papilloedema	Cranial nerves palsy	Ataxia/ unsteadiness	Cognitive dysfunction[a]	Decreased consciousness/ coma	Visual field defects
Baskin and Wilson [13] $n = 74$ 38 <18 years	50%	–	–	–	–	–	–	72%
Hetelekidis et al. [19] $n = 61$ all <21 years	77%	43%	29%	20%	18%	–	3%	–
Weiner et al. [10] $n = 56$ 26 <19 years	7% adults 15% children	–	–	8% children	–	20% adults 8% children	–	60% adults 54% children
Duff et al. [14] $n = 121$ 32 ≤16 years	74%	21%	10%	2%	–	10%	29%	62%
Van Effenterre and Boch [17] $n = 122$ 29 <16 years	53%	–	16%	3%	–	13%	8%	79%
Karavitaki et al. [11] $n = 119$ 41 <16 years	56% adults 78% children	26% adults 54% children	6% adults 29% children	9% adults 27% children	3% adults 7% children	17% adults 10% children	4% adults 10% children	60% adults 46% children

Gautier et al. [15] $n = 171$ 65 <18 years	61% adults 61% children	–	–	–	–	–	–	–	
Nielsen et al. [16] $n = 189$ 39 <15 years	59% adults 71.8% children	17% adults 51.3% children	–	7% adults 13% children	–	13% adults 13% children	3% adults 3% children	–	
Range	7–81%	21–54%	6–29%	2–27%	3–18%	8–17%	3–29%	46–79%	
Reference, number and age of patients	**Decreased visual acuity or visual deterioration**	**Growth failure[b]**	**Failure of sexual development[b]**	**Hypogonadism (adults)**	**Poor energy**	**Somnolence**	**Anorexia/poor weight gain or weight loss**	**Obesity or weight gain**	**Polyuria/polydipsia**
Baskin and Wilson [13] $n = 74$ 38 <18 years	–	93%	–	85%	–	–	–	–	23%
Hetelekidis et al. [19] $n = 61$ all <21 years	–	25%	–	–	–	–	–	–	13%
Weiner et al. [10] $n = 56$ 26 <19 years	47% adults 50% children	8%	4%	10%	–	–	–	4% children	3% adults
Duff et al. [14] $n = 121$ 32 ≤16 years	62%	17%	14%	–	23%	–	–	8%	12%

(continued)

Table 1 (continued)

Reference, number and age of patients	Headache	Nausea/ vomiting	Papilloedema	Cranial nerves palsy	Ataxia/ unsteadiness	Cognitive dysfunction[a]	Decreased consciousness/ coma	Visual field defects	
Van Effenterre and Boch [17] $n = 122$ 29 <16 years	80%	45%	–	40%	–	–	–	–	18%
Karavitaki et al. [11] $n = 119$ 41 <16 years	40% adults 39% children	32%	24%	28%	32% adults 22% children	10% adults 5% children	8% adults 20% children	13% adults 5% children	15% adults 15% children
Gautier et al. [15] $n = 171$ 55 <18 years	71% adults 75% children	–	–	–	–	–	–	–	12% adults 14% children
Nielsen et al. [16] $n = 189$ 39 <15 years	65% adults 49% children	–	–	–	47% adults 31% children	–	10% adults 10% children	17% adults 8% children	19% adults 10% children
Range	39–80%	8–93%	4–24%	10–85%	22–47%	5–10%	8–20%	4–17%	3–23%

[a]Memory loss, impaired concentration, confusion, disorientation
[b]Children

Table 2 Pituitary hormone deficits and hyperprolactinaemia in adults and children with craniopharyngioma at the time of presentation (number of patients with the deficit/total number of patients checked and percentages)

Reference, number and age of patients	GH deficiency	FSH/LH deficiency[a]	ACTH deficiency	TSH deficiency	Hyperprolactinaemia	Diabetes insipidus
Hetelekides et al. [20] $n = 61$ all <21 years	12/34 (35%)	–	7/34 (21%)	7/34 (21%)	–	8/34 (23%)
Paja et al. [36] $n = 35$ 13 <19 years	27/35 (77%)	27/33 (82%)	12/35 (34%)	13/35 (37%)	7/29 (24%)	13/35 (38%)
De Vile et al. [37] $n = 75$ all <16.3 years	13/15 (87%)	3/6 (50%)	16/50 (32%)	20/62 (32%)	12/37 (32%)	22/75 (29%)
Honegger et al. [28] $n = 143$ 30 <16 years	59/82 (72%) 17/23 in children (74%)	96/143 (77%) 10/11 in children (91%)	45/143 (32%) 8/30 in children (27%)	35/143 (25%) 6/30 in children (20%)	59/143 (41%) 5/30 in children (17%)	23/143 (16%) 3/30 in children (10%)
Karavitaki et al. [11] $n = 121$ 42 <16 years	21/22 (95%) 15/15 in children (100%)	40/54 (74%)[b]	40/65 (62%) 15/22 in children (68%)	29/81 (36%) 7/28 in children (25%)	24/44 (55%)[b]	19/104 (18%) 7/32 in children (22%)
Nielsen et al. [16] $n = 189$ 39 <15 years	23/45 (51%) in adults 5/9 (56%) in children	89/109 (82%)	40/123 (33%) in adults 8/20 (40%) in children	31/121 (26%) in adults 7/27 (26%) in children	–	20/131 (15%) in adults 2/32 (6%) in children
Range	35–100%	50–91%	21–68%	20–37%	17–55%	6–38%

[a]Pre-pubertal children were excluded from the evaluations
[b]Only adults were included in the evaluations

Hypothalamic involvement affecting several medial and posterior hypothalamic nuclei has been related with higher body mass index (BMI) at diagnosis [34]. Structural damage of these nuclei leads to hyperphagia, rapid weight gain, central insulin and leptin resistance, decreased sympathetic activity, low energy expenditure and increased energy storage in adipose tissue with adverse long-term sequelae [35].

Summary

The clinical presentation of patients with human adamantinomatous craniopharyngioma at the time of diagnosis shows a wide variety and relates to pressure effects the mass exerts on vital structures of the brain (visual pathways, brain parenchyma, ventricular system, major blood vessels and hypothalamo-pituitary system). The presenting signs and symptoms can be classified into three main groups: neurological, visual and hypothalamo-pituitary necessitating thorough neurological, ophthalmological and endocrine assessment. Headaches, nausea/vomiting, visual disturbances, growth failure (in children) and hypogonadism (in adults) are the most frequently reported. The pituitary function may be severely affected already at the time of presentation contributing to the reported signs and symptoms and hypothalamic involvement with structural damage of the medial and posterior nuclei can lead to obesity, one of the most challenging problems seen in these patients.

References

1. Raimondi AJ, Rougerie J (1994) A critical review of personal experiences with craniopharyngioma: clinical history, surgical technique and operative results. Pediatr Neurosurg 21:134–115
2. Karavitaki N, Cudlip S, Adams CBT, Wass JAH (2006) Craniopharyngiomas. Endocr Rev 27:371–397
3. Roderick E, Karavitaki N, Wass JA (2008) Craniopharyngiomas. Historical aspects on their management. Hormones 7(3):271–274
4. Bunin GR, Surawicz TS, Witman PA, Preston-Martin S, Davis F, Bruner JM (1998) The descriptive epidemiology of craniopharyngioma. J Neurosurg 89:547–551
5. Bailey W, Freidenberg GR, James HE, Hesselink JR, Jones KL (1990) Prenatal diagnosis of a craniopharyngioma using ultrasonography and magnetic resonance imaging. Prenat Diagn 10(1):623–629
6. Müller-Scholden J, Lehrnbecker T, Müller HL, Bensch J, Hengen RH, Sörensen N, Stockhausen HB (2000) Radical surgery in a neonate with craniopharyngioma—report of a case. Pediatr Neurosurg 33:265–269
7. Sorva R, Heiskanen O (1986) Craniopharyngioma in Finland. A study of 123 cases. Acta Neurochir 81(3–4):85–89
8. Adamson TE, Wiestler OD, Kleihues P, Yasargil MG (1990) Correlation of clinical and pathological features in surgically treated craniopharyngiomas. J Neurosurg 73:12–17
9. Kahn EA, Gosch HH, Seeger JF, Hicks SP (1973) Forty-five years experience with the craniopharyngiomas. Surg Neurol 1:5–12
10. Weiner HL, Wisoff JH, Rosenberg ME, Kupersmith MJ, Cohen H, Zagzag D, Shiminski-Maher T, Flamm ES, Epsten FJ, Miller D (1994) Craniopharyngiomas: a clinicopathological analysis of factors predictive of recurrence and functional outcome. Neurosurgery 35(6):1001–1011
11. Karavitaki N, Brufani C, Warner JT, Adams CBT, Richards P, Ansorge O, Shine B, Turner HE, Wass JAH (2005) Craniopharyngiomas in children and adults: systematic analysis of 121 cases with long-term follow-up. Clin Endocrinol (Oxf) 62:397–409
12. Petito CK, De Girolami U, Earle K (1976) Craniopharyngiomas. A clinical and pathological review. Cancer 37:1944–1952

13. Baskin DS, Wilson CB (1986) Surgical management of craniopharyngiomas. J Neurosurg 65:22–27
14. Duff JM, Meyer FB, Ilstrup DM, Laws Jr ER, Scleck CD, Scheithauer BW (2000) Long-term outcomes for surgically resected craniopharyngiomas. Neurosurgery 46(2):291–305
15. Gautier A, Godbout A, Grosheny C, Tejedor I, Coudert M, Courtillot C, Jublanc C, De Kerdanet M, Poirier JY, Riffaud L, Sainte-Rose C, Van Effenterre R, Brassier G, Bonnet F, Touraine P, Craniopharyngioma Study Group (2012) Markers of recurrence and long-term morbidity in craniopharyngioma: a systematic analysis of 171 patients. J Clin Endocrinol Metab 97(4):1258–1267
16. Nielsen EH, Jørgensen JO, Bjerre P, Andersen M, Andersen C, Feldt-Rasmussen U, Poulsgaard L, Kristensen LO, Astrup J, Jørgensen J, Laurberg P (2013) Acute presentation of craniopharyngioma in children and adults in a Danish national cohort. Pituitary 16:528–535
17. Van Effenterre R, Boch AL (2002) Craniopharyngioma in adults and children: a study of 122 surgical cases. J Neurosurg 97:3–11
18. Hoffmann A, Boekhoff S, Gebhardt U, Sterkenburg AS, Daubenbüchel AM, Eveslage M, Müller HL (2015) History before diagnosis in childhood craniopharyngioma: associations with initial presentation and long-term prognosis. Eur J Endocrinol 173(6):853–862
19. Hetelekidis S, Barnes PD, Tao ML, Fischer EG, Schneider L, Scott RM, Tarbell NJ (1993) 20-year experience in childhood craniopharyngioma. Int J Radiat Oncol Biol Phys 27:189–195
20. Rosenfeld A, Arrington D, Miller J, Olson M, Gieseking A, Etzl M, Harel B, Schembri A, Kaplan A (2014) A review of childhood and adolescent craniopharyngiomas with particular attention to hypothalamic obesity. Pediatr Neurol 50(1):4–10
21. Chen C, Okera S, Davies PE, Selva D, Crompton JL (2003) Craniopharyngioma: a review of long-term visual outcome. Clin Exp Ophthalmol 31:220–228
22. Müller HL, Emser A, Faldum A, Bruhnken G, Etavard-Gorris N, Gebhardt U, Oeverink R, Kolb R, Sörensen N (2004) Longitudinal study on growth and body mass index before and after diagnosis of childhood craniopharyngioma. J Clin Endocrinol Metab 89(7):3298–3305
23. Banna M, Hoare RD, Stanley P, Till K (1973) Craniopharyngioma in children. J Pediatr 83(5):781–785
24. Hoffman HJ, DeSilva M, Humphreys RP, Drake JM, Smith ML, Blaser SI (1992) Aggressive surgical management of craniopharyngiomas in children. J Neurosurg 76:47–52
25. Gonzales-Portillo G, Tomita T (1998) The syndrome of inappropriate secretion of antidiuretic hormone: an unusual presentation for childhood craniopharyngioma: report of three cases. Neurosurgery 42(4):917–992
26. Patrik BS, Smith RR, Bailey TO (1974) Aseptic meningitis due to spontaneous rupture of craniopharyngioma cysts: case report. J Neurosurg 41:387–390
27. Honegger J, Buchfelder M, Fahlbusch R (1999) Surgical treatment of craniopharyngiomas: endocrinological results. J Neurosurg 90:251–257
28. Kanungo N, Just N, Black M, Mohr G, Glikstein R, Rochon L (1995) Nasopharyngeal craniopharyngioma in an unusual location. Am J Neuroradiol 16:1372–1374
29. Ahsan F, Rashid H, Chapman A, Ah-See KW (2004) Infrasellar craniopharyngioma presenting as epistaxis, excised via Denker's medial maxillectomy approach. J Laryngol Otol 118(11):895–898
30. Kawasaki A, Purvin VA (2002) Photophobia as the presenting visual symptom of chiasmal compression. J Neuroophthalmol 22(1):3–8
31. Miyoshi Y, Yunoki M, Yano A, Nishimoto K (2003) Diencephalic syndrome of emaciation in an adult associated with a third ventricle intrinsic craniopharyngioma: case report. Neurosurgery 52(1):224–227
32. Waga S, Handa H (1976) Radiation-induced meningioma: with review of literature. Surg Neurol 5:215–219
33. Van den Bergh P, Dom R (1983) Wallenberg's syndrome caused by a craniopharyngioma "en plaque". J Neurol 229(1):61–64

34. Daubenbuchel AM, Hoffmann A, Gebhardt U, Warmuth-Metz M, Sterkenburg AS, Muller HL (2015) Hydrocephalus and hypothalamic involvement in pediatric patients with craniopharyngioma or cysts of rathke's pouch: impact on long-term prognosis. Eur J Endocrinol 172:561–569
35. Roth CL (2015) Hypothalamic obesity in craniopharyngioma patients: disturbed energy homeostasis related to extent of hypothalamic damage and its implication for obesity intervention. J Clin Med 4(9):1774–1797
36. Paja M, Lucas T, Garcia-Uria F, Salame F, Barcelo B, Estrada J (1995) Hypothalamic-pituitary dysfunction in patients with craniopharyngioma. Clin Endocrinol (Oxf) 42:467–473
37. De Vile CJ, Grant DB, Hayward RD, Stanhope R (1996) Growth and endocrine sequelae of craniopharyngioma. Arch Dis Child 75:108–114

Endocrine Deficits in Patients with Human Craniopharyngioma

Hoong-Wei Gan, Helen A. Spoudeas, and Mehul T. Dattani

Abstract Despite their benign histology, craniopharyngiomas are the archetypal suprasellar tumour for causing hypothalamo-pituitary endocrine dysfunction both from tumour- and treatment-related effects. Endocrine deficits are frequently present at diagnosis but are often under-recognised; thus a detailed pre-treatment endocrine assessment is essential, particularly to exclude adrenocorticotrophic hormone deficiency and central diabetes insipidus prior to any surgery. Management of fluid balance in the perioperative period is particularly complex and requires the expertise of a paediatric endocrinologist familiar in managing patients with multiple endocrinopathies. Post-operatively, virtually all survivors face a lifetime risk of evolving endocrine deficits that may be extremely complex to manage, therefore requiring careful long-term follow-up and monitoring into adulthood with appropriate plans for transition between paediatric and adult care.

Keywords Craniopharyngioma • hypopituitarism • hypothalamic disease • diabetes insipidus

Abbreviations

ACTH Adrenocorticotrophic hormone
ADH Antidiuretic hormone (arginine-vasopressin)
AFP α-fetoprotein

H.-W. Gan (✉) • M.T. Dattani
Genetics and Genomic Medicine Programme, UCL Great Ormond Street Institute of Child Health, London, UK

The London Centre for Paediatric Endocrinology and Diabetes, Great Ormond Street Hospital for Children NHS Foundation Trust, London, UK
e-mail: hoong.gan.11@ucl.ac.uk

H.A. Spoudeas
The London Centre for Paediatric Endocrinology and Diabetes, Great Ormond Street Hospital for Children NHS Foundation Trust, London, UK

J.P. Martinez-Barbera, C.L. Andoniadou (eds.), *Basic Research and Clinical Aspects of Adamantinomatous Craniopharyngioma*,
DOI 10.1007/978-3-319-51890-9_5

BMI	Body mass index
DI	Diabetes insipidus
FSH	Follicle-stimulating hormone
GH	Growth hormone
GHRH	Growth hormone-releasing hormone
GnRH	Gonadotrophin-releasing hormone
IGF-1	Insulin-like growth factor 1
IGF-BP3	Insulin-like growth factor binding protein 3
LH	Luteinising hormone
PRL	Prolactin
SD	Standard deviation
SIADH	Syndrome of inappropriate ADH secretion
TRH	Thyrotrophin-releasing hormone
TSH	Thyroid-stimulating hormone
β-hCG	β-human chorionic gonadotrophin

Introduction

More than 80% of all central nervous system tumour survivors develop at least one endocrinopathy in the long-term [1]. Despite the benign histology of craniopharyngiomas, their tendency for multiple progression and their suprasellar position in close proximity to vital hypothalamo-pituitary structures makes them the archetypal tumour for causing such endocrine dysfunction, not only from direct tumour invasion or mass effects, but also from treatments targeted to this area such as surgery and radiotherapy. When tested, the prevalence of hypothalamo-pituitary endocrine dysfunction can therefore be as high as 87–100% at diagnosis [2, 3], and is practically universal post-treatment [2, 4]. In the longer-term, survivors also have a high incidence of obesity, with its attendant metabolic and endocrine consequences (Chap. 7). The high overall survival rate seen with these tumours is therefore accompanied by significant long-term neuroendocrine morbidity faced by a rapidly accruing cohort of survivors. Management of craniopharyngioma as a chronic disease with lifelong follow-up of patients and clear plans for transition to adult services is therefore essential.

Endocrine Dysfunction at Diagnosis

Symptoms related to disruption of the hypothalamo-pituitary axes are generally under-recognised in craniopharyngioma patients, but are the third commonest group of clinical features at diagnosis after that of raised intracranial pressure and visual deterioration [5], with up to 20% of patients having panhypopituitarism [6, 7].

There is a degree of discrepancy between the proportion of patients spontaneously reporting endocrine symptoms such as linear growth failure, weight loss or gain, lethargy, polyuria and polydipsia as the initial presenting complaint versus the proportion who truly have manifestations of deficiencies in growth hormone (GH), thyroid-stimulating hormone (TSH), adrenocorticotrophic hormone (ACTH) or antidiuretic hormone (ADH) on direct questioning or biochemical testing [2]. Hypothalamo-pituitary dysfunction can also arise as a result of hormone excess, such as that occurring in central precocious puberty.

GH Deficiency

In a cross-sectional study of 138 paediatric craniopharyngioma patients, Muller et al. [8] demonstrated that decelerations in height velocity and increases in weight and BMI SDS often occur for a significant period prior to diagnosis (height deceleration median interval 33 (range 12–96) months, weight and BMI increase median interval 24 (range 24–48) months). In some cases, this means that auxological changes can be found from as early as 6–7 months of age [9]. Indeed, changes in height and/or weight have been found to occur in up to two-thirds of patients with a variety of suprasellar tumours prior to presenting with neuro-ophthalmological symptoms [10]. On testing, between 35% and 100% of craniopharyngioma patients have GH deficiency at diagnosis [2, 3, 6, 7, 11–13], depending on whether dynamic testing was performed on all patients or only those exhibiting growth failure, and it is probably the commonest endocrinopathy at presentation. Serum insulin-like growth factor 1 (IGF-1) and its binding protein, IGF-BP3, are useful but less accurate surrogate markers of GH deficiency [14], particularly if formal GH stimulation testing involving insulin-induced hypoglycaemia or glucagon stimulation is not possible pre-treatment or is contraindicated [15].

Gonadotrophin Dysfunction

Gonadotrophin dysfunction can manifest in two ways at presentation—either as gonadotrophin [luteinising hormone (LH)/follicle-stimulating hormone (FSH)] deficiency, or as central precocious puberty (i.e. LH/FSH excess). The former is more common, with 5–24% of patients presenting with pubertal delay or arrest [2, 3, 6, 7, 16], and up to 85% having evidence of hypogonadotropic hypogonadism on biochemical testing [2, 6, 11, 12, 16]. In contrast, precocious puberty, which is thought to arise from hypothalamic (rather than pituitary) disruption leading to gonadotrophin-releasing hormone (GnRH) excess [17], is rarer and occurs in only 3% of patients at diagnosis [7, 16, 18]. Assessment of this axis is difficult as it is quiescent through most of childhood, being active only during the first 6 months of life (a period known as “mini-puberty”) and during puberty [19]. Measurement of basal and GnRH-stimulated gonadotrophin concentrations outside these windows is therefore unlikely to be informative, so children presenting during this period require careful follow-up through puberty as discussed below.

ACTH Deficiency

ACTH deficiency or secondary hypoadrenalism is present in 21–71% of patients at the time of diagnosis, and can be associated with symptoms of lethargy, somnolence or prolonged recovery from recurrent infections [2, 3, 6, 7, 12, 16, 20]. Such non-specific symptoms may be easily missed during history-taking or attributed to other non-endocrine causes, but are important to diagnose, particularly prior to any surgical intervention as this may precipitate a fatal adrenal crisis in the absence of adequate hydrocortisone replacement. It is usually seen in the presence of multiple pituitary hormone deficiencies. Due to the permissive effect of cortisol on renal free water clearance via the inhibition of ADH secretion, it is also worth remembering that hypocortisolaemia may therefore mask coexistent diabetes insipidus and patients with combined ACTH and ADH deficiencies may not have symptoms of polyuria or polydipsia until the glucocorticoid deficiency is corrected. Comprehensive assessment of this axis can be logistically difficult, especially if dexamethasone has been commenced to reduce peritumoural oedema prior to surgery, and is discussed further below.

Thyrotropin-Releasing Hormone/TSH Deficiency

Secondary and/or tertiary hypothyroidism, like secondary hypoadrenalism, presents with non-specific symptoms such as lethargy and cold intolerance and occurs in 13–32% of patients at diagnosis [2, 3, 6, 7, 12]. Biochemically, it usually presents as a low free thyroxine (T_4) in the presence of an inappropriately normal TSH (a subnormal TSH at diagnosis is uncommon). Differentiation between TSH and thyrotropin-releasing hormone (TRH) deficiency is a purely academic exercise and a TRH stimulation test is often unhelpful [21].

Hyperprolactinaemia

Mild hyperprolactinaemia (<2000 mU/l) is present in 11–52% of craniopharyngioma patients at diagnosis [2, 6, 16], but is usually clinically insignificant and does not lead to galactorrhoea. Its presence indicates disruption of inhibitory hypothalamic dopaminergic neurons by pituitary stalk compression but must be differentiated from true prolactinomas where prolactin (PRL) concentrations are usually >5000 mU/l.

Posterior Pituitary Dysfunction

Although less common, posterior pituitary dysfunction can manifest in several ways at diagnosis of a craniopharyngioma. The commonest of these is central diabetes insipidus (DI) and occurs in 17–29% of patients [3, 7, 11, 12, 22], presenting as a dilute polyuria (or in some cases secondary enuresis) and polydipsia. Missing this diagnosis pre-operatively can prove fatal, as the combination of a general anaesthetic and further hypothalamic damage sustained during surgery concurrently deprives the patient of free access to water and can render them effectively adipsic. As mentioned above, treatment of concurrent secondary hypoadrenalism can result in unmasking of occult DI.

More rarely, posterior pituitary dysfunction may manifest as the syndrome of inappropriate antidiuretic hormone (SIADH) secretion, resulting in a concentrated oliguria, or cerebral salt-wasting syndrome, leading to concentrated polyuria. The pathophysiology of the latter syndrome is not well understood, and is thought to be secondary to an increased secretion of brain natriuretic peptide [23] Therefore, paired early morning plasma and urine osmolalities, electrolytes and glucose are mandatory investigations at diagnosis to determine fluid management during the perioperative period.

Endocrine Investigations and Management at Diagnosis

Given the frequency of abnormalities in growth patterns at presentation, accurate auxology (height, weight and BMI and their corresponding SD scores) is mandatory prior to any treatment (Table 1). If available, prior growth curves from community screening may indicate linear growth failure or rapid weight gain in the months preceding diagnosis. Pre-treatment BMI SD scores in particular are predictive of future hypothalamic obesity and may indicate the need for earlier intervention and screening for complications [9]. Pubertal assessment for precocious, delayed or arrested puberty supported by an inappropriate delayed or advanced bone age help predict future growth potential and determine the need for treatments to maximise this. A delayed bone age may also be a manifestation of longstanding, untreated hypothyroidism.

Endocrine tests prior to any definitive treatment are crucial for several reasons. Firstly, differentiating craniopharyngiomas from other suprasellar tumours in this region such as germinomas (some of which secrete α-fetoprotein (AFP) and/or β-human chorionic gonadotrophin (β-hCG)), prolactinomas (PRL) and somatotrophinomas (GH) is important as tumours such as germinomas and prolactinomas can be treated medically by chemotherapy and dopamine agonists, respectively, without necessitating biopsy or surgical intervention. Measurement of these plasma hormones (AFP, β-hCG, PRL, and GH or IGF-1), and occasionally cerebrospinal fluid AFP and β-hCG concentrations is therefore mandatory in all such cases

Table 1 Recommended pre-treatment endocrine assessment for craniopharyngiomas

First-line	Second-line (if first-line abnormal)
Clinical assessment	
Height	
Weight	
Sitting height	
Body mass index (BMI)	
Tanner pubertal stage	
Bone age	
Endocrine biochemistry	
IGF-1/IGF-BP3	GH provocation test (e.g. insulin tolerance test, glucagon stimulation test)
LH, FSH, oestradiol/testosterone (if <2 years, and >8 years in girls, >9 years in boys)	GnRH stimulation test
TSH, free T_4 ± free T_3	
7, 8, 9 a.m. cortisol and ACTH (before dexamethasone)	Adrenal stimulation test (e.g. insulin tolerance test, standard or physiological synacthen test)
Early morning paired urine and plasma osmolalities and electrolytes	Water deprivation test
Tumour markers	
PRL	
AFP	
β-hCG	

IGF-1 insulin-like growth factor 1, *IGF-BP3* insulin-like growth factor binding protein 3, *LH* luteinising hormone, *FSH* follicle-stimulating hormone, *TSH* thyroid-stimulating hormone, T_4 thyroxine, T_3 triiodothyronine, *ACTH* adrenocorticotrophic hormone, *PRL* prolactin, *AFP* α-fetoprotein, *β-hCG* β-human chorionic gonadotrophin, *GH* growth hormone, *GnRH* gonadotrophin-releasing hormone

A basal pituitary hormone screen helps identify which patients need prioritising for dynamic endocrine testing prior to any definitive surgery. Further dynamic testing should prioritise establishing the status of the hypothalamo-pituitary-adrenal and ADH axes prior to embarking on any surgical intervention in order to avoid the fatal consequences of uncorrected cortisol insufficiency and/or central diabetes insipidus. Pre-operative early morning paired urine and plasma electrolytes and osmolalities, as well as early morning cortisol and ACTH concentrations, should be measured particularly prior to the administration of high-dose dexamethasone for peritumoural oedema as this will subsequently suppress the hypothalamo-pituitary-adrenal axis. Morning cortisol values <200 nmol/l may indicate ACTH deficiency and in the absence of dexamethasone prophylaxis should be confirmed by insulin-induced hypoglycaemia (which will concurrently assess GH reserve) or standard synacthen stimulation, with a peak cortisol concentration of >500 nmol/l being normal.

Further investigation for other hypothalamo-pituitary endocrine deficits and their treatment must be supervised by a paediatric endocrinologist. Dynamic testing and treatment of GH and gonadotrophin deficiency can usually be delayed to the post-treatment period. It would, however, seem sensible to render a hypothyroid patient euthyroid pre-operatively. The pattern of endocrine dysfunction at diagnosis may also be important—for instance, isolated central DI is unusual as a presenting feature in craniopharyngiomas (usually only being present with other pituitary hormone deficiencies), and is usually a feature of germinomas and Langerhans cell histiocytosis [24, 25].

Endocrine Considerations in Perioperative Management

In the absence of dexamethasone prophylaxis and where a patient's adrenal status has not been established, intramuscular or intravenous hydrocortisone in stress doses (1–2 mg/kg) should be administered at induction of anaesthetic, with repeated doses every 6–8 h thereafter. This should be gradually weaned to oral maintenance doses of hydrocortisone (10–15 mg/m^2/day) over 5–10 days until adrenal status can be assessed.

Pre-operative central DI warrants desmopressin administration prior to surgery (0.1–0.2 μg intramuscularly/subcutaneously) with careful attention to fluid balance monitoring with an indwelling urinary catheter intra-operatively. Fluid losses should be replaced ml for ml with an additional 300 ml/m^2/day to account for insensible losses. Uncontrolled posterior pituitary dysfunction is a significant risk factor for mortality and was found to be associated with nearly 50% of deaths from hypothalamo-chiasmatic low-grade gliomas, another suprasellar tumour [26]. Close monitoring by an experienced paediatric endocrinologist is essential as dangerous post-operative swings in salt and water balance resulting in rapid shifts from hyper- to hyponatraemia can occur for several reasons leading to the attendant risks of seizures and cerebral oedema.

Firstly, a well-described triphasic response of transient DI (up to 48 h), SIADH (up to 2 weeks) and a second phase of DI may evolve over the first 2–3 post-operative weeks. The latter is more likely to be permanent if it persists beyond 21 days, is preceded by a prolonged or severe period of SIADH, or if extensive surgery has been performed [27, 28]. This triphasic phenomenon is more frequent in children than adults (23% vs. 14% in craniopharyngiomas) [29]. The three phases may also occur independently, with or without the coexistence of cerebral salt-wasting syndrome which further complicates diagnosis and management. Additionally, changes in dosing of glucocorticoid replacement therapy, particularly when weaning from supraphysiological doses, may result in changes in requirement for desmopressin replacement. Other factors that increase the risk of posterior pituitary dysfunction include the use of anticonvulsants such as carbamazepine and lamotrigine, which can potentiate the effect of desmopressin and cause an SIADH-like syndrome. Lastly, surgically-induced damage to the thirst centre in the hypothalamus can lead to hypothalamic adipsia, where the child may need to be restricted to a set amount of fluid per day to maintain normal fluid balance.

Long-Term Endocrine Dysfunction in Craniopharyngioma Survivors

The pattern of post-treatment endocrine dysfunction seen in survivors of craniopharyngioma is remarkably very similar in frequency and timing, regardless of the treatment modalities used. Deficiency of GH is commonest (20–99%), followed by TSH/TRH (39–97%), LH/FSH (30–95%), ACTH (39–89%) and finally ADH (42–94%), and 31–84% of patients have panhypopituitarism [2–4, 6, 7, 30–38]. Longitudinal comparisons indicate that the hierarchical evolution of deficits mimics that of many other suprasellar tumours, with the GH axis being the most susceptible, and the ACTH and posterior pituitary axes being the most robust [3, 5, 26, 33, 39]. The frequency of deficits also mimics that of congenital neurodevelopmental malformations in this region such as septo-optic dysplasia [40]. Although it is well known that the various hypothalamo-pituitary axes are differentially sensitive to irradiation [41], the similar pattern seen even in its absence as well as with other suprasellar tumours and congenital disorders suggests that this observation is more to do with tumour position and the inherent susceptibility of various hypothalamo-pituitary networks to damage rather than the therapeutic modality *per se*. The continued possibility of new endocrine deficits emerging over time (Fig. 1) means that lifelong endocrine follow-up with regular clinical and biochemical assessments is crucial, with at least 6-monthly follow-up appointments until adult height is achieved, after which annual reviews should suffice with clear plans for transition to adult care.

GH Deficiency

GH deficiency occurs in nearly all craniopharyngioma survivors at some stage. If not already present at diagnosis, it is virtually guaranteed to occur after any therapy to the hypothalamo-pituitary region. It has been shown that the GH axis is particularly sensitive to irradiation [26, 41, 42], and neurosecretory dysfunction with disturbances to spontaneous GH pulsatility may be present before eventual evolution to frank GH deficiency [43]. Diagnosis requires dynamic endocrine testing with the gold standard being the insulin tolerance test, but this is contraindicated in patients with a history of seizures resulting from tumour- or treatment-related damage due to the unavoidable side effect of hypoglycaemia. If performed, we would recommend routine glucose rescue at 25–30 min supported by an age-appropriate dose of emergency hydrocortisone at the end of the test, regardless of whether hypoglycaemia is achieved or not, due to the risk of concurrent ACTH deficiency. Other alternatives include the glucagon stimulation test, clonidine stimulation test and arginine-GHRH stimulation test, although the latter will not detect GH deficiency resulting from hypothalamic damage and disruption of GHRH release. As mentioned previously, serum IGF-1 and IGF-BP3 are less accurate markers of GH deficiency, with the

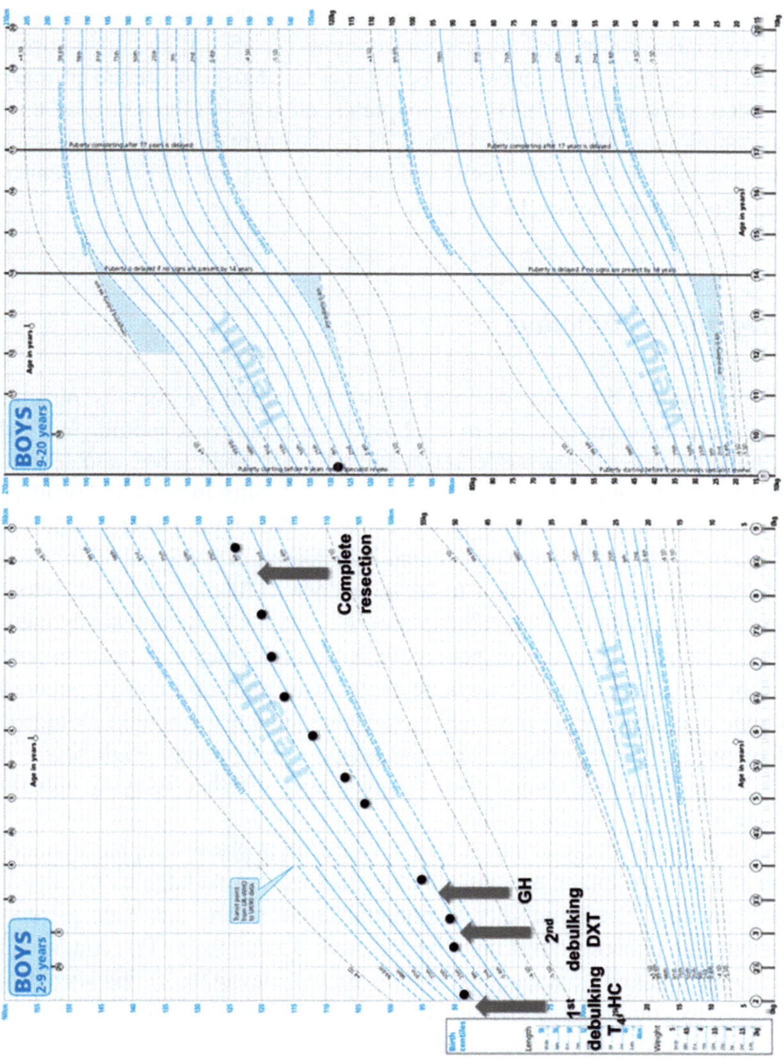

Fig. 1 Typical growth chart of a patient with craniopharyngioma who had multiple progressions and required multiple surgical interventions and radiotherapy. Not noted here for clarity are the multiple cystic progressions which required intracystic drain and reservoir insertions. Note the deceleration in height velocity prior to each progression. T_4, levothyroxine; *HC*, hydrocortisone; *GH*, growth hormone

former being also dependent on nutritional status and liver function, but may be useful in cases where dynamic testing is considered too hazardous.

Paradoxically normal growth may also occur despite GH deficiency, either due to the phenomenon of "growth without growth hormone", which is poorly understood and thought to be secondary to hyperinsulinaemia occurring from rapid post-treatment weight gain [44], or precocious puberty. It is worth noting that GH therapy in replacement doses is not associated with an increased risk of tumour recurrence or progression in both craniopharyngiomas [2, 45, 46] and other suprasellar lesions [26]. There may, however, be an increased risk of second primary neoplasms but the evidence for this is not completely clear [47]. Given this, we tend to advocate "minimum effective doses" for GH replacement, starting at a dose of 0.5–0.7 mg/m^2/day (10–15 IU/m^2/week) and aiming for normal IGF-1 concentrations and an age-appropriate height velocity.

Gonadotrophin Dysfunction

Gonadotrophin dysfunction can manifest either as central precocious puberty (defined as a testicular volume of ≥4 ml in a boy <9 years or breast budding in a girl <8 years), pubertal delay (failure to enter puberty by age 14 in a boy and 13 in a girl), or arrest (manifesting as sexual dysfunction or infertility in boys and primary amenorrhoea or oligomenorrhoea in girls). Central precocious puberty tends to occur in the setting of irradiation exposure and is increased in younger children and with lower irradiation doses (thought to be secondary to disruption of inhibitory hypothalamic pathways on the GnRH pulse generator) [43, 48]. As previously mentioned, the coexistence of precocious puberty with GH deficiency may mask growth failure, and height velocity must always be assessed in light of a child's pubertal stage and bone age. Treatment with GnRH analogues such as triptorelin alongside GH supplementation in the presence of deficiency can help restore adult height if commenced early, but consideration must be given as to whether delaying puberty is appropriate for each individual child.

Conversely, pubertal delay or arrest as a result of hypogonadotrophic hypogonadism can result from either tumour- or treatment-related damage to the hypothalamus or pituitary. The possibility of pubertal arrest means that even patients with a spontaneous pubertal onset must be followed up carefully. The concurrent presence of untreated GH deficiency (which is usually inevitable in this scenario) can be managed by commencing GH supplementation at least 6 months prior to sex steroid supplementation. The timing of sex steroid initiation, however, must be tailored appropriately to the individual child, bearing in mind that delaying commencement beyond the usual pubertal age (~12 years in girls and 14–15 years in boys) is probably inappropriate, particularly given the long-term benefits of sex steroid replacement on bone mineral accretion and long-term bone health [49–51].

In both cases, a GnRH stimulation test provides supporting information but is not always necessary. It is also worth noting that central precocious puberty and later hypogonadotrophic hypogonadism are not mutually exclusive entities. Indeed, in a retrospective longitudinal analysis of survivors of childhood hypothalamo-chiasmatic low-grade gliomas, Gan et al. [26] demonstrated that children with central precocious puberty were potentially at increased risk for future hypogonadotrophic hypogonadism. Consequently, children who have been treated for the former with GnRH analogues should be monitored carefully on the cessation of such treatment, looking out for failure of subsequent normal pubertal progress.

TSH/TRH Deficiency

Unlike the thyroid gland itself, TRH-secreting neurons and thyrotrophs in the pituitary gland are relatively resistant to irradiation [41]. Secondary and/or tertiary hypothyroidism is therefore only likely to arise in the setting of hypothalamic involvement of the tumour or from direct surgical damage. As previously mentioned, TRH stimulation tests do not aid in the differentiation between secondary and tertiary hypothyroidism and serial thyroid function tests are more specific and easier to perform [21]. In this setting, dose titration of levothyroxine supplementation should be based entirely on free thyroxine concentrations and not TSH. Some clinicians favour keeping the free thyroxine concentration in the upper half of the normal range, given the predisposition of many craniopharyngioma survivors to obesity.

ACTH Deficiency

Out of all the anterior pituitary hormones, the hypothalamo-pituitary-adrenal axis is fortunately most robust to irradiation. Post-operative assessment of this axis after the use of perioperative dexamethasone and subsequent maintenance hydrocortisone replacement is probably best left for at least 3 months. If neither dexamethasone nor hydrocortisone has been commenced, then serial monitoring of early morning cortisol concentrations (aiming for >200 nmol/l) should be performed.

The gold standard insulin tolerance test assesses the entire hypothalamo-pituitary-adrenal pathway and concurrently tests the integrity of the GH axis as well, but may not always be possible as discussed above. Given the risks associated with hypoglycaemia resulting from this test, it should be performed in a dedicated paediatric endocrinology centre used to managing patients with multiple pituitary hormone deficiencies. Other methods of assessment such as the standard (high-dose) synacthen test, low-dose synacthen test and 24-h cortisol day curves are less reliable but may be useful in cases where the insulin tolerance test is contraindicated. It is worth remembering that the standard synacthen test involves a supraphysiological stimulus of synthetic ACTH and may be less sensitive at detecting more subtle degrees of deficiency [52–54]. However, the sensitivity of the low-dose synacthen test is also widely debated, not least because the protocols used in different studies vary considerably [55–57]. Regardless of the method used, hydrocortisone should be omitted on the morning of the test as it interferes with most commonly used biochemical assays for cortisol.

Hydrocortisone replacement therapy is generally given at a dose of 10–15 mg/m^2/day, with parents and/or patients advised to double or even treble their doses in the face of illness. Doses can be titrated to trough concentrations, or more accurately using a 24-h cortisol day curve. More recently, we have also advised patients to take an extra double dose of hydrocortisone at 4 a.m. when they are unwell, as this coincides with the beginning of the physiological cortisol morning surge. Patients should also be educated on how to administer emergency intramuscular hydrocortisone (doses <1 year 25 mg, 1–5 years 25–50 mg, >5 years 100 mg) and correct hypoglycaemia.

Hyperprolactinaemia and Prolactin Deficiency

Hyperprolactinaemia is usually asymptomatic in craniopharyngioma survivors but may be a sign of tumour progression due to stalk compression disrupting the hypothalamic inhibitory dopaminergic pathways regulating pituitary prolactin secretion. In this setting prolactin concentrations are usually <2000 mU/l and galactorrhoea is rare [58]. Conversely, prolactin deficiency, although increasingly recognised, is usually seen in the setting of multiple pituitary hormone deficiencies but is of unknown clinical significance [59].

Posterior Pituitary Dysfunction

Central DI is the least common of all hypothalamo-pituitary hormone deficits and is the only endocrinopathy that has been recognised to spontaneously resolve in this situation. As part of the triphasic response seen in the perioperative period, the final phase of DI may be transient as described above. As a rule, central DI does not develop after cranial irradiation alone, and is only seen after direct hypothalamic tumour invasion or surgery [26]. It is notable that central DI can occur even after relatively minor surgical procedures such as biopsies or ventriculoperitoneal shunt insertions [26], and therefore all such patients should be treated as being at risk of this. The combination of central DI with hypothalamic adipsia and ACTH deficiency is particularly dangerous and can prove fatal, and patients undergoing radical resection of their tumour are at particularly high risk of this combination [2, 7, 22, 32, 33, 37].

Confirmation of diagnosis of central DI in this situation does not always warrant a water deprivation test, particularly in the presence of frank, dilute polyuria (specific gravity <1.010, urine osmolality <300 mOsm/kg) despite evidence of dehydration (plasma osmolality >295 mOsm/kg). This is usually best assessed on early morning paired urine and plasma samples. A random urine osmolality of >600 mOsm/kg or a urine: plasma osmolality ratio of >1.5 also effectively excludes DI. A water deprivation test result confirming DI should be immediately followed by the DDAVP test to confirm it is truly central and not nephrogenic in origin.

In the management of patients with central DI and an intact thirst axis, it is worth bearing in mind that untreated patients are able to maintain themselves euvolaemic and eunatraemic by adjusting their oral intake appropriately. DDAVP treatment therefore aims to provide the patient with a relatively normal quality of life by reducing the burden of polyuria and polydipsia. We would normally advise that DDAVP be commenced as an inpatient and titrated to stable doses prior to discharge, guided by pre-DDAVP paired urine and plasma osmolalities. The aim of treatment should generally be to err on the side of underdosing, as overdosing DDAVP can result in hyponatraemia which is difficult to correct safely and carries the significant risk of cerebral oedema and death. Patients with hypothalamic adipsia are even more difficult to manage and require a strict fluid intake regimen to maintain euvolaemia.

The Hypothalamic Syndrome

The hypothalamic syndrome is a loosely defined constellation of features attributed to hypothalamic dysfunction. The cardinal feature of this disorder is hypothalamic obesity, a syndrome of morbid, inexorably escalating obesity resulting in a BMI of >+3 SDS, occurring in up to 77% of craniopharyngioma survivors [60] with attendant risks of type 2 diabetes and cardiovascular mortality. It is often associated with other features of hypothalamic dysfunction, including impaired thermoregulation, disruption of the sleep-wake cycle and behavioural disturbances which are often autistic in nature. Risk factors include younger age at diagnosis, hypothalamic injury by tumour, irradiation or surgery and multiple pituitary hormone deficiencies [61]. Its pathophysiology is poorly understood, and although initially thought to be caused by disruption of the ventromedial hypothalamus which acts as an appetite-regulating centre, studies have now shown that hyperphagia is not necessarily the cause of weight gain [62]. Instead, the concept of an imbalance in energy homeostasis involving anorexigens (leptin, pro-opiomelanocortin, α-melanocyte-stimulating hormone, oxytocin) and orexigens (ghrelin, neuropeptide Y, agouti-related peptide) has now been put forward [60]. Further discussion of the management of obesity in craniopharyngioma survivors is covered in Chap. 7.

Conclusions

Endocrine dysfunction in craniopharyngioma patients is remarkably common, and is already often present at the time of diagnosis, although in some cases this may not be recognised unless specifically tested for. Pre-treatment endocrine clinical assessment and biochemical testing is essential not just to exclude other differential diagnoses for tumours in this region, but also to provide a baseline assessment of hypothalamo-pituitary function. Careful pre-operative assessment of the ACTH and ADH axes is particularly crucial to avoid the life-threatening risk of untreated glucocorticoid deficiency and diabetes insipidus. Survivors of craniopharyngioma require lifelong follow-up due to the ongoing risk of evolving hypothalamo-pituitary endocrine dysfunction which is present in virtually all patients and the long-term risk of hypothalamic obesity with its effects on cardiovascular and metabolic morbidity.

References

1. Brignardello E, Felicetti F, Castiglione A, Chiabotto P, Corrias A, Fagioli F, et al (2013) Endocrine health conditions in adult survivors of childhood cancer: the need for specialized adult-focused follow-up clinics. Eur J Endocrinol 168(3):465–472. European Federation of Endocrine Societies
2. DeVile CJ, Grant DB, Hayward RD, Stanhope R (1996) Growth and endocrine sequelae of craniopharyngioma. Arch Dis Child 75(2):108–114
3. Karavitaki N, Brufani C, Warner JT, Adams CB, Richards P, Ansorge O et al (2005) Craniopharyngiomas in children and adults: systematic analysis of 121 cases with long-term follow-up. Clin Endocrinol (Oxf) 62(4):397–409
4. Crom DB, Smith D, Xiong Z, Onar A, Hudson MM, Merchant TE, et al (2010) Health status in long-term survivors of pediatric craniopharyngiomas. J Neurosci Nurs 42(6):323–328; quiz 9–30. Journal of the American Association of Neuroscience Nurses
5. Gan H-W, Bulwer C, Spoudeas HA (2014) Chapter 11a: Pituitary & hypothalamic tumour syndromes in childhood. In: Grossman A (ed) Pituitary disease & neuroendocrinology. MDText.Com Inc, Massachusetts
6. Caldarelli M, Massimi L, Tamburrini G, Cappa M, Di Rocco C (2005) Long-term results of the surgical treatment of craniopharyngioma: the experience at the Policlinico Gemelli, Catholic University, Rome. Childs Nerv Syst 21(8–9):747–757. Official Journal of the International Society for Pediatric Neurosurgery
7. Hetelekidis S, Barnes PD, Tao ML, Fischer EG, Schneider L, Scott RM et al (1993) 20-year experience in childhood craniopharyngioma. Int J Radiat Oncol Biol Phys 27(2):189–195
8. Muller HL, Kaatsch P, Warmuth-Metz M, Flentje M, Sorensen N (2003) Kraniopharyngeom im Kindes-und Jugendalter: Diagnostische und therapeutische Strategien (Childhood craniopharyngioma - diagnostic and therapeutic strategies). Monatsschr Kinderheilkd 151:1056–1063
9. Muller HL, Emser A, Faldum A, Bruhnken G, Etavard-Gorris N, Gebhardt U et al (2004) Longitudinal study on growth and body mass index before and after diagnosis of childhood craniopharyngioma. J Clin Endocrinol Metab 89(7):3298–3305
10. Taylor M, Couto-Silva AC, Adan L, Trivin C, Sainte-Rose C, Zerah M et al (2012) Hypothalamic-pituitary lesions in pediatric patients: endocrine symptoms often precede neuro-ophthalmic presenting symptoms. J Pediatr 161(5):855–863
11. Halac I, Zimmerman D (2005) Endocrine manifestations of craniopharyngioma. Childs Nerv Syst 21(8–9):640–648
12. Muller HL (2012) Childhood craniopharyngioma. Pituitary 16:56–67
13. Sorva R (1988) Children with craniopharyngioma. Early growth failure and rapid postoperative weight gain. Acta Paediatr Scand 77(4):587–592

14. Hindmarsh PC, Swift PG (1995) An assessment of growth hormone provocation tests. Arch Dis Child 72(4):362–367 discussion 7-8
15. Shah A, Stanhope R, Matthew D (1992) Hazards of pharmacological tests of growth hormone secretion in childhood. BMJ 304(6820):173–174
16. de Vries L, Lazar L, Phillip M (2003) Craniopharyngioma: presentation and endocrine sequelae in 36 children. J Pediatr Endocrinol Metab 16(5):703–710
17. Fuqua JS (2013) Treatment and outcomes of precocious puberty: an update. J Clin Endocrinol Metab 98(6):2198–2207
18. Puget S, Garnett M, Wray A, Grill J, Habrand JL, Bodaert N et al (2007) Pediatric craniopharyngiomas: classification and treatment according to the degree of hypothalamic involvement. J Neurosurg 106(1 Suppl):3–12
19. Rey RA (2014) Mini-puberty and true puberty: differences in testicular function. Ann Endocrinol 75(2):58–63
20. Sklar CA (1994) Craniopharyngioma: endocrine abnormalities at presentation. Pediatr Neurosurg 21(Suppl 1):18–20
21. Mehta A, Hindmarsh PC, Stanhope RG, Brain CE, Preece MA, Dattani MT (2003) Is the thyrotropin-releasing hormone test necessary in the diagnosis of central hypothyroidism in children. J Clin Endocrinol Metab 88(12):5696–5703
22. De Vile CJ, Grant DB, Kendall BE, Neville BG, Stanhope R, Watkins KE et al (1996) Management of childhood craniopharyngioma: can the morbidity of radical surgery be predicted? J Neurosurg 85(1):73–81
23. Oh JY, Shin JI (2014) Syndrome of inappropriate antidiuretic hormone secretion and cerebral/renal salt wasting syndrome: similarities and differences. Front Pediatr 2:146
24. Phi JH, Kim SK, Lee YA, Shin CH, Cheon JE, Kim IO, et al (2013) Latency of intracranial germ cell tumors and diagnosis delay. Childs Nerv Syst 29(10):1871–1881. Official Journal of the International Society for Pediatric Neurosurgery
25. Donadieu J, Rolon MA, Thomas C, Brugieres L, Plantaz D, Emile JF et al (2004) Endocrine involvement in pediatric-onset Langerhans' cell histiocytosis: a population-based study. J Pediatr 144(3):344–350
26. Gan HW, Phipps K, Aquilina K, Gaze MN, Hayward R, Spoudeas HA (2015) Neuroendocrine morbidity after pediatric optic gliomas: a longitudinal analysis of 166 children over 30 years. J Clin Endocrinol Metab 100(10):3787–3799
27. Finken MJ, Zwaveling-Soonawala N, Walenkamp MJ, Vulsma T, van Trotsenburg AS, Rotteveel J (2011) Frequent occurrence of the triphasic response (diabetes insipidus/hyponatremia/diabetes insipidus) after surgery for craniopharyngioma in childhood. Horm Res Paediatr 76(1):22–26
28. Ghirardello S, Hopper N, Albanese A, Maghnie M (2006) Diabetes insipidus in craniopharyngioma: postoperative management of water and electrolyte disorders. J Pediatr Endocrinol Metab 19(Suppl 1):413–421
29. Pratheesh R, Swallow DM, Rajaratnam S, Jacob KS, Chacko G, Joseph M et al (2013) Incidence, predictors and early post-operative course of diabetes insipidus in paediatric craniopharygioma: a comparison with adults. Childs Nerv Syst 29(6):941–949
30. Habrand JL, Ganry O, Couanet D, Rouxel V, Levy-Piedbois C, Pierre-Kahn A et al (1999) The role of radiation therapy in the management of craniopharyngioma: a 25-year experience and review of the literature. Int J Radiat Oncol Biol Phys 44(2):255–263
31. Hoffman HJ, De Silva M, Humphreys RP, Drake JM, Smith ML, Blaser SI (1992) Aggressive surgical management of craniopharyngiomas in children. J Neurosurg 76(1):47–52
32. Lin LL, El Naqa I, Leonard JR, Park TS, Hollander AS, Michalski JM et al (2008) Long-term outcome in children treated for craniopharyngioma with and without radiotherapy. J Neurosurg Pediatr 1(2):126–130
33. Merchant TE, Kiehna EN, Sanford RA, Mulhern RK, Thompson SJ, Wilson MW et al (2002) Craniopharyngioma: the St. Jude Children's Research Hospital experience 1984–2001. Int J Radiat Oncol Biol Phys 53(3):533–542

34. Muller HL, Gebhardt U, Teske C, Faldum A, Zwiener I, Warmuth-Metz M et al (2011) Post-operative hypothalamic lesions and obesity in childhood craniopharyngioma: results of the multinational prospective trial KRANIOPHARYNGEOM 2000 after 3-year follow-up. Eur J Endocrinol 165(1):17–24
35. Poretti A, Grotzer MA, Ribi K, Schonle E, Boltshauser E (2004) Outcome of craniopharyngioma in children: long-term complications and quality of life. Dev Med Child Neurol 46(4):220–229
36. Schulz-Ertner D, Frank C, Herfarth KK, Rhein B, Wannenmacher M, Debus J (2002) Fractionated stereotactic radiotherapy for craniopharyngiomas. Int J Radiat Oncol Biol Phys 54(4):1114–1120
37. Stripp DC, Maity A, Janss AJ, Belasco JB, Tochner ZA, Goldwein JW et al (2004) Surgery with or without radiation therapy in the management of craniopharyngiomas in children and young adults. Int J Radiat Oncol Biol Phys 58(3):714–720
38. Van Effenterre R, Boch AL (2002) Craniopharyngioma in adults and children: a study of 122 surgical cases. J Neurosurg 97(1):3–11
39. Armstrong GT, Conklin HM, Huang S, Srivastava D, Sanford R, Ellison DW et al (2011) Survival and long-term health and cognitive outcomes after low-grade glioma. Neuro Oncol 13(2):223–234
40. Webb EA, Dattani MT (2010) Septo-optic dysplasia. Eur J Hum Genet 18(4):393–397
41. Littley MD, Shalet SM, Beardwell CG, Robinson EL, Sutton ML (1989) Radiation-induced hypopituitarism is dose-dependent. Clin Endocrinol (Oxf) 31(3):363–373
42. Spoudeas HA (2002) Growth and endocrine function after chemotherapy and radiotherapy in childhood. Eur J Cancer 38(13):1748–1759 discussion 60-1
43. Talbot L, Spoudeas H (2010) Late effects in relation to childhood cancer. In: Estlin EJ, Gilbertson RJ, Wynn RF (eds) Pediatric hematology and oncology: scientific principles & clinical practice. Wiley-Blackwell, Oxford, pp 367–391
44. Phillip M, Moran O, Lazar L (2002) Growth without growth hormone. J Pediatr Endocrinol Metab 15(Suppl 5):1267–1272
45. Moshang T Jr, Rundle AC, Graves DA, Nickas J, Johanson A, Meadows A (1996) Brain tumor recurrence in children treated with growth hormone: the National Cooperative Growth Study experience. J Pediatr 128(5 Pt 2):S4–S7
46. Muller HL, Gebhardt U, Schroder S, Pohl F, Kortmann RD, Faldum A et al (2010) Analyses of treatment variables for patients with childhood craniopharyngioma–results of the multicenter prospective trial KRANIOPHARYNGEOM 2000 after three years of follow-up. Horm Res Paediatr 73(3):175–180
47. Raman S, Grimberg A, Waguespack SG, Miller BS, Sklar CA, Meacham LR et al (2015) Risk of neoplasia in pediatric patients receiving growth hormone therapy–a report from the pediatric endocrine society drug and therapeutics committee. J Clin Endocrinol Metab 100(6):2192–2203
48. Ogilvy-Stuart AL, Clayton PE, Shalet SM (1994) Cranial irradiation and early puberty. J Clin Endocrinol Metab 78(6):1282–1286
49. Carel JC (2006) Management of short stature with GnRH agonist and co-treatment with growth hormone: a controversial issue. Mol Cell Endocrinol 254–255:226–233
50. Gan H-W, Spoudeas HA (2014) Long-term follow-up of survivors of childhood cancer (SIGN Clinical Guideline 132). Arch Dis Child Educ Pract Ed 99(4):138–143
51. Scottish Intercollegiate Guidelines Network (SIGN) (2013) SIGN 132: long term follow up of survivors of childhood cancer. SIGN, Edinburgh
52. Crowley S, Hindmarsh PC, Holownia P, Honour JW, Brook CG (1991) The use of low doses of ACTH in the investigation of adrenal function in man. J Endocrinol 130(3):475–479
53. Patterson BC, Truxillo L, Wasilewski-Masker K, Mertens AC, Meacham LR (2009) Adrenal function testing in pediatric cancer survivors. Pediatr Blood Cancer 53(7):1302–1307
54. Schmiegelow M, Feldt-Rasmussen U, Rasmussen AK, Lange M, Poulsen HS, Muller J (2003) Assessment of the hypothalamo-pituitary-adrenal axis in patients treated with radiotherapy and chemotherapy for childhood brain tumor. J Clin Endocrinol Metab 88(7):3149–3154

55. Maghnie M, Uga E, Temporini F, Di Iorgi N, Secco A, Tinelli C et al (2005) Evaluation of adrenal function in patients with growth hormone deficiency and hypothalamic-pituitary disorders: comparison between insulin-induced hypoglycemia, low-dose ACTH, standard ACTH and CRH stimulation tests. Eur J Endocrinol 152(5):735–741
56. Maguire AM, Biesheuvel CJ, Ambler GR, Moore B, McLean M, Cowell CT (2008) Evaluation of adrenal function using the human corticotrophin-releasing hormone test, low dose Synacthen test and 9am cortisol level in children and adolescents with central adrenal insufficiency. Clin Endocrinol (Oxf) 68(5):683–691
57. O'Grady MJ, Hensey C, Fallon M, Hoey H, Murphy N, Costigan C et al (2013) Lack of sensitivity of the 1-mug low-dose ACTH stimulation test in a paediatric population with suboptimal cortisol responses to insulin-induced hypoglycaemia. Clin Endocrinol (Oxf) 78(1):73–78
58. Karavitaki N, Thanabalasingham G, Shore HC, Trifanescu R, Ansorge O, Meston N et al (2006) Do the limits of serum prolactin in disconnection hyperprolactinaemia need redefinition? A study of 226 patients with histologically verified non-functioning pituitary macroadenoma. Clin Endocrinol (Oxf) 65(4):524–529
59. Toledano Y, Lubetsky A, Shimon I (2007) Acquired prolactin deficiency in patients with disorders of the hypothalamic-pituitary axis. J Endocrinol Invest 30(4):268–273
60. Lustig RH (2011) Hypothalamic obesity after craniopharyngiomas: mechanisms, diagnosis and treatment. Front Endocrinol 2:60
61. Lustig RH, Post SR, Srivannaboon K, Rose SR, Danish RK, Burghen GA et al (2003) Risk factors for the development of obesity in children surviving brain tumors. J Clin Endocrinol Metab 88(2):611–616
62. Harz KJ, Muller HL, Waldeck E, Pudel V, Roth C (2003) Obesity in patients with craniopharyngioma: assessment of food intake and movement counts indicating physical activity. J Clin Endocrinol Metab 88(11):5227–5231

Obesity and Metabolic Disturbances in Adamantinomatous Craniopharyngioma Patients

Eva Marie Erfurth

Abstract An adamantinous craniopharyngioma (CP) is rare benign pituitary tumour often growing invasively and thereby affecting the hypothalamus. The recurrence rate in CP is high and therapy of choice is surgery, followed by cranial radiotherapy (CRT) in about half of the patients. Patients with CP have a 3–19 fold higher cardiovascular mortality in comparison to the general population. The majority of patients have an extended hypopituitarism that needs balanced hormone supplementation. About half of the patient population suffer from hypothalamic damage due to the tumour or operation. These patients suffer from hypothalamic obesity together with increased vascular morbidity and compromised bone health, thirst disorders as well as somnolence and sleep apnea. Cardiovascular risk factors need up front therapy. There is a need for centralising this rare aggressive and difficult tumour to only a few centres and that it will be managed by a multidisciplinary team with experienced neurosurgeons, endocrinologists, neurooncologists, neuroradiologists, and neuroopthamologists for a better prognosis.

Keywords Childhood craniopharyngioma • Cardiovascular mortality • Cardiovascular morbidity • Cardiovascular risk factors • Cardiovascular diseases • Hypothalamic damage • Leptin resistance • Bone mineral density • Insulin

E.M. Erfurth (✉)
Department of Endocrinology, Skånes University Hospital, SE 221 85 Lund, Sweden

Department of Clinical Sciences, Medical Faculty of Lund University, Lund, Sweden
e-mail: Eva_Marie.Erfurth@med.lu.se

J.P. Martinez-Barbera, C.L. Andoniadou (eds.), *Basic Research and Clinical Aspects of Adamantinomatous Craniopharyngioma*,
DOI 10.1007/978-3-319-51890-9_6

Introduction

A craniopharyngioma (CP) is a benign pituitary tumour often growing invasively and thereby affecting the hypothalamus. The tumour is a partly cystic embryonic malformation of the sellar and para-sellar region. An adamantinomatous craniopharyngioma is the most common CP in childhood. The overall incidence of a CP is 0.5–2.0 cases/million per year [1–3] and among childhood CP in Denmark the incidence was 2.14 (1.53–2.92) [2]. The peak age distribution in childhood CP in Sweden was highest at 5–10 years and 10–15 years of age [3] which is similar to the Danish age distribution [4]. Due to its growth and/or treatment, hypopituitarism is seen in the vast majority of cases and obesity due to hypothalamic involvement in up to 50% of the patients [5, 6]. Women and men are equally affected [7].

Typical clinical manifestations at diagnosis in children are visual problems, elevated intracranial pressure (headache, nausea) followed by endocrine symptoms [1]. The recurrence rate in CP is high and therapy of choice is surgery, followed by cranial radiotherapy (CRT) in about half of the patients. The long-term morbidity in patients with CP is substantial and is mainly based on the tumour location and size, recurrence rate, and its treatment. The morbidity is affecting the life span of these patients with hypopituitarism, increased cardiovascular risk, hypothalamic damage, visual and neurological deficits, reduced bone health, and compromised quality of life and cognitive function [8].

The aim of this chapter is to highlight metabolic disturbances and obesity particularly in childhood onset adamantinomatous CP patients. Further, to discuss the background to these metabolic disturbances.

Mortality in Craniopharyngioma

CPs have the highest mortality of all pituitary tumour and women are more affected than men [1, 3, 9]. The overall survival rates in recent years in CPs range from 89 to 94% at 5-years and from 85 to 90% at 10-years follow-up [8, 10], and an average of 62–76% at 20 years [11, 12]. The cause specific late mortality, after 20 years, was multi-factorial, but rarely due to disease progression [11].

Long-term follow-up of CPs include often mixed populations of children and adults and the mortality risk varied in different countries. The standardised overall mortality rate (SMR) from two Swedish studies was 5.5 (CI% 3.68–8.22) and 3.8 (2.9–5.0), respectively [1, 3], the Netherlands SMR 2.88 (95% CI of 1.35–4.99) [13], and from the UK 9.28 (95% CI 5.84–14.75) [9]. In a study from Ireland including 70 CP patients the increased overall mortality was confirmed (SMR 8.75; 95% CI 5.4–13.3), and again with a somewhat higher mortality in women (SMR 10.51, 95% CI 5.04–19.3) than among men (SMR 7.55 95% CI 3.77–13.52) [11]. The percentage of childhood onset CP was mentioned in three of the cohorts and was between 34% and 43% [1, 3, 13].

Table 1 Hazard ratios (HR) and 95% confidence intervals (CI) for death after surgery for craniopharyngioma with respect to tumor recurrence, radicality at surgery, and radiotherapy, with and without age stratification (<29 and ≥29 year at operation)

	No age stratification		With age stratification	
Cox regression models	HR 95% CI	*P*	HR 95% CI	*P*
All patients ($n=60$)				
Recurrence (time dependent; yes *vs.* no)	2.7 (1.0–7.3)	0.05	4.4 (1.4–14)	0.01
Radicality at surgery (subtotal *vs.* total)	2.0 (0.9–4.5)	0.09	1.1 (0.5–2.5)	0.9
Radiotherapy (yes *vs.* no)	0.5 (0.2–1.2)	0.1	0.3 (0.1–0.8)	0.01
Patients surviving the first 6 months after surgery ($n=50$)				
Recurrence (time dependent: yes *vs.* no)	1.6 (0.5–4.7)	0.4	2.2 (0.6–7.6)	0.2
Radicality at surgery (subtotal *vs.* total)	1.6 (0.6–4.5)	0.4	0.9 (0.3–2.5)	0.8
Radiotherapy (yes *vs.* no)	1.4 (0.4–4.1)	0.6	0.7 (0.2–2.4)	0.6

Survival is, however, related to choice of therapy which is depending on tumour size, extension of the tumour, and of recurrence of the tumour. Karavitaki et al. [8] recorded no significant difference in the 10-year survival rates between patients treated with gross total removal (GTR) (100%) and partial removal (PR) (86%). Stripp et al. [14] found also comparable 10-year survival rates in patients treated by surgery alone (GTR or PR) or by surgery followed with CRT (83%). Finally, Rajan et al. [15] found no difference if patients were treated with CRT alone or with surgery and no influence were seen of the degree of surgery. Bulow et al. [1] showed in a multivariate model, including radicality at surgery, radiotherapy (yes vs no), and recurrence as time dependent factor with a broad age stratification (<29 years and ≥29 years), a protective effect of radiotherapy (HR 0.3; 95% CI, 0.1–0.8). Furthermore, increased risk of death after recurrence (Hazard Ratio 4.4; 95% CI 1.4–14) was shown, but no obvious positive effect of radicality at surgery (Table 1).

Cardiovascular Mortality

Patients with CP have a 3–19 fold higher cardiovascular mortality in comparison to the general population [1, 9, 13]. The cohort from UK had the highest SMR of 19.4 (95% CI 8.08–46.7) for cerebrovascular deaths [9]. In the first Swedish cohort the cardiovascular (including cerebrovascular) mortality was three-fold enhanced (SMR 3.21, 95% CI 1.29–6.61) [1] and in a recent study the cerebrovascular mortality was 5.1 (1.7–12) [3]. Female gender seemed to be at particular risk for cardiovascular mortality (SMR 11.4 and 4.9, respectively) compared to male gender (SMR 4.79 and 3.2, respectively). This gender difference was also seen in the Dutch cohort with a higher SMR among females (SMR 3.84; 95% CI 1.47–7.22) compared to males (1.84; 95% CI 0.33–4.58) [13].

Morbidity in Craniopharyngioma

The long-term morbidity in patients with CP is substantial and is mainly based on the tumour location and size, recurrence rate, and its treatment. The morbidity includes hypopituitarism, hypothalamic involvement (obesity, thirst disorders, thermoregulatory disorders, somnolence and sleep apnoea, and cardiac arrhythmia), cardiovascular risk factors, visual and neurological problems, and reduced quality of life and cognitive function [6]. The treatment of a CP may be achieved using either primary gross total resection, which may increase the risk of hypothalamic, neurological, endocrine, and visual damage [6, 16], or partial resection of the tumour followed by CRT, as an effective means of preventing recurrences ([17]).

Cardiovascular Morbidity

The increased cardiovascular mortality is also related to cardiovascular morbidity in CP patients. Childhood obesity due to CP is leading to severe atherosclerotic cardiovascular disease, type II diabetes mellitus, and metabolic syndrome [13, 18]. A study based on The Swedish National Patient Registry showed an increased risk for type II DM (six-fold) [3]. Recently, a total of 19 childhood onset CPs patients with hypothalamic tumour involvement agreed to participate in a CT scan investigation of the liver [19, 20]. A nonalcoholic fatty liver disease (NAFLD) was recorded in about 50% of CP patients with hypothalamic involvement which was associated with elevated liver enzymes and homeostasis model assessment index. NAFLD is considered a feature in the metabolic syndrome.

Holmer et al. [21] evaluated the prevalence of cardiovascular risk factors after long-term complete hormone replacement in childhood onset CP patients and showed increased cardiovascular risk factors in particularly CP women and in patients with hypothalamic involvement by the tumour [21]. Typical risk assessments for cardiovascular disease showed increased levels of hormone sensitive C-reactive protein and low density lipoproteins [21]. In addition, significantly more treatment for cardiovascular diseases (anti-hypertensive and anti-diabetes treatments and lipid lowering drugs,) and/or manifestations of the metabolic syndrome were present [21]. This is in accordance with another study also showing an increased prevalence of hypertension and of other cardiovascular morbidities in CP patients [13].

Hypopituitarism

Complete hypopituitarism is encountered in a majority of CP patients. At least three pituitary hormone deficiencies have been reported in 54–100% of patients with CP [22]. In a study from the Netherlands the long-term prevalence rate of total anterior pituitary insufficiency was 89% [13], and from Ireland the percentage of GH-, gonadotropin-, adrenocorticotropin-, and TSH-deficiency it was 91%, 93.5%, 92%, and 86%, respectively [11]. The prevalence of diabetes insipidus (DI) was 81% [11].

Pituitary deficiency per se and its treatment might through various metabolic effects contribute to the enhanced cardiovascular morbidity and mortality seen in epidemiological studies. However, panhypopituitarism is interrelated to recurrences of the CP, cranial radiotherapy, and with hypothalamic involvement by the tumour, thus it is almost impossible to identify hypopituitarism as an independent risk factor. GH deficiency is probably the most frequent pituitary hormone deficiency and associated with increased levels of cardiovascular risk factors [23]. GH therapy was shown to have beneficial effects on lean and body fat mass, total and low density lipoprotein cholesterol, and diastolic blood pressure in hypopituitary patients [24]. Only few studies have evaluated the effect GH therapy in CP patients and in comparison to patients with non-functioning pituitary adenoma, CP patients had higher prevalence of pituitary deficiencies, were more obese, and had more dyslipidaemia [25]. Two years of GH replacement showed the same effect on fat-free mass, lipids, but CPs were less likely to lose body fat, which is possibly a hypothalamic effect.

Also ACTH deficiency and glucocorticoid supplementation is of importance. This was highlighted by patients with Addison's disease, who suffer from increased premature mortality in comparison to the general population [26]. In addition, high substitution doses of cortisone is associated with increased cardiovascular mortality in patients with previous acromegaly [27]. Subclinical hypothyroidism causes increased cardiovascular risk [28]. A study from Klose et al. [29] showed that low levels of serum free T4 were associated with higher BMI, waist, and hip circumference, but lower levels of the good -HDL- cholesterol.

Hypothalamic Obesity

The hypothalamus (HT) is a small, complex area of the brain and is involved in endocrine function and energy homeostasis. Hypothalamic damage due to tumour or its treatment is the most devastating consequence of a CP. Approximately 50% of children treated for CP are obese at follow-up and hypothalamic damage seems to be a major cause [30, 31]. The HT consists of distinct nuclei including the infundibular nucleus (INF), the paraventricular nucleus (PVN), the dorsomedial nucleus (DMN), and the ventromedial nucleus (VMN) [32, 33]. The peripheral afferent hormones leptin, ghrelin, and insulin all have receptors located on neurons within VMH. Damage to these neurons leads to defective transduction of afferent signal of fat accumulation, satiety, and metabolism. In the postoperative phase of a CP destruction of these nuclei induces hyperphagia, hyperinsulinaemia, and weight gain [30, 33]. The resultant lack of anorexigenic pressure on the melanocortin receptor-4 results in increased feeding behaviour and energy efficiency (with reduced fat oxidation) to store energy substrate as fat. Overt hyperphagia is very prevalent in the first period after operation for about 12 months [34] but seem to be conspicuously absent in the long-term follow-up of these patients [35, 36].

Adipose tissue is richly innervated by sympathetic nerve fibres that control lipolysis. Lipogenesis is also controlled by parasympathetic innervation of adipose tissue originating from separate sympathetic and parasympathetic neurons in the periventricular nucleus (PVN) and supra chiasmatic nucleus (SCN) [37]. Thus, SCN has a unique position to balance both branches of the autonomous nervous system. A large proportion of the CP patients have supra sellar involvements and thus damage to the SCN with an altered regulation of central clock mechanism, which predispose to alterations in metabolism [14, 37]. A disturbance in the circadian rhythms may cause daytime sleepiness demonstrated in childhood onset CPs patients with hypothalamic obesity, which correlate with low morning and nocturnal melatonin levels in saliva [38]

The primary pathogenesis of hypothalamic damage seems to start with hyperinsulinaemia, due to destruction of the VMH nuclei causing imbalance of autonomic nervous system, resulting in suppression of the sympathetic nervous system and stimulation of the vagus [33]. Indeed CP patients have hyperinsulinaemia, with increased levels in the relation to tumour growth (Fig. 1, [36]). Hyperinsulinaemia increases lipogenesis in liver and adipose tissue and also lipoprotein lipase activity accelerating endogenous (very low density lipoproteins and triglycerides) lipid production [32]. Fat mass is accumulated with an increase in weight and BMI [32].

There is a high degree of correlation between total adiposity and leptin levels in obese CP patients [33] similar to levels found in normal obese subjects, but among the CP patients higher leptin/kg fat mass were found, suggesting lack of sensitivity

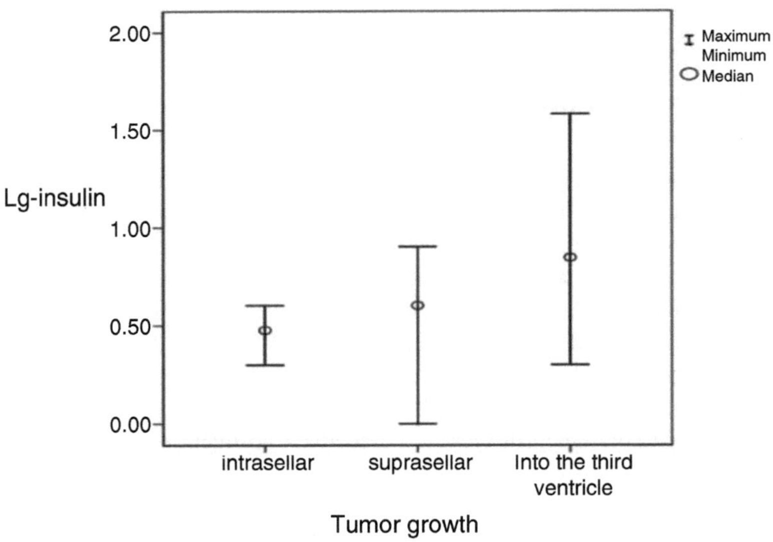

Fig. 1 Correlation between tumour growth and the logarithm for serum insulin for all patients, $r = 0.57$, $P < 0.001$; for men, $r = 0.53$, $P = 0.011$; and for women, $r = 0.66$, $P = 0.002$ [36]

to endogenous leptin. Elevated leptin levels in relation to BMI, thus leptin resistance, have been recorded in both children [39] and adult CP [36] with hypothalamic involvement.

The gastric hormone, ghrelin, is a peripheral produced and centrally active orexigenic hormone that is considered to be an important gut-brain signal for appetite control and energy balance. It has been proposed that ghrelin acts as an initiator for food consumption in humans, with a pre-prandial rise and a post-prandial fall in plasma [40]. In obese normal individuals there is an inverse relation between ghrelin and obesity and BMI [41], which is also confirmed in CP patients [36]. Further, in CPs there is an inverse correlation between plasma ghrelin with tumour growth into the hypothalamus [36]. Previously a weaker post-meal ghrelin response was shown in obese CP patients with hypothalamic involvement, probably due to insulin resistance [42]. It is possible that a low ghrelin response together with uncoupling of feeding-related signals such as leptin with sympathetic activity on the weight regulatory loop reinforces obesity in acquired HO and not hyperphagia. In addition, chronic pro-obesity effects of vagally induced hyperinsulinaemia contribute to persistence of obesity in adulthood [43]. A cross talk between signaling events downstream of insulin receptors and leptin receptors occurs and thus cross-desensitisation between insulin and leptin pathways can occur. Insulin resistance at the intracellular signaling levels causes a relative resistance to leptin thereby enabling the body to tolerate new adiposity. And indeed CP patients with HO have relative hyperinsulinaemia for their obesity and with a significant correlation to tumour growth [36].

Importantly a reduction in energy expenditure has been recorded in a study of children with hypothalamic obesity (HO) suffering from tumour in the hypothalamic region, i.e. CPs [44]. Pediatric and adult patients with childhood onset CPs were found to have a lower resting energy expenditure when compared to controls [45]. Further, Holmer et al. [36] have shown that adults with childhood onset CP had significantly lower basal metabolic rate (BMR) compared to controls. After adjustment, patients had lower BMR compared to controls (−90 kcal/24 h, $P = 0.02$), and the ratio (Energy Intake) EI/BMR was significantly lower in patients with hypothalamic damage (Fig. 2). The total caloric intake was not significantly different between CP patients and controls. This accords with a self-assessment by nutritional diary from a previous study on childhood CP and BMI matched controls [35]. Further, Holmer et al. [36] also showed that similar dietary macronutrient composition was found in patients and controls, and interestingly a lower EI (1778 vs 2094 kcal/24 h; $P = 0.008$) was recorded among patients with hypothalamic involvement by the tumour. With the use of questionnaire only significantly higher scales in restricting food intake, but no difference in disinhibition or hunger, were recorded in patients [35]. In addition, reduced physical activity compounded by neurological deficits and visual disturbances worsens the obesity in these patients [35]. Harz et al. [35] suggested that reduced physical activity rather than increased energy intake is responsible for the obesity in these subjects.

Recent data show that hypothalamic involvement was associated with lower overall survival at diagnosis and follow-up when compared with sellar mass without hypothalamic involvement [46]. Hoffmann et al. [19] analysed eating disorders and

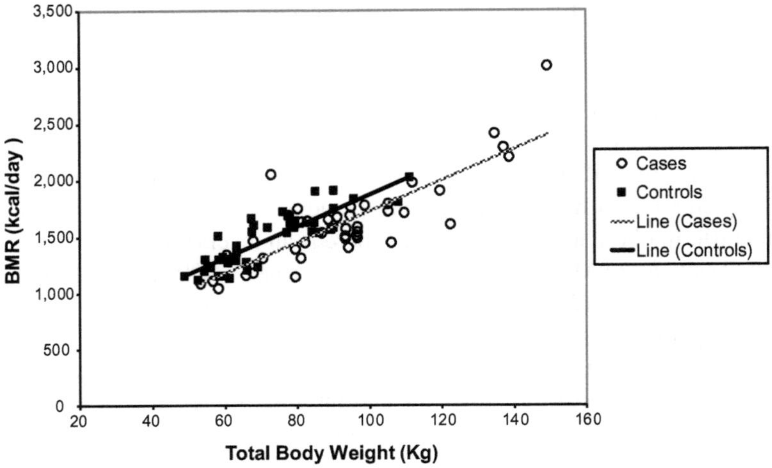

Fig. 2 Linear association between total body weight (kg) and BMR (kilocal/day) among patients and controls. Patients had significantly lower BMR compared with controls after adjustment for sex and total body weight in linear regression analysis (mean difference, −90 kcal/24 h; 95% CI, −160 to −10; $P = 0.02$) [36]

behaviours of 101 childhood onset CPs survivors and 85 BMI matched controls. Severely obese patients presented with pathological eating behaviours and more weight problems than obese and normal or overweight patients. However, childhood onset CPs with different degrees of obesity showed similar or even less pathological findings. The conclusion was that although disturbed eating was present this was not the root cause of their hypothalamic obesity [19, 20]. Roth et al. assessed the pre- and post-meal response to visual food cues in childhood onset CPs' brain regions of interest using functional magnetic imaging (fMRI) [47]. Following a test meal, BMI-matched controls showed suppression of activation by high-caloric food cues while in childhood onset CPs patients showed trends toward higher activation. These altered perceptions of food cues indicated on fMRI, especially after food intake, support the body of evidence that disturbed eating behaviour is not the root cause of obesity in childhood onset CPs with hypothalamic obesity.

Bone Health

The hypothalamus regulates bone and adipose tissue via a complex and fine-tuned interplay of endocrine mechanisms (of which neuropeptide Y is a key regulator) together with the sympathetic nervous system [48]. Obesity in healthy subject is protective against low BMD [49] possibly through increased leptin levels [50]. However, in obese CP patients with hypothalamic damage leptin levels were shown to be higher in relation to body mass index, indicating leptin resistance [39], and

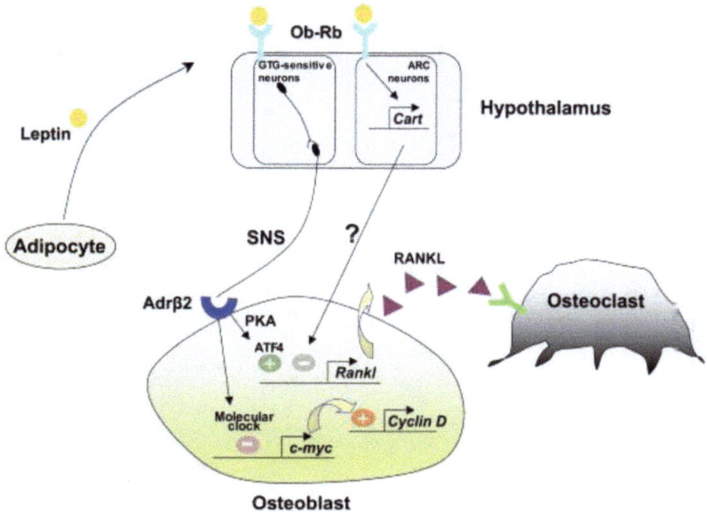

Fig. 3 Current model of the Leptin-dependent regulation of bone mass [52]

among adolescents leptin resistance was shown to be negatively correlated to BMD [51]. In animal models leptin resistance has been correlated to bone loss [52].

Leptin binds to its hypothalamic receptors and induces two cascades to control bone mass ([52], Fig. 3). In the arcuate nuclei, it increases Cart expression that in turn regulates, via an unknown mechanism, Rankl expression by osteoblasts, and thereby bone resorption. Leptin also binds to receptors on the GTG-sensitive neurons of the VMH nuclei, inducing an increase in sympathetic activity which signals to osteoblasts via the β2 adrenergic receptors present at their surface. Two distinct molecular cascades downstream of this receptor are then activated. One inhibits osteoblast proliferation via the molecular clock regulation of c-myc and Cyclin-D expression. The other, mediated by PKA phosphorylation of ATF4, promotes Rankl expression and thereby bone resorption [52]. Upon inactivation of the molecular clock, the leptin/SNS arm regulating osteoblast proliferation via c-myc and Cyclin-D is overactivated, leading to an increase in bone formation and thus a high bone mass phenotype. Cart absence leads to a decrease in bone mass because of an isolated increase in osteoclast activity following Rankl overexpression by osteoblasts [52].

Whereas many of the effects occur via direct actions on osteoblasts or adipocytes, sex hormones can also mediate effects on bone and adipose tissue via interaction with neuronal pathways. In CP children BMD has been investigated with the use of volumetric BMD (vBMD), showing a lower radial *Z*-score, which was most obvious in lean male CP patients [53]. In contrast, female gender and severe obesity seemed to be protective against low vBMD in childhood [53]. A study in adults with a childhood onset CP, females, but not CP males showed significantly lower BMD than matched population controls, despite similarities between gender in pituitary hormone deficiencies and substitution therapies [54]. Furthermore, Holmer et al. [54]

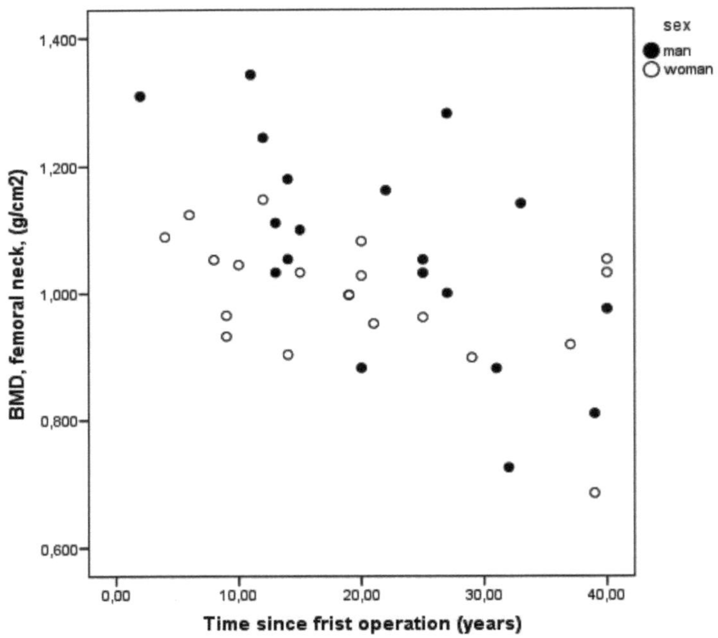

Fig. 4 Correlation between bone mineral density (BMD) femoral neck and time since first operation (years) ($r = -0.45$, $P = 0.004$) in 39 patients with a childhood onset craniopharyngioma [54]

showed that serum leptin levels were significantly increased in both CP women and men and correlated significantly negatively with BMD. Sixty-seven per cent of all CP women, and 75% of the whole patient cohort with *Z*-scores of ≤−2.0 (SD) at L2–L4 were affected by hypothalamic damage from the tumour and the vast majority of these latter patients had also high BMI (>30 kg/m^2). Thus, high BMI was not protective against low BMD.

BMD at femoral neck was significantly negatively correlated to time since first operation (Fig. 4). About 45% of CP women had *Z*-scores ≤−2.0 SDS, despite numerous factors known to be positive for bone formation, e.g. increased BMI, fat mass, and insulin levels. Suggested contributing factors were late-onset puberty resulting in later than optimal introduction of sex steroids, as late-onset puberty is associated with low peak bone mass in both genders [54, 55], but is more deleterious to the female skeleton [56]. Reduced BMD has been shown in healthy adolescents after use of oral contraceptives with an average ethinyl estradiol dose of 20–35 μg [56], which could indicate insufficient sex steroid replacement. Glucocorticoid-induced osteoporosis is well known [57], but long-term glucocorticoid replacement is not reported to affect BMD [58].

Thus, both early and persistent hypothalamic dysfunction and insufficient sex hormone replacement at disease onset are likely to have attributed to the observed long-term effects on bone in CP patients. Thus, the interaction between the fat-related endocrine system and bone seems to be complex and may be modulated by central and peripheral mechanisms as well as local resistance to the putative protective effects of insulin and leptin on bone.

Thirst Disorder, Thermoregulatory Disorders, Somnolence, Sleep Apnoea, and Cardiac Arrhythmias

Diabetes insipidus (DI) with absent or impaired sense of thirst is one of the most difficult complications to manage [5, 59] in CP. Smith et al. [59] reported absence of thirst in 19% of adults with DI treated with surgery combined or not with CRT. A loss of temperature control (very low temperature) was recorded in adult CP patients after surgery for a CP [60]. Somnolence is common in adult CP patients and in the majority due to sleep apnoea [61]. Sleep disorder and excessive daytime sleepiness have been recorded in up to one third of adult CP patients [62]. Fatal outcome was recorded in an adult CP patient [63] after GTR together with CRT of a large hypothalamic CP. The patient had sleep disturbance together with hypothermia and also cardiac arrhythmia, needing a pacemaker. Sudden cardiac deaths have been reported in CP children [64]. In a prospective cardiac screening of 12 survivors, nearly a third was identified with significant QTc prolongation [64]. This emphasises that we need electrocardiogram screening to identify those CP patients at risk.

Conclusion

CP has a special status among pituitary tumours as an aggressive causing high morbidity and increased mortality in comparison to the general population. The management of the tumour is complex and lifelong surveillance by a multidisciplinary team (experienced neurosurgeons, endocrinologists, neurooncologists, neuroradiologists, and neuroopthamologists) is required for better prognosis. Careful surgery to prevent further complications from the hypothalamus is essential. Indeed cardiovascular risk factors need up front therapy. In addition, pituitary hormone deficiencies need balanced replacement therapies. Early postoperative intervention with strict dieting and physical activity is the main stay if hypothalamic damage is present pre- or post-operative. Surveillance of BMD is recommended in patients with a hypothalamic tumour involvement

Acknowledgment This work was supported by the Swedish Children's Cancer Foundation, and the Medical Faculty, Lund University, Sweden.

The South Medical Research Council and the Medical Faculty, Lund University, Sweden and the Swedish Children's Cancer Foundation.

Conflict of Interest Statement EM Erfurth has no disclosure.

References

1. Bulow B, Attewell R, Hagmar L, Malmstrom P, Nordstrom CH, Erfurth EM (1998) Postoperative prognosis in craniopharyngioma with respect to cardiovascular mortality, survival, and tumor recurrence. J Clin Endocrinol Metab 83:3897–3904
2. Nielsen EH, Feldt-Rasmussen U, Poulsgaard L, Kristensen LO, Astrup J, Jørgensen JO, Bjerre P, Andersen M, Andersen C, Jørgensen J, Lindholm J, Laurberg P (2011) Incidence of craniopharyngioma in Denmark (n = 189) and estimated world incidence of craniopharyngioma in children and adults. J Neuro-Oncol 104(3):755–763
3. Olsson DS, Andersson E, Bryngelsson IL, Nilsson AG, Johannsson G (2015) Excess mortality and morbidity in patients with craniopharyngioma, especially in patients with childhood onset: a population-based study in Sweden. J Clin Endocrinol Metab 100(2):467–474
4. Nielsen EH, Lindholm J, Laurberg P (2007) Excess mortality in women with pituitary disease: a meta-analysis. Clin Endocrinol 67(5):693–697
5. De Vile CJ, Grant DB, Hayward RD, Stanhope R (1996) Growth and endocrine sequelae of craniopharyngioma. Arch Dis Child 75:108–114
6. Müller HL, Emser A, Faldum A, Bruhnken G, Etavard-Gorris N, Gebhardt U, Oeverink R, Kolb R, Sorensen N (2004) Longitudinal study on growth and body mass index before and after diagnosis of childhood craniopharyngioma. J Clin Endocrinol Metab 89:3298–3305
7. Bunin GR, Surawicz TS, Witman PA, Preston-Martin S, Davis F, Bruner JM (1998) The descriptive epidemiology of craniopharyngioma. J Neurosurg 89(4):547–551
8. Karavitaki N, Brufani C, Warner JT, Adams CB, Richards P, Ansorge O, Shine B, Turner HE, Wass JA (2005) Craniopharyngiomas in children and adults: systematic analysis of 121 cases with long-term follow-up. Clin Endocrinol 62(4):397–409
9. Tomlinson JW, Holden N, Hills RK, Wheatley K, Clayton RN, Bates AS, Sheppard MC, Stewart PM (2001) Association between premature mortality and hypopituitarism. Lancet 357:425–431
10. Van Effenterre R, Boch AL (2002) Craniopharyngioma in adults and children: a study of 122 surgical cases. J Neurosurg 97(1):3–11
11. Crowley RK, Hamnvik OP, O'Sullivan EP, Behan LA, Smith D, Agha A, Thompson CJ (2010) Morbidity and mortality in patients with craniopharyngioma after surgery. Clin Endocrinol 73(4):516–521
12. Visser J, Hukin J, Sargen M, Steinbok P, Goddard K, Fryer C (2010) Late mortality in pediatric patients with craniopharyngioma. J Neuro-Oncol 100:105–111
13. Pereira AM, Schmid EM, Schutte PJ, Voormolen JH, Biermasz NR, van Thiel SW, Corssmit EP, Smit JW, Roelfsema F, Romijn JA (2005) High prevalence of long-term cardiovascular, neurological and psychosocial morbidity after treatment for craniopharyngioma. Clin Endocrinol 62:197–204
14. Stripp DC, Maity A, Janss AJ, Belasco JB, Tochner ZA, Goldwein JW, Moshang T, Rorke LB, Phillips PC, Sutton LN, Shu HK (2004) Surgery with or without radiation therapy in the management of craniopharyngiomas in children and young adults. Int J Radiat Oncol Biol Phys 58(3):714–720
15. Rajan B, Ashley S, Gorman C, Jose CC, Horwich A, Bloom HJ, Marsh H, Brada M (1993) Craniopharyngioma-a long-term results following limited surgery and radiotherapy. Radiother Oncol 26(1):1–10
16. Müller HL (2014) Craniopharyngioma. Endocr Rev 35(3):513–543
17. Kalapurakal JA (2005) Radiation therapy in the management of pediatric craniopharyngiomas-a review. Childs Nerv Syst 21:808–816
18. Srinivasan S, Ogle GD, Garnett SP, Briody JN, Lee JW, Cowell CT (2004) Features of the metabolic syndrome after childhood craniopharyngioma. J Clin Endocrinol Metab 89:81–86
19. Hoffmann A, Postma FP, Sterkenburg AS, Gebhardt U, Müller HL (2015b) Eating behavior, weight problems and eating disorders in 101 long-term survivors of childhood-onset craniopharyngioma. J Pediatr Endocrinol Metab 28(1–2):35–43

20. Hoffmann A, Bootsveld K, Gebhardt U, Daubenbüchel AM, Sterkenburg AS, Müller HL (2015a) Nonalcoholic fatty liver disease and fatigue in long-term survivors of childhood-onset craniopharyngioma. Eur J Endocrinol 173(3):389–397
21. Holmer H, Ekman B, Björk J, Nordstöm CH, Popovic V, Siversson A, Erfurth EM (2009) Hypothalamic involvement predicts cardiovascular risk in adults with childhood onset craniopharyngioma on long-term GH therapy. Eur J Endocrinol 161(5):671–679
22. Karavitaki N, Cudlip S, Adams CB, Wass JA (2006) Craniopharyngiomas. Endocr Rev 27:371–397
23. Molitch ME, Clemmons DR, Malozowski S, Merriam GR, Vance ML (2011) Endocrine Society. Evaluation and treatment of adult growth hormone deficiency: an Endocrine Society clinical practice guideline. J Clin Endocrinol Metab 96(6):1587–1609
24. Maison P, Griffin S, Nicoue-Beglah M, Haddad N, Balkau B, Chanson P (2004) Impact of Growth Hormone (GH) Treatment on Cardiovascular Risk Factors in GH-Deficient Adults: A Meta-analysis of Blinded, Randomized, Placebo-Controlled Trials. J Clin Endocrinol Metab 89:2192–2199
25. Verhelst J, Kendall-Taylor P, Erfurth EM, Price DA, Geffner M, Koltowska-Häggström M, Jönsson PJ, Wilton P, Abs R (2005) Baseline characteristics and response to 2 years of growth hormone (GH) replacement of hypopituitary patients with GH deficiency due to adult-onset craniopharyngioma in comparison with patients with nonfunctioning pituitary adenoma: data from KIMS (Pfizer International Metabolic Database). J Clin Endocrinol Metab 90(8):4636–4643
26. Bergthorsdottir R, Leonsson-Zachrisson M, Odén A, Johannsson G (2006) Premature mortality in patients with Addison's disease: a population-based study. J Clin Endocrinol Metab 91(12):4849–4853
27. Sherlock M, Reulen RC, Alonso AA, Ayuk J, Clayton RN, Sheppard MC, Hawkins MM, Bates AS, Stewart PM (2009) ACTH deficiency, higher doses of hydrocortisone replacement, and radiotherapy are independent predictors of mortality in patients with acromegaly. J Clin Endocrinol Metab 94(11):4216–4223
28. Owen PJ, Lazarus JH (2003) Subclinical hypothyroidism: the case for treatment. Trends Endocrinol Metab 14:257–261
29. Klose M, Marina D, Hartoft-Nielsen ML, Klefter O, Gavan V, Hilsted L, Rasmussen AK, Feldt-Rasmussen U (2013) Central hypothyroidism and its replacement have a significant influence on cardiovascular risk factors in adult hypopituitary patients. J Clin Endocrinol Metab 98(9):3802–3810
30. Duff J, Meyer FB, Ilstrup DM, Laws ER Jr, Schleck CD, Scheithauer BW (2000) Long-term outcomes for surgically resected craniopharyngiomas. Neurosurgery 46(2):291–302
31. Hetelekidis S, Barnes PD, Tao ML, Fischer EG, Schneider L, Scott RM, Tarbell NJ (1993) 20-year experience in childhood craniopharyngioma. Int J Radiat Oncol Biol Phys 27:189–195
32. Bray GA, Gallagher TF Jr (1975) Manifestations of hypothalamic obesity in man: a comprehensive investigation of eight patients and a reveiw of the literature. Medicine (Baltimore) 54(4):301–330
33. Lustig RH (2002) Hypothalamic obesity: the sixth cranial endocrinopathy. Endocrinologist 12:210–217
34. Roth CL (2015) Hypothalamic obesity in craniopharyngioma patients: disturbed energy homeostasis related to extent of hypothalamic damage and its implication for obesity intervention. J Clin Med 4(9):1774–1797
35. Harz KJ, Müller HL, Waldeck E, Pudel V, Roth C (2003) Obesity in patients with craniopharyngioma: assessment of food intake and movement counts indicating physical activity. J Clin Endocrinol Metab 88:5227–5231
36. Holmer H, Pozarek G, Wirfält E, Popovic V, Ekman B, Björk J, Erfurth EM (2010) Reduced energy expenditure and impaired feeding-related signals but not high energy intake reinforces hypothalamic obesity in adults with childhood onset craniopharyngioma. J Clin Endocrinol Metab 95(12):5395–5402

37. Kreier F, Fliers E, Voshol PJ, Van Eden CG, Havekes LM, Kalsbeek A, Van Heijningen CL, Sluiter AA, Mettenleiter TC, Romijn JA, Sauerwein HP, Buijs RM (2002) Selective parasympathetic innervation of subcutaneous and intra-abdominal fat-functional implications. J Clin Invest 110(9):1243–1250
38. Müller HL (2010) Increased daytime sleepiness in patients with childhood craniopharyngioma and hypothalamic tumor involvement: review of the literature and perspectives. Int J Endocrinol 2010:519607
39. Roth C, Wilken B, Hanefels F, Schröter W, Leonhardt U (1998) Hyperphagia in children with craniopharyngioma is associated with hyperleptinaemia and a failure in the downregulation of appetite. Eur J Endocrinol 138:89–91
40. Cummings DE, Purnell JQ, Frayo RS et al (2001) A preprandial rise in plasma ghrelin levels suggests a role in meal initiation in humans. Diabetes 50:1714–1719
41. Hillman JB, Tong J, Tschop M (2011) Ghrelin biology and its role in weight-related disorders. Discov Med 11:521
42. Goldstone AP, Patterson M, Kalingag N, Ghatei MA, Brynes AE, Bloom SR, Grossman AB (2005) Korbonits M 2005 Fasting and postprandial hyperghrelinemia in Prader-Willi syndrome is partially explained by hypoinsulinemia, and is not due to peptide YY3-36 deficiency or seen in hypothalamic obesity due to craniopharyngioma. J Clin Endocrinol Metab 90(5):2681–2690
43. Bray GA, Inoue S, Nishizawa Y (1981) Hypothalamic obesity. The autonomic hypothesis and the lateral hypothalamus. Diabetologia 20(Suppl):366–377
44. Shaikh MG, Grundy RG, Kirk JM (2008) Reductions in basal metabolic rate and physical activity contribute to hypothalamic obesity. J Clin Endocrinol Metab 93:2588–2593
45. Bomer I, Saure C, Caminiti C, Ramos JG, Zuccaro G, Brea M, Bravo M, Maza C (2015) Comparison of energy expenditure, body composition, metabolic disorders, and energy intake between obese children with a history of craniopharyngioma and children with multifactorial obesity. Pediatr Endocrinol Metab 28(11–12):1305–1312
46. Daubenbüchel AM, Hoffmann A, Gebhardt U, Warmuth-Metz M, Sterkenburg AS, Müller HL (2015) Hydrocephalus and hypothalamic involvement in pediatric patients with craniopharyngioma or cysts of Rathke's pouch: impact on long-term prognosis. Eur J Endocrinol 172(5):561–569
47. Roth CL, Eslamy H, Werny D, Elfers C, Shaffer ML, Pihoker C, Ojemann J, Dobyns WB (2015) Semiquantitative analysis of hypothalamic damage on mri predicts risk for hypothalamic obesity. Obesity (Silver Spring) 23:1226–1233
48. Karsenty G, Oury F (2010) The central regulation of bone mass, the first link between bone remodeling and energy metabolism. J Clin Endocrinol Metab 95(11):4795–4801 Review
49. Felson DT, Zhang Y, Hannan MT, Anderson JJ (1993) Effects of weight and body mass index on bone mineral density in men and women: the Framingham study. J Bone Miner Res 8:567–573
50. Hamrick MW, Ferrari SL (2008) Leptin and the sympathetic connection of fat to bone. Osteoporos Int 19:905–912 Review
51. Do Prado WL, de Piano A, Lazaretti-Castro M, de Mello MT, Stella SG, Tufik S, do Nascimento CM, Oyama LM, Lofrano MC, Tock L, Caranti DA, Dâmaso AR (2009) Relationship between bone mineral density, leptin and insulin concentration in Brazilian obeses adolscents. J Bone Miner Metab 27:613–619
52. Karsenty G (2006) Convergence between bone and energy homeostases: leptin regulation of bone mass. Cell Metab 4(5):341–348
53. Müller HL, Schneider P, Bueb K, Etavard-Gorris N, Gebhardt U, Kolb R, Sörensen N (2003) Volumetric bone mineral density in patients with childhood craniopharyngioma. Exp Clin Endocrinol Diabetes 111:168–173
54. Holmer H, Popovic V, Ekman B, Follin C, Siversson A-B, Erfurth EM (2011) Hypothalamic involvement and insufficient sex steroid supplementation are associated with low bone mineral density in women with childhood onset craniopharyngioma. Eur J Endocrinol 165:25–31

55. Finkelstein JS, Neer RM, Biller BM, Crawford JD, Klibanski AN (1992) Osteopenia in men with a history of delayed puberty. N Engl J Med 326:600–604
56. Almstedt Shoepe H, Snow CM (2005) Oral contraceptive use in young women is associated with lower bone mineral density than that of controls. Osteoporos Int 16:1538–1544
57. Saag KG (2003) Glucocorticoid-induced osteoporosis. Endocrinol Metab Clin N Am 32:135–157
58. Mora S, Saggion F, Russo G, Weber G, Bellini A, Prinster C, Chiumello G (1996) Bone density in young patients with congenital adrenal hyperplasia. Bone 18:337–340
59. Smith D, Finucane F, Phillips J, Baylis PH, Finucane J, Tormey W, Thompson CJ (2004) Abnormal regulation of thirst and vasopressin secretion following surgery for craniopharyngioma. Clin Endocrinol 61:273–279
60. Griffiths AP, Henderson M, Penn ND, Tindall H (1988) Haematological, neurological and psychiatric complications of chronic hypothermia following surgery for craniopharyngioma. Postgrad Med J 64(754):617–620
61. Crowley RK, Woods C, Fleming M, Rogers B, Behan LA, O'Sullivan EP, Kane T, Agha A, Smith D, Costello RW, Thompson CJ (2011) Somnolence in adult craniopharyngioma patients is a common, heterogeneous condition that is potentially treatable. Clin Endocrinol 74(6):750–755
62. van der Klaauw AA, Biermasz NR, Pereira AM, van Kralingen KW, Dekkers OM, Rabe KF, Smit JW, Romijn JA (2008) Patients cured from craniopharyngioma or nonfunctioning pituitary macroadenoma (NFMA) suffer similarly from increased daytime somnolence despite normal sleep patterns compared to healthy controls. Clin Endocrinol 69(5):769–774
63. Rehman HU, Atkin SL (1999) Sleep disturbances and cardiac arrhythmia after treatment of a craniopharyngioma. J R Soc Med 92(11):585–586
64. Mong S, Pomeroy SL, Cecchin F, Juraszek A, Alexander ME (2008) Cardiac risk after craniopharyngioma therapy. Pediatr Neurol 38(4):256–260

Radiology and Radiotherapy of Craniopharyngioma

Babita Jyoti, Daniel J. Indelicato, Julie A. Bradley, and Ronny L. Rotondo

Abstract Craniopharyngioma is a histologically benign tumour commonly seen in the paediatric population. This tumour develops in the central region of the brain surrounded by several critical structures, including the optic pathway, pituitary, hypothalamus, circle of Willis, temporal lobes, and brainstem. Its central intracranial location poses a challenge to surgical resection and radiotherapy. Radiotherapy following limited surgery offers high rates of local control and long-term survival but may be associated with significant side effects. Proton therapy in this setting offers a dosimetric advantage over conventional radiotherapy because of the unique physical properties of proton particles, which deposit the dose in a narrow region called the Bragg peak before stopping. This phenomenon reduces the dose to adjacent critical structures and could translate into a reduction in late treatment-related sequelae and an improvement in survivors' quality of life.

Keywords Craniopharyngioma • Proton therapy • Brain tumour • Paediatric radiotherapy • Late effects • Radiation toxicity • Survivorship • Dosimetry • Intensity-modulated radiotherapy

Abbreviations

3D	3-dimensional
CT	computed tomography
CTV	clinical target volume
GTR	gross total resection
IMPT	intensity-modulated proton therapy
IMRT	intensity-modulated radiotherapy
IQ	intelligence quotient

B. Jyoti • D.J. Indelicato (✉) • J.A. Bradley • R.L. Rotondo
University of Florida Health Proton Therapy Institute,
2015 North Jefferson St., Jacksonville, FL, USA
e-mail: dindelicato@floridaproton.org

J.P. Martinez-Barbera, C.L. Andoniadou (eds.), *Basic Research and Clinical Aspects of Adamantinomatous Craniopharyngioma*,
DOI 10.1007/978-3-319-51890-9_7

MRI	magnetic resonance imaging
PFS	progression-free survival
PTV	planning target volume
RBE	relative biologic effectiveness
STR	subtotal resection

Radiology of Adamantinomatous Craniopharyngioma

The presence of craniopharyngioma can be identified on plain radiographs of the skull. Sellar changes such as enlargement or erosion of the clinoid process or dorsum sella, as well as calcification may be evident on plain radiographs. In a large retrospective review of patients with craniopharyngioma, plain skull X-ray results were available in 190 patients; of these radiographs, 25% were normal, 47% displayed suprasellar calcification, and 44% demonstrated enlarged and/or eroded sella turcica. The presence of calcification was highly significant in younger patients [1].

Computed tomography (CT) and magnetic resonance imaging (MRI) have largely replaced plain radiographs of the skull as an initial investigation for the diagnosis of craniopharyngioma. On CT, the presence of a mixed solid/cystic enhancing suprasellar mass with calcification is strongly suggestive of an adamantinomatous craniopharyngioma [2]. These calcifications, most clearly identified on CT, are a characteristic feature of craniopharyngioma [2–4]. Morphologically, these can be described as punctate, globular, or curvilinear calcifications [5]. Bone erosion may be seen secondary to the effect of pressure from long-standing large tumours (Fig. 1).

MRI is the diagnostic imaging of choice. It is useful to confirm the presence of cystic components and accurately measure the size and extent of the tumour.

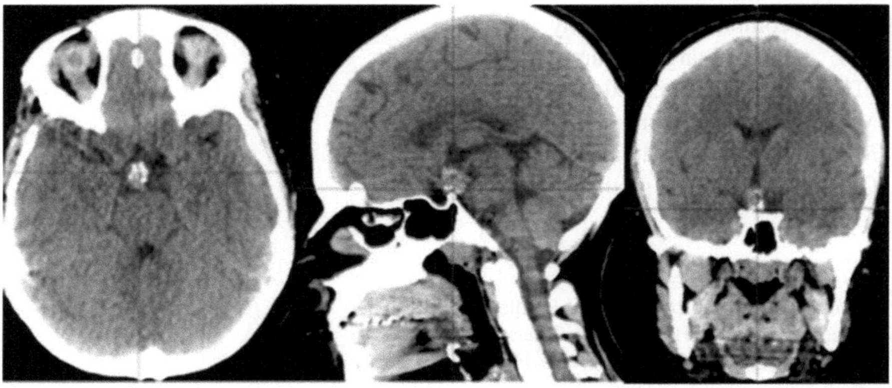

Fig. 1 Computed tomography scan with calcification in the suprasellar region indicative of craniopharyngioma

Craniopharyngiomas are typically located in the parasellar region, with a prominent suprasellar component in the majority. These tumours are described in relation to the optic chiasm (pre-, retro-, or subchiasmatic). Prechiasmatic tumours extend anteriorly displacing the chiasm postero-superiorly and they may also extend toward the frontal lobes and cause displacement of the A-1 segment of the anterior cerebral artery. Retrochiasmatic tumours tend to displace the optic chiasm anteriorly. Growth of retrochiasmatic tumours into the third ventricle can result in obstructive hydrocephalus. Posterior and inferior extension can result in displacement of the basilar artery and compression of the midbrain. Compression of the temporal lobes occurs with lateral growth. Approximately 30% of craniopharyngiomas extend anteriorly, 25% extend into the middle cranial fossa, and up to 20% have retroclival extension to the foramen magnum or the cerebellar pontine angle [6] (Fig. 2).

On T2-weighted sequences, including those with fluid-attenuated inversion recovery (FLAIR), the solid portion is heterogeneous in intensity, and the cysts are mostly or partly hyperintense. Fluid-attenuated inversion recovery is useful in differentiating the cystic portions of the tumour from loculated portions of the third ventricle or other craniospinal fluid spaces that tend to be isointense. Calcifications are hypointense on T2 sequences. Susceptibility sequences, such as gradient recalled echo (GRE), are helpful in demonstrating calcifications and haemorrhage [4]. On T1 sequences, the cysts tend to be iso- to hyperintense. The solid component is

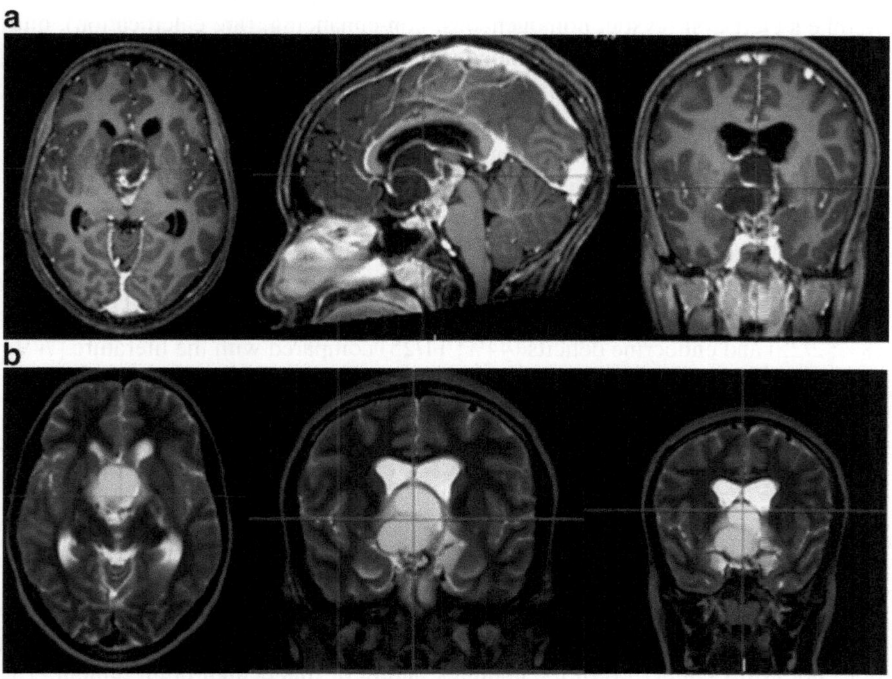

Fig. 2 Axial, sagittal, and coronal slices of (**a**) T1-weighted and (**b**) T2-weighted sequences of a magnetic resonance imaging scan demonstrating a solid/cystic suprasellar lesion

generally smaller in proportion to the cystic component and enhances with the administration of contrast. Magnetic resonance angiography is an important study in identifying patients with baseline vessel stenosis or collateralisation. On magnetic resonance spectroscopy, cysts demonstrate a significant lipid content [4].

Differentiation between the more common childhood adamantinomatous craniopharyngioma and the adult squamous-papillary craniopharyngioma based on MRI features was reported by Sartoretti-Schefer et al. Of the 42 cases examined, 25 were adamantinomatous, 15 had squamous-papillary, and 2 mixed subtypes. Adamantinomatous craniopharyngiomas were seen mainly in children but also occasionally in adults. They were predominantly cystic and suprasellar in location. On T1-weighted images, the cysts were typically hyperintense. The tumour was mostly lobulated with a tendency to encase blood vessels. Calcifications were less commonly seen in the squamous-papillary subtype. The squamous-papillary type occurred in adults and was predominantly solid and occasionally demonstrated calcifications. When this subtype contained cysts, they appeared hypointense on T1-weighted images. There was no encasement of blood vessels and these tumours were mostly spherical in shape. Some of these features achieved statistical significance: squamous-papillary were predominantly solid and spherical with rare hypointense cysts; adamantinomatous type were lobulated, with frequent hyperintense cysts and vessel encasement [3].

The types of tumours in this area can often be distinguished based on imaging characteristics. Differential diagnoses of sellar and suprasellar mass lesions include Rathke's cleft cyst (cystic, homogeneous, non-enhancing, rare calcification), pituitary adenoma (enlarged sella, homogeneous, and less cystic), meningioma (isointense on T1 and T2 and rarely cystic), and optic pathway glioma (may be cystic but are usually non-calcified) [5].

While the role of diagnostic imaging in craniopharyngioma is well established, and in some cases can obviate the need for biopsy, imaging can also be used to assist in the therapeutic management of craniopharyngiomas. Intraoperative MRI provides guidance for dissection. Hofmann et al. report a 16% increase in the rate of total tumour resection (TTR) using intraoperative MRI, resulting in a gross total resection (GTR) rate of 80% in their series with similar rates of new ophthalmologic (8%; 2/25) and endocrine deficits (44%; 11/25) compared with the literature [7–9].

Radiotherapy in the Management of Craniopharyngioma

Adamantinomatous craniopharyngiomas most commonly arise in the paediatric population, with a peak incidence between ages 5 and 14 years. The management of craniopharyngiomas has been debated for decades based on the theory that surgical resection of a benign tumour should achieve cure. Nevertheless, this simple solution does not always apply given the complex nature of this benign brain tumour. The location and intimate association of the tumour with critical structures such as the visual pathway, pituitary, hypothalamus, intra-cerebral vasculature, brainstem, and temporal lobes, and the morbidity associated with aggressive surgery have led to a multimodality approach using limited, safe surgery followed by radiotherapy with the goal of local control while minimising treatment-related sequelae.

According to recently published data from the Surveillance, Epidemiology, and End Result Program database [10], 644 patients were identified with craniopharyngioma between 2004 and 2008. On multivariate analysis, factors associated with prolonged survival included younger age, smaller tumour size, subtotal resection (STR), and radiotherapy. Black race was associated with an incidence rate of 2.1 cases per 1,000,000 person-years versus 1.7 cases per person-years and an age-adjusted relative risk for craniopharyngioma of 1.26 (95% confidence interval, 0.98–1.59) compared with white race. Survival rates at 1 and 3 years for the entire cohort were 91.5% and 86.2%, respectively. Black race was also associated with worse overall survival compared to white patients; 87.4% and 74.8% at 1, and 3 years versus 92.2% and 89.9%, respectively.

A significant risk of progression is present in cases without gross total resection and aggressive surgery carries a higher risk of surgical morbidity. Individualised therapeutic management in a multidisciplinary approach involving neurosurgery, radiation oncology, paediatric oncology, neuro-ophthalmology, neuro-psychology, and endocrinology is invaluable to achieve optimal outcomes with maximal tumour control and minimal morbidity [11–13]. Overall recurrence-free survival (RFS) rates at 5 and 10 years were 62% and 49% according to a study by investigators at the Children's Memorial Hospital (Chicago, IL) [13]. The extent of tumour resection influenced RFS. With GTR, RFS rates were 83% and 70% at 5 and 10 years, while the RFS decreased to 71% and 36%, at 5 and 10 years, respectively, with STR and radiotherapy. Authors of this study concluded that recurrence rates and surgical complications remained high following radical resection.

In a retrospective analysis of 86 children who underwent GTR (103 operations), three children (3%) experienced perioperative deaths. The tumour recurred in 14 of 17 (20%) after GTR [14]. Factors negatively affecting overall survival and progression-free survival (PFS) included STR, tumour size >5 cm, and the presence of hydrocephalus or a ventriculoperitoneal shunt. Authors conclude that radical resection performed by experienced surgeons at presentation offers good chance of disease control and, while GTR does not preclude recurrence, it is more difficult to achieve in recurrent tumours [14].

The role of limited surgery followed by radiotherapy to doses of 5000–6550 Roentgens leading to improved disease control was reported as early as 1961 by Kramer et al. and was updated by Rajan et al. in 1993 [15, 16]. In 2010, based on an extensive literature review of 32 published reports, Merchant et al. concluded that conservative surgery and radiotherapy resulted in disease control rates ranging from 44 to 100%, with modern studies showing at least 90% disease control with 5-year follow-up [17]. In addition, other studies [18, 19] have shown similar rates of PFS (86% and 65%) with postoperative radiotherapy.

A large retrospective analysis of 122 patients treated between 1980 and 2009 included 46 patients who were ≤18 years old. In patients treated with GTR versus STR + radiotherapy no difference in overall survival or PFS rates was identified

($p = 0.544$ and $p = 0.735$, respectively) [20]. As expected, STR alone resulted in a significantly inferior PFS compared to STR + radiotherapy or GTR ($p = <0.001$ for both). STR was associated with significantly decreased overall survival compared to STR + radiotherapy ($p = 0.050$) with a trend to shorter overall survival compared to GTR ($p = 0.066$). GTR was also significantly associated with a greater risk of developing diabetes insipidus (56.3% vs 13.3%, $p < 0.001$) and panhypopituitarism (54.8 vs 26.7%, $p = 0.014$) compared with STR + radiotherapy.

The effect of modern treatment techniques and the use of combined-modality therapy with limited surgery and radiotherapy were evaluated in a retrospective analysis of children with craniopharyngioma treated from 1976 to 2003 [21]. Recurrence rates after limited surgery and radiotherapy were 21% in the pre-1988 era and 5% in the post-1988 era. Investigators also noted that patients treated with surgery alone were 7.7 times more likely to recur compared to those treated with the combined approach (95% confidence interval: 2.0 to 28.7). Although there was no difference in the 10-year overall survival rate between patients treated with surgery alone versus those treated with combined-modality therapy (97% vs 93%; $p = 0.53$), the 10-year local control rate was higher for patients treated with combined-modality therapy compared to surgery alone (84% vs 52%; $p = 0.006$). Radiotherapy in this series varied from a conventional three-field, 3D conformal technique, stereotactic fractionated radiotherapy, arc therapy, and intensity-modulated radiotherapy (IMRT).

The timing of radiotherapy invites debate. Lin et al. [22] in 2008 reported overall survival and PFS rates in a retrospective study of 31 children; early radiotherapy showed a 100% local control rate compared to 32% when radiation was delivered at relapse. The data is countered, however, by several other studies [13, 23–26] that have indicated comparable overall survival and PFS with early versus delayed or salvage radiotherapy.

Acceptable local control and overall survival achieved with combined-modality therapy appropriately raise concerns of the late effects of radiotherapy, a highly relevant survivorship issue as it can permanently affect function and quality of life. Young age is an established risk factor for neurocognitive decline in children receiving cranial radiotherapy [27–29]. With improved dose conformality and decreased exposure of adjacent critical structures, technological advancements in radiotherapy may maintain disease control while reducing in the long-term toxicity of treatment.

Radiotherapy

External-beam radiotherapy involves directing beams of high-energy radiation from outside the body using various modalities such as X-rays (photons), gamma rays, and electron beams, and particles such as protons or heavy ions like carbon. Another form of therapeutic radiation known as brachytherapy is delivered via a radioactive source placed within a body cavity or tissue, as in the treatment of uterine or prostate cancer. Radiation oncologists exploit the different physical characteristics of each type of ionising radiation to produce an individual treatment plan for each patient. Ionising radiation causes damage to cell DNA, leading to cell death. Unlike abnormal tumour cells, normal cells have the ability to repair this damage.

Preparation and Delivery of Radiotherapy

Patient Set-up and Data Acquisition

Patient position, immobilisation, and reproducibility of position for daily treatment are important considerations when planning external-beam radiation. For the treatment of a craniopharyngioma, patients are typically in the supine position and immobilised using a face/head mask with or without a bite plate to reduce intra- and interfraction motion (Fig. 3). Patients undergo a CT simulation scan with a slice thickness of 1–2 mm. Marks on the mask in conjunction with triangulating lasers in the room are used for the daily alignment of the patient for treatment. Daily image guidance is recommended to minimise set-up uncertainties. Cone-beam CT, when available, provides robust image guidance. For proton therapy, daily image guidance is currently performed with orthogonal kV X-rays typically aligned to bony landmarks of the skull (Fig. 4). Digital radiographic reconstructions (DRRs) generated from the CT simulation images are used to reproduce the position set during CT simulation. As bone development is incomplete in the prepubertal age group, bony landmarks may be suboptimal for daily patient alignment in this age group [30]. As a result, some institutions recommend placement of four small titanium fiducial markers prior to proton radiotherapy for these young patients to assist in optimal daily alignment. In addition, daily patient set-up can be aligned to existing surgical hardware when present.

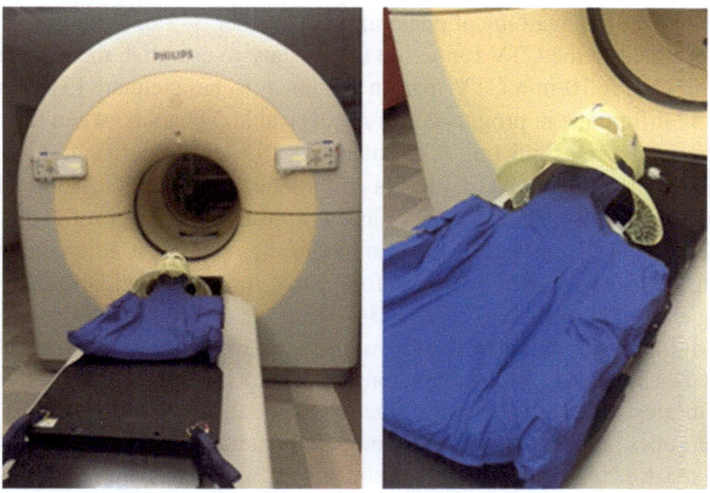

Fig. 3 Computed tomography simulation with face mask and body mould

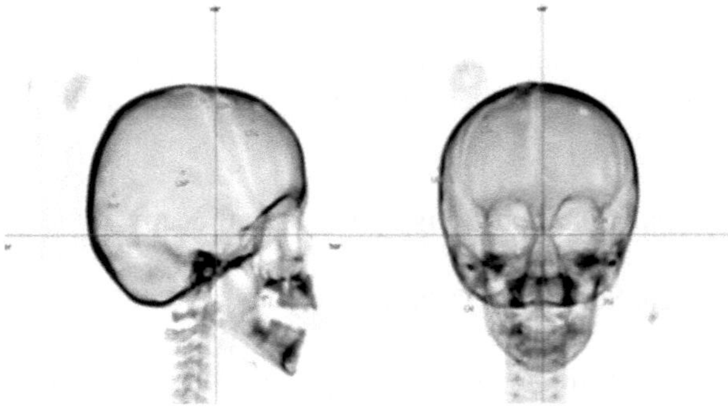

Fig. 4 Orthogonal images using bony landmarks for daily patient alignment in proton therapy

In the delivery of proton therapy, patients undergo a CT simulation scan with a slice thickness of 1 mm. A similar set-up is used for fractionated stereotactic radiotherapy and stereotactic radiosurgery. Young children may require a general anesthetic for CT simulation as well as for each fraction of radiation treatment delivery.

Target Volume Delineation

Target volumes are defined on the CT simulation images. Diagnostic high resolution MR (1–1.5 mm slice thickness) T1 and T2 weighted post contrast and FLAIR sequences are used, with image fusion to the planning CT images for accurate delineation of targets and organs at risk. Calcifications seen on the CT scan also help define the gross tumour volume (GTV). The gross tumour is expanded by 5 mm to obtain the clinical target volume (CTV), which may be subsequently enlarged to encompass any surface originally in contact with the tumour, or decreased based on anatomical constraints. A further 3–5 mm institution specific margin is added to the CTV for set-up uncertainty to establish the planning target volume (PTV). A 10-mm CTV margin is known to be adequate [31]. Other data showed no difference in progression-free survival at 5 years between those treated with a >5 mm versus ≤5 mm CTV margin [32]. An ongoing phase II protocol (RT2CR) at the University of Florida in collaboration with St. Jude Children's Research Hospital is examining the efficacy of limited surgery, proton therapy, and a limited margin (5 mm CTV) for craniopharyngioma. Cyst enlargement during radiation is well recognised and, therefore, early, regular surveillance and adaptive planning are recommended (Fig. 5) [33–35].

In addition to the treatment target volumes, normal tissues or organs at risk are contoured as avoidance targets, including the optic chiasm, optic nerves, brainstem surface, brainstem core, bilateral retinae, lens, bilateral cochlea, lacrimal glands, temporal lobes, hippocampal head and tail bilaterally, pituitary, hypothalamus, bilateral mastoid air cells, posterior nasopharynx, scalp, and supratentorial brain.

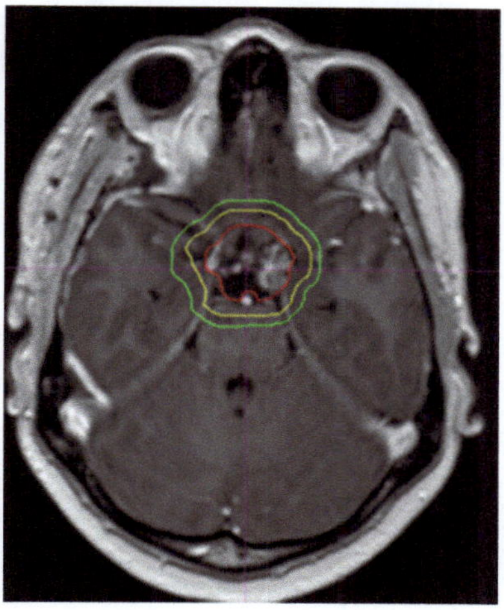

Fig. 5 Representative slice of craniopharyngioma target definition. Gross tumour volume in *red*; clinical target volume in *yellow*; planning target volume in *green*

Dose and Fractionation

The standard dose for the treatment of craniopharyngioma with fractionated external beam radiotherapy is 54 Gy at 1.8 Gy per fraction, and a range from 50 to 60 Gy has been reported in the literature [16, 18, 24, 36].

Types of Radiotherapy

Conventional External-beam Radiotherapy and 3-Dimensional Conformal Radiotherapy

Conventional external-beam radiotherapy, the most common type of radiation used worldwide, is delivered in the form of photon beams (X-rays) using linear accelerators. The total dose of radiation is usually delivered over a fractionated course of 1.8–2 Gy per fraction, each day, allowing for normal tissue repair between fractions. Normal tissue tolerance to radiotherapy along with the therapeutic dose required to adequately treat the tumours guide radiotherapy design. The fundamental principle of radiotherapy is to deliver a therapeutic dose to the target volume while minimising the dose to adjacent normal structures. The tolerance dose of the brainstem is 54–59.4 Gy and that of the optic nerves and chiasm is 55–60 Gy [37]. Sensitivity of the pituitary's endocrine function varies with growth hormone demonstrating the highest sensitivity (QUANTEC) [38]. These goals are achieved by incorporating modern techniques in the radiotherapy planning process, such as the use of three-dimensional (3D) imaging including CT and MRI to assist with target volume and normal structure delineation, reproducible patient immobilisation, and enhanced precision delivery techniques like image guidance.

A 3D conformal plan to treat craniopharyngioma is typically delivered using non-coplanar beams shaped by a multileaf collimator. A common beam arrangement

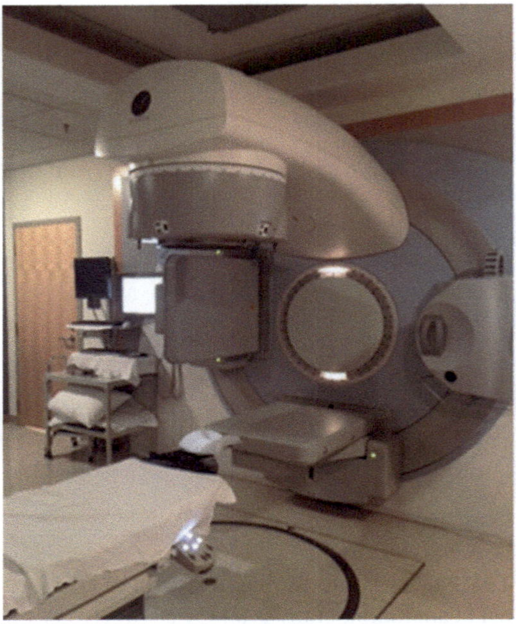

Fig. 6 Linear accelerator ± 3-dimensional conformal craniopharyngioma plan

consists of a superior oblique beam arrangement and opposing lateral or posterior oblique beams (Fig. 6).

Rajan et al. reported one of the largest and earliest series of patients treated with limited surgery and external-beam radiotherapy for craniopharyngioma [16]. Of the 173 patients, 22 received radiotherapy alone. At a median follow-up of 12 years, the 10- and 20-year PFS rates were 83% and 79%. The 10- and 20-year overall survival rates were 77% and 66%. Age (<40 years old) and use of modern techniques were found to be significant independent positive prognostic factors for survival. Other prognostic factors that influenced outcome was extent of surgery, radiation dose of <55 Gy, and treatment prior to the routine use of diagnostic CT scans (between 1961 and 1974) [19].

A smaller series from Philadelphia [39] with 19 paediatric patients showed a higher 20-year survival rate for those treated for primary disease (78% versus 25%) compared to those treated for a recurrence [24]. Overall, the 20-year survival rate was 62%. In addition, surgical extent, radiation dose (≤54 Gy vs ≥54 Gy with a recurrence rate of 50% in those receiving <54 Gy and 15% in those receiving >54 Gy), and treatment prior to the routine use of CT scans (between 1961 and 1974) appeared to have a major influence on patient outcomes.

Another large retrospective analysis included 61 children treated between 1979 and 1990 [18]. Among them, nine children were treated with radiotherapy alone, 15 with surgery alone, and 37 with both surgery and radiotherapy to a median dose of 5464 cGy. At a median follow-up of 10 years (range, 2–20.5 years), the 10-year actuarial overall survival rate was 91% for all patients. The 10-year actuarial freedom from progression rate for the surgery group was 31% compared with 100% for the radiation-alone group and 86% for the patients treated with surgery and radiotherapy at diagnosis. This study established tumour size greater than 5 cm to be an important predictor of local recurrence.

Intensity-Modulated Radiotherapy

IMRT uses multiple small beamlets of different intensities to treat the target volume by precisely sculpting the high-dose volume around complex target shapes, thereby significantly reducing the dose to adjacent organs at risk. IMRT is delivered using either a dynamic or static multileaf collimator or by a multileaf intensity-modulating collimator using rotational arc therapy (such as RapidArc or volumetric arc therapy). IMRT can significantly reduce the dose to adjacent organs at risk due to its ability to conform the high dose to complex shapes. However, the cost of the conformal high dose is an increased low dose exposure of surrounding normal tissue due to the multiple beams or arcs used to achieve the conformal high target dose (Fig. 7).

In a retrospective review of 24 children treated with IMRT, the 5- and 10-year PFS rates were 65.8% and 60.7% [40]. The 5- and 10-year cystic PFS rates were 70.2% and 65.2%, respectively. The solid PFS rates were no different at 90.7%. Half of these patients had received radiotherapy as salvage therapy for disease progression on imaging following initial surgical treatment. Dynamic cyst changes, both expansion and reduction (range, −20.7% to 82%), can occur during radiotherapy [33] and warrant adaptive planning based on surveillance weekly MRI. Adaptive targeting allows for both target volume coverage and the limiting of dose to adjacent critical structures such as the optic chiasm, which is particularly relevant when a smaller CTV is used with modern techniques. Compared to traditional conformal RT (CRT) using a 5-mm CTV margin and 3-mm PTV margin, IMRT with similar margins reduced the mean dose to the cochlea from 18.2 to13.3 Gy ($p < 0.001$), temporal lobes from 14.3 to 7.9 Gy ($p < 0.001$) and hippocampus from 26.8 to17.6 Gy ($p < 0.001$). Although IMRT plans are more susceptible to the effects of dynamic cyst changes compared to 3D–conformal plans, IMRT is favoured to reduce the dose to critical structures when weekly surveillance imaging is available [33].

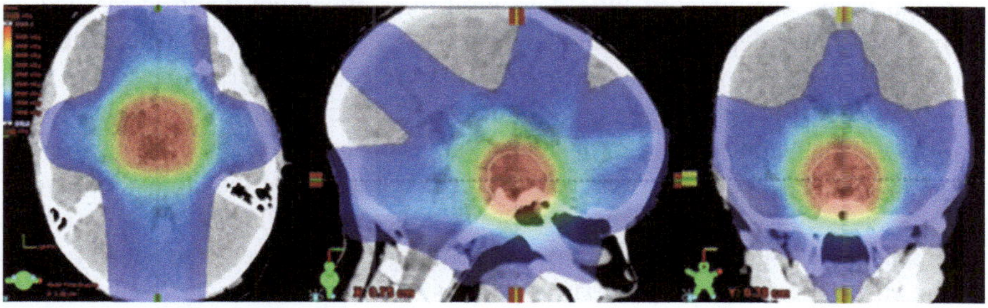

Fig. 7 Axial, sagittal, and coronal slices demonstrating a low dose entry and exit doses in an IMRT plan

Fractionated Stereotactic Radiotherapy

Fractionated stereotactic radiotherapy combines the precise dose delivery of stereotactic radiosurgery through noninvasive immobilisation with the radiobiological advantage of fractionation, enabling the treatment of tumours that may be too large for stereotactic radiosurgery [41]. A study on stereotactic fractionated radiotherapy included 39 patients with a median age of 18 years (range, 3–68 years) treated with stereotactically guided conformal radiotherapy of 30–33 daily fractions over 6 to 6.5 weeks to a total dose of 50 Gy using 6-MV photons. All patients had previously undergone one or more surgical procedures. The 3- and 5-year PFS rates were 97% and 92%. The 3- and 5-year overall survival rates were 100%. Two patients required debulking surgery for progressive disease at 8 and 41 months. Twelve patients (30%) experienced acute clinical deterioration secondary to cyst enlargement and required cyst aspiration. Seven patients with normal pituitary function prior to radiotherapy had no deficits following radiotherapy at a median follow-up of 40 months. One patient with severe visual impairment before radiotherapy deteriorated and lost vision after radiotherapy.

Long-term outcomes from another study report local control rates of 95.3%, 92.1%, and 88.1% at 5, 10, and 20 years, respectively, with a median follow-up time of 128 months (range, 2–276 months) after stereotactic fractionated radiotherapy to a dose of 52.2 Gy (range, 50–57.6 Gy) at 1.8 Gy per fraction [42]. The overall survival rate at 10 and 20 years was 83.3% and 67.8%, respectively. This study included 55 patients with a median age of 37 years (range, 6–70 years) treated between 1989 and 2012, 8 of whom were younger than 18 years old. Toxicities reported included 1 patient with complete anosmia and 1 patient with further deterioration in vision. Of note, neuropsychological testing was not performed.

Proton Therapy

Protons are positively charged particles with a unique property of depositing their maximum energy in tissue shortly before they come to rest [43]. Therapeutic proton beams are generated from a cyclotron, a type of particle accelerator, to attain the desired energy and thus the desired depth of penetration. As protons enter and travel through tissue, they deposit little dose until their maximum depth is reached, at which point they deposit dose in a characteristic sharp, narrow distribution called the Bragg peak. As protons come to rest abruptly after the Bragg peak, the "exit dose," or dose distal to the target, is eliminated. This lack of dose deposition beyond the target lies in contrast to photons, where additional dose is deposited beyond the target as the beam exits the body beyond the distal aspect of the target. A range of energies is used to create a spread-out Bragg peak (SOBP). The SOBP allows dose deposition over a greater length of tissue, resulting in high dose to the entirety of the target volume. This unique property of protons results in the primary clinical advantage of decreased normal tissue exposure and thereby reduced toxicity (Fig. 8).

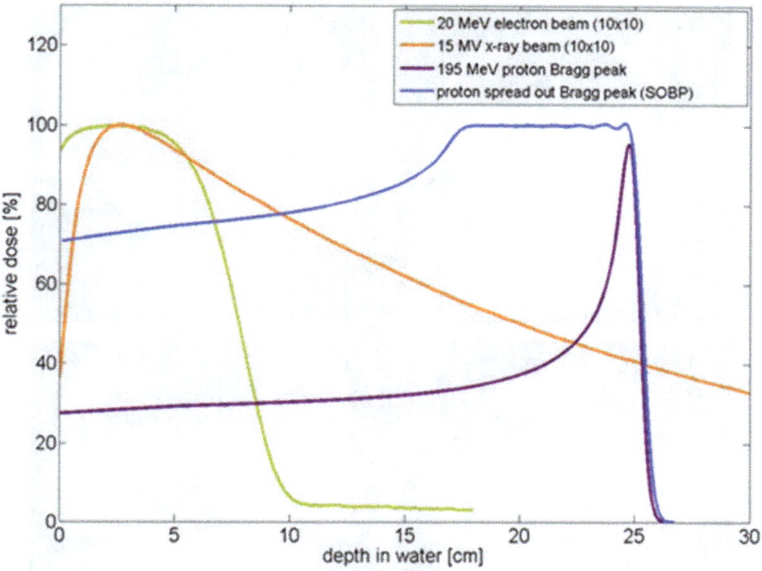

Fig. 8 Depth dose distribution for 195 MeV proton displaying the low entrance dose, Bragg peak with a sharp distal drop off; spread-out Bragg peak (SOBP); 15 MV X-ray (photons) demonstrating an initial buildup of dose followed by a gradual decline with depth; 20 MeV electron beam displaying the higher entrance dose and low dose at depth

Protons, like photons, are an ionising radiation with a low linear energy transfer. The relative biologic effectiveness (RBE) of protons is approximately 1.1; therefore, unlike heavy ion therapy, which also eliminates the exit dose but carries a high linear energy transfer, protons and photons have a similar and more predictable impact on normal tissue.

The primary benefit of proton therapy in the treatment of craniopharyngioma is a decrease in the undesired dose to adjacent critical structures and lower integral dose while maintaining adequate dose to the tumour. This decreased dose to surrounding critical structures is expected to translate into fewer acute and late toxicities from radiotherapy. The reduction in late toxicity is particularly relevant in this predominantly childhood tumour where durable local control is achievable (Fig. 9).

This advantage of proton therapy possible through exploiting the Bragg peak has been evaluated in multiple dosimetric studies. A study of ten paediatric craniopharyngioma patients compared three-dimensional conformal proton therapy, IMRT, and intensity-modulated proton therapy (IMPT) [44]. Coverage of the PTV was adequate for all modalities. The IMRT and IMPT plans were the most conformal, but offered a less homogenous dose distribution. 3D conformal proton and IMPT plans allowed a relative reduction in the integral dose to the hippocampus, dentate gyrus, subventricular zone, vasculature, infratentorial and supratentorial brain, brainstem, and whole brain. Another study on 14 patients with craniopharyngioma compared double-scatter proton therapy, IMPT, and IMRT [45]. Conformality index

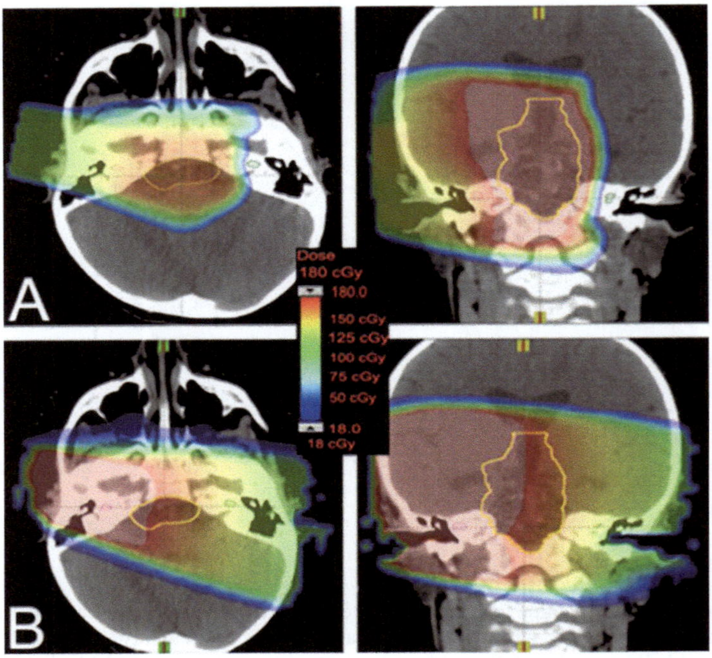

Fig. 9 (**a**) Dose distribution from one axial and coronal proton beam compared to (**b**) one axial and coronal photon beam demonstrating the higher entrance dose for photon beams and the lack of exit dose with the proton beam

was higher with IMPT and IMRT. Heterogeneity and the dose to organs at risk, such as the cochlea, optic chiasm, and brain, was lower with double-scatter proton therapy.

Decreasing the integral dose with proton therapy should decrease the incidence of radiation-induced tumours. According to the Childhood Cancer Survivor Study assessing all types of tumours, the cumulative incidence of a second malignancy at 30 years after diagnosis was 7.9% [46]. The relative risk of a second malignancy in patients treated with radiotherapy was 2.7. A 2- to 15-fold decrease in the rate of second malignancies with proton therapy compared to photon radiotherapy was demonstrated [47].

Another potential means to reducing the integral dose with the use of proton therapy is through the development of cognitive dose models to allow predictive comparison of outcomes. Based on one such longitudinal model in patients with craniopharyngioma, a diminution in intelligence quotient (IQ) was noted in younger patients (5 years vs 9 years) and with the use of photons compared to double-scatter proton therapy [48]. A large prospective study of 70 craniopharyngioma patients who underwent cognitive testing at baseline, 6 months, after radiotherapy, and then annually for 10 years demonstrated an association of longitudinal IQ with mean whole-brain radiation dose and time with a loss of 0.0027 points/cGy/year [32].

Table 1 Evidence of outcomes from proton therapy for patients with craniopharyngioma

Study; no. of patients	Median follow-up (years)	Treatment modality	Actuarial 5-year local control rate	Acute toxicity	Late toxicity	Mortality (absolute number)
Fitzek et al. 2006; $N = 15$	13.1	Surgery/biopsy + proton-photon	93% 5-year	none,7; nausea, 1; fatigue,3; headaches,4	Visual deficits,2; endocrinopathy,15; learning difficulty,1	PD, 2; vascular complications, 1; treatment-related hypothalamic syndrome, 1
Luu et al. 2006; $N = 16$	5	Surgery + proton or proton alone (Surgery+RT, 4; recurrent after surgery, 12; re-resection + RT, 7; RT, 4)	94%[b] (recurrence at 80 months, 1)	NR	Panhypopituitarism, 1; CVA, 1 with full recovery; meningioma, 1 following reirradiation	3 at 12, 52, and 120 months after re-resection and RT (PD, 1; sepsis, 1; MCA infarct, 1)
Chang et al. 2009[a]; $N = 14$	1.3	Surgery/biopsy + proton therapy	100%[b]	NR	Vision, stable or improved; endocrinopathy, 11 (of 11 with results)	0
Alapetite et al. 2012[a]; $N = 49$	4.4	Surgery + proton-photon, 10; surgery + proton, 39	90%[b]	NR	Altered short-term memory, social and emotional functioning and significant school difficulties in children who had RT after several surgeries. Behavioural disorder rates lower after STR+RT	NR

(continued)

Table 1 (continued)

Study; no. of patients	Median follow-up (years)	Treatment modality	Actuarial 5-year local control rate	Acute toxicity	Late toxicity	Mortality (absolute number)
Confer et al. 2012[a]; *N* = 13	0.7	Surgery/biopsy + proton	85%[b]	Grade 2 headache, 1	NR	0
Indelicato et al. 2013[a]; *N* = 40	0.7	Surgery/biopsy + proton therapy	100%[b]	Emesis, 1; headache, 1; presyncope, 2; nausea, 9	None to date	0
Bishop et al., 2013[a]; N = 19	2.7	Surgery/biopsy + proton, 15; proton alone, 4	91%	NR	Vasculopathy, 2; endocrinopathy, 17, of which 15 were post-op and 2 new after RT	secondary to surgically induced DI, 1
Winkfield et al. 2008[a]; *N* = 24	3.7	Surgery/biopsy + proton therapy	100%	NR	NR	intracranial haemorrhage at 1 year, 1 (in a child with 3 previous surgeries followed by RT)

Abbreviations: *CVA* cerebrovascular accident, *DI* diabetes insipidus, *MCA* middle cerebral artery, *NR* not reported, *PD* progressive disease, *RT* radiotherapy
[a]abstract only
[b]Crude rate at time reported; cerebral artery

Table 1 summarises current evidence of outcomes from proton therapy for patients with craniopharyngioma.

Early reports on the use of proton therapy in craniopharyngioma primarily include patients treated with a combined proton-photon technique. In one analysis of 15 patients treated between 1981–1988, one-third of the cohort was treated with proton therapy alone and two-thirds with a combined technique [49]. The analysis included five children and ten adults. Due to a change in the calibration for proton dosimetry in 1997, the dose delivered was recalculated and found to be 6.5% greater than the prescribed dose. The median delivered dose was 55.6 Gy (RBE) for the paediatric patients and 62.7 Gy (RBE) for the adult patients. Proton dose contribution to the total dose in those who underwent a combined treatment was 47% (median dose, 26.9 Gy [RBE]). All treatments were delivered using the same non-coplanar three-field technique with two lateral beams and one superior oblique field. The actuarial 5- and 10-year survival rates observed were 93% and 72%, respectively. The 5- and 10-year local control rates were 93% and 85%, respectively. No paediatric patients experienced tumour progression. These findings were further substantiated by data from the same institution in a report of 24 children who underwent proton therapy for biopsy-proven craniopharyngioma to a median dose of 52.2 Gy (RBE) (range, 52.2–54 Gy [RBE]) between 2001 and 2007 [50]. The local control, PFS, and overall survival rates were 100%, 100%, and 94%, respectively, at a median follow-up of 45 months (range, 3–95 months).

In a study by Luu et al., proton therapy was utilised as salvage therapy in 16 patients who had undergone primary radical resection [51]. Of these, seven patients underwent more than one surgical resection. Nine others underwent one resection. Four of these received adjuvant proton therapy, five were treated with proton therapy at recurrence. One patient was lost to follow-up. At a mean follow-up of 60.2 months (range, 12–121 months), local control was achieved in 14 patients and only one patient had a relapse at 80 months.

Alapetite et al. reported on 49 patients with craniopharyngioma, 39 of whom were treated with protons alone and ten with a combination of protons and photons between 1994 and 2009. At a median follow-up of 53 months (range, 24–156 months), five patients experienced disease relapse, one of which occurred along the surgical access route after 56 months; the other four were cystic relapses at a median time of 43 months (range, 3–68 months) [52].

Indelicato et al. from the University of Florida reported early clinical outcomes in a retrospective analysis of 40 children with a median age of 7.8 years (range, 1.5–19.3 years) treated with double-scatter proton therapy from 2008 to 2012. Of these children, 35 received 54 Gy (RBE) (range, 50.4–54 Gy [RBE]). At a median follow-up of 0.7 years (range, 0.1–3.7 years), the solid tumour control and survival rates were 100% [53].

Early outcomes in 13 children treated with biopsy, cyst decompression, STR, or a combination of these, followed by proton therapy to a median dose of 50.4 Gy (RBE) (range, 50.4–54 Gy [RBE]) suggest this treatment is well tolerated with minimal acute toxicity [54]. In this series, 2 patients required cyst decompression for symptomatic growth at a median of 4.7 months after radiation. At a median follow-up of 8.4 months, all patients were alive and 11 (85%) had not shown any evidence of recurrence or cyst expansion.

Chang et al. reported on a cohort of 14 patients treated with proton therapy between 2004 and 2008 to a median dose of 54.0 Gy (RBE) (range, 50.4–57.6 Gy [RBE]) [55]. In this group, one patient started treatment urgently with photons due to deteriorating vision and then switched to proton therapy. Another patient received proton therapy for a recurrence 8 years after receiving photon irradiation (55.4 Gy). The local control rate at a median follow-up of 16.5 months (range, 6–42 months) was 100%. In another cohort of 19 paediatric patients with craniopharyngioma treated with proton therapy between 2007 and 2012, the 5-year actuarial overall survival rate was 93% and the 5-year PFS rate was 91% [56].

Outcomes for this cohort were updated in 2014 in an analysis of 52 children, 21 treated with proton beam therapy, and 31 with IMRT at two institutions from 1996 and 2012 [35]. Failure-free survival rates were reported separately for nodular and cystic disease. The median follow-up was 59.6 months (33 months for the proton therapy group and 106 months for the IMRT group). At 3 years, the overall survival rate was 96%, the nodular failure-free survival rate was 95%, and the cystic failure-free survival rate was 76%. No difference was seen in overall survival or disease control between the two treatment groups. For the proton therapy group, the cystic failure-free survival rate was 67% and the nodular failure-free survival rate was 91.7% at 3 years. As the follow-up for patients treated with proton therapy was shorter, data points were censored at 3 years and the data were re-analysed. There was no statistical difference between the two arms for any disease control measure.

Passive Scattering and Uniform Scanning

3D conformal passive double-scattering proton therapy is the most widely used technique and uses a second high-Z scattering device to spread the beam laterally. Custom brass apertures and Lucite compensators are designed for each field for each patient to conform the beam laterally and distally (Fig. 10).

3D conformal proton therapy is also delivered using uniform scanning with a scanning magnet that expands the pencil beam laterally. The aperture is designed according to the PTV.

Spot Scanning and Intensity-modulated Proton Therapy

Spot scanning uses small mono-energetic pencil beams to cover the target area using magnetic fields, without the use of apertures or compensators. The advantage to spot scanning over passive-scatter techniques is the improved conformality allowed by smaller spot sizes. The disadvantage to spot scanning is increased time to treatment delivery and increased susceptibility to target motion.

In a clinical case study, Yeung et al. created IMPT plans for eight patients who had been treated with double-scatter proton therapy [57]. IMPT increased the therapeutic ratio through improved target coverage conformality and normal-tissue

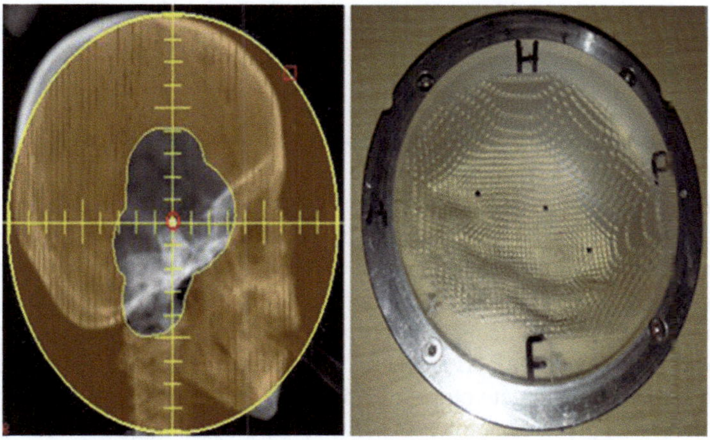

Fig. 10 Custom brass aperture and Lucite compensator designed individually for each proton field

avoidance. Similar to the transition to IMRT, the lack of mature data on spot-scanning techniques, concern for marginal misses from enhanced conformality and precision, and the range uncertainties at the distal margin will continue to be debated by skeptics until more mature data become available.

Stereotactic Radiosurgery

Stereotactic radiosurgery delivers high doses of radiation to a small, precise target in a single or limited fractions, and can either involve the use of a stereotactic frame or be frameless. The use of SRS requires a tumour diameter of <2.5 cm and distance to the optic pathway of >3 mm from the tumour edge. These parameters may not be achievable for many craniopharyngiomas. Nevertheless, one of the largest reported series by Kobayashi et al. analysed 98 patients with craniopharyngioma at a mean follow-up time of 65.5 months (range, 6–148 months) [58]. The mean tumour diameter was 18.8 mm and median tumour volume was 3.5 ml. The maximum dose was 21.8 Gy and the tumour margin dose was 11.5 Gy. The complete response rate was 19.4% and partial response rate was 67.4%, the tumour control rate was 79.6%, and the tumour progression rate was 20.4%. The 5- and 10-year PFS rates were 60.8% and 53.6%, respectively. A deterioration in visual and endocrine function was reported in 6.1%.

In 2008, Lee et al. reported on 16 patients with a residual or recurrent craniopharyngioma treated with frameless cyberknife stereotactic radiosurgery [59]. The mean marginal dose was 21.6 Gy and the mean maximum dose was 29.9 Gy. Outcomes for 11 of these 16 patients were available at a mean follow-up time of 15.4 months. Patient age ranged from 13 to74 years. There was no deterioration in visual or endocrine function. Seven patients experienced tumour shrinkage while three others showed tumour control. Optic apparatus constraint was <5 Gy. Patients received treatment in multiple sessions of 3–10 Gy fractions. The mean volume treated was 6 cm^3.

Cyst Dynamics and Pattern of Response to Radiotherapy on MRI

Dynamic changes in cyst size during radiotherapy call for surveillance imaging and adaptive planning to ensure adequate target coverage and avoid overdosing adjacent critical structures. Winfield et al. reported on 17 patients treated with proton therapy who underwent imaging during radiotherapy; six patients experienced significant cyst growth. Of the 17 patients, four (24%) required revised radiotherapy plans to cover cyst enlargement [34]. As a general rule, adaptive planning should be considered when a change in target volume of ≥5% for IMPT, ≥10% for IMRT, and ≥25% for double-scatter proton therapy is observed [45]. Replanning for proton therapy is more complex than for conventional radiation. In double-scattered treatments, extending the CTV and PTV may require a change in the aperture and compensator,

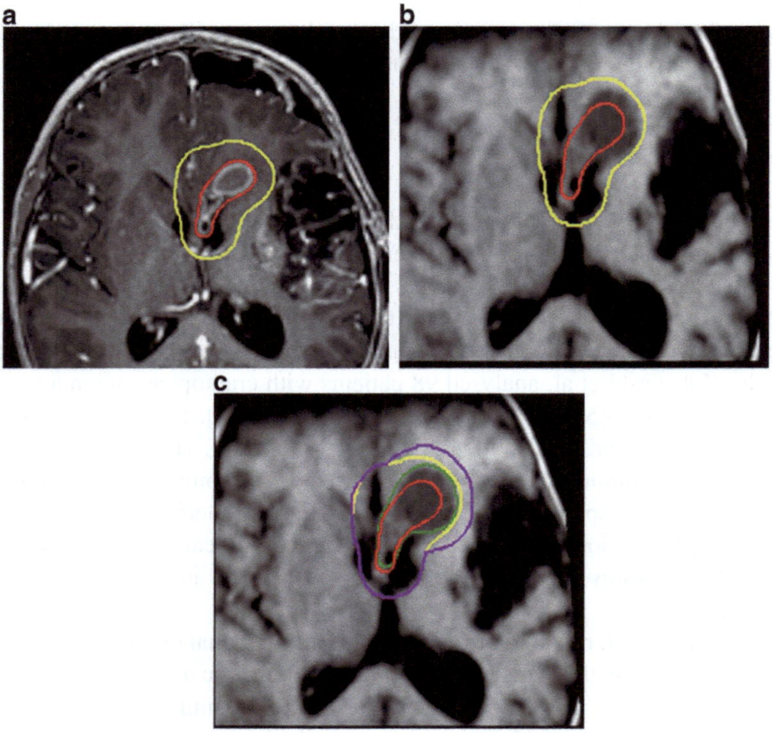

Fig. 11 Image demonstrating cyst growth during radiotherapy requiring replanning. (**a**) Planning magnetic resonance imaging scan with initial gross tumour volume outlined in red and the initial clinical target volume in *yellow*. (**b**) Week 1 magnetic resonance imaging scan with the initial gross tumour volume in red, the initial clinical target volume in *yellow*, and (**c**) the new gross tumour volume outlined in green, and the new clinical tumour volume in *purple*

on the planning computer, and with manufacture of the hardware. The distal margin of one or more beams may also need adjustment (Fig. 11).

Bishop et al. compared IMRT to proton therapy for the treatment of craniopharyngioma and reported that 40% of patients experienced cyst growth with 20% requiring intervention [35]. Immediately after treatment (<3 months after radiotherapy), 33% experienced cyst growth, the majority of which were transient (14/17). Late cyst growth (>3 months after radiotherapy) was seen in 27% (32% in the IMRT group versus 19% in the proton group; $p = 0.354$) with 40% requiring intervention. On multivariate analysis, cyst growth was related to visual and hypothalamic toxicity ($p = 0.009$ and 0.04, respectively). Visual and endocrine outcomes were worse in patients treated at recurrence compared to those receiving adjuvant therapy.

Response to treatment was evaluated with serial MRI scans in 21 children treated with radiotherapy for craniopharyngioma [60]. Early postradiotherapy scans showed tumour or cyst expansion in 52.4% of patients (11/21), which can persist for up to 5–6 months after completing radiotherapy. The median time to initial response after radiation was 4.2 months (range, 0.6–9.2 months). The median time to maximal response after radiation was 9.5 months (range, 3.5–39.9 months). Time to maximum response radiologically was up to 3.3 years. Salvage surgery or cyst aspiration soon after radiation in the absence of overriding surgical indications could expose patients to unnecessary risk.

Side Effects of Radiotherapy

The typical location of craniopharyngiomas and its close proximity to critical intracranial regions may result in both tumour and treatment-related dysfunction of these structures, with a considerable proportion of patients presenting with such dysfunction. While there is substantial data regarding long-term sequelae in the current literature, it is difficult to discern the direct effects of radiation from those attributable to surgery or the combined effect of surgery and radiation. It is important to establish baseline neuroendocrine, ophthalmological, auditory, and neurocognitive function prior to the start of radiation treatment. A baseline magnetic resonance angiogram (MRA) is useful given the high rate of vasculopathy present in craniopharyngioma patients prior to radiation.

The effects of radiation therapy are related to the damage caused to the local tissues and organs that are within the radiation fields. The development of these injuries is dependent on the dose received, and the sensitivity of that tissue to radiation. Fractionated radiotherapy allows for normal tissue repair between fractions. Toxicities are typically graded using the National Cancer Institute Common Terminology Criteria for Adverse Events (CTCAE) or the Radiation Therapy Oncology Group/European Organisation for Research and Treatment of Cancer (RTOG/EORTC) scoring system, with Grade 1 representing the most mild and Grade 5 the most severe, or death related to the toxicity.

Acute effects are defined as those that occur during and within several weeks after radiotherapy (commonly <12 weeks). These effects are the result of damage to early responding tissues. Common acute effects in the treatment of craniopharyngioma include fatigue, skin reaction (dermatitis), nausea, anorexia, and alopecia. Less commonly patients report headaches or vomiting requiring treatment with medications, including oral steroids in some cases. The use of proton therapy in the treatment of central nervous system tumours and specifically for craniopharyngioma is generally well tolerated with mild acute toxicities that can be managed with supportive care [53, 61].

Late effects of radiation can occur months to years after treatment and are usually of greater concern in this group of patients due to the young age at treatment and large number of long-term survivors as well as the permanence of these effects.

Table 2 summarises available data on late toxicity from treatment of craniopharyngioma.

Disease and Treatment-related Mortality

Mortality following treatment of craniopharyngioma may be linked directly to tumour progression or to treatment-related morbidity. The Winkfield cohort of 79 patients had four deaths; two were secondary to disease progression, one was caused by a second malignancy, and one death was unrelated [21].

Among 173 patients followed by Raja et al., 46 patients died; half died from disease progression [16]. Two patients died of endocrine deficiency without disease progression, which were considered disease-related deaths. Three patients died of a cerebrovascular accident at 30, 26, and 15 years after treatment. Thirteen patients died of unrelated causes (including intercurrent illness and glottic edema following peritonsillar abscess). The cause of death was unknown in five patients.

Of the 19 patients in the Regine et al. series, four patients died of disease recurrence and one from a second brain tumour 12 years after treatment [24]. Schoenfeld et al. reported two deaths among 48 patients who received STR and radiotherapy: one died from disease progression 60 months after initial treatment and one from leukemia 151.4 months after initial treatment [20]. In the Weiss et al. series, five patients underwent aggressive surgery and radiotherapy as initial treatment [36]. One of these patients died at 88 months following initial treatment after undergoing salvage surgery for a recurrence at 42 months. Hetelekidis et al. reported three deaths in a cohort of 37 patients treated with surgery and radiotherapy [18]. One patient progressed following initial treatment and died at 31 months. The second patient developed severe metabolic complications following total resection and died 13 months later without evidence of recurrence. The third patient died of a biopsy-proven brainstem glioma 93 months after surgery and radiotherapy.

In their comparison of proton therapy to IMRT, Bishop et al. reported four deaths in 52 patients [35]. One death occurred from cyst progression after STR and adju-

Table 2 Available data on late toxicity from treatment of craniopharyngioma

Study; no. of patients	Mortality	Endocrine	Vision	Cerebrovascular	2nd malignancy	Neurocognitive	Radiation necrosis
Rajan et al. 1993; *N* = 173	46/173 (PD, 24; endocrine def secondary to disease, 2; CVA, 3; unrelated, 13; NK, 5)	Required GH supplementation, 41%; hypothyroid, 66% (99/150); hypogonadism, 39% (59/150); corticosteroid sup, 59%; ADH sup, 25%	VF defects improved, 36% (38/106); VA improved, 30% (27/91); VF +VA deterioration, 1	NR	None at 12 year median FU	Overall = excellent normal, independent life, 46% (80/173); mild disability, active useful life, 37% (52/173); serious disability but capable of self-care, 4%	NR
Winkfield et al. 2011; *N* = 79 (initial Surgery+RT = 43)	4 (PD, 2; 2nd malignancy, 1; unrelated, 1)	Pre-RT deficits, 37% (29/79); post-RT, 72% (57/79); in Surgery+RT arm, 23% (10/43); APHD alone, 66.7% (28/43) DI+APHD	Blindness, 5% (2/43) VF deficit, 53% (23/43)	Moyamoya, 5/79	4 (malignant astrocytoma at field edge, 1; spindle cell sarcoma at CP angle, 1; meningiomas, 2)	NR	NR
Regine et al. 1992; *N* = 19 (Surgery+RT)	5 (PD, 4; 2nd malignancy, 1)	endocrine sequelae, 9/19; thyroid, 7; GH, 6; steroid, 7; panhypopituitarism, 3; no DI	Progressive unilateral amblyopia/oculomotor dysfunction, 1/19	Mild CVA with full recovery, 1/19	Glioblastoma multiforme, 1/19	Mild mental retardation, 1/19	NR
Schonfeld et al. 2012; *N* = 122 (GTR+RT, 3; STR+RT, 48)	2/48 (PD, 1; leukemia1)	DI, 6/48; panhypopituitarism, 12/48	Pre-RT deficits, 25/48 (improved post-RT, 60%; stable, 20%; deteriorated, 20%)	3/48 (infarcts, 2; multiple strokes, 1)	Parathyroid cancer, 1/48	NR	NR

(continued)

Table 2 (continued)

Study; no. of patients	Mortality	Endocrine	Vision	Cerebrovascular	2nd malignancy	Neurocognitive	Radiation necrosis
Hetelekidis et al., 1993; $N = 61$ (RT alone, 9; Surgery+RT, 37)	3/37 (PD, 1; metabolic complications post op at recurrence, 1; 2nd malignancy, 1)	Abnormal tests, 51 (30/51 and 20/33 who received RT = panhypopituitarism, or 3 hormones deficient; 16/51 = 2 abnormalities; 5/51 = 1 abnormality) Significant DI in surgery group (RT, 22%; S, 79%; Surgery+RT group, 38%)	Visual deficits, 11/37 (RT alone, no blindness; Surgery+RT, 2 stable unilateral blindness; bilateral blindness at recurrence, 1; blind after surgery for recurrence, 2)	Moyamoya, 1/37	Glioma, 1/37	RT alone, 1/9 (mild disability); surgery+RT: mild, 7/31; major disability, 2/31; and recurrence had major disability, 4/31	NR
Merchant et al. 2002; $N = 30$ (initial S, 15; RT at relapse, 8; initial Surgery+RT, 15)	NR	Required hormone replacement, 29/30	Post treatment: normal VA bilaterally, 60% (18/30); normal VF bilaterally, 37% (11/30)	NR	NR	Mean loss of IQ points greater in surgery group Mean health utility index better in surgery+RT group	NR
Liu et al. 2009; $N = 20$	NR	NR	NR	Vasculopathy, 6/20 (bilateral temporal cavernomas, 1; moyamoya, 1; aneurysm of ICA, 1; asymptomatic decrease in ICA caliber, 3)	NR	NR	NR

Bishop et al. 2014; *N* = 52	4 (PD, 1; treatment-related morbidity, 3)	At least 1 additional endocrinopathy after RT, 77% (40/52); panhypopituitarism, 46% (24/52); hypothyroid, 13% (7/52); hypogonadism, 8% (4/52); hypothalamic obesity, 13/52 (2 directly related to surgical toxicity)	Visual function influenced by persistent cyst growth	Vascular injury, 5 (symptomatic strokes, 3 radiologic vascular malformation, 1; radiologic moyamoya, 1)	NR	NR	NR
Weiss et al. 1989; *N* = 31 (initial surgery+RT, 5; STR alone, 7; TR alone, 19)	4 (Surgery+RT group, 1 at recurrence; surgery-alone group, 3)	Panhypopituitarism post op and pre-RT, 26/31	Visual loss after surgery+RT, 1; VA loss post-op and pre-RT, 6	NR	NR	Normal intelligence test, 8/11; mild impairment, 2/11; moderate impairment, 1/11	NR

Abbreviations: *APHD* anterior pituitary hormone deficiency, *CMT* combined modality treatment, *CP* cerebellopontine, *CVA* cerebrovascular accident, *DI* diabetes insipidus, *GTR* gross total resection, *ICA* internal carotid artery, *NR* not recorded, *PD* progressive disease, *RT* radiotherapy, *RT* surgery and radiotherapy, *STR* subtotal resection, *TR* total resection, *VA* visual acuity, *VF* visual fields vant radiotherapy; three resulted from treatment-related morbidity, including uncontrolled diabetes insipidus and postoperative neurologic injury.

Neurocognitive Sequelae and Effects on Quality of Life

A decline in cognition, including IQ, is the single-most concerning late effect in long-term survivors. Injury to brain tissue, specifically the temporal lobes and hippocampi, can lead to deficits in learning, memory, and executive functioning and affect quality of life. Demyelination and axonal degradation are postulated to contribute to neurocognitive deficits [62].

Additionally, the presence of secondary obstructive hydrocephalus from tumour and shunt insertion are known to negatively affect neurocognitive outcome and may be reflective of a more aggressive natural history in these patients. Other variables predictive of a negative outcome on learning and memory are treatment with chemotherapy, younger age at the time of treatment, and female sex [63].

Assessment of adaptive function, or the age-appropriate ability to perform tasks of daily living, is a meaningful tool to estimate the effects of treatment on neurocognition. In a prospective study of 62 patients with craniopharyngioma and 77 patients with low-grade glioma, Netson et al. observed a significant decline over time in communication and socialisation in craniopharyngioma patients, but there was no correlation between variation in adaptive functioning indices and IQ scores except for a change in the Daily Living Skills index [64]. The investigators also found that, although participants with severe visual impairment obtained lower scores on many IQ and adaptive indices across the 5-year follow-up period, the only statistically significant difference occurred in the Daily Living Skills and Adaptive Behaviour Composite at 12 and 48 months after treatment, respectively. No deficit in the socialisation index was observed with visual impairment.

Quality of life in long-term survivors is dictated by the effects of the tumour, surgery, and radiotherapy on neurologic function and cognition as well as endocrine effects. Merchant et al. analysed 30 patients with craniopharyngioma (median age, 8.6 years), half of whom were treated initially with surgery alone (8 subsequently received salvage radiotherapy at relapse) and the other half with limited surgery and radiotherapy [65]. The mean loss of IQ points was greater in the surgery group than the combined-modality group (9.8 vs. 1.25; $p < 0.067$). The mean health utility index, a functional quality of life index, was also higher for the combined group (0.85 vs 0.71; $p < 0.063$, one-sided t test). Rajan et al. reported excellent general functional outcomes in patients treated with primary radiotherapy [16]. Eighty patients (46%) lived normal independent lives (grade I), 37% (52) had a grade II status (mild disability, leading an active useful life), and 4% had a grade III functional status (serious disability but capable of self-care). The four-point functional scale included grade IV status, which is total disability.

In the study by Weiss et al., intelligence testing was performed in 11 out of 31 patients after achieving stable disease status; eight patients were within normal limits, two were mildly impaired (one was treated with STR alone and one with STR followed by re-resection and radiation), and one was moderately impaired (treated with initial total removal followed by salvage re-resection and radiation) [36]. In the study by Fitzek et al., all five children in the study completed high school and three (at age 8.2, 16.7, and 16.8 years) attended or were attending college [49]. All were able to live independently and support themselves. One of the five had some learning difficulties (in verbal skills, math, and writing), but was able to pass college courses despite these impediments. The patients treated in adulthood all resumed their previous professional duties.

Cerebral white matter degradation, detected by diffusion tensor MRI (DTI) may predict late delayed cognitive decline after radiotherapy. A correlation between early changes in diffusivity and a late decline in verbal recall has suggested the use of DTI as a potential biomarker for late cognitive decline [66].

In a study by Hetelekidis et al., degree of disability was graded from 1 to 4 (1, no disability; two, mild handicap; three, major disability; and four, incapable of self-care) [18]. Of the nine patients who received radiotherapy alone, only one patient had a mild handicap. In the group who received surgery and radiotherapy ($n = 31$), four patients with recurrent disease developed a major disability but were still capable of self-care; of those whose disease did not recur, 18 had no disability, seven had mild disability, and two had major disability.

Effects on Hypothalamus and Pituitary Function

Endocrine effects from damage to the hypothalamus, the pituitary, or both can vary from isolated deficiency of a single hormone to panhypopituitarism. Endocrinology evaluation prior to radiotherapy and continued life-long follow-up is important in all patients. Hormone replacement or supplementation may be required life-long. Growth failure may be the consequence of growth hormone deficiency, central hypothyroidism, gonadotropin deficiency, or all three combined.

Hypothalamic injury can cause a wide range of disturbances. Hypothalamic obesity occurs from the disruption of appetite and energy homeostasis through interference with complex neuronal and hormonal pathways connecting the hypothalamus to the gastrointestinal system and peripheral adipose tissue. It also causes disturbances of sleep-wake pattern and patients may experience increased daytime somnolence further contributing to an inadequate energy homeostasis [67].

Diabetes insipidus occurs more commonly following surgery. Merchant et al. examined the differences in neurologic, cognitive, and endocrine outcomes between patients undergoing surgery alone and those treated with limited surgery and radiotherapy [65]. Overall, the combined-modality group had fewer or comparable sequelae; diabetes insipidus was the only endocrine deficiency that differed significantly between the two cohorts (75% vs 5%). Almost all patients in this combined cohort (29/30) required hormone replacement therapy.

The study by Hetelekidis et al. revealed baseline endocrine abnormalities in 13 of the 34 patients tested (total of 61 patients in all three arms: 9, radiotherapy alone;

15, surgery alone; and 37, surgery and radiotherapy combined) [18]. Growth hormone was abnormal in 35% of patients, and thyroid and cortisol abnormalities were present in 21%. Following treatment, 51 of the 57 patients who underwent postradiotherapy testing tested had abnormal tests, with panhypopituitarism in 30 patients (three abnormal hormone levels), two abnormal hormones in 16 patients, and one abnormal hormone level in five patients. Among the three groups in this study, a significant difference was seen only for patients with symptomatic diabetes insipidus after treatment: 22%, 79%, and 38% for the radiotherapy alone, surgery alone, and surgery + radiotherapy groups, respectively ($p = 0.024$).

Rajan et al. reported on 19 children with growth hormone deficiency at presentation and following radiotherapy 26 of an evaluable 64 children (41%) who required growth hormone supplementation [16]. Hypothyroidism was present in 15 patients initially and in 99 of 150 patients (66%) after radiotherapy. Hypogonadism at presentation was seen in 18 patients and rose to 59 of 150 (39%) patients after radiotherapy. Corticosteroid supplements were required in 59% of patients after radiation. At presentation, clinical or biochemical diabetes insipidus was evident in 28 patients and eventually 25% (35/150) of patients required long-term anti-diuretic hormone supplementation. In the 48 patients treated with STR and radiotherapy reported by Schoenfeld, six developed diabetes insipidus and 12 developed panhypopituitarism [20].

Weiss et al. reported panhypopituitarism in 26 of 31 patients following initial surgery and prior to radiotherapy. In this cohort, five children developed a severe diencephalic syndrome with rage attacks, hyperphagia, and limited attention span [36]. All patients had undergone surgery (two for total resection, one for STR, and one for total resection followed by re-resection and radiation at the time of recurrence).

In the study by Winkfield et al., 37% (29/79) presented with endocrinopathies ranging from growth hormone deficiency, hypothyroidism, and sexual hormone dysfunction among others [21]. On treatment completion, 72% (57/79) had a documented endocrine disorder. In the group with initial surgery and radiotherapy (43/79) there were no patients who developed diabetes insipidus alone. Isolated cases were however seen in the surgery alone group (5.6%). Overall, 23% (10/43) had anterior pituitary hormone deficiency alone, while 66.7% (28/43) had both diabetes insipidus and anterior pituitary hormone deficiency.

Maximum doses > 6100 cGy were thought to have contributed to the endocrine toxicity seen in the Regine et al. study [24]. Nine (9/19) patients developed endocrine sequelae; 4 of whom had received maximum doses of >6100 cGy. The abnormalities included thyroid ($n = 7$), growth hormone ($n = 6$), and steroid hormone ($n = 7$) deficiency. Three patients had panhypopituitarism. There was no radiation attributable to diabetes insipidus in this group.

In the Bishop et al. study comparing proton therapy with IMRT after surgery, over 50% of patients had some perioperative morbidity, with nearly 40% (10/52) developing diabetes insipidus or panhypopituitarism [35]. More than 75% (40/52) of patients had at least one additional endocrinopathy after radiotherapy, of which 46% (24/40) of cases were new onset hypopituitarism. The most common hormone deficiencies in the remaining 16 patients were hypothyroidism (13%; 7/52) and hypogonadism (8%; 4/52).

Effects on Visual Function

In the large cohort reported by Rajan et al. of patients treated with limited surgery and radiotherapy for craniopharyngioma, over half presented with visual disturbance (95 patients; 55%) and 75% (131 patients) had visual impairment on formal testing [16]. Visual field defect improved after radiotherapy in 36% (38/106) and visual acuity improved in 30% (27/91) of patients. Only one patient experienced deterioration of both visual field and acuity, secondary to recurrent tumour. No patient developed radiation-induced optic neuropathy.

Weiss et al. reported seven survivors with marked loss of visual acuity; in six of these patients, the deficits were present preoperatively [36]. All patients underwent aggressive initial surgery. One case of visual loss following treatment occurred, but the authors were unable to attribute it to a single treatment modality. Winkfield et al. reported blindness in 5% (2/43) and visual deficits ranging from field cuts to decrements in visual acuity in 53% (23/43) of patients after surgery and radiotherapy [21]. Merchant et al. reported that 23 of 30 patients in their cohort received radiotherapy (15 as initial treatment and eight for a recurrence after initial surgery); of these, 20 patients were treated with external-bream radiotherapy and three received ^{32}P [65]. Following radiotherapy, 60% (18/30) had normal visual acuity in both eyes and 37% (11/30) had normal visual fields bilaterally.

Of the 48 patients treated with STR + radiotherapy reported by Schoenfeld et al., 25 patients had initial preradiotherapy deficits; 60% of these improved 20% were stable, and 20% showed deterioration after radiotherapy [20]. Hetelekidis et al. reported 11 visual deficits (field defects or decreased acuity) among the 37 patients in the surgery and radiotherapy arm of their cohort [18]. There was no blindness reported in the radiotherapy-alone group. In the surgery + radiotherapy group, two patients had persistent unilateral blindness present before radiotherapy. One patient developed bilateral blindness at the time of recurrence and two patients were blind after salvage surgery for recurrent disease.

Authors of the Regine et al. series record one case of progressive unilateral amblyopia/oculomotor dysfunction 12 years after total resection and radiation, which was attributed to combined toxicity from surgery and radiation [24].

Finally, the Bishop et al. study found a relationship between persistent cyst growth and visual function that was statistically significant (hazard ratio, 6.01; $p = 0.009$); specific visual outcomes were not reported [35].

Cerebrovascular Complications

Moyamoya disease, cavernous haemangiomas, and stroke are well-documented radiation-induced vascular abnormalities and are an uncommon occurrence in children receiving brain radiotherapy. Moyamoya disease is a progressive occlusive cerebrovascular disorder at the base of the brain usually involving the circle of Willis and the arteries that feed it. The subsequent development of collateral vessels is visible on angiography as a "puff of smoke." Risk factors for the development of Moyamoya disease include radiation dose, presence of neurofibromatosis type I, and younger age at treatment [68]. Additionally, stenosis and aneurysm formation may occasionally occur.

In a retrospective analysis of 20 patients who received radiotherapy for craniopharyngioma from 1995 to 2008, six patients developed some type of radiation-induced vasculopathy at a median imaging follow-up of 44 months (range, 4–85 months) [68]. Abnormalities noted were aneurysm of the internal carotid, bilateral temporal cavernomas, moyamoya disease, aneurysm of the internal carotid artery, and asymptomatic decrease in the caliber of the carotid or cerebral arteries ($n = 3$). Two of the six patients with complications had also received intracystic bleomycin, which the authors thought contributed to the development of vasculopathy.

Winkfield et al. reported five cases of Moyamoya vasculopathy in a cohort of 79 children [21]. Four of these five patients had received radiotherapy as part of their initial treatment and did not have any risk factors like neurofibromatosis type 1; 3 of these patients required surgical intervention and the remaining patient was managed medically. The fifth patient presented with right-upper-extremity weakness and was diagnosed by arteriogram with Moyamoya disease 6 years after receiving RT for recurrent disease after surgery and was managed successfully with bilateral encephaloduroarteriosynangiosis.

In the Schoenfeld et al. cohort of 48 patients who underwent STR + radiotherapy, one patient developed a left middle cerebral artery infarct 2 years after treatment, another had a left parietal infarct after 7 years, and a third patient had multiple strokes at the last follow-up (16.5 years) [20].

One patient in the surgery and radiotherapy group (1/37) developed moyamoya with resultant hemiparesis, seizure disorder, visual field defect, hearing loss, and learning disabilities in the Hetelekidis series.

Regine et al. reported one patient (1/19) treated with surgery and RT with mild mental retardation 5 years after treatment; the patient then had a mild cerebrovascular event 18 years after radiotherapy from which there was complete recovery [39]. A correlation with maximum doses of >6100 cGy was suggested as a cause by the authors.

Bishop et al. detected vascular injury in 5 of 52 patients who received either IMRT or proton therapy. Of these, three patients had symptomatic strokes, one had a radiological vascular malformation, and one had radiological evidence of Moyamoya disease [35].

Radiation Necrosis

Radiation-induced cerebral necrosis is an uncommon complication of radiotherapy and can occur a few months to several years after treatment. The diagnosis is challenging and can be difficult to distinguish from a tumour recurrence. The effects of necrosis can range from asymptomatic radiographic changes to symptoms related to tumour recurrence, seizures, stroke, and death in some cases. Total dose, irradiated volume, treatment duration, dose per fraction, chemotherapy, previous radiation, and male sex have been identified as potential risk factors. Various treatments have been used, including observation when asymptomatic, steroids, hyperbaric oxygen therapy, and bevacizumab [69]. A recent study comparing the predictive risk of radiation-induced cerebral necrosis in paediatric brain cancer patients has shown a reduced risk with the use of passively scattered proton therapy and IMPT compared to photons using volumetric modulated arc therapy [70]. Merchant et al. report one case of parenchymal brain necrosis with subsequent motor and mental dysfunction in a patient treated with external-bream radiotherapy (to 5580 cGy using an arc technique) [65].

Second Malignancies

Benign or malignant tumours may develop within a previously treated radiation field years after completion of radiotherapy. One of the largest cohorts of patients treated with limited surgery and radiotherapy between 1950 and 1986 did not report any second malignancies at a median follow-up of 12 years [16]. Weiss et al. reported four cases of second malignancy, including one case of intracranial sarcoma after 14,400 cGy over 3 years, 2 cases of malignant astrocytomas after 6000 cGy, and one case of glioblastoma multiforme after 5400 cGy [36]. Winkfield et al. also reported four cases of second malignancy: one patient had a malignant astrocytoma at the edge of the treatment field 8 years after receiving 55 Gy postoperatively, one patient had spindle cell neoplasm at the cerebellopontine angle 14 years after receiving treatment at the age of 9 years old, and two patients had meningiomas, one of which was diagnosed more than 20 years after initial therapy [21]. Schoenfeld et al. reported 1 case of parathyroid cancer (out of 43 patients) 16 years after STR + radiotherapy. Radiation-induced sequelae reported by Regine et al. included one case of glioblastoma multiforme of the right temporal lobe 12 years after treatment [20]. Hetelekidis et al. reported one case of an in-field glioma 7.75 years after radiotherapy [18].

A lower rate of second malignancies was noted in patients treated with proton therapy compared to those receiving photon therapy (5.2% vs 7.5%, respectively; hazard ratio, 0.52; 95% confidence interval, 0.32–0.85; $p = 0.009$) at a median follow-up of 6.7 years [71]. The 10-year cumulative incidence rates were 5.4% for proton patients versus 8.6% for photon patients. Less than 10% of patients in both groups were in the paediatric age group and none of them developed a malignancy in the follow-up period. Longer follow-up and the inclusion of paediatric patients may help reveal potential significant differences.

Summary

Although surgery continues to play an important role in the management of craniopharyngiomas, the multimodality approach with safe resection and radiation therapy helps address the challenges posed by the vital normal tissues surrounding the tumour. Advanced radiation techniques may spare critical areas of the brain while maintaining local tumour control, especially in the paediatric population, leading to reduced late effects and improved quality of life.

References

1. Petito CK, DeGirolami U, Earle KM (1976) Craniopharyngiomas: a clinical and pathological review. Cancer 37:1944–1952
2. Fitz CR, Wortzman G, Harwood-Nash DC, Holgate RC, Barry JF, Boldt DW (1978) Computed tomography in craniopharyngiomas. Radiology 127:687–691
3. Sartoretti-Schefer S, Wichmann W, Aguzzi A, Valavanis A (1997) MR differentiation of adamantinous and squamous-papillary craniopharyngiomas. AJNR Am J Neuroradiol 18:77–87
4. Curran JG, O'Connor E (2005) Imaging of craniopharyngioma. Childs Nerv Syst 21:635–639
5. Donovan J, Nesbit G (1996) Distinction of masses involving the sella and suprasellar space: specificity of imaging features. AJR Am J Roentgenol 167:597–603
6. Harwood-Nash DC (1994) Neuroimaging of childhood craniopharyngioma. Pediatr Neurosurg 21(Suppl 1):2–10
7. Hofmann BM, Nimsky C, Fahlbusch R (2011) Benefit of 1.5-T intraoperative MR imaging in the surgical treatment of craniopharyngiomas. Acta Neurochir 153:1377–1390 discussion 1390
8. Chakrabarti I, Amar AP, Couldwell W, Weiss MH (2005) Long-term neurological, visual, and endocrine outcomes following transnasal resection of craniopharyngioma. J Neurosurg 102:650–657
9. Shirane R, Ching-Chan S, Kusaka Y, Jokura H, Yoshimoto T (2002) Surgical outcomes in 31 patients with craniopharyngiomas extending outside the suprasellar cistern: an evaluation of the frontobasal interhemispheric approach. J Neurosurg 96:704–712
10. Zacharia BE, Bruce SS, Goldstein H, Malone HR, Neugut AI, Bruce JN (2012) Incidence, treatment and survival of patients with craniopharyngioma in the surveillance, epidemiology and end results program. Neuro-Oncology 14:1070–1078
11. Komotar RJ, Roguski M, Bruce JN (2009) Surgical management of craniopharyngiomas. J Neuro-Oncol 92:283–296
12. Steinbok P (2015) Craniopharyngioma in children: long-term outcomes. Neurol Med Chir (Tokyo) 55:722–726
13. Tomita T, Bowman RM (2005) Craniopharyngiomas in children: surgical experience at Children's Memorial Hospital. Childs Nerv Syst 21:729–746
14. Elliott RE, Hsieh K, Hochm T, Belitskaya-Levy I, Wisoff J, Wisoff JH (2010) Efficacy and safety of radical resection of primary and recurrent craniopharyngiomas in 86 children. J Neurosurg Pediatr 5:30–48
15. Kramer S, McKissock W, Concannon JP (1961) Craniopharyngiomas. Treatment by combined surgery and radiation therapy. J Neurosurg 18:217–226
16. Rajan B, Ashley S, Gorman C, Jose CC, Horwich A, Bloom HJ et al (1993) Craniopharyngioma--a long-term results following limited surgery and radiotherapy. Radiother Oncol 26:1–10

17. Kiehna EN, Merchant TE (2010) Radiation therapy for pediatric craniopharyngioma. Neurosurg Focus 28:E10
18. Hetelekidis S, Barnes PD, Tao ML, Fischer EG, Schneider L, Scott RM et al (1993) 20-year experience in childhood craniopharyngioma. Int J Radiat Oncol Biol Phys 27:189–195
19. Habrand JL, Ganry O, Couanet D, Rouxel V, Levy-Piedbois C, Pierre-Kahn A et al (1999) The role of radiation therapy in the management of craniopharyngioma: a 25-year experience and review of the literature. Int J Radiat Oncol Biol Phys 44:255–263
20. Schoenfeld A, Pekmezci M, Barnes MJ, Tihan T, Gupta N, Lamborn KR et al (2012) The superiority of conservative resection and adjuvant radiation for craniopharyngiomas. J Neuro-Oncol 108:133–139
21. Winkfield KM, Tsai HK, Yao X, Larson E, Neuberg D, Pomeroy SL et al (2011) Long-term clinical outcomes following treatment of childhood craniopharyngioma. Pediatr Blood Cancer 56:1120–1126
22. Lin LL, El Naqa I, Leonard JR, Park TS, Hollander AS, Michalski JM et al (2008) Long-term outcome in children treated for craniopharyngioma with and without radiotherapy. J Neurosurg Pediatr. 1:126–130
23. Sung D, Chang C, Harisiadis L, Carmel P (1981) Treatment results of craniopharyngiomas. Cancer 47:847–852
24. Regine WF, Mohiuddin M, Kramer S (1993) Long-term results of pediatric and adult craniopharyngiomas treated with combined surgery and radiation. Radiother Oncol 27 13–21
25. Stripp DC, Maity A, Janss AJ, Belasco JB, Tochner ZA, Goldwein JW et al (2004) Surgery with or without radiation therapy in the management of craniopharyngiomas in children and young adults. Int J Radiat Oncol Biol Phys 58:714–720
26. Moon SH, Kim IH, Park SW, Kim I, Hong S, Park CI et al (2005) Early adjuvant radiotherapy toward long-term survival and better quality of life for craniopharyngiomas--a study in single institute. Childs Nerv Syst 21:799–807
27. Jannoun L, Bloom HJ (1990) Long-term psychological effects in children treated for intracranial tumors. Int J Radiat Oncol Biol Phys 18:747–753
28. Bleyer WA, Fallavollita J, Robison L, Balsom W, Meadows A, Heyn R et al (1990) Influence of age, sex, and concurrent intrathecal methotrexate therapy on intellectual function after cranial irradiation during childhood: a report from the Children's Cancer Study Group. Pediatr Hematol Oncol 7:329–338
29. Merchant TE, Conklin HM, Wu S, Lustig RH, Xiong X (2009) Late effects of conformal radiation therapy for pediatric patients with low-grade glioma: prospective evaluation of cognitive, endocrine, and hearing deficits. J Clin Oncol 27:3691–3697
30. Aldana PR, MacGregor TL, James HE, Marcus RB Jr, Indelicato DJ (2010) Titanium screws as fiducial markers for image-guided stereotactic proton irradiation in pediatric central nervous system tumors: a technical report. Pediatr Neurosurg 46:227–231
31. Merchant TE, Kiehna EN, Kun LE, Mulhern RK, Li C, Xiong X et al (2006) Phase II trial of conformal radiation therapy for pediatric patients with craniopharyngioma and correlation of surgical factors and radiation dosimetry with change in cognitive function. J Neurosurg 104:94–102
32. Merchant TE, Kun LE, Hua CH, Wu S, Xiong X, Sanford RA et al (2013) Disease control after reduced volume conformal and intensity modulated radiation therapy for childhood craniopharyngioma. Int J Radiat Oncol Biol Phys 85:e187–e192
33. Beltran C, Naik M, Merchant TE (2010) Dosimetric effect of target expansion and setup uncertainty during radiation therapy in pediatric craniopharyngioma. Radiother Oncol 97:399–403
34. Winkfield KM, Linsenmeier C, Yock TI, Grant PE, Yeap BY, Butler WE et al (2009) Surveillance of craniopharyngioma cyst growth in children treated with proton radiotherapy. Int J Radiat Oncol Biol Phys 73:716–721
35. Bishop AJ, Greenfield B, Mahajan A, Paulino AC, Okcu MF, Allen PK et al (2014) Proton beam therapy versus conformal photon radiation therapy for childhood craniopharyngioma:

multi-institutional analysis of outcomes, cyst dynamics, and toxicity. Int J Radiat Oncol Biol Phys 90:354–361
36. Weiss M, Sutton L, Marcial V, Fowble B, Packer R, Zimmerman R et al (1989) The role of radiation therapy in the management of childhood craniopharyngioma. Int J Radiat Oncol Biol Phys 17:1313–1321
37. Lawrence YR, Li XA, el Naqa I, Hahn CA, Marks LB, Merchant TE et al (2010) Radiation dose-volume effects in the brain. Int J Radiat Oncol Biol Phys 76:S20–S27
38. Toogood AA (2004) Endocrine consequences of brain irradiation. Growth Hormon IGF Res 14(Suppl A):S118–S124
39. Regine WF, Kramer S (1992) Pediatric craniopharyngiomas: long term results of combined treatment with surgery and radiation. Int J Radiat Oncol Biol Phys 24:611–617
40. Greenfield BJ, Okcu MF, Baxter PA, Chintagumpala M, Teh BS, Dauser RC et al (2015) Long-term disease control and toxicity outcomes following surgery and intensity modulated radiation therapy (IMRT) in pediatric craniopharyngioma. Radiother Oncol 114:224–229
41. Minniti G, Saran F, Traish D, Soomal R, Sardell S, Gonsalves A et al (2007) Fractionated stereotactic conformal radiotherapy following conservative surgery in the control of craniopharyngiomas. Radiother Oncol 82:90–95
42. Harrabi SB, Adeberg S, Welzel T, Rieken S, Habermehl D, Debus J et al (2014) Long term results after fractionated stereotactic radiotherapy (FSRT) in patients with craniopharyngioma: maximal tumor control with minimal side effects. Radiat Oncol 9:203
43. Durante M, Loeffler JS (2010) Charged particles in radiation oncology. Nat Rev Clin Oncol 7:37–43
44. Boehling NS, Grosshans DR, Bluett JB, Palmer MT, Song X, Amos RA et al (2012) Dosimetric comparison of three-dimensional conformal proton radiotherapy, intensity-modulated proton therapy, and intensity-modulated radiotherapy for treatment of pediatric craniopharyngiomas. Int J Radiat Oncol Biol Phys 82:643–652
45. Beltran C, Roca M, Merchant TE (2012) On the benefits and risks of proton therapy in pediatric craniopharyngioma. Int J Radiat Oncol Biol Phys 82:e281–e287
46. Friedman DL, Whitton J, Leisenring W, Mertens AC, Hammond S, Stovall M et al (2010) Subsequent neoplasms in 5-year survivors of childhood cancer: the Childhood Cancer Survivor Study. J Natl Cancer Inst 102:1083–1095
47. Miralbell R, Lomax A, Cella L, Schneider U (2002) Potential reduction of the incidence of radiation-induced second cancers by using proton beams in the treatment of pediatric tumors. Int J Radiat Oncol Biol Phys 54:824–829
48. Merchant TE, Hua CH, Shukla H, Ying X, Nill S, Oelfke U (2008) Proton versus photon radiotherapy for common pediatric brain tumors: comparison of models of dose characteristics and their relationship to cognitive function. Pediatr Blood Cancer 51:110–117
49. Fitzek MM, Linggood RM, Adams J, Munzenrider JE (2006) Combined proton and photon irradiation for craniopharyngioma: long-term results of the early cohort of patients treated at Harvard Cyclotron Laboratory and Massachusetts General Hospital. Int J Radiat Oncol Biol Phys 64:1348–1354
50. Winkfield KM, Linsenmeier C, Yeap BY, Tarbell NJ, Yock TI (2008) Proton radiotherapy for childhood craniopharyngioma: initial clinical outcomes [abstract]. Int J Radiat Oncol Biol Phys 72:S496
51. Luu QT, Loredo LN, Archambeau JO, Yonemoto LT, Slater JM, Slater JD (2006) Fractionated proton radiation treatment for pediatric craniopharyngioma: preliminary report. Cancer J 12:155–159
52. Alapetite C, Puget S, Ruffier A, Habrand J-L, Bolle S, Noel G et al (2010) CR-09: proton therapy for craniopharyngioma in children: update of the the orsay proton center experience [abstract]. Pediatr Blood Cancer 14(Suppl 1):i22–i25
53. Indelicato DJ, Rotondo R, Flampouri S, Kowalchik K, Sandler E, Aldana P et al (2012) Proton therapy for craniopharyngioma: early clinical outcomes [abstract]. Int J Radiat Oncol Biol Phys 84:S634

54. Confer ME, McNall-Knapp R, Krishnan S, Gross N, Keole S (2012) Proton radiation therapy for pediatric craniopharyngiomas: initial results [abstract]. Int J Radiat Oncol Biol Phys 84:S635
55. Chang AL, Fitzek MM, Kruter LE, Roach K, Thornton AF (2009) Outcomes of pediatric craniopharyngioma treated with proton radiation therapy [abstract]. Int J Radiat Oncol Biol Phys 75:S513
56. Bishop AJ, Mahajan A, Okcu MF, Allen PK, Kahalley LS, McAleer MF et al (2013) P054: proton therapy for the treatment of childhood craniopharyngiomas: cyst dynamics and initial outcomes. In: 95th Annual Meeting American Radium Society, Scottsdale, Arizona, 2013
57. Yeung D, McKenzie C, Indelicato DJ (2014) A dosimetric comparison of intensity-modulated proton therapy optimization techniques for pediatric craniopharyngiomas: a clinical case study. Pediatr Blood Cancer 61:89–94
58. Kobayashi T, Kida Y, Mori Y, Hasegawa T (2005) Long-term results of gamma knife surgery for the treatment of craniopharyngioma in 98 consecutive cases. J Neurosurg 103:482–488
59. Lee M, Kalani MY, Cheshier S, Gibbs IC, Adler JR, Chang SD (2008) Radiation therapy and CyberKnife radiosurgery in the management of craniopharyngiomas [abstract]. Neurosurg Focus 24:E4
60. Shi Z, Esiashvili N, Janss AJ, Mazewski CM, MacDonald TJ, Wrubel DM et al (2012) Transient enlargement of craniopharyngioma after radiation therapy: pattern of magnetic resonance imaging response following radiation. J Neuro-Oncol 109:349–355
61. Suneja G, Poorvu PD, Hill-Kayser C, Lustig RA (2013) Acute toxicity of proton beam radiation for pediatric central nervous system malignancies. Pediatr Blood Cancer 60:1431–1436
62. Khong PL, Leung LH, Fung AS, Fong DY, Qiu D, Kwong DL et al (2006) White matter anisotropy in post-treatment childhood cancer survivors: preliminary evidence of association with neurocognitive function. J Clin Oncol 24:884–890
63. Di Pinto M, Conklin HM, Li C, Merchant TE (2012) Learning and memory following conformal radiation therapy for pediatric craniopharyngioma and low-grade glioma Int J Radiat Oncol Biol Phys 84:e363–e369
64. Netson KL, Conklin HM, Wu S, Xiong X, Merchant TE (2013) Longitudinal investigation of adaptive functioning following conformal irradiation for pediatric craniopharyngioma and low-grade glioma. Int J Radiat Oncol Biol Phys 85:1301–1306
65. Merchant TE, Kiehna EN, Sanford RA, Mulhern RK, Thompson SJ, Wilson MW et al (2002) Craniopharyngioma: the St. Jude Children's Research Hospital experience 1984–2001. Int J Radiat Oncol Biol Phys 53:533–542
66. Chapman CH, Nagesh V, Sundgren PC, Buchtel H, Chenevert TL, Junck L et al (2012) Diffusion tensor imaging of normal-appearing white matter as biomarker for radiation-induced late delayed cognitive decline. Int J Radiat Oncol Biol Phys 82:2033–2040
67. Cohen M, Guger S, Hamilton J (2011) Long term sequelae of pediatric craniopharyngioma - literature review and 20 years of experience. Front Endocrinol (Lausanne) 2:81
68. Liu AK, Bagrosky B, Fenton LZ, Gaspar LE, Handler MH, McNatt SA et al (2009) Vascular abnormalities in pediatric craniopharyngioma patients treated with radiation therapy. Pediatr Blood Cancer 52:227–230
69. Chao ST, Ahluwalia MS, Barnett GH, Stevens GH, Murphy ES, Stockham AL et al (2013) Challenges with the diagnosis and treatment of cerebral radiation necrosis. Int J Radiat Oncol Biol Phys 87:449–457
70. Freund D, Zhang R, Sanders M, Newhauser W (2015) Predictive risk of radiation induced cerebral necrosis in pediatric brain cancer patients after VMAT versus proton therapy. Cancers (Basel) 7:617–630
71. Chung CS, Yock TI, Nelson K, Xu Y, Keating NL, Tarbell NJ (2013) Incidence of second malignancies among patients treated with proton versus photon radiation. Int J Radiat Oncol Biol Phys 87:46–52

Surgical Treatment of Human ACP

Kristian Aquilina and Michael Buchfelder

Abstract Adamantinomatous craniopharyngiomas are complex cystic-solid lesions, typically located within the suprasellar cistern. Their growth causes symptoms by pressure and distortion of the optic nerves and chiasm, the hypothalamus and the pituitary gland. Involvement of the third ventricle causes hydrocephalus. Initial management is surgical and directed towards preservation of vision and control of raised intracranial pressure. Further surgical management is dictated by the extent of hypothalamic involvement, evident on clinical presentation and magnetic resonance imaging. Limited surgery, followed by radiotherapy, is as effective at long-term local control as aggressive surgery aimed at complete resection. The former approach is superior at protecting hypothalamic function and is preferable whenever hypothalamic involvement by the tumour is significant.

Surgery may be carried out transcranially or transsphenoidally. The development of endoscopic transsphenoidal surgery has allowed larger and higher craniopharyngiomas to be resected safely through this route. The management of recurrent craniopharyngiomas remains challenging and may include revisional surgery or administration of interferon into cystic components.

Keywords Adamantinomatous craniopharyngioma • Hypothalamus • Trans sphenoidal surgery • Transcranial resection • Obesity • Recurrence

K. Aquilina (✉)
Department of Neurosurgery, Great Ormond Street Hospital for Children,
Great Ormond Street, London WC1N 3JH, UK
e-mail: Kristian.aquilina@gosh.nhs.uk

M. Buchfelder
Department of Neurosurgery, University Hospital Erlangen,
Schwabachanlage 6, Erlangen 91054, Germany
e-mail: michael.buchfelder@uk-erlangen.de

J.P. Martinez-Barbera, C.L. Andoniadou (eds.), *Basic Research and Clinical Aspects of Adamantinomatous Craniopharyngioma*,
DOI 10.1007/978-3-319-51890-9_8

Introduction

Surgery is an important treatment modality in the management of adamantinomatous craniopharyngiomas. Most of these tumours are large on presentation and require surgical decompression to reduce their mass effect, aspirate cystic components, relieve hydrocephalus and alleviate pressure on the optic apparatus. Surgery is sometimes also necessary to confirm diagnosis. The location of craniopharyngiomas within the suprasellar space, and their intimate relationship to the optic nerves and chiasm, pituitary gland, stalk, hypothalamus, critical vascular structures, and occasionally the brainstem ensures that these tumours remain some of the most challenging lesions paediatric neurosurgeons have to deal with.

In this chapter we review aspects of the surgical management of craniopharyngiomas. This has evolved extensively over the last hundred years; the first successful removal of a craniopharyngioma was reported by Halstead in 1910, using a sublabial gingival incision and transsphenoidal approach [1]. Concerns about operative mortality led to a phase of limited surgery from the 1950's to the 1970's, followed by a period of more aggressive attempts at gross total resection from the 1970s onwards. We explore the currently prevalent approach of limited surgery and adjuvant radiotherapy, advances in approaches to craniopharyngiomas and also review how magnetic resonance imaging facilitates selection of those tumours where gross total resection may still be possible. We also address the adverse effects of surgery, with particular emphasis on hypothalamic deficits, and the particular difficulties related to the management of recurrent craniopharyngiomas.

Investigations

The presentation of adamantinomatous craniopharyngiomas is variable. Most commonly they present with non-specific symptoms related to raised intracranial pressure. This may be related to the direct mass effect of a large craniopharyngioma cyst, a large solid component or to hydrocephalus occurring as a result of obstruction to cerebrospinal fluid (CSF) flow at the third ventricle. Craniopharyngiomas may also present with visual deficit, hypothalamic symptoms such as obesity, and endocrine disturbance. As most adamantinomatous craniopharyngiomas grow within the suprasellar region, disturbance of vision is a frequent initial manifestation, occurring in more than half of the patients [2]. At diagnosis, over 80% of patients have a deficit of at least one hormone, most commonly growth hormone [2]. Most children would have gained weight and demonstrated impaired growth over the year preceding diagnosis [3]. In addition to obesity, other evidence of hypothalamic dysfunction, such as daytime sleepiness, hyperphagia and disturbed circadian rhythms, have been found in over a third of children at diagnosis [4, 5].

Initial investigations of craniopharyngioma must include detailed imaging studies. Computed tomography (CT) imaging is useful in determining the diagnosis preoperatively. Specks of calcification on CT scans are evident in up to 90% of craniopharyngiomas and effectively exclude other suprasellar pathology such as germ cell tumours, pituitary adenomas and Rathke's cleft cysts. Acute hydrocephalus, evident on CT imaging, is an independent predictor of hypothalamic involvement [6, 7].

Magnetic resonance imaging is essential to allow definition of the anatomy of the craniopharyngioma. The relationship of the tumour to the walls of the third ventricle defines hypothalamic involvement and the risk associated with surgical resection. In their case series, De Vile et al. described the association of pre-operative hypothalamic involvement on MRI with obesity [8]. Topographical characteristics that are associated with pre-operative hyperphagia and obesity include absence of the pituitary stalk, displacement of the optic chiasm and peritumoural hypothalamic oedema on T2 and FLAIR sequences [5].

Pre-operative scoring systems have been developed to define the degree of hypothalamic involvement and predict the clinical outcome of complete resection [9–11]. Van Gompel et al. demonstrated that hypothalamic signal changes on T2-weighted MRI, as well as irregular contrast enhancement, predicted increased hypothalamic risk [10]. Increased risk was also associated with tumours that extended posteriorly beyond the mammillary bodies [12]. Tumours with a retrochiasmatic growth pattern and an incompetent diaphragm were associated with a higher post-operative BMI [5]. In addition, craniopharyngiomas extending into the ventricles are associated with a higher hypothalamic complication rate than those that do not [13].

Puget et al. reviewed their results in 66 paediatric craniopharyngiomas and concluded that outcome was related to MRI grade, pre-operative clinical hypothalamic status and experience of the surgeon [9]. The MRI scoring system they proposed has now been widely adopted. A simple three point grading system defined hypothalamic involvement as grade 0 (no involvement), grade 1 (tumour abutting or displacing the hypothalamus) and grade 2 (hypothalamus significantly involved and no longer identifiable) [9]. Another grading system defined the location of the tumour as involving one of four areas: section 1 (limited superiorly by the diaphragma sellae), section 2 (below the optic chiasm and mammillary bodies) and section 3 (above the latter two structures) [6]. This is then divided further into the areas anterior and posterior to the mammillary bodies, 3a and 3b. Involvement of higher and more posterior structures was associated with increased hypothalamic risk on attempt of complete resection.

Surgical Management

Acute Management and Cyst Drainage

The initial priorities in the management of craniopharyngiomas are to treat raised intracranial pressure and restore vision. This often involves drainage of a large tumour cyst obstructing the third ventricle and causing hydrocephalus, or visual disturbance caused by displacement of the chiasm. Decisions regarding definitive management of the tumour are often considered once acute symptoms are under control [14].

As cysts extend into the third ventricle, a neuro-endoscopic approach through the lateral ventricle is ideal for drainage. A catheter may be placed accurately into the cyst under direct vision, permitting recurrent aspiration of the cyst and allowing longer term instillation of intracavitary ablative therapy such as interferon alpha [15]. Such cysts often occlude the foramina of Monro, and therefore a septum pellucidotomy at the same time facilitates the long-term management of hydrocephalus.

Image guidance is useful to help select the ideal trajectory. A perpendicular approach to the surface of the cyst wall is necessary to facilitate safe penetration. The burr hole is typically placed on the coronal suture. The side selected depends on the degree of deformation of the foramina of Monro. The drainage catheter may be passed through the same burr hole and same transcortical tract, using a 3.5 mm rigid endoscope within a 4.7 mm peel-apart sheath. The catheter can then be advanced adjacent to the endoscope within the sheath. Speckled calcifications on the cyst surface are usually evident. The surface of the cyst is coagulated and the catheter is advanced under direct vision. It is important to minimise leakage of cyst fluid into the ventricular system, as this may cause chemical meningitis or, rarely, obstructive hydrocephalus. A transparent catheter allows visualisation of the cyst fluid and monitoring of the progress of aspiration. Failure of this catheterisation technique has been reported when the cyst capsule is heavily calcified. Pre-operative CT scan to exclude extensive wall calcification is helpful [15]. The distal end of the catheter is then attached to an Ommaya reservoir that is placed subcutaneously at the burr hole.

This technique is also possible, although more challenging, when the ventricles are small. Under image guidance the lateral ventricle is first cannulated with a ventricular catheter and the ventricular system is insufflated with 2–3 mL aliquots of saline [16]. Vigilance for hypertension and bradycardia is maintained. Once the ventricles expand the catheter is replaced by the endoscope.

If a large cyst does not extend into the ventricular system, direct image-guided implantation of a catheter into the cyst is preferred. Using electromagnetic or optic image guidance, the catheter is sheathed over the guidance probe and passed along a pre-determined optimal trajectory. Significant resistance, exerted by the tough cyst capsule, is often felt at the point of entry into the cyst. The optimal length of the catheter within the cyst is determined from the pre-operative MRI scan. As in the endoscopic technique, an Ommaya reservoir attached to the distal end of the catheter allows repeated cyst aspiration.

Drainage of a large cyst reduces the size of the tumour and renders the breach in the third ventricular floor, as well as the degree of hypothalamic involvement, easier to evaluate [14]. In this series, nine of the ten patients undergoing endoscopic drainage were reclassified to a lower Paris grade, leading to a significant change in surgical strategy. The authors prefer to shrink the cyst capsule rather than insert an Ommaya reservoir, arguing that the shrinkage leads to better re-grading than simple drainage. The development of a wide stoma in the cyst capsule, with resection of

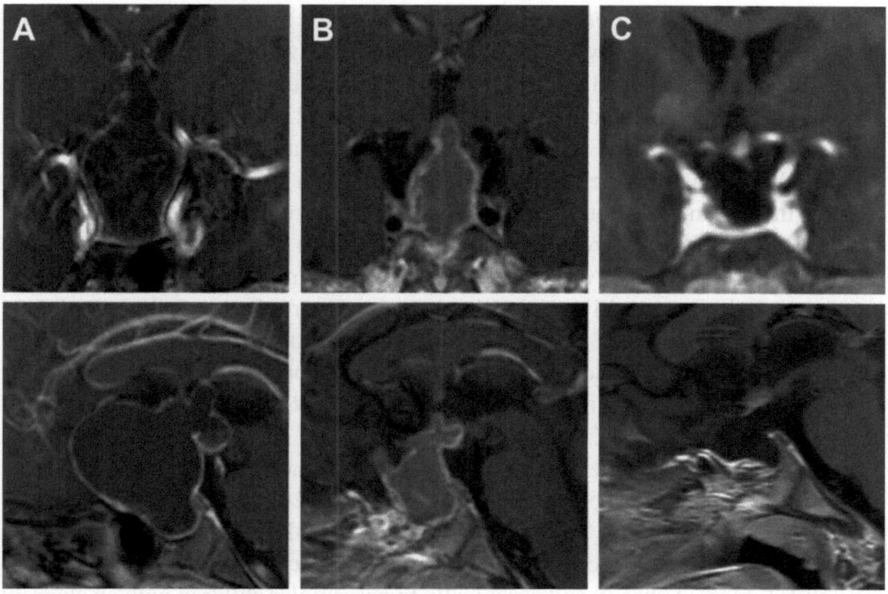

Fig. 1 Staged therapy in a 10 year-old boy presenting with a large cystic craniopharyngioma (**A**). He underwent transsphenoidal decompression, with improvement in vision. The residual craniopharyngioma was evident on an MRI scan at three months post-operatively (**B**). This was subsequently resected via a right fronto-temporal craniotomy with very small tumor residuals which where adherent to the hypothalamus three months postoperatively (**C**)

much of the capsule and reduction of the solid component through the endoscopic approach, has also been described [16]. However, consideration should be given to the fact that residual cysts often enlarge during radiotherapy, and an implanted catheter-reservoir system facilitates repeated aspiration without interruption of the radiotherapy programme [17]. In a similar way, cysts presenting low in the sellar and suprasellar region may be drained transsphenoidally and subsequently evaluated for definitive management (Fig. 1).

Definitive Surgical Management

Current management of craniopharyngiomas is based upon an individualised risk-adapted strategy, where the extent of resection is tailored to the degree of hypothalamic involvement, as determined clinically and on pre-operative MR imaging. Although complete resection leads to optimal long-term disease control, the benign nature of these tumours and high survival rates mandate that low hypothalamic morbidity remains an important objective. The rarity and complexity of these tumours also requires that they be managed in institutions with enough experience and volume.

In the early series, the primary treatment objectives were preservation of vision and complete tumour removal [18–21]. Awareness of a significant number of recurrences despite documented complete resections, as well as insight into hypothalamic morbidity, led to a more conservative approach [22]. Accurate replacement of deficient pituitary hormones did not necessarily lead to a 'normal' life. In one of the early studies discussing the role of aggressive surgical management, 45 out of 50 children underwent gross total resection [18]. 34% recurred at a mean of 32.6 months. In addition, out of 28 children who were followed up in the long term, 16 had some impairment of memory, 24 were obese and 24 attended regular school, some requiring special help for cognitive and vision reasons. Over a 7-year follow-up, Sainte Rose et al. observed a 36% recurrence rate after previously confirmed 'gross total resection' [22]. In 70 children undergoing an attempt at complete resection by Yasargil et al., good outcomes were only reported in 60% [23]. In a series of 37 children treated at Boston Children's Hospital between 1972 and 1981, none of the four patients who had undergone radical excision of their tumour and reached the end of high school, were able to work independently [24]. In another study involving long-term follow-up of 75 patients, 59 of whom underwent complete resection, 56% demonstrated evidence of hypothalamic dysfunction and in 16% this was severe [21]. Morbid obesity and chronically hyperosmolar children with an aberrant sense of thirst were seen almost exclusively after radical surgery.

Acknowledgement of long-term survival, increased understanding of the cost of hypothalamic morbidity, an awareness of a significant risk of tumour recurrence even after apparently complete resection, and an appreciation of the efficacy of radiotherapy at local control have led to the development of a tailored risk-adapted individualised strategy for this tumour. The relatively unknown long-term effects of radiotherapy, and the potential risks of managing recurrent tumours, dampen the enthusiasm for a universally minimal approach [22]. Most experienced units currently consider limited surgery only for those high risk cases where hypothalamic involvement is evident from pre-operative imaging and examination [9, 25, 26].

Operative Approaches

Selection of Operative Approach

The optimal choice of approach to adamantinomatous craniopharyngiomas remains controversial. However, there are a few rules of thumb, which are widely accepted by expert surgeons. Tumours of different size, location and extension require different strategies and individualised treatment concepts. As discussed above, there is unique agreement that a child who clearly exhibits signs of hypothalamic disturbance pre-operatively is not a candidate for extensive procedures such as attempts at total or subtotal excisions. These patients should only undergo procedures that are as minimally invasive as possible, and restricted to measures that decompress and clarify diagnosis.

The authors consider any patient with an intra- and suprasellar craniopharyngioma that is similar in location and shape to a pituitary adenoma an ideal candidate for transsphenoidal surgery; this is irrespective of whether a microscopic or endoscopic approach is used. The degree of sellar enlargement is the most decisive criterion on the suitability of a transsphenoidal approach. In contrast to the conventional, relatively conservative, indication for selection of the transsphenoidal route, those who favour extended endoscopic approaches see almost no limits to using this technique even for large lesions that are exclusively outside the sella. Enlargement of the sella by drilling the skull base bone and incising the dura until sufficient space is available allows dissection and resection of tumours located entirely outside the sella. The classification of Kassam et al. helps to find the ideally suitable patient, provided that sufficient experience with extended transsphenoidal approaches is available in the respective neurosurgical centre [27].

Most surgeons consider a fronto-temporal craniotomy and a pterional approach suitable for most suprasellar craniopharyngiomas, as long as their posterior extent does not expand beyond the clivus behind the dorsum sellae. For those who have a considerably retrosellar component, a fronto-basal midline approach, which requires dissection of the olfactory nerves, allows sufficient visualisation of the blood vessels of the posterior circulation, at least once the lamina terminalis is opened. Moreover, this approach is also suitable for tumours that have significant extension within the third ventricle. Transventricular approaches to the third ventricle are most appropriate for tumours which are entirely located within the third ventricle. Whether a lateral transventricular approach is used or a transcallosal paramedian route is selected depends mostly on the size of the foramen of Monro. However, every surgeon has his own preference and his own experience with the various approaches, and one usually selects a technique which one feels most comfortable with because of familiarity with the operative view.

Cyst punctures and tumour biopsies can be equally performed with frame-based or frameless navigation-guided procedures. Intraoperative MRI facilitates all surgical techniques, by demonstrating the extent of decompression or resection, and updating the image-guidance data set (Fig. 2).

Pterional Approach

The pterional approach is the most frequently used approach in some large series [23, 28]. This represents the shortest route to the suprasellar region and allows dissection of the tumour through the opticocarotid triangle and the prechiasmatic space. The lower and anterior third ventricle are accessed through the lamina terminalis. The primary limitations of this approach relate to poor visualisation of a retrosellar tumour component and the small perforating arteries within the opticocarotid triangle [28]. This approach is also particularly difficult in the context of a prefixed chiasm, or if the tumour extends posteriorly into the third ventricle beyond its anterior third. The approach remains excellent for small craniopharyngiomas; the

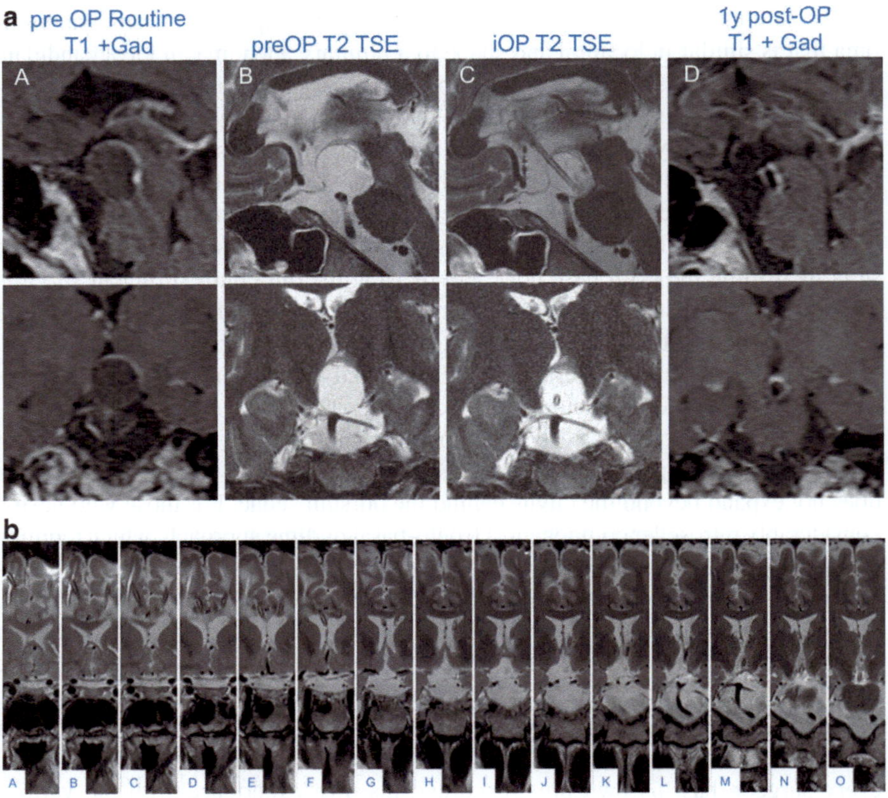

Fig. 2 (**a**) 29 year-old male patient presented with an increasing cystic remnant 10 years (September 2009) after bifrontal craniotomy for a adamantinomatous craniopharyngioma. Preoperative contrasted T1 and T2 images are shown in (*A*) and (*B*). A cysto-ventricular drain was inserted and its tip position was controlled on intraoperative MR-imaging (*C*). The cyst was aspirated successfully, as demonstrated on a post-contrast T1 MRI scan one year later (*D*). (**b**) Standard post-operative serial coronal T2-weighted imaging (2 mm distance) demonstrates the catheter running through the left lateral ventricle (*F–J*) with the tip ending in the cystic residual craniopharngioma (*N* and *O*) above the pons. It serves as a continuous drainage of the content of cystic residual into the lateral ventricle. Additionally a right sided ventriculoperitoneal shunt is visible, (*A–D*); this was implanted prior to the bifrontal craniotomy due to obstructive hydrocephalus

subfrontal or anterior interhemispheric approach is preferable for larger lesions, particularly if there is a retrosellar component.

Early decompression of the tumour facilitates dissection at the plane between the capsule and the arachnoid. Division of the falciform ligament and opening the optic foramen protects the ipsilateral optic nerve during dissection. Residual tumour is most likely located below the ipsilateral nerve and this area may be inspected with a 'dental' mirror towards the end of the procedure.

Preservation of the pituitary stalk is challenging as it often blends into the posterior capsule of the tumour, often eccentrically to one side. As the tumour is debulked and the capsule is freed off Liliequist's membrane posteriorly, the stalk may become visible as it courses towards the infundibulum behind the chiasm.

Subfrontal and Interhemispheric Midline Approach

In the bifrontal interhemispheric approach, a low craniotomy, involving the frontal sinus, allows minimal basal frontal retraction. Frontal lobe trauma has been shown to be a significant predictor of morbidity, both immediately after surgery and at long-term follow-up [21]. The olfactory nerves are microsurgically freed from the basal frontal lobes [28]. Opening the chiasmatic and interhemispheric cisterns, allowing time for CSF to egress and the frontal lobes to relax, allows exposure of the optic nerves, chiasm, proximal optic tracts, lamina terminalis, anterior communicating artery and both A2 vessels. Initial tumour dissection begins in the prechiasmatic region. Division of the lamina terminalis allows approach to the retrochiasmatic tumour. This is usually distended by tumour and can be divided safely. Division of the capsule and internal decompression of the tumour is followed by careful dissection of the capsule from the reactive gliotic layer or arachnoid membranes; perforating vessels must be protected. This exposes the retrosellar space, and allows dissection of the tumour from the premesencephalic space as well as the basilar artery and proximal posterior cerebral arteries. Dissection within the retrochiasmatic space carries a slightly increased risk of visual deterioration through manipulation of the optic chiasm [28].

Frontolateral Approach

The frontolateral approach represents a combination of the pterional and the unilateral subfrontal craniotomy. It offers a short distance to the suprasellar region and allows good visualisation of the anatomy within the basal cisterns. It allows early decompression through CSF release from the basal cisterns, reduces risk to the olfactory nerves and minimises injury to the basal frontal lobes which results from bifrontal retraction in the bilateral subfrontal approach. The head is fixed in extension and rotated 30 degrees to the opposite side [29]. After release of CSF from the basal cisterns and internal decompression of the tumour, the anatomical structures are recognised and their relationship to the tumour is evaluated.

Using this approach, tumour below the ipsilateral optic nerve often represents the most difficult dissection. Opening of the opticocarotid triangle facilitates complete removal in this region. When the chiasm is post-fixed and the optic nerves are long, dissection of the suboptic component of the tumour can be accomplished through the interoptic and opticocarotid spaces. When the chiasm is prefixed dissection through the lamina terminalis is required.

The pituitary stalk may be identified as a thin, vascular structure attached to the capsule of the tumour and the diaphragma sellae, running behind the chiasm towards the infundibulum. It may be possible to dissect tumour from it.

Careful inspection of the tumour bed with the operating microscope is the principal method of verification of the extent of resection [29]. Endoscopic inspection and intraoperative MRI are useful adjuncts.

Transventricular and Transcallosal Approaches

These approaches are useful for craniopharyngiomas that extend into, or are almost entirely within, the upper third ventricle towards the foramina of Monro. In selected cases, this approach may avoid the need to violate the lamina terminalis or floor of the third ventricle, potentially preserving pituitary function. Neuronavigation is used to select and plan the optimal trajectory in relation to the position of the tumour.

Some authors prefer the transcortical route in the presence of large ventricles [30]. Resection can be achieved through a 1–2 cm corticotomy. The third ventricle is explored through the foramen of Monro. After opening the tumour capsule the core of the tumour is debulked in piecemeal fashion. A cottonoid placed in the posterior third ventricle prevents egress of blood towards and into the aqueduct. Caution needs to be exercised when peeling the capsule from the walls and floor of the third ventricle; the authors prefer to leave residual capsule if it appears tightly adherent on exploration [30]. If the tumour extends into the posterior third ventricle beyond the trajectory through the foramen of Monro, a 30 degree endoscope is useful to evaluate the extent of resection and facilitate additional resection. The primary concern with this approach remains a small risk of causing epilepsy, recently estimated at around 8% [31].

If the ventricles are small, the transcallosal approach is preferred. The tumour is dissected and resected through the foramen of Monro unilaterally; this avoids potential bilateral injury to the fornices. The foramen may be extended posteriorly by dividing the anterior septal vein and the tela choroidea, allowing access to the posterior half of the third ventricular cavity [32].

Transsphenoidal Approaches

Craniopharyngiomas typically extend to various levels from the sella through the suprasellar space posterior to the optic chiasm. This suggests that they are excellent candidates for the transsphenoidal approach, as their long axis lies along the axis of visualisation [33]. Formal comparison between transsphenoidal and transcranial approaches is difficult, as tumour morphology is highly variable and the surgeon's personal experience and preference is particularly relevant. Published meta-analyses have reached different conclusions. One study concluded that transsphenoidal approaches have led to higher rates of complete resection and lower recurrence rates, as well as lower rates of post-operative diabetes insipidus [34]. Another study restricted to paediatric craniopharyngioma concluded that although transsphenoidal surgery is associated with higher resection rates and lower morbidity, baseline differences lead to preferred selection of one or other approach and generate bias which may itself account for the differences [35]. The traditional microsurgical approach has more recently been supplemented by the endoscopic approach as well as the development of extended transsphenoidal and cranial base approaches. It remains the belief of some authors that microscope-guided transsphenoidal surgery allows higher stereoscopic visualisation and maintains better surgical maneuverability and tactile feedback when dissecting delicate structures off a tough tumour capsule [36].

The microscopic transsphenoidal approach proceeds in distinct phases, from the nasal to the sphenoidal and sellar stage. After wide exposure of the sellar floor, the bone is resected to the planum sphenoidale anteriorly and to the cavernous sinuses laterally. The dura is opened and the pituitary gland may be mobilised inferiorly or laterally as necessary. For intrasellar craniopharyngiomas, tumour dissection and resection is begun inferiorly and laterally, allowing the more superior component to descend [37]. For suprasellar lesions, division of the diaphragm at the tuberculum allows visualisation of the tumour capsule behind the arachnoid. After tumour cysts are decompressed, the capsule can be carefully dissected from the optic nerve, chiasm and stalk. The interface with the hypothalamus needs very cautious dissection [37]. In the final phase, the CSF barriers are reconstructed, usually with a fat graft harvested from the abdominal wall and a pedicled nasoseptal flap based on the sphenopalatine artery [27].

The development of endoscopic and extended skull base approaches has led to a higher use of the transsphenoidal route for craniopharyngiomas beyond the sellar region. With the endoscopic approach, the opening into the skull base is tailored to the extent of the tumour, and may include not just the pituitary sella but also the tuberculum, the planum sphenoidale and the posterior ethmoidal air cells [38]. The dura is opened in the midline from the posterior aspect of the planum sphenoidale to a point anterior to the pituitary gland, through the coagulated intercavernous sinus. On opening the arachnoid, the tumour capsule is seen anterior and superior to the pituitary gland and below the chiasm. The capsule is opened and systematic internal tumour debulking is carried out. Cystic components are drained. After tumour decompression, the plane around the capsule is explored. The capsule may be tightly adherent to neurovascular structures, including the optic nerves and the hypothalamus. If it is not possible to develop an adequate plane, it would be best to leave parts of the capsule behind and avoid neurological injury. A combination of angled endoscopes, including 30, 45 and 70 degree scopes, are useful to inspect the tumour cavity and achieve maximal safe resection. As in the microscopic transsphenoidal approach, a CSF leak is averted by combinations of a fat and/or fascial graft and a nasoseptal flap, harvested in the nasal phase of the procedure.

Results

The variation in size and extent of craniopharyngiomas, as well as the multiple surgical options and lack of standardised clinical trials, render a study of surgical outcomes difficult. This is complicated by the fact that in most publications results pertaining to adamantinomatous and papillary craniopharyngiomas are not presented separately. Nevertheless, a number of meta-analyses have attempted to comprehensively present the important outcome measures related to local control as well as endocrine and visual outcome.

A large meta-analysis evaluated the efficacy of local tumour control with various treatment strategies [39]. From a total of 442 patients, reported on from 1990, including adult and paediatric age groups, who underwent tumour resection, gross total resection (GTR) was achieved in 256 patients and subtotal resection (STR) in 186. In 85 patients STR was followed by adjuvant radiotherapy. Two and five-year progression free survival (PFS) were similar for the GTR and STR and radiotherapy group (81 and 91% at 2 years, respectively, and similarly 67 and 66%). The five and 10-year overall survival (OS) rates were also similar (98 and 99% at 5 years, and 98 and 95% at 10 years). The authors conclude that the addition of radiotherapy to STR is a 'reasonable' approach to achieving local control while limiting the well-reported hypothalamic morbidity associated with radical resection.

In another systematic review restricted to paediatric craniopharyngioma a total of 109 studies detailed extent of resection and long-term follow-up, with recurrence data available for a cohort of 377 patients [40]. The study found no significant difference in 1 and 5 year PFS between those children who underwent GTR (89% and 77%, respectively) and those who underwent STR combined with radiotherapy (77% and 73%, respectively). STR alone was significantly worse than STR with radiotherapy; one-year PFS was 84% for STR and radiotherapy and 76% for STR alone; 5 year values were 73% and 43%, respectively [40]. Twenty-two patients in the study underwent biopsy only followed by intracystic chemotherapy; there was no difference in PFS between these patients and those who underwent STR and radiotherapy.

The importance of adjuvant radiotherapy has been emphasised in several publications. Iannalfi et al. conducted a large literature review including 43 studies and 1716 patients between 1990 and 2012 who underwent radiotherapy for craniopharyngioma [41]. The reported 10 year local control rates varied from 77% to 100% and the long-term toxicity of combined limited surgery and radiotherapy was less than that of radical surgery [41].

Although meta-analyses provide a good overall statistical overview, they do not reflect the heterogeneity in treatment philosophies evident in the published series. STR may mean different things to different surgeons, and is likely to represent a more extensive resection for surgeons who aim for a GTR [39]. The data is derived from series of self-reported outcomes and it is impossible to control for biases at meta-analysis level [39, 42, 43].

Prospective multicentre studies remain useful. Recent publications have reported on three-year follow-up of 117 pediatric patients recruited prospectively across centres in Germany, Austria and Switzerland [44, 45]. The reported three-year overall and event free survival were 97% and 46%, respectively; recurrence rates after complete resection were high, up to 36%, in 3 years [44]. Children who underwent irradiation reduced their risk of progression or recurrence by 88%. This study also observed a trend towards reduced surgical aggressiveness throughout its recruitment period. Children recruited within the second half of the study demonstrated far fewer hypothalamic complications (13% vs 35%) than those from the earlier half [45].

A recent study underlined the impact of minimising hypothalamic injury to overall outcomes in 65 children and compared them to a historical cohort of 59 children treated between 1985 and 2002 [46]. Surgical resection was complete in 60% of controls while 37% of children, recruited prospectively, underwent 'hypothalamus-sparing' surgery. Radiotherapy was given to 20 controls and 34 prospective patients. Post-operatively, patients who underwent 'hypothalamus-sparing' surgery had lower rates of obesity and endocrinopathies, particularly for corticotropic deficiency, gonadotropic hypogonadism and diabetes insipidus; there was no difference in growth hormone or thyroid stimulating hormone deficiency. The authors report no difference in the recurrence rate between the two groups, but acknowledge significantly shorter follow-up in the prospective group [46].

In a similar Canadian study, three cohorts of children undergoing treatment for craniopharyngioma between 1975 and 2011 were compared [47]. The frequency of GTR decreased over this period, whereas that of limited surgery, adjuvant radiotherapy and intracystic chemotherapy increased. Recurrence rates were not significantly different across the three time periods, changing from 34% and 30% in the first two decades to 52% in the last decade. There was, however, a significant reduction in the rate of diabetes insipidus (69% and 79% in the first two decades and 49% in the last decade) and ACTH deficiency (89% and 93% in the first two decades and 64% in the last decade). The prevalence of severe obesity decreased by 36%, although this was not statistically significant [47]. Similarly, in a combined review involving 122 adult and paediatric patients, the rate of local control with STR and radiotherapy was similar to those undergoing aggressive resection but the clinical outcome was superior [48].

Visual impairment is evident on presentation in more than half of children with craniopharyngioma [49]. Up to 50% of these improve post-operatively [35]. In a large meta-analysis that included adult patients, visual decline was reported in 3.7% of patients undergoing resective surgery and up to 8.6% in those who underwent radiotherapy alone [43]. This study also showed a statistical trend towards worse visual outcome in patients receiving radiotherapy after resection, compared to those undergoing resection, whether gross or subtotal, alone.

Children who have significant visual impairment on presentation and those with prechiasmatic tumour have a worse visual prognosis [35]. Mixed adult and paediatric series have demonstrated that visual prognosis is better in children [28]. In this series, post-operative visual deterioration did not occur in any of the 21 children who underwent transcranial surgery; however, across the whole series, this occurred in up to 15%. In this large series, visual outcome was best after transsphenoidal surgery [28]. This approach allows decompression of the tumour from below before any manipulation of the optic apparatus [50]. Elliott et al. found that in 86 children treated transcranially, the commonest new visual deficit, almost always transient, was contralateral homonymous hemianopia, related to dissection of the retrochiasmatic tumour component from a pterional approach [35].

Complications

Surgical complications are primarily related to the suprasellar location of craniopharyngiomas and include hypothalamic, vascular, cranial nerve, pituitary gland and stalk, as well as brainstem, injury. Although most complications are transient, craniopharyngioma and its treatment has a significant impact on children's quality of life. Children with craniopharyngioma rated their health-related quality of life as considerably lower than healthy controls; the domains of social and emotional functioning were particularly affected [51]. Parents rated the impact of the disease higher than the children themselves.

Older publications describing radical surgery have underlined the consequence of tumour adherence to the carotid artery and its major branches; in a study of 75 children who underwent radical surgery between 1973 and 1994, intraoperative arterial bleeding, usually from the carotid bifurcation, was reported in nine [21]. Profuse venous haemorrhage occurred in six patients. However, in a large meta-analysis, vascular injury resulting in cerebral infarction occurred in only two of 540 cases [43]. In this review, new neurological deficits occurred in 5.1% of patients undergoing surgery. Fusiform dilatation of the carotid artery may occur when adherent tumour tissue is dissected off the vessel, causing adventitial injury [52]. None of the nine patients who developed carotid fusiform dilatation developed haemorrhagic complications and the author recommended careful surveillance only.

Endocrine deficiencies are common after treatment of craniopharyngioma, and some authors argue that a degree of endocrinopathy is unavoidable after craniopharyngioma treatment [53]. Transient diabetes insipidus occurs in 80 to 100% of cases post-operatively; the rate for permanent diabetes insipidus has been quoted at between 40 and 93% [54]. The most prevalent post-operative endocrine deficiency is growth hormone, followed by TSH and ACTH [28]. Growth hormone deficiency is common at diagnosis. In a meta-analysis, the overall rate of new endocrinopathy for all patients undergoing surgical resection was 37%, with anterior panhypopituitarism reported in up to 12% of patients undergoing GTR [43].

Elliott et al. compare the incidence of pituitary dysfunction following transcranial and transsphenoidal surgery in children [35]. In their review, only 36% of patients treated transsphenoidally developed DI, in comparison to 69% of those treated transcranially. They attribute this difference to better visualisation of the pituitary stalk through the transsphenoidal approach and to the distal involvement of the stalk in tumours selected for transsphenoidal surgery. Even when, during transcranial surgery, the integrity of the stalk is maintained, the rate of diabetes insipidus is still high; the authors therefore conclude that stalk preservation should not preclude GTR [35]. Similarly, anterior pituitary injury was commoner in transcranial compared to transsphenoidal approaches.

In one of the largest published meta-analyses, 540 patients underwent surgery, with gross total resection (GTR) in 289 and subtotal resection (STR) in 251 [43]. An additional 138 patients received biopsy only, followed by radiotherapy. Over a mean follow-up of 54 months, the overall rate of new endocrine disturbance was 37% in patients undergoing surgical resection. Patients receiving GTR had over 2.5 times the rate of developing at least one endocrinopathy compared to patients having STR alone or STR with radiotherapy. Even after controlling for the extent of hypothalamic involvement on multivariate analysis, GTR conferred a significant increase in the risk of endocrinopathy compared to STR with radiotherapy, with an odds ratio of 3.54 [43]. It was also associated with an increase in neurological deficit, with an odds ratio of 5.05. This study also demonstrated a statistical trend towards worse visual outcome in patients receiving radiotherapy after STR compared to STR or GTR alone. The authors express concern that a larger study would have been likely to show a significant visual risk with radiotherapy and consider recommending a therapeutic strategy in which surgery aims to increase the distance between the tumour residuum and the optic apparatus. This study also found a significant reduction in most complications, neurological and endocrinological with increasing size of the series.

In a large meta-analysis restricted to paediatric craniopharyngioma, reviewing a total of 531 patients across 109 studies, GTR was associated with higher risks of new endocrine dysfunction (OR 5.4), panhypopituitarism (OR 7.8) and new neurological deficit (OR 9.9) when compared with biopsy procedures [40]. STR was also associated with new endocrine deficits compared to biopsy (OR 4.6), but not increased risk of diabetes insipidus or new neurological deficit. GTR is also associated with increased risk of post-operative diabetes insipidus compared to STR with radiotherapy or STR alone. This is presumably related to the surgical dissection incurred in attempting a complete resection. The addition of adjunctive radiotherapy to STR increases the risk of panhypopituitarism compared with STR alone (OR 9.1); there are no other significant differences related to the addition of radiotherapy [40].

Hypothalamic dysfunction remains one of the most important complications of craniopharyngioma surgery and has been one of the primary causes of the shift in surgical treatment paradigms over the last 30 years [22]. It is manifested by obesity, excessive daytime sleepiness and disturbance of circadian rhythm [6]. Other effects include antisocial behaviour, depression and other degrees of psychosocial problems [55]. There may also be changes in temperature regulation, blood pressure and heart rate [56]. About a third of children with craniopharyngiomas demonstrate hypothalamic disturbance on presentation [4]. Factors associated with high risk of hypothalamic injury include hypothalamic symptoms at presentation, large tumour size, extensive involvement of the third ventricle, the presence of obstructive hydrocephalus and the pursuit of radical resection in tumours adherent to the hypothalamus [8, 57]. Surgical resection posterior to the mammillary bodies is an additional factor [56].

Children who sustain hypothalamic damage after surgery for craniopharyngioma gain weight rapidly over the first 6–12 months post-operatively [58]. This is related both to abnormal energy management, mediated by the hypothalamus, and reduced levels of activity [45]. Parasympathetic innervation of adipose tissue is mediated by the periventricular and suprachiasmatic nuclei in the hypothalamus;

disturbance of this system leads to increased lipogenesis [6]. Obese children with craniopharyngiomas do not have a higher calorie intake when compared to other children of similar body mass index [59]. Reduction of melatonin secretion by a damaged suprachiasmatic nucleus affects daytime activity and reduces energy expenditure; this has been reversed by melatonin administration [56]. Hypothalamic obesity is poorly responsive to diet, physical activity and most pharmacological therapies. However, in a recent study, glucagon-like peptide 1 analogues were successful at causing sustained weight loss in obese adult patients with hypothalamic obesity [60].

When reviewing the quality of life of these children it is difficult to distinguish the effects of the treatment from those of the tumour itself. Patients typically rate their psychosocial health below their physical health, with particular difficulties related to social and emotional functioning [51]. Hypothalamic dysfunction had profound effects on physical ability, social functioning and body image [51, 58]. In a study of 16 children who underwent surgery without radiotherapy, the overall level of cognitive functioning was within the normal range, but their ability to retrieve learned information was poor [61]. This was evident in both the language and visuospatial domains. In a similar study of children who underwent gross total resection, retrochiasmatic tumour location was associated with reduced psychosocial quality of life, social-emotional and behavioural functioning [62]. This tended to improve as post-operative time increased. In a study comparing children who underwent radical surgery with others who had limited surgery and radiation therapy, the surgery group lost a mean of 9.8 points in full scale IQ compared to 1.25 in the combined modality group; this, however, was not statistically significant [63]. The authors conclude that limited surgery and radiotherapy cause lesser or comparable sequelae. The only significant difference was the rate of post-operative diabetes insipidus at 73% in the radical surgery group compared to 33% in the limited surgery and radiation therapy group [63].

Recurrent Craniopharyngioma

Craniopharyngiomas that recur or regrow after initial treatment are challenging. Most children would have already incurred complications related to the tumour and its primary treatment. Management of recurrent craniopharyngiomas requires a concerted multidisciplinary effort, as well as full discussions with the child and family to define the risk of treatment and to determine the additional complications they may be willing to tolerate.

Up to 80% of recurrent craniopharyngiomas present with new symptoms and are identified on imaging [64]. About half develop headaches or new visual deficits. In this study reviewing 54 adult patients treated for recurrent craniopharyngiomas, only 30% were identified on routine surveillance imaging [64].

The recurrence rates after GTR and after STR with adjuvant radiotherapy are similar [35, 39, 63]. Risk factors for recurrence include large or heavily calcified

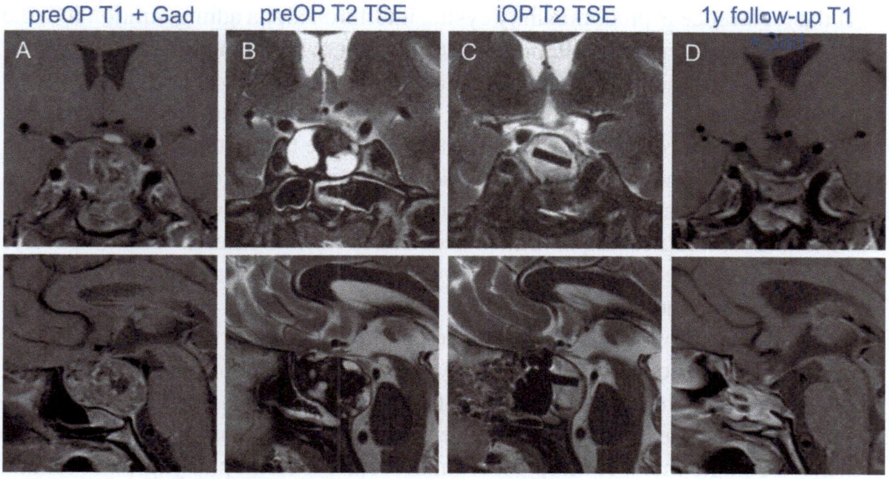

Fig. 3 Progressive adamantinomatous craniopharyngioma in a 50-year-old male patient (after two transcranial resections 1989 and 1995)

tumours, those that extend beyond the suprasellar compartment and presentation with hydrocephalus or visual disturbance [28, 64, 65]. In a paediatric series where radiotherapy was only administered after surgical resection and subsequent regrowth, the extent of the first surgical resection had a significant impact on time to regrowth [66]. Primary treatment strategies involving STR or cyst aspiration without radiotherapy, or radiotherapy alone, do not provide the same level of protection from recurrence than GTR or STR with adjuvant radiotherapy [28, 53, 65].

Treatment options for recurrent craniopharyngiomas include further surgery, further radiotherapy, stereotactic radiosurgery and intracystic treatment, alone or in combination (Fig. 3). Surgery should be considered whenever possible, particularly if the recurrence is predominantly solid [67]. Previous surgery or radiotherapy affects the plane between the tumour and surrounding neural and vascular structures, rendering revisional surgery more difficult and increasing the risk of complications [53, 67]. If the same approach is used, adhesions and scar tissue need to be dissected; this is particularly true for the subfrontal approach. The pterional approach may still leave some previously unopened routes, such as the sylvian fissure. The transsphenoidal approach is useful for recurrences in the sella or along the anterior skull base.

In a study from St Jude Children's Hospital, evaluating 18 out of 97 patients who sustained tumour progression following limited surgery and radiotherapy, the median time to progression was 4.6 years [68]. The five and ten year treatment free survival rates for the whole cohort were 89% and 76.3%, respectively. Seven patients had a single episode of progression and 11 had more than one; moreover, the time interval between episodes of progression became shorter. The authors conclude that decompression of new cysts alone has a low rate of long-term success, and optimal management involves an attempt at GTR at the first recurrence [68].

If the recurrence is predominantly cystic, interferon alpha administration into the cyst may be considered. Instillation of radiological contrast medium into the catheter confirms the absence of leakage and lack of communication of the cyst with the subarachnoid space. Interferon alpha 2A is currently considered the safest agent, with low risk of neurological injury or irritation outside the cyst wall [69]. It is given for at least three times a week for 4 weeks in the first instance. In the largest prospective paediatric study reported to date, 60 children underwent treatment [69]. Clinical and radiological improvement occurred in 76% of cases, with mild adverse effects, mostly related to headache and eyelid oedema, occurring in a third. A mean of five cycles were administered.

Stereotactic radiosurgery has also been employed for recurrent craniopharyngioma. The main difficulty is the proximity of these tumours to the optic apparatus. However, in as much as these are well-defined, contrast-enhancing lesions, with no intracranial dissemination, and no concern about change of shape or size in the context of a single therapeutic episode, small recurrent craniopharyngiomas remain a suitable target for this modality [70]. In a large series involving 137 patients, 54% demonstrated shrinkage of the tumour and 16.5% showed continued tumour growth; in 8% the cysts enlarged [71]. Local control was worse for cystic or mixed tumours compared to solid tumours, and for larger tumour volumes [71].

Reported outcomes for the treatment of recurrent craniopharyngioma vary widely. The likelihood of achieving GTR is lower and the risks to vision, endocrine function and overall functional outcome is lower [53, 64]. In addition, second recurrences occur in 14 to 33% of patients after reoperation for a recurrent tumour [53, 64].

Conclusion

Craniopharyngiomas remain challenging tumours. Their long-term survival and slow growth mandates a surgical approach that preserves as much function as possible, ensuring good quality of life in the long term. The changes in surgical attitudes that have occurred over the last 40 years reflect this objective. The difficult balance between safe surgery and long-term local control require individualised patient care and multidisciplinary expertise.

References

1. Barkhoudarian G, Laws ER (2013) Craniopharyngioma: history. Pituitary 16:1–8
2. Muller HL (2008) Childhood craniopharyngioma. Horm Res 69:193–202
3. Ahmet A, Blaser S, Stephens D, Guger S, Rutkas JT, Hamilton J (2006) Weight gain in craniopharyngioma--a model for hypothalamic obesity. J Pediatr Endocrinol Metab 19:121–127
4. Elliott RE, Wisoff JH (2010) Surgical management of giant pediatric craniopharyngiomas. J Neurosurg Pediatr 6:403–416
5. Mortini P, Gagliardi F, Bailo M, Spina A, Parlangeli A, Falini A, Losa M (2016) Magnetic resonance imaging as predictor of functional outcome in craniopharyngiomas. Endocrine 51:148–162

6. Müller HL (2014) Childhood craniopharyngioma: treatment strategies and outcomes. Expert Rev Neurother 14:187–197
7. DeVile CJ, Grant DB, Hayward RD, Stanhope R (1996) Growth and endocrine sequelae of craniopharyngioma. Arch Dis Child 75:108–114
8. De Vile CJ, Grant DB, Hayward RD, Kendall BE, Neville BG, Stanhope R (1996) Obesity in childhood craniopharyngioma: relation to post-operative hypothalamic damage shown by magnetic resonance imaging. J Clin Endocrinol Metab 81:2734–2737
9. Puget S, Garnett M, Wray A et al (2007) Pediatric craniopharyngiomas: classification and treatment according to the degree of hypothalamic involvement. J Neurosurg 106:3–12
10. Van Gompel JJ, Nippoldt TB, Higgins DM, Meyer FB (2010) Magnetic resonance imaging-graded hypothalamic compression in surgically treated adult craniopharyngiomas determining postoperative obesity. Neurosurg Focus 28:E3
11. Müller HL (2015) Preoperative staging in childhood craniopharyngioma: standardization as a first step towards improved outcome. Endocrine. doi:10.1007/s12020-015-0800-x
12. Müller HL, Gebhardt U, Teske C et al (2011) Post-operative hypothalamic lesions and obesity in childhood craniopharyngioma: results of the multinational prospective trial KRANIOPHARYNGEOM 2000 after 3-year follow-up. Eur J Endocrinol 165:17–24
13. Šteňo J, Bízik I, Šteňo A, Matejčík V (2011) Craniopharyngiomas in children: how radical should the surgeon be? Childs Nerv Syst 27:41–54
14. Mallucci C, Pizer B, Blair J, Didi M, Doss A, Upadrasta S, Newman W, Avula S, Pettorini B (2012) Management of craniopharyngioma: the Liverpool experience following the introduction of the CCLG guidelines. Introducing a new risk assessment grading system. Childs Nerv Syst 28:1181–1192
15. Pettorini BL, Tamburrini G, Massimi L, Caldarelli M, Di Rocco C (2009) Endoscopic transventricular positioning of intracystic catheter for treatment of craniopharyngioma. Technical note. J Neurosurg Pediatr 4:245–248
16. Margetis K, Souweidane MM (2013) Endoscopic treatment of intraventricular cystic tumors. World Neurosurg 79:S19.e1–S19.11
17. Merchant TE, Kun LE, Hua C-H, Wu S, Xiong X, Sanford RA, Boop FA (2013) Disease control after reduced volume conformal and intensity modulated radiation therapy for childhood craniopharyngioma. Int J Radiat Oncol Biol Phys 85:e187–e192
18. Hoffman HJ, De Silva M, Humphreys RP, Drake JM, Smith ML, Blaser SI (1992) Aggressive surgical management of craniopharyngiomas in children. J Neurosurg 76:47–52
19. Caldarelli M, Massimi L, Tamburrini G, Cappa M, Di Rocco C (2005) Long-term results of the surgical treatment of craniopharyngioma: the experience at the Policlinico Gemelli, Catholic University, Rome. Childs Nerv Syst 21:747–757
20. Weiner HL, Wisoff JH, Rosenberg ME, Kupersmith MJ, Cohen H, Zagzag D, Shiminski-Maher T, Flamm ES, Epstein FJ, Miller DC (1994) Craniopharyngiomas: a clinicopathological analysis of factors predictive of recurrence and functional outcome. Neurosurgery 35:1001–10 discussion 1010–1
21. De Vile CJ, Grant DB, Kendall BE, Neville BG, Stanhope R, Watkins KE, Hayward RD (1996) Management of childhood craniopharyngioma: can the morbidity of radical surgery be predicted? J Neurosurg 85:73–81
22. Sainte-Rose C, Puget S, Wray A, Zerah M, Grill J, Brauner R, Boddaert N, Pierre-Kahn A (2005) Craniopharyngioma: the pendulum of surgical management. Childs Nerv Syst 21:691–695
23. Yaşargil MG, Curcic M, Kis M, Siegenthaler G, Teddy PJ, Roth P (1990) Total removal of craniopharyngiomas. Approaches and long-term results in 144 patients. J Neurosurg 73:3–11
24. Fischer EG, Welch K, Shillito J, Winston KR, Tarbell NJ (1990) Craniopharyngiomas in children. Long-term effects of conservative surgical procedures combined with radiation therapy. J Neurosurg 73:534–540
25. Elliott RE, Sands SA, Strom RG, Wisoff JH (2010) Craniopharyngioma Clinical Status Scale: a standardized metric of preoperative function and posttreatment outcome. Neurosurg Focus 28:E2

26. Müller HL (2016) Risk-adapted treatment and follow-up management in childhood-onset craniopharyngioma. Expert Rev Neurother. doi:10.1586/14737175.2016.1166959
27. Kassam AB, Gardner PA, Snyderman CH, Carrau RL, Mintz AH, Prevedello DM (2008) Expanded endonasal approach, a fully endoscopic transnasal approach for the resection of midline suprasellar craniopharyngiomas: a new classification based on the infundibulum. J Neurosurg 108:715–728
28. Fahlbusch R, Honegger J, Paulus W, Huk W, Buchfelder M (1999) Surgical treatment of craniopharyngiomas: experience with 168 patients. J Neurosurg 90:237–250
29. Gerganov V, Metwali H, Samii A, Fahlbusch R, Samii M (2014) Microsurgical resection of extensive craniopharyngiomas using a frontolateral approach: operative technique and outcome. J Neurosurg 120:559–570
30. Chamoun R, Couldwell WT (2013) Transcortical-transforaminal microscopic approach for purely intraventricular craniopharyngioma. Neurosurg Focus 34:Video 4
31. Milligan BD, Meyer FB (2010) Morbidity of transcallosal and transcortical approaches to lesions in and around the lateral and third ventricles: a single-institution experience. Neurosurgery 67:1483–1496
32. Patel P, Cohen-Gadol AA, Boop F, Klimo P (2014) Technical strategies for the transcallosal transforaminal approach to third ventricle tumors: expanding the operative corridor. J Neurosurg Pediatr 14:365–371
33. Laws ER (2010) Surgical outcome in 90 patients with craniopharyngiomas: an evaluation of the transsphenoidal approach. World Neurosurg 74:254–255
34. Komotar RJ, Starke RM, Raper DMS, Anand VK, Schwartz TH (2012) Endoscopic endonasal compared with microscopic transsphenoidal and open transcranial resection of craniopharyngiomas. World Neurosurg 77:329–341
35. Elliott RE, Jane JA, Wisoff JH (2011) Surgical management of craniopharyngiomas in children: meta-analysis and comparison of transcranial and transsphenoidal approaches. Neurosurgery 69:630–643 discussion 643
36. Ikeda H, Gotoh H, Watanabe K (2012) Outcome of endoscopy-assisted microscopic extended transsphenoidal surgery for suprasellar adult craniopharyngiomas. Front Endocrinol (Lausanne) 3:25
37. Laws ER (1997) Craniopharyngioma: transsphenoidal surgery. Curr Ther Endocrinol Metab 6:35–38
38. Kenning TJ, Beahm DD, Farrell CJ, Schaberg MR, Rosen MR, Evans JJ (2012) Endoscopic endonasal craniopharyngioma resection. J Neurosurg 32(Suppl):E5
39. Yang I, Sughrue ME, Rutkowski MJ, Kaur R, Ivan ME, Aranda D, Barani IJ, Parsa AT (2010) Craniopharyngioma: a comparison of tumor control with various treatment strategies. Neurosurg Focus 28:E5
40. Clark AJ, Cage TA, Aranda D, Parsa AT, Sun PP, Auguste KI, Gupta N (2013) A systematic review of the results of surgery and radiotherapy on tumor control for pediatric craniopharyngioma. Childs Nerv Syst 29:231–238
41. Iannalfi A, Fragkandrea I, Brock J, Saran F (2013) Radiotherapy in craniopharyngiomas. Clin Oncol (R Coll Radiol) 25:654–667
42. Cage TA, Clark AJ, Aranda D, Gupta N, Sun PP, Parsa AT, Auguste KI (2013) A systematic review of treatment outcomes in pediatric patients with intracranial ependymomas. J Neurosurg Pediatr 11:673–681
43. Sughrue ME, Yang I, Kane AJ, Fang S, Clark AJ, Aranda D, Barani IJ, Parsa AT (2011) Endocrinologic, neurologic, and visual morbidity after treatment for craniopharyngioma. J Neuro-Oncol 101:463–476
44. Müller HL, Gebhardt U, Schröder S et al (2010) Analyses of treatment variables for patients with childhood craniopharyngioma--results of the multicenter prospective trial KRANIOPHARYNGEOM 2000 after three years of follow-up. Horm Res Paediatr 73:175–180
45. Hoffmann A, Warmth-Metz M, Gebhardt U, Pietsch T, Pohl F, Kortmann R-D, Calaminus G, Müller HL (2014) Childhood craniopharyngioma – changes of treatment strategies in the trials KRANIOPHARYNGEOM 2000/2007. Klin Padiatr 226:161–168

46. Elowe-Gruau E, Beltrand J, Brauner R et al (2013) Childhood craniopharyngioma: hypothalamus-sparing surgery decreases the risk of obesity. J Clin Endocrinol Metab 98:2376–2382
47. Cohen M, Bartels U, Branson H, Kulkarni AV, Hamilton J (2013) Trends in treatment and outcomes of pediatric craniopharyngioma, 1975–2011. Neuro-Oncology 15:767–774
48. Schoenfeld A, Pekmezci M, Barnes MJ et al (2012) The superiority of conservative resection and adjuvant radiation for craniopharyngiomas. J Neuro-Oncol 108:133–139
49. Müller HL (2014) Craniopharyngioma. Endocr Rev 35:513–543
50. Laws ER, Trautmann JC, Hollenhorst RW (1977) Transsphenoidal decompression of the optic nerve and chiasm. Visual results in 62 patients. J Neurosurg 46:717–722
51. Poretti A, Grotzer MA, Ribi K, Schönle E, Boltshauser E (2004) Outcome of craniopharyngioma in children: long-term complications and quality of life. Dev Med Child Neurol 46:220–229
52. Sutton LN (1994) Vascular complications of surgery for craniopharyngioma and hypothalamic glioma. Pediatr Neurosurg 21(Suppl 1):124–128
53. Elliott RE, Hsieh K, Hochm T, Belitskaya-Levy I, Wisoff J, Wisoff JH (2010) Efficacy and safety of radical resection of primary and recurrent craniopharyngiomas in 86 children. J Neurosurg Pediatr 5:30–48
54. Müller HL (2015) Childhood craniopharyngioma – current status and recent perspectives in diagnostics and treatment. J Pediatr Endocrinol Metab 28:1–2
55. Pierre-Kahn A, Recassens C, Pinto G, Thalassinos C, Chokron S, Soubervielle JC, Brauner R, Zerah M, Sainte-Rose C (2005) Social and psycho-intellectual outcome following radical removal of craniopharyngiomas in childhood. A prospective series. Childs Nerv Syst 21:817–824
56. Müller HL (2016) Craniopharyngioma and hypothalamic injury: latest insights into consequent eating disorders and obesity. Curr Opin Endocrinol Diabetes Obes 23:81–89
57. Curtis J, Daneman D, Hoffman HJ, Ehrlich RM (1994) The endocrine outcome after surgical removal of craniopharyngiomas. Pediatr Neurosurg 21(Suppl 1):24–27
58. Müller HL, Bueb K, Bartels U et al (2001) Obesity after childhood craniopharyngioma--German multicenter study on pre-operative risk factors and quality of life. Klin Padiatr 213:244–249
59. Harz KJ, Müller HL, Waldeck E, Pudel V, Roth C (2003) Obesity in patients with craniopharyngioma: assessment of food intake and movement counts indicating physical activity. J Clin Endocrinol Metab 88:5227–5231
60. Zoicas F, Droste M, Mayr B, Buchfelder M, Schöfl C (2013) GLP-1 analogues as a new treatment option for hypothalamic obesity in adults: report of nine cases. Eur J Endocrinol 168:699–706
61. Carpentieri SC, Waber DP, Scott RM, Goumnerova LC, Kieran MW, Cohen LE, Kim F, Billett AL, Tarbell NJ, Pomeroy SL (2001) Memory deficits among children with craniopharyngiomas. Neurosurgery 49:1053–1057 discussion 1057–8
62. Sands SA, Milner JS, Goldberg J, Mukhi V, Moliterno JA, Maxfield C, Wisoff JH (2005) Quality of life and behavioral follow-up study of pediatric survivors of craniopharyngioma. J Neurosurg 103:302–311
63. Merchant TE, Kiehna EN, Sanford RA, Mulhern RK, Thompson SJ, Wilson MW, Lustig RH, Kun LE (2002) Craniopharyngioma: the St. Jude Children's Research Hospital experience 1984-2001. Int J Radiat Oncol Biol Phys 53:533–542
64. Liubinas SV, Munshey AS, Kaye AH (2011) Management of recurrent craniopharyngioma. J Clin Neurosci 18:451–457
65. Duff J, Meyer FB, Ilstrup DM, Laws ER, Schleck CD, Scheithauer BW (2000) Long-term outcomes for surgically resected craniopharyngiomas. Neurosurgery 46:291–302 discussion 302–5
66. Kalapurakal JA, Goldman S, Hsieh YC, Tomita T, Marymont MH (2003) Clinical outcome in children with craniopharyngioma treated with primary surgery and radiotherapy deferred until relapse. Med Pediatr Oncol 40:214–218

67. Vinchon M, Dhellemmes P (2008) Craniopharyngiomas in children: recurrence, reoperation and outcome. Childs Nerv Syst 24:211–217
68. Klimo P, Venable GT, Boop FA, Merchant TE (2015) Recurrent craniopharyngioma after conformal radiation in children and the burden of treatment. J Neurosurg Pediatr 15:499–505
69. Cavalheiro S, Di Rocco C, Valenzuela S et al (2010) Craniopharyngiomas: intratumoral chemotherapy with interferon-alpha: a multicenter preliminary study with 60 cases. Neurosurg Focus 28:E12
70. Murphy ES, Chao ST, Angelov L, Vogelbaum MA, Barnett G, Jung E, Recinos VR, Mohammadi A, Suh JH (2015) Radiosurgery for pediatric brain tumors. Pediatr Blood Cancer 63:398–405. doi:10.1002/pbc.25831
71. Lee C-C, Yang H-C, Chen C-J, Hung Y-C, Wu H-M, Shiau C-Y, Guo W-Y, Pan DH-C, Chung W-Y, Liu K-D (2014) Gamma Knife surgery for craniopharyngioma: report on a 20-year experience. J Neurosurg 121(Suppl):167–178

Intracystic Administration of Interferon-Alpha for Reduction of Cystic Tumour Burden

John-Paul Kilday and Ute Bartels

Abstract Despite technical advances in the neurosurgical and radiotherapeutic foundations of craniopharyngioma treatment, clinical management remains controversial with persisting high morbidity rates for those affected. As such, there continues to be a scope for therapeutic improvements and innovation in the field. The cystic composition of adamantinomatous craniopharyngiomas (ACPs) makes them an ideal candidate to evaluate one such development; the instillation of intracystic agents. This mode of drug delivery aims to obtain durable cyst shrinkage with minimum consequent morbidity and toxicity for patients. In this chapter we will highlight common agents used to date, paying particular attention to the drug currently gaining interest, interferon-alpha.

Keywords Craniopharyngioma • Adamantinomatous • Intracystic • Therapy • Interferon-alpha

Introduction

Defining the most favourable management strategy for craniopharyngiomas continues to prove a controversial subject of debate amongst neurosurgeons, neuro-oncologists and endocrinologists. Evidence-based support for a particular therapy is often limited by insufficient cohort sizes. While the evolving field of molecular analysis suggests aberrant signalling pathways exist in craniopharyngioma subtypes that could be targeted by novel biological agents in the future [1], translating this into successful personalised therapy for affected patients remains elusive.

Currently, complete surgical resection is often recommended for smaller craniopharyngiomas without associated obstructive hydrocephalus or hypothalamic dysfunction [2] and may offer the best chance of long-term cure. However, disease

J.-P. Kilday
Department of Oncology, Royal Manchester Children's Hospital, Manchester, UK

U. Bartels (✉)
Department of Oncology, The Hospital for Sick Children, Toronto, ON, Canada
e-mail: ute.bartels@sickkids.ca

J.P. Martinez-Barbera, C.L. Andoniadou (eds.), *Basic Research and Clinical Aspects of Adamantinomatous Craniopharyngioma*,
DOI 10.1007/978-3-319-51890-9_9

recurrence after an assumed complete resection has been reported [3–6] and significant post-operative hypothalamo-pituitary axis injury and/or visual decline remain major threats [7, 8]. Alterations in operational approaches (transsphenoidal, transventricular), advances in imaging (intra-operative ultrasound or magnetic resonance imaging (MRI)) and improvements in equipment (endoscopy) have attempted to counteract this risk, achieving varying levels of success. However despite this, post-operative morbidity represents a considerable ongoing concern.

Alternative management strategies combining subtotal resections with additional post-operative therapies have been devised with the aim of preserving vision and maintaining patient quality of life. Unfortunately, this is typically offset by a higher risk of disease progression. Conformal radiotherapy remains the current standard adjuvant therapy for older children, adolescents and adults, with evidence that focal radiotherapy combined with incomplete resection can provide similar progression-free survival rates as a gross tumour removal [9], with potentially less neurological deficits [7]. However, concerns persist regarding the longer-term neurocognitive decline, endocrinological, visual, metabolic, vascular and secondary neoplastic sequelae arising from radiotherapy use, particularly when administered to younger children [6, 8, 10–12]. In addition, the efficacy of systemic chemotherapy as an alternative adjuvant therapy in this scenario has proved limited and disappointing [13, 14].

The toxicity of radiotherapy is of particular importance to the developing brain and thereby paediatric patients who predominantly present with cystic craniopharyngiomas of the adamantinomatous histological subtype. Led by work focussing on the younger age, management strategies for such lesions have shifted in recent years to focus on either repeated cyst aspiration or the instillation of intracystic therapy as a means of delaying or avoiding the harmful effects of radiotherapy or definitive tumour resection with minimal clinical consequences for the patient.

While cyst aspiration is advocated by some groups as a preferred therapeutic option [15, 16], this is not echoed by others as it is not a definitive treatment, its need is unpredictable and therefore no standardised scheduling for use exists [14]. The regularity of the procedure is often dictated by symptomatology which can be unreliable or misdiagnosed in the younger age range, risking significant and potentially irreversible morbidity to the patient before action is taken. As an alternative, research in recent years has aimed to identify an effective and safe intracystic agent for instillation in adamantinomatous craniopharyngiomas. This chapter will particularly focus on paediatric therapies and a relatively novel drug currently gaining most interest, interferon-alpha.

Intracystic Catheter Insertion for Drug Delivery

Insertion of a catheter into a cystic adamantinomatous craniopharyngioma has the advantage over simple cyst fenestration in that, if re-accumulation is observed, repetitive aspiration or the administration of intracystic substances can be attempted

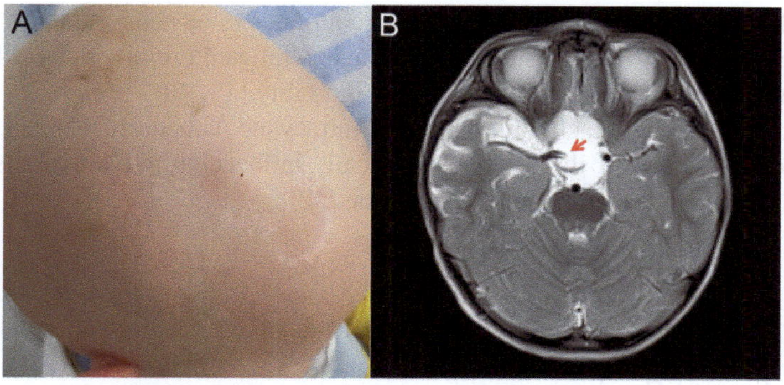

Fig. 1 Ommaya catheter insertion demonstrated in an infant with the access point protruding from the scalp (**a**). T2-axial MRI brain scan of an unrelated craniopharyngioma patient demonstrating catheter tip placement in the cystic component (*red arrow*) (**b**)

(Fig. 1). Placement of the catheter can be achieved using different neurosurgical techniques; the most frequently employed involving open, stereotactic or open approaches [17–20].

A combined multicentre review of such surgical approaches from 1990 to 2000 in France (Paris, Marseille) and Brazil (Belo Horizonte, Goiânia, São Paulo) was undertaken for 50 cystic craniopharyngioma patients [17]. The cohort was predominantly paediatric with complications reported for three distinct patient groups, categorised by surgical technique. The first group (n = 26) underwent catheter placement via a frontal or pterional craniotomy under direct vision. Catheter placement was stereotactic for group two (n = 14), while group three (n = 11) had free hand catheter placement through a preceding intracranial burr hole. All catheters were connected to either a Rickham (n = 23) or an Ommaya reservoir subcutaneously. Post-operative permeability imaging (the injection of radio-opaque contrast into the cyst via a catheter) revealed either catheter misplacement or contrast leakage in eight children (16%). However, this was not significantly associated with a particular group/technique and no additional catheter placement complications were identified.

More recently, neuroendoscopic positioning of intracystic catheters has been advocated as a safer approach that open or stereotactic techniques. An Italian case series of eight patients (five paediatrics) with cystic craniopharyngioma employed this method using a Storz rigid ventriculoscope [18]. No complications were reported except for a technical failure in one child (12%) secondary to a heavily calcified cyst wall. As a consequence, the authors advocate performing a pre-operative computerised tomography (CT) head scan as opposed to the preferred MRI scan as a means of detecting such calcification. Neuroendoscopic positioning also has the advantage of enabling ventricular septostomy in cases of associated hydrocephalus to enable improved cerebrospinal fluid circulation.

The location of a craniopharyngioma and proximity to eloquent structures such as the visual apparatus and hypothalamo-pituitary axis may limit the stereotactic accessibility of a lesional cyst [21], however, combining a stereotactic endoscopic approach for intracystic catheter placement continues to be supported as a safe and effective technique, with two studies reporting operative morbidity rates of less than three percent and no peri- or post-operative mortality [19, 20].

Ultimately, the approach undertaken for intracystic catheter placement, and the subsequent outcome for the patient, is typically influenced by the expertise of the operating surgeon with a particular technique [22].

Historical Intracystic Agents Used in Cystic Craniopharyngioma

The objective behind the instillation of intracystic therapy in craniopharyngiomas is both to reduce tumour burden for the patient through cyst shrinkage and sclerose the secretive elements of the epithelial-lined cyst walls in order to prevent re-accumulation. The substances used should demonstrate efficacy at low toxicity. Prior to the introduction of interferon-alpha, agents historically used for adamantinomatous craniopharyngiomas therapy included several beta-emitting radioisotopes and the chemotherapeutic agent bleomycin (Table 1).

Beta-Emitting Radioisotopes

Over the last 40 years several radioisotopes, particularly phosphorus^{32}, yttrium^{90} and rhenium^{186} have been reported to show efficacy as intracystic therapy in craniopharyngioma [23–26]. The use of these agents is, however, now diminished, limited to a select number of global centres with the appropriate equipment and experience in administration [27–29].

Of the small clinical studies published, no particular radioisotope has produced consistently superior results compared to that of other agents; randomised comparison studies are lacking. It could be that the choice of β-emitting radionucleotide used is influenced by accessibility, since phosphorus^{32} is the only isotope licensed for this purpose in the USA, whereas phosphorus^{32}, yttrium^{90} and rhenium^{186} are available across certain institutions in Europe and East Asia [27–30]. Likewise, the radio-physical attributes of the different radionucleotides may determine preference. For instance, phosphorus^{32} has a longer half-life, emits less energy with a lower resulting tissue penetration than other isotopes [31]. However, yttrium^{90} is favoured by other groups because of its paradoxical higher dose release and penetration over a shorter time period [32].

While tumour responses following administration of β-emitting radionucleotides have been reported (Table 1), a definition of the *degree* of such a response has not been standardised for use across these craniopharyngioma studies, therein leading to potential bias. The durability of such responses is also often lacking and limited

Table 1 Agents other than interferon-alpha used for intracystic administration in adamantinomatous craniopharyngioma—English language series involving paediatric patients

Agent	Author (year)	Number (paeds)	Median/Mean age (range–years)	CR	PR	Median/Mean F/U (range–years)	Median/Mean PFS (range–years)	Complications or toxicity
Radioisotope P^{32}	Pollock (1995) [31]	30 (10)	35 (3–70)	0.1	0.83	3.1 (0.6–9.7)	NR	3—new behavioural and visual problem 3—new onset diabetes insipidus
	Hasegawa (2004) [26]	49 (15)	29 (3–74)	0.17	0.59	4.1	NR	5—worsened pituitary deficiency 3—visual deterioration
	Barriger (2011) [35]	19 (NR)	20 (3–54)	0.05	0.26	5.2 (0.7–11.3)	0.8 (0.1–4.5)	6—worsened pituitary deficiency 1—new visual field deficit
	Kickingereder (2012) [27]	53 (19)	31.1 (7.6–75.5)	0.1	0.5	4.4 (0.3–16.3)	0.8 (0.7–0.9)	10—endocrinological progression 8—neurological deterioration 5—visual field cut deterioration 4—visual acuity deterioration 1—septic meningitis 1—cranial nerve palsy 1—hemiparesis
	Yu (2015) [28]	20—all paeds	Xx (0.2–14)	0.6	0.4	4 (0.6–11.5)	2.8 (0.5–7)	8—other cranial nerve injury 2—optic nerve injury
	Ansari (2015) [33]	9—all paeds	6.7 (3–15)	0.23	0.67	6.7 (1–12)	1.85	1—hyponatraemia
Radioisotope Y^{90}	Voges (1997) [25]	62 (32)	17 (4–71)	0.45	0.35	11.9 (1.5–16.4)	NR	3—amaurosis fugax 3—new endocrinological dysfunction 1— new visual field cut 1—death
	Julow[b] (2013) [29]	78 (26)	28 (2.9–73)	0.2	0.6	NR	NR	59—worsened pituitary deficiency 4—chiasmal injury 2—paramedian infarct (1 death) 1—carotid radiation injury 1—memory loss 1—new onset diabetes insipidus

(continued)

Table 1 (continued)

Agent	Author (year)	Number (paeds)	Median/Mean age (range–years)	CR	PR	Median/Mean F/U (range–years)	Median/Mean PFS (range–years)	Complications or toxicity
Radioisotope R^{186}	Derrey (2008) [30]	42 (11)	38.7 (5–85)	0.44	0.44	3.6 (0.7–12.3)	NR	3—new visual field cut or acuity decline 2—septic meningitis 2—chemical meningitis 2—central hyperthermia 1—intracranial hypertension 1—memory loss
Bleomycin	Hader (2000) [40]	9—all paeds	8.4 (2.5–14)	0.14	0.71	3 (0.5–5)	NR	2—transient headaches and fever 1—panhypopituitarism
	Park (2002) [44]	10 (5)	30.2 (3–65)	NR	NR	2.8 (1–6.6)	NR	1—visual disturbance 1—cerebellar infarction/death 1—hypersomnia/memory loss 1—bedridden 1—mental impairment
	Mottolesse[b] (2005) [43]	24—all paeds	NR (6–16)	0.5	0.25	6.7 (1–14)	NR	11—endocrine deficiency 4—visual decline 3—new onset of diabetes insipidus 1—blindness (after toxic dose)
	Takahashi[b] (2005) [45]	11—all paeds	NR (2–14)	0.27	0.64	NR (3–16)	NR	1—hypothalamic insufficiency/death
	Hukin[b] (2007) [42]	17—all paeds	6 (1–14)	0.29	0.35	5 (0.5–10.2)	0.7 (0.3–6.1)[a]	2—panhypopituitarism 1—cranial nerve palsies and hemiparesis 1—decreased LOC 1—precocious puberty
	Kim (2007) [46]	11 (5)	27.9 (3–67)	0.36	0.64	2.6 (0.8–6.6)	1.4 (0.5–3)	3—transient headache, vomiting and elevation of body temperature
	Hsu (2009) [41]	9—all paeds	7.8 (3.3–11.8)	NR	NR	3.7 (0.8–6.6)	2.4 (0.2–6.6)	NR

Key: *NR* not recorded, *CR* complete response, *PR* partial response, *F/U* follow-up, *PFS* progression-free survival, P^{32} phosphporous[32], Y^{90} Yttrium isotope, R^{186} Rhenium isotope, *LOC* level of consciousness.

aRefers to 12 patients treated at the point of initial diagnosis

bRefers to updated case series encompassing previous, smaller case series by same author. Note: This table is an updated version of that published by Bartels et al. in 2012 [14] to the cystic portion receiving therapy without an effect on any neighbouring cysts or solid tumour components if present [14, 27, 33]. In addition, the dose of isotope used between patients is not consistent, reflecting dose adjustments to match treated cyst volume. There is evidence that cyst volumes exceeding 100 ml will not respond well to this therapy, requiring higher dose and potential increased toxicity [21, 30, 34]. Indeed, visual, endocrine, vascular and infective complications have been reported in as many as 30% of exposed cohorts, sadly including fatalities following certain cases [25–31, 33, 35]. Prospective, randomised-controlled evaluations of efficacy are lacking.

Bleomycin

In 1974, the antitumour antibiotic bleomycin was reported to demonstrate in vitro cytotoxicity against cultured craniopharyngioma cells [36]. This finding, in conjunction with the more widespread clinical availability of bleomycin compared to β-emitting radionucleotides, made the chemotherapeutic agent an attractive alternative candidate for use as intracystic therapy.

Over the following 35 years, several case reports [37–39] and retrospective institutional case series reported the efficacy of bleomycin in inducing sustained craniopharyngioma cyst shrinkage for paediatric patients [40–52] (Table 1). However, as was the case for β-emitting radionucleotides, bleomycin did not appear to influence extrinsic craniopharyngioma cyst or solid component growth [14, 46]. The retrospective analyses undertaken are hindered by small cohort sizes, variable definitions of disease response and inconsistent drug dosing and scheduling. Indeed total bleomycin doses administered to patients ranged from 14.5 to 185 mg in one treatment cycle [44, 46, 50, 53]. Moreover, whilst bleomycin appeared to be generally well tolerated with side effects limited typically to headaches, pyrexia and emesis, concerns of significant toxicity were voiced, likely related to drug leakage from the cyst into surrounding brain structures. Post-therapeutic seizures, hemiparesis [42, 44, 53, 54], panhypopituitarism [40, 42], hypothalamic injuries [44, 55], blindness [52, 54] and mortality [45, 56] were all attributed to bleomycin leakage, in some instances despite verification of correct catheter placement.

In 2012 and 2014, two Cochrane systematic literature reviews of intracystic bleomycin use in childhood cystic craniopharyngioma failed to identify any randomised-controlled trials in which the only differing factor was the instillation of bleomycin [57, 58]. As a result, no conclusion on its efficacy could be drawn and the Cochrane review team was unable to recommend bleomycin as a legitimate treatment in this setting. This finding, together with concerns regarding its toxicity following administration [55], has now rendered intracystic bleomycin obsolete as a therapeutic option for cystic craniopharyngiomas.

Intracystic Interferon-alpha: Biological Rationale for Use

Craniopharyngiomas are thought to arise from rests of pharyngeal epithelium during embryonic development of Rathke's pouch [59]. The secretive squamous epithelium in turn contributes to the formation of cysts containing fluid, epithelial cells and cholesterol-rich crystals [39, 60]. To date, little is known about the biological composition of the cystic fluid extracted from adamantinomatous craniopharyngiomas [61–67] but recent proteomic work has detected an increase in the concentration of anti-inflammatory peptides called alpha-defensins in cyst fluid from untreated children [60]. Explanations for the presence of these alpha-defensins remain elusive, but hypotheses include involvement of the innate immune response system [60], a proposal reinforced by evidence of increased immunoglobulin production in craniopharyngioma cyst fluid [63] and the association of pro-inflammatory Natural Killer (NK) cells with craniopharyngiomas [67].

Interferon-alpha belongs to the large interferon family of cell signalling proteins that are produced and secreted by cells in response to a biological induction from pathogens such as viruses, bacteria and tumour cells [68]. The interferons exert their effect by binding to specific cell receptors, initiating a range of intracellular effects. In tumours, the direct mechanism of action is unclear. It is postulated these actions may include activation of the immunological response system via immune cell activation, cytokine induction and the inhibition of neoplastic vascularisation [69], or direct promotion of cellular differentiation and the inhibition of proliferation through modulating signalling mechanisms such as the JAK-STAT and PI3K pathways [70, 71]. As a consequence, interferon therapy has been used for a range of haematological and solid cancers [72–77].

One such malignancy where interferon-alpha has demonstrated clinical activity is squamous cell carcinoma of the skin [77–79]. Since squamous cell carcinomas and cystic craniopharyngiomas share the same cell of origin, clinical analyses were developed to investigate a potential therapeutic role for this drug in craniopharyngiomas.

A Phase II study of subcutaneous interferon-alpha injections in 12 evaluable patients (aged 4.2–19.8 years) with progressive or recurrent craniopharyngiomas revealed radiological improvement in three patients (25%). Therapy involved daily injections at a dose of 8 million units/m^2 for 16 weeks, followed by the same dose three times per week for an additional 32 weeks. All 12 patients developed a transient flu-like illness while seven children required dose reduction or temporary stoppage of the drug due to toxicity [80]. A further, smaller study of five children (aged 9–15 years) with recurrent craniopharyngiomas evaluated the weekly, subcutaneous administration of pegylated interferon-alpha with the theory that the amended formulation would enable a sustained drug exposure, thereby improving efficacy [81]. Four patients (80%) had protracted partial or complete responses to treatment with minimal toxicity. This experience prompted a prospective phase II trial by the Paediatric Brain Tumour Consortium which is still ongoing (ClinicalTrials.gov Identifier: NCT01964300).

These results lend credence to a potential role for interferon-alpha as an intralesional therapeutic agent in cystic craniopharyngioma. Moreover, whilst concern surrounding the toxicity from CSF leakage of intracystic β-emitting radionucleotides and bleomycin often precluded their use in this setting, interferon-alpha has been used successfully when administered as intrathecal therapy for subacute sclerosing panencephalitis and neoplastic meningitis [82, 83], suggesting potentially less risk if leakage into surrounding CSF spaces and tissue parenchyma was to occur.

Studies evaluating interferon-alpha as an intracystic therapy in childhood craniopharyngioma have been performed and are described following an overview of the administration schedule that has been typically followed.

Administration of Intracystic Interferon-alpha

On overview of the most widely used administration schedule used in recent clinical studies is shown in Table 2 [14, 84–86]. Two weeks after insertion of the intracystic catheter, a permeability imaging study (either a CT or MRI brain with contrast injection and saline flush into the catheter) is recommended to ensure the correct position of the catheter tip in the cyst (Fig. 2), establish an approximate calculation of cyst volume and to confirm no leakage of contrast occurs from the catheter following instillation. Baseline pituitary functioning and visual assessment is recommended before commencing therapy.

One cycle of therapy encompasses tri-weekly cystic aspirations immediately followed by the intracystic administration of three million units (1 ml) of interferon-alpha (IntronA®—recombinant interferon-alpha-2b ready to use solution, Merck, USA or Roferon-A®—recombinant interferon-alpha-2a, Roche, Switzerland) and 1 ml of normal saline solution flush via the catheter over a 4 week time period, equivalent to 12 doses of interferon-alpha. On the first day of the cycle, the maximum amount of cyst fluid should be slowly aspirated from the patient before injecting interferon-alpha, taking into account the estimated volume of the cyst calculated from the permeability study. This aspiration can be curtailed if the patient acutely complains of any new symptoms such as headache or visual disturbance.

Table 2 Schedule for administration of intracystic interferon-alpha-2b

Drug	Route	Dose	Weeks	Days
Interferon-alpha-2b (IntronA®—ready to use solution)	Intracystic injection	3,000,000 IU	Weeks 1–4	Monday, Wednesday and Friday of each week

Note: This schedule is in accordance with the intracystic interferon-alpha-2b standard of care protocol for cystic craniopharyngiomas as published by The Hospital for Sick Children, Toronto, Canada [14]. A single cycle represents 12 doses, for a duration of four weeks. The total dose administered in one cycle is 36,000,000 International Units (IU). Subsequent cycles can be repeated depending on clinical and radiological response. There is no dose reduction for different ages or weights of patients. Following aspiration of the cyst, 3 million units (1 millilitre) of interferon-alpha is instilled, followed by an intracystic 1 millilitre of normal saline flush

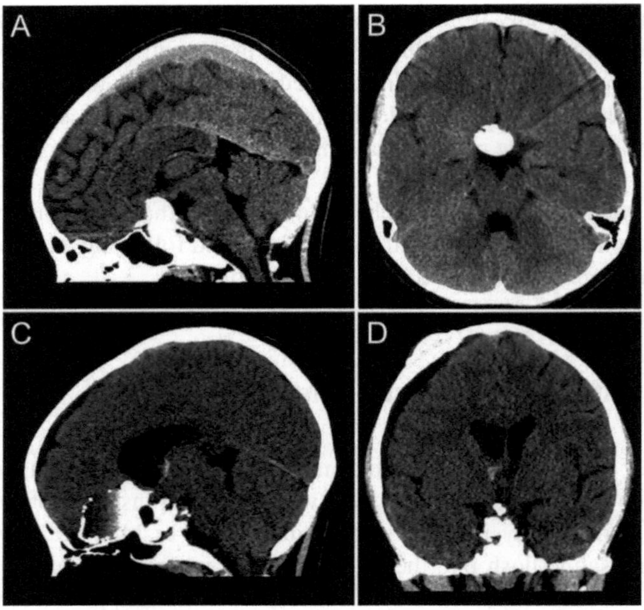

Fig. 2 Examples of a permeability study. Sagittal CT (**a**) and axial CT (**b**) of the brain post instillation of contrast through an Ommaya reservoir ruling out leakage. This is clearly distinct from a sagittal CT (**c**) and a coronal CT (**d**) of the brain post-contrast instillation through an Ommaya reservoir demonstrating leakage of contrast into the body of the right lateral ventricle and adjacent Foramen of Monro

For all subsequent aspirations, at least 2 ml of fluid should attempted to be aspirated to enable volume equilibrium in the cyst as this corresponds to the total volume of each instillation (drug and flush). More cyst fluid can be aspirated if calculations suggest a significant residual volume remains and it is clinically tolerated.

A permeability imaging study should be performed before the first therapeutic cycle of 12 interferon-alpha doses is commenced or if there is clinical concern the aspirated fluid is suspicious for CSF (clear as opposed to proteinaceous) as cyst volume alterations and increased cyst wall permeability during treatment may influence the intracystic catheter tip location predisposing to leakage of instilled products [55].

Studies of Intracystic Interferon-alpha in Paediatric Craniopharyngioma

The first intracystic instillation of interferon-alpha for children with cystic craniopharyngioma was published by Brazilian colleagues, describing nine children (mean age 10.8 years; range 1.8–18) treated between 2000 and 2004 [85]. Whilst the dose of interferon-alpha administered per treatment cycle was standardised to 36 million units, the total dose administered to children ranged from 36 to 108 million units. The agent used was recombinant Roferon-A®. A significant reduction in cyst volume was observed in all nine children with seven (78%) achieving complete disappearance. Six children (67%) developed transient pyrexia with one patient suffering arthralgia and chronic fatigue symptoms that resolved with medication. The mean follow-up period was under 2 years and, although the assumption is that no progression had been observed, no comment was made whether cyst-re-accumulation, new cyst formation or solid tumour growth had been reported for any treated patient.

A follow-on multicentre analysis led by the Brazilian group, collaborating with Italian and Chilean clinical teams, resulted in the largest prospective study of intracystic interferon-alpha use in childhood craniopharyngioma published to date [86]. Sixty children (mean age 11 years, range 1.7–18) were treated with Roferon-A® over a 9 year period according to an agreed protocol akin to that described above in this chapter. Thirty-one patients (52%) received this therapy as their primary treatment, while the remaining cohort had already undergone either unsuccessful surgery (n = 18) or bleomycin therapy (n = 3). For patients with measurable cysts, mean volume reduced following therapy from 27.7 ml (3.3–134.5) to 9.5 ml (0.14–70.7). One third of the cohort experienced mild, transient side effects including headaches (n = 6), palpebral oedema (n = 5) and eight patients developed new endocrinopathies. One patient required optic nerve decompression due to visual decline despite cyst shrinkage, while chronic fatigue and arthralgia were again recorded for the child treated in the original study. None of these adverse events required discontinuation of therapy and no mortalities were reported. The vast majority of children (78%) attained more than 50% cyst shrinkage with therapy, although a mean follow-up period of 3.7 years revealed 13 children (22%) had progressed requiring subsequent surgical intervention (Table 3).

The most recent published craniopharyngioma case series by the Hospital for Sick Children, Toronto also suggests an improved outcome with minimal adverse effects from the administration of intracystic interferon-alpha [14]. Six children (mean age 10.8 years, range 4–18) underwent a compatible schedule of administration to that used in the previous studies, albeit using a different interferon-alpha formulation (IntronA®). Four patients (67%) achieved more than 50% cell shrinkage with sustained response observed after 2–4 therapeutic cycles at a relatively short median follow-up of 17 (range 1–30) months. Again, transient headaches were common amongst all six patients, yet no new endocrinopathies, obesity, significant morbidities or deaths were reported (Table 3). Indeed, one such female patient maintained a complete response, with intact vision and hypothalamo-pituitary activity, 30 months after two cycles of interferon-alpha treatment. To date, 7 years after therapy, this patient continues to remain disease free with normal vision and pituitary function (Fig. 3).

Biological work has also supported these clinical findings of potential efficacy with evidence of increased Fas-induced cell apoptosis in craniopharyngioma cyst fluid following the administration of interferon-alpha on children treated within the

Table 3 Intracystic interferon-alpha in childhood adamantinomatous craniopharyngioma—retrospective case series

Agent	Author (year)	Number (paeds)	Median/Mean age (range–years)	CR	PR	Median/Mean F/U (range–years)	Median/Mean PFS (range–years)	Complications or toxicity
Interferon-alpha	Cavalheiro (2005)[a] [85]	9—all paeds	10 (1.8–18)	0.78	0.22	1.8 (1–3.5)	NR	6—pyrexia 1—arthralgia/chronic fatigue
	Ierardi (2007)[a] [61]	21—all paeds	10 (1–19)	0.5	0.5	2.25 (0.5–4)	NR	NR
	Cavalheiro (2010)[a] [86]	60—all paeds	11 (1.6–18)	NR	0.78	3.7 (0.3–7)	NR	18—transient headache/fever/fatigue/arthritis 8—new endocrinological dysfunction
	Bartels (2012) [14]	6—all paeds	10.8 (4–18)	0.17	0.5	1.4 (0.08–2.5)	NR	6—manageable headache 1—nausea and vomiting

Key: *NR* not recorded, *CR* complete response, *PR* partial response, *F/U* follow-up, *PFS* progression-free survival.
[a]likely patient overlap between studies but not documented as such in manuscripts

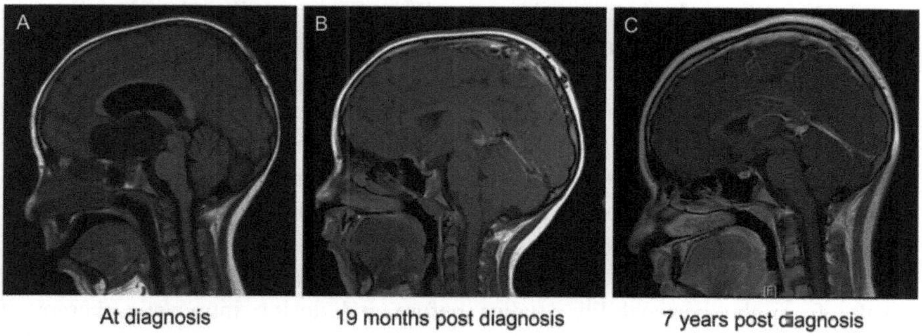

Fig. 3 Four-year-old female child diagnosed with a cystic craniopharyngioma (**a**), who was subsequently treated with two cycles of intracystic interferon-alpha therapy. The patient has demonstrated a sustained complete response at 19 months post-diagnosis (**b**) and again now, aged 11 years; 7 years post-diagnosis (**c**). No further treatment has been required. Clinically the patient has normal visual acuity and fields, and other than precious puberty diagnosed at the age of 8 years, intact pituitary function

Brazilian clinical studies [61, 85, 86]. Likewise, the aforementioned increased levels of alpha-defensins described within cystic craniopharyngiomas demonstrated a sequential decrease in concentration following interferon-alpha instillation [60].

Whilst evidence from these paediatric craniopharyngioma cohort studies would suggest intracystic interferon-alpha is a promising therapeutic agent that could potentially supersede its predecessors, β-emitting radionucleotides and bleomycin, with respect to disease response and tolerability, a note of caution comes from the fact that the analyses to date have been non-randomised studies performed on relatively small cohort sizes with brief follow-up periods. They have often lacked standardised response criteria and precise data on progression-free survival times and the nature of any resulting progressions, similar to the limitations labelled at the other intracystic agent studies.

In addition, although not observed in the larger craniopharyngioma case series described above, case reports of more significant toxicity following intracystic interferon-alpha administration are emerging [87–89], postulated to result from leakage of the drug into surrounding CSF spaces. One study described a craniopharyngioma patient who suffered a focal neurological event with subsequent permeability imaging revealing suspected leakage beyond the cyst [87]. Another report described an 8-year-old boy who was commenced on intracystic interferon-alpha after suffering a cystic craniopharyngioma recurrence with associated hydrocephalus despite previous surgery and focal radiotherapy. Although the patient's pre-therapeutic permeability imaging revealed widespread leakage of contrast outside the cyst into the ventricular system, interferon-alpha treatment was continued due to pressing clinical need and the drug's reassuring neurological safety profile [89]. Two weeks after completing one treatment cycle, the child presented with confusion, electrolyte imbalance and incontinence. Imaging revealed extensive parenchymal atrophy, no improvement in cyst size and worsening obstructive hydrocephalus. Clinical improvement was seen after insertion of a ventriculo-peritoneal shunt, making it difficult to ascribe the deterioration observed to raised intracranial pressure or interferon-alpha therapy. A third study reported the case of a heavily pre-treated 13-year-old patient who presented with recurrent, cystic craniopharyngioma and developed permanent visual field loss and reversible acuity decline following the concomitant administration of intracystic *and* subcutaneous *pegylated* interferon-alpha, a novel mode of both administration and drug formulation in this context [88]. Retrospective permeability imaging revealed a cyst wall leak.

Results from these case reports reinforce the importance of taking into account results from the permeability studies before and during intracystic therapy if required. Even for an agent such as interferon-alpha, with an apparent acceptable safety profile from intrathecal use [82, 83], its relatively recent introduction as an intra-lesional therapeutic agent for craniopharyngiomas means that when radiological evidence of cyst leakage is observed, this should be considered an indication not to proceed until more is understood about its acute and chronic safety profile. Furthermore, this data highlights several outstanding issues regarding intracystic interferon use that require elucidation, such as the most appropriate dosing, schedule and drug formulation, or whether specific craniopharyngiomas and patients would benefit from intracystic interferon therapy compared to others [84]. For instance, it is plausible that, as seems the case with other intracystic therapy, lesions with cystic and solid components beyond the treated cyst may not respond, while preceding treatments such as focal radiotherapy or previous intracystic therapies may alter the permeability of recurrent cyst walls increasing the likelihood of treatment failure.

Currently, any decision on which patients would potentially benefit from intracystic-interferon therapy should be made within the confines of a multidisciplinary team setting incorporating the patient and family where possible. For instance, based on the above current knowledge, infants and young patients with single isolated cysts, minimal endocrinopathies and visual difficulties may be considered more appropriate candidates for intracystic interferon therapy than older adolescents with a multicystic or mixed cystic-solid craniopharyngioma but this decision may be influenced by multi-professional discussion. As stated, there should be consensus, however, that all proposed candidates must undergo cyst wall pre-therapeutic permeability screening with cancellation of treatment if cyst wall leakage is encountered.

Conclusion

Developing an intracystic craniopharyngioma treatment strategy for widespread use continues to remain challenging since supporting data purporting a particular agent is often derived retrospectively from different institutions adopting a non-randomised range of schedules, doses and response definitions. Effectiveness in this context has often been determined by either radiological reduction in size or clinical improvement. However, studies have shown that these two features may not necessarily correlate with clinical deterioration observed in the face of radiological shrinkage [86]. Furthermore, the benefit: risk value of repeated intracystic treatments in postponing more definitive surgery or radiotherapy has not been fully evaluated.

Nevertheless, interpretation of current literature suggests that introduction of interferon-alpha as intracystic therapy for craniopharyngioma holds promise, at the very least for a subset of patients, with evidence of sustained tumour responses in conjunction with fewer episodes of significant morbidity to date than its predecessors, β-emitting radionucleotides and bleomycin. More multi-national, randomised studies would be required to define appropriate scheduling, formulation, patient selection, sustainability of response, effect on neighbouring cysts or solid craniopharyngioma components if present, and toxicity profile.

Consequently, in an initial attempt to obtain more of this information on intracystic interferon-alpha in paediatric cystic craniopharyngioma, the authors of this chapter have embarked on an global review of institutional experience with the non-pegylated formulation of interferon-alpha, collaborating with over 20 paediatric brain tumour centres from the European division of the International Society of Paediatric Oncology (SIOP-E) and the International Society for Paediatric Neurosurgery (ISPN). It is hoped that results from analyses such as this may influence appropriate patient selection and future collaborative prospective clinical trial work. The latter is clearly needed to enhance patient outcome and quality of life while the development of personalised, targeted therapies based on the molecular profile of craniopharyngioma subtypes remains an evolving work in progress.

References

1. Holsken A et al (2016) Adamantinomatous and papillary craniopharyngiomas are characterized by distinct epigenomic as well as mutational and transcriptomic profiles. Acta Neuropathol Commun 4(1):20
2. Mallucci C et al (2012) Management of craniopharyngioma: the Liverpool experience following the introduction of the CCLG guidelines. Introducing a new risk assessment grading system. Childs Nerv Syst 28(8):1181–1192
3. Steinbok P, Hukin J (2010) Intracystic treatments for craniopharyngioma. Neurosurg Focus 28(4):E13
4. Duff J et al (2000) Long-term outcomes for surgically resected craniopharyngiomas. Neurosurgery 46(2):291–302 discussion 302-5
5. Fahlbusch R et al (1999) Surgical treatment of craniopharyngiomas: experience with 168 patients. J Neurosurg 90(2):237–250
6. Villani RM et al (1997) Long-term results of treatment for craniopharyngioma in children. Childs Nerv Syst 13(7):397–405
7. Clark AJ et al (2012) Treatment-related morbidity and the management of pediatric craniopharyngioma: a systematic review. J Neurosurg Pediatr 10(4):293–301
8. Muller HL (2010) Childhood craniopharyngioma: current controversies on management in diagnostics, treatment and follow-up. Expert Rev Neurother 10(4):515–524

9. Yang I et al (2010) Craniopharyngioma: a comparison of tumor control with various treatment strategies. Neurosurg Focus 28(4):E5
10. Kalapurakal JA et al (2003) Clinical outcome in children with craniopharyngioma treated with primary surgery and radiotherapy deferred until relapse. Med Pediatr Oncol 40(4):214–218
11. Srinivasan S et al (2004) Features of the metabolic syndrome after childhood craniopharyngioma. J Clin Endocrinol Metab 89(1):81–86
12. Merchant TE et al (2006) Phase II trial of conformal radiation therapy for pediatric patients with craniopharyngioma and correlation of surgical factors and radiation dosimetry with change in cognitive function. J Neurosurg 104(2 Suppl):94–102
13. Hargrave DR (2006) Does chemotherapy have a role in the management of craniopharyngioma? J Pediatr Endocrinol Metab 19(Suppl 1):407–412
14. Bartels U et al (2012) Intracystic therapies for cystic craniopharyngioma in childhood. Front Endocrinol (Lausanne) 3:39
15. Moussa AH, Kerasha AA, Mahmoud ME (2013) Surprising outcome of ommaya reservoir in treating cystic craniopharyngioma: a retrospective study. Br J Neurosurg 27(3):370–373
16. Spaziante R et al (1989) Management of primary or recurring grossly cystic craniopharyngiomas by means of draining systems. Topic review and 6 case reports. Acta Neurochir (Wien) 97(3–4):95–106
17. Zanon N, Cavalheiro S, da Silva MC (2008) Does the choice of surgical approach to insert an intratumoral catheter influence the results of intratumoral cystic treatment? Surg Neurol 70(1):66–69 discussion 69
18. Pettorini BL et al (2009) Endoscopic transventricular positioning of intracystic catheter for treatment of craniopharyngioma. Technical note. J Neurosurg Pediatr 4(3):245–248
19. Joki T et al (2002) Neuroendoscopic placement of Ommaya reservoir into a cystic craniopharyngioma. Childs Nerv Syst 18(11):629–633
20. Hellwig D, Bauer BL, List-Hellwig E (1995) Stereotactic endoscopic interventions in cystic brain lesions. Acta Neurochir Suppl 64:59–63
21. Floyd JR et al (2009) Endoscopic, image-guided, transnasal instillation of (32)P for recurrent infrachiasmatic cystic craniopharyngioma. Minim Invasive Neurosurg 52(3):137–140
22. Sanford RA (1994) Craniopharyngioma: results of survey of the American Society of Pediatric Neurosurgery. Pediatr Neurosurg 21(Suppl 1):39–43
23. Backlund EO (1973) Studies on craniopharyngiomas. IV. Stereotaxic treatment with radiosurgery. Acta Chir Scand 139(4):344–351
24. Musolino A et al (1985) Stereotaxic treatment of expanding cysts in craniopharyngiomas by endocavitary beta-irradiation (Re-186; Au-198; Y-90). Neurochirurgie 31(3):169–178
25. Voges J et al (1997) Cystic craniopharyngioma: long-term results after intracavitary irradiation with stereotactically applied colloidal beta-emitting radioactive sources. Neurosurgery 40(2):263–269 discussion 269–70
26. Hasegawa T et al (2004) Management of cystic craniopharyngiomas with phosphorus-32 intracavitary irradiation. Neurosurgery 54(4):813–820 discussion 820–2
27. Kickingereder P et al (2012) Intracavitary brachytherapy using stereotactically applied phosphorus-32 colloid for treatment of cystic craniopharyngiomas in 53 patients. J Neurooncol 109(2):365–374
28. Yu X et al (2015) Interstitial radiotherapy using phosphorus-32 for giant posterior fossa cystic craniopharyngiomas. J Neurosurg Pediatr 15(5):510–518
29. Julow JV (2013) Intracystic irradiation for craniopharyngiomas. Pituitary 16(1):34–45
30. Derrey S et al (2008) Management of cystic craniopharyngiomas with stereotactic endocavitary irradiation using colloidal 186Re: a retrospective study of 48 consecutive patients. Neurosurgery 63(6):1045–1052 discussion 1052–3
31. Pollock BE et al (1995) Phosphorus-32 intracavitary irradiation of cystic craniopharyngiomas: current technique and long-term results. Int J Radiat Oncol Biol Phys 33(2):437–446
32. Blackburn TP, Doughty D, Plowman PN (1999) Stereotactic intracavitary therapy of recurrent cystic craniopharyngioma by instillation of 90yttrium. Br J Neurosurg 13(4):359–365

33. Ansari SF et al (2016) Efficacy of phosphorus-32 brachytherapy without external-beam radiation for long-term tumor control in patients with craniopharyngioma. J Neurosurg Pediatr 17(4):439–445
34. Albright AL et al (2005) Individualized treatment of pediatric craniopharyngiomas. Childs Nerv Syst 21(8–9):649–654
35. Barriger RB et al (2011) Phosphorus-32 therapy for cystic craniopharyngiomas. Radiother Oncol 98(2):207–212
36. Kubo O et al (1974) Intracystic therapy of bleomycin for craniopharyngioma--effect of bleomycin for cultured craniopharyngioma cells and intracystic concentration of bleomycin (author's transl). No Shinkei Geka 2(10):683–688
37. Cavalheiro S et al (1996) Use of bleomycin in intratumoral chemotherapy for cystic craniopharyngioma. Case report. J Neurosurg 84(1):124–126
38. Savas A et al (1999) Intracavitary chemotherapy of polycystic craniopharyngioma with bleomycin. Acta Neurochir (Wien) 141(5):547–548 discussion 549
39. Alen JF et al (2002) Intratumoural bleomycin as a treatment for recurrent cystic craniopharyngioma. Case report and review of the literature. Neurocirugia (Astur) 13(6):479–485 discussion 485
40. Hader WJ et al (2000) Intratumoral therapy with bleomycin for cystic craniopharyngiomas in children. Pediatr Neurosurg 33(4):211–218
41. Hsu TR, Chang KP, Wong TT (2009) Intracystic bleomycin therapy for craniopharyngioma in children. Neuro-oncology 11:902
42. Hukin J et al (2007) Intracystic bleomycin therapy for craniopharyngioma in children: the Canadian experience. Cancer 109(10):2124–2131
43. Mottolese C et al (2005) Craniopharyngiomas: our experience in Lyon. Childs Nerv Syst 21(8–9):790–798
44. Park DH et al (2002) Outcome of postoperative intratumoral bleomycin injection for cystic craniopharyngioma. J Korean Med Sci 17(2):254–259
45. Takahashi H, Yamaguchi F, Teramoto A (2005) Long-term outcome and reconsideration of intracystic chemotherapy with bleomycin for craniopharyngioma in children. Childs Nerv Syst 21(8–9):701–704
46. Kim SD et al (2007) Radiological findings following postsurgical intratumoral bleomycin injection for cystic craniopharyngioma. Clin Neurol Neurosurg 109(3):236–241
47. Yang SH, Kim IS, Hong JT, Sung JH, Son BC, Lee SW (2009) Intratumoral bleomycin injection as a primary therapy for cystic craniopharyngioma. Neuro-oncology 11:903
48. Hernandez J, Morel C, Gonzalez A, Garcia S, Hernandez M, Zarate A (2002) Use of bleomycin through a local reservoir in patients with cystic craniopharyngiomas. [Spanish] Uso de la bleomicina mediante uso reservorio local en pacientes con craneofaringiomas quisticos. Medicina Interna de Mexico 18:273–277
49. Franzini A, Leocata F, Pintucci M, Broggi G (2001) Cystic craniopharyngioma: stereotactic intracavitary Bleomycin administration. [Italian] La somministrazione di Bleomicina intracavitaria nel trattamento chirurgico del craniofaringioma cistico. Revista Medica 7:41–45
50. Takahashi H, Nakazawa S, Shimura T (1985) Evaluation of postoperative intratumoral injection of bleomycin for craniopharyngioma in children. J Neurosurg 62(1):120–127
51. Hukin J et al (2005) Childhood craniopharyngioma: Vancouver experience. Childs Nerv Syst 21(8–9):758–765
52. Mottolese C et al (2001) Intracystic chemotherapy with bleomycin in the treatment of craniopharyngiomas. Childs Nerv Syst 17(12):724–730
53. Jiang R, Liu Z, Zhu C (2002) Preliminary exploration of the clinical effect of bleomycin on craniopharyngiomas. Stereotact Funct Neurosurg 78(2):84–94
54. Belen D et al (2007) Delayed neurotoxic complication of intracavitary bleomycin therapy for craniopharyngioma in a child who had previously undergone radiosurgery. Case report. J Neurosurg 106(5 Suppl):391–393

55. Lafay-Cousin L et al (2007) Neuroradiological findings of bleomycin leakage in cystic craniopharyngioma. Report of three cases. J Neurosurg 107(4 Suppl):318–323
56. Savas A et al (2000) Fatal toxic effect of bleomycin on brain tissue after intracystic chemotherapy for a craniopharyngioma: case report. Neurosurgery 46(1):213–216 discussion 216–7
57. Zheng J et al (2014) Intracystic bleomycin for cystic craniopharyngiomas in children. Cochrane Database Syst Rev 9:CD008890
58. Fang Y et al (2012) Intracystic bleomycin for cystic craniopharyngiomas in children. Cochrane Database Syst Rev 4:CD008890
59. May JA et al (2006) Craniopharyngioma in childhood. Adv Pediatr 53:183–209
60. Pettorini BL et al (2010) The role of inflammation in the genesis of the cystic component of craniopharyngiomas. Childs Nerv Syst 26(12):1779–1784
61. Ierardi DF et al (2007) Apoptosis in alpha interferon (IFN-alpha) intratumoral chemotherapy for cystic craniopharyngiomas. Childs Nerv Syst 23(9):1041–1046
62. Hirano A, Ghatek NR, Zimmerman HM (1973) Fenestrated blood vessels in craniopharyngioma. Acta Neuropathol 26(2):171–177
63. Arefyeva IA et al (2002) Analysis of fluid in craniopharyngioma-related cysts in children: proteins, lactate and pH. Acta Neurochir (Wien) 144(6):551–554 discussion 554
64. Petito CK (1996) Craniopharyngioma: prognostic importance of histologic features. AJNR Am J Neuroradiol 17(8):1441–1442
65. Szeifert GT et al (1991) Secretory component of cystic craniopharyngiomas: a mucinohistochemical and electron-microscopic study. Surg Neurol 36(4):286–293
66. Vaquero J et al (1999) Expression of vascular permeability factor in craniopharyngioma. J Neurosurg 91(5):831–834
67. Stevens A, Kloter I, Roggendorf W (1988) Inflammatory infiltrates and natural killer cell presence in human brain tumors. Cancer 61(4):738–743
68. Le Page C et al (2000) Interferon activation and innate immunity. Rev Immunogenet 2(3):374–386
69. Ferrantini M, Capone I, Belardelli F (2007) Interferon-alpha and cancer: mechanisms of action and new perspectives of clinical use. Biochimie 89(6–7):884–893
70. Stark GR et al (1998) How cells respond to interferons. Annu Rev Biochem 67:227–264
71. Platanias LC (2005) Mechanisms of type-I- and type-II-interferon-mediated signalling. Nat Rev Immunol 5(5):375–386
72. Ahmed S, Rai KR (2003) Interferon in the treatment of hairy-cell leukemia. Best Pract Res Clin Haematol 16(1):69–81
73. Strander H (1986) Interferon treatment of human neoplasia. Adv Cancer Res 46:1–265
74. Goldstein D, Laszlo J (1988) The role of interferon in cancer therapy: a current perspective. CA Cancer J Clin 38(5):258–277
75. Kirkwood JM et al (2001) High-dose interferon alfa-2b significantly prolongs relapse-free and overall survival compared with the GM2-KLH/QS-21 vaccine in patients with resected stage IIB-III melanoma: results of intergroup trial E1694/S9512/C509801. J Clin Oncol 19(9):2370–2380
76. Salles G et al (2008) Rituximab combined with chemotherapy and interferon in follicular lymphoma patients: results of the GELA-GOELAMS FL2000 study. Blood 112(13): 4824–4831
77. Edwards L et al (1992) Treatment of cutaneous squamous cell carcinomas by intralesional interferon alfa-2b therapy. Arch Dermatol 128(11):1486–1489
78. Shin DM et al (2001) Combined interferon-alfa, 13-cis-retinoic acid, and alpha-tocopherol in locally advanced head and neck squamous cell carcinoma: novel bioadjuvant phase II trial. J Clin Oncol 19(12):3010–3017
79. Kim KH et al (2004) Intralesional interferon alpha-2b in the treatment of basal cell carcinoma and squamous cell carcinoma: revisited. Dermatol Surg 30(1):116–120
80. Jakacki RI et al (2000) Phase II evaluation of interferon-alpha-2a for progressive or recurrent craniopharyngiomas. J Neurosurg 92(2):255–260

81. Yeung JT et al (2012) Pegylated interferon-alpha-2b for children with recurrent craniopharyngioma. J Neurosurg Pediatr 10(6):498–503
82. Gascon GG (2003) Randomized treatment study of inosiplex versus combined inosiplex and intraventricular interferon-alpha in subacute sclerosing panencephalitis (SSPE): international multicenter study. J Child Neurol 18(12):819–827
83. Chamberlain MC (2002) A phase II trial of intra-cerebrospinal fluid alpha interferon in the treatment of neoplastic meningitis. Cancer 94(10):2675–2680
84. Bailey S, Parkes J (2015) Intracystic interferon therapy in childhood craniopharyngioma: who, when and how? Clin Endocrinol (Oxf) 82(1):29–34
85. Cavalheiro S et al (2005) Use of interferon alpha in intratumoral chemotherapy for cystic craniopharyngioma. Childs Nerv Syst 21(8–9):719–724
86. Cavalheiro S et al (2010) Craniopharyngiomas: intratumoral chemotherapy with interferon-alpha: a multicenter preliminary study with 60 cases. Neurosurg Focus 28(4):E12
87. Sharma J et al (2015) Intracystic interferon-alpha treatment leads to neurotoxicity in craniopharyngioma: case report. J Neurosurg Pediatr 16(3):301–304
88. Tiedemann LM et al (2016) Visual field loss in a case of recurrent cystic craniopharyngioma during concomitant treatment with pegylated interferon alpha-2b. J Pediatr Hematol Oncol 38(1):e26–e28
89. Mohammed KEA, Mike KRA, Parkes J (2013) Unexpected brain atrophy following administrationof intratumoral interferon alpha 2-b for cystic cranioparyngioma: a case report. International Journal of Case Reports and Images 4(12):719–722

Long-Term Management and Clinical Trials in Adamantinomatous Craniopharyngioma

Hermann L. Müller

Abstract This report is a review of findings on the diagnosis, treatment, clinical course, follow-up, and prognosis of craniopharyngioma patients with special regard to clinical trials and long-term management. Craniopharyngiomas are rare, embryonic malformations of the sellar/parasellar region with low histological grade. Clinical manifestations are related to increased intracranial pressure, visual impairment, and hypothalamic/pituitary deficiencies. If the tumour is favourably localised, therapy of choice is complete resection, with care taken to preserve hypothalamic and optic functions. In patients with unfavourable tumour location (i.e. involvement of hypothalamic areas), recommended therapy is limited hypothalamus-sparing surgical strategy followed by irradiation. Irradiation has proven effective in treatment of recurrences and progression. Surgical lesions and/or anatomical involvement of posterior hypothalamic areas can result in serious sequelae, mainly hypothalamic syndrome. It is crucial that craniopharyngioma can be managed as a frequently chronic disease, providing ongoing care of pediatric and adult patients by experienced multidisciplinary teams in the context of multicentre trials.

Keywords Craniopharyngioma • Neurosurgery • Hypothalamic obesity • Sleep • Pituitary deficiency • Neurocognitive • Recurrence • Quality of life • Brain tumours • Irradiation

Introduction

Craniopharyngiomas are rare embryonic malformations of the sellar/parasellar area with low histological grade (WHO I°). Despite high overall survival rates (87–95% 20-year overall survival in childhood-onset adamantinomatous craniopharyngioma), patients with hypothalamic involvement have reduced survival rates. Quality

H.L. Müller (✉)
Department of Pediatrics and Pediatric Hematology and Oncology, Klinikum Oldenburg AöR, Medical Campus University Oldenburg, Rahel-Straus-Strasse 10, 26133 Oldenburg, Germany
e-mail: mueller.hermann@klinikum-oldenburg.de

J.P. Martinez-Barbera, C.L. Andoniadou (eds.), *Basic Research and Clinical Aspects of Adamantinomatous Craniopharyngioma*,
DOI 10.1007/978-3-319-51890-9_10

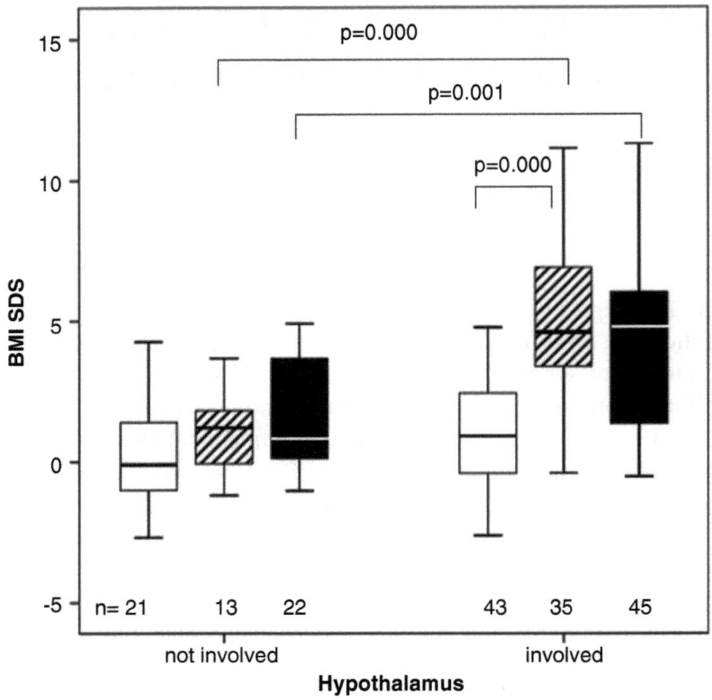

Fig. 1 Weight development in childhood-onset adamantinomatgous craniopharyngioma patients recruited in HIT Endo according to hypothalamic involvement. Body mass index (BMI) SDS is shown at time of diagnosis and at two intervals after diagnosis (8–12 years and more than 12 years). White boxes: BMI at diagnosis; hatched: 8–12 years follow-up; black: more than 12 years follow-up. The horizontal line in the middle of the box depicts the median. The top and bottom edges of the box, respectively, mark the 25th and 75th percentiles. Whiskers indicate the range of values that fall within 1.5 box-lengths. Reproduced from Sterkenburg et al. [11] with kind permission of Oxford University Press

of life is frequently impaired in long-term survivors due to sequelae caused by the anatomical proximity of craniopharyngiomas to hypothalamic–pituitary axes and the optic chiasma [1–9]. Severe obesity has major impact on long-term quality of survival and frequently occurs in craniopharyngioma with hypothalamic involvement [10] (Fig. 1). Adult-onset craniopharyngioma patients have higher mortality compared to the general population, namely an up to 19 fold higher cerebrovascular mortality [12]. Any clinical improvement in the prognosis of craniopharyngioma patients will require the development of risk-adapted radiooncological and neurosurgical treatment strategies in a multidisciplinary setting that provides medical as well as psychosocial support for these patients [3, 13]. Recent multicentre cooperation has already led to beneficial results [2, 14, 15].

Epidemiology

Craniopharyngioma is rare, with an incidence of 0.5–2 cases per million persons per year, 30–50% of all cases presenting during childhood and adolescence [16, 17]. Craniopharyngioma represents 1.2–4% of all childhood intracranial tumours. In the pediatric age group, its histological type is usually adamantinomatous with cyst formation [18–21]. Craniopharyngioma can be detected at any age, even in the prenatal and neonatal periods [18]. A bimodal age distribution has been shown, with peak incidence rates in children of ages 5–14 years and adults of ages 50–74 years. No gender differences have been observed in population-based studies. Craniopharyngioma cases have been reported within two families but an underlying genetic susceptibility has not been verified [22, 23]. Metastases are rare events in childhood-onset adamantinomatous craniopharyngioma, mainly occurring in the corridor of surgical approach [24].

The German Pediatric Cancer Registry (GPCR) systematically documents cases of childhood-onset adamantinomatous craniopharyngioma [25]. The GPCR data from 1980 to 2007 obtained for 496 childhood-onset craniopharyngioma patients diagnosed ≤18 years reveal most patients (451; 91%) were younger than 15 years of age at initial diagnosis, with a 1:1 sex ratio and a median age at initial diagnosis of 8.8 years. The 1980–2007 overall survival rate was 97% after 3 years from diagnosis, 96% after 5 years, and 93% after 10 years. Patients diagnosed with childhood-onset craniopharyngioma in the 1980s had a lower overall survival rate than patients diagnosed in the 1990s (overall survival at 5 years, 91% vs. 98%) [25, 26].

Pathogenesis and Molecular Pathology

Craniopharyngiomas are non-glial intracranial tumours derived from a malformation of embryonic tissue [27]. There are two differing hypotheses on its embryonic origin: originating from residual embryonal epithelium of the anterior infundibulum and the anterior pituitary gland versus originating from ectodermal remnants of Rathke's pouch [27].

Recent observations have given insight into biological markers of pituitary tumours [28]. Beta-catenin is highly expressed in adamantinomatous craniopharyngioma, whereas in papillary craniopharyngioma beta-catenin expression is absent [29]. Beta-catenin has been identified as a downstream component of the Wnt signal pathway that regulates cellular proliferation, development, and morphology [28]. Mouse studies have shown that only pituitary embryonic precursors or adult stem cells are able to generate tumours when targeted with oncogenic β-catenin, which suggests that the cell context is critical in order for mutant β-catenin to exert its oncogenic effect. Interestingly, mutant stem cells do not generate the bulk of the tumour cells; instead, they induce tumours in a paracrine manner. Combining basic studies in mice and humans will provide further insights into the biology of these neoplasms and will reveal pathogenic pathways that could be targeted with specific inhibitors for the benefit of patients [30].

Presenting Clinical Manifestations

The diagnosis of childhood-onset craniopharyngioma is often made late—sometimes years after initial appearance of symptoms [25, 31]—with a clinical picture often dominated by manifestations of intracranial pressure (e.g. headache and nausea) [32, 33]. Further primary manifestations are endocrine deficits (52–87%) and visual impairment (62–84%). Hormonal deficits are frequently caused by disturbances to the hypothalamic–pituitary axes that affect growth hormone secretion (75%), gonadotropins (40%), thyroid-stimulating hormone (TSH) (25%), and adrenocorticotropic hormone (ACTH) (25%). At the time of initial diagnosis, 40–87% of patients present with at least one hormonal deficit [34], and other endocrine symptoms such as central diabetes insipidus are present preoperatively in 17–27% of all patients [34, 35]. An analysis of anthropometric data obtained in routine checkups before the diagnosis of childhood-onset adamantinomatous craniopharyngioma in 90 patients [36] revealed that a pathologically reduced growth velocity—an early manifestation of the disease—presents in patients as young as 12 months, but that significant weight gain, predictive of hypothalamic obesity, tends to occur as a later manifestation, shortly before diagnosis.

Hoffmann et al. [37] concluded from their analysis of clinical presentation at diagnosis and duration of history in 411 patients with childhood-onset adamantinomatous craniopharyngioma that childhood-onset craniopharyngioma is frequently diagnosed after very long duration of history, especially in older children and adolescents. However, long duration of history before initial diagnosis was not a risk factor for large tumour size or impaired outcome in terms of survival or functional capacity. Visual impairment and neurological deficits in history should be considered as symptoms necessitating rapid diagnostic work-up. Growth failure and weight gain are early symptoms in history, which should lead to early consideration of craniopharyngioma in differential diagnosis.

Risk-Adapted Treatment Strategies

Novel risk-adapted treatment strategies are focusing on the following main goals: (a) reversal of visual compression symptoms, (b) relief of raised intracranial pressure, (c) prevention of tumour regrowth/progression, and (d) restoration or substitution of pituitary hormone deficits plus all other supplement-supportive measures, while minimising acute and long-term mortality and morbidity [2–4].

Surgery

De Vile et al. [38, 39] published the first reports on the association between attempts at radical gross-total resection in case of hypothalamic involvement and long-term morbidity. Elowe-Gruau et al. [40] published an algorithm for surgical treatment of childhood-onset adamantinomatous craniopharyngioma patients, which recommends a hypothalamus-sparing strategy based on a grading of hypothalamic tumour involvement in preoperative magnetic resonance imaging (MRI) [41]. The authors reported that patients neurosurgically treated according to this algorithm using a hypothalamus-sparing approach had similar relapse rates and a lower prevalence of severe obesity than patients treated by gross-total resection (28% versus 54%, respectively). This was the first report in the literature proving the tolerability and efficacy of a hypothalamus-sparing strategy by comparing cohorts treated by the same experienced surgical team at a single institution, and thus eliminating the bias of surgical experience on outcome analysis. However, it is important to note that although the "hypothalamus-sparing surgery" increased the percentage of "normal" body mass index (BMI) from 17% to 38%, the likelihood of clinically significant weight gain remained 62% with nearly half of all patients developing morbid obesity. Based on this grading system, Garré et al. [27] published a modified algorithm for risk-adapted treatment strategies in childhood-onset craniopharyngioma. The authors emphasised the treatment by experienced neurosurgical teams and suggested proton beam therapy, especially for young patients (< 5 years of age) after limited hypothalamus-sparing surgery. Müller et al. [42, 43] published studies on a risk-adapted treatment strategy based on pre- and post-surgical grading of hypothalamic involvement/damage in MRI. The assessment of the suprasellar tumour extension towards the mammillary bodies is considered essential for their grading into anterior or posterior hypothalamic involvement/lesion (Fig. 2). According to their report, patients with post-surgical lesions affecting posterior hypothalamic structures presented with an increased BMI and reduced self-assessed quality of survival during long-term prospective follow-up. Mallucci et al. [44] published a treatment algorithm, suggesting a two-staged surgical approach with initial relief of cystic pressure and thereby down-staging the risk grade in appropriate cases. Van Gompel et al. [45] developed a MRI-based grading system to predict the risk of postoperative obesity in adults treated for craniopharyngioma. Irregular contrast enhancement and hypothalamic T2 signal change correlated and predicted more severe hypothalamic involvement. Larger postoperative weight gains correlated with progressive hypothalamic involvement. Steno et al. [46] reported an increased hypothalamic complication rate, in case of tumours with intraventricular extension as compared with purely extraventricularly located tumours. To preserve the hypothalamic function, the authors recommended to distinguish between extraventricular and intra- and extraventricular located suprasellar craniopharyngiomas. In the first group, the hypothalamus can be simply compressed and not infiltrated. Elliott et al. [47] used a newly designed Craniopharyngioma Clinical Status Scale (CCSS) to show that preoperative CCSS scores predicted postoperative outcome better than

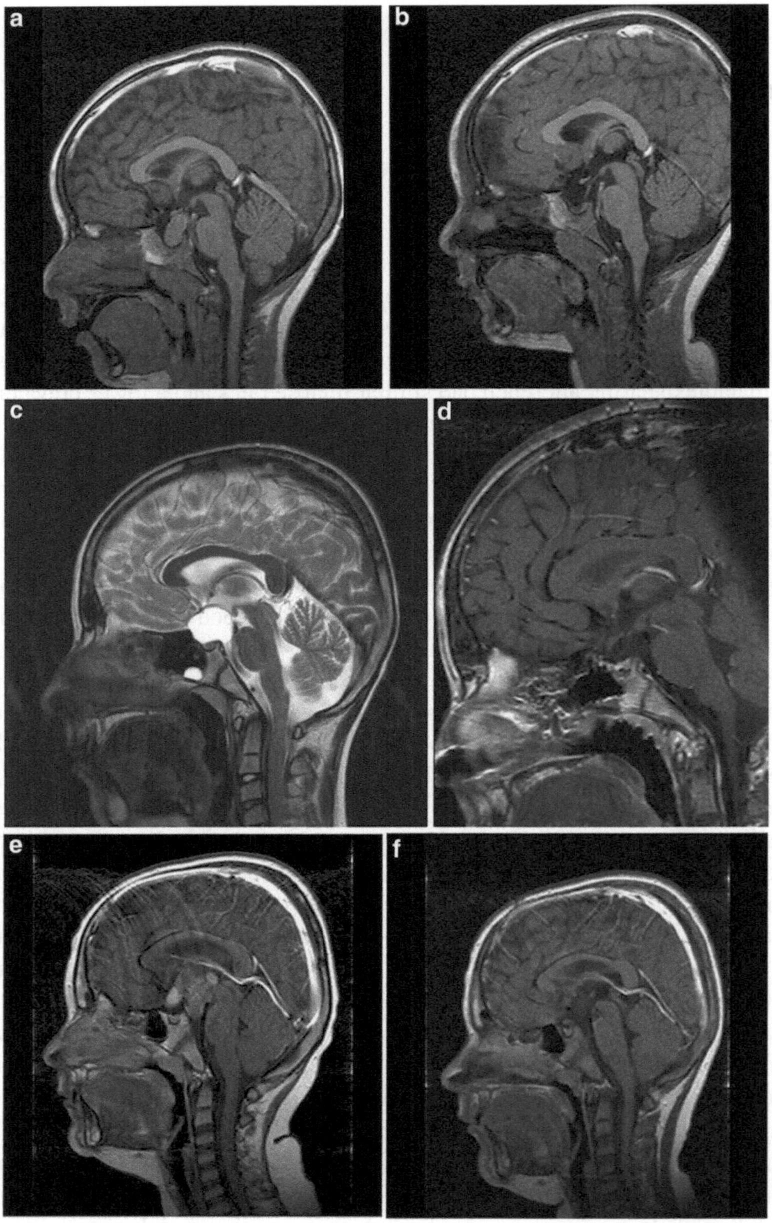

Fig. 2 Body mass index (BMI) and magnetic resonance imaging (MRI) at diagnosis and 36 months after surgery in three cases of childhood-onset adamantinomatous craniopharyngioma with different grade of hypothalamic involvement/lesion. (**a, b**) Patient with craniopharyngioma confined to the intrasellar space [0° = no hypothalamic involvement (**a**)/surgical lesion (**b**)]. BMI at diagnosis: −0.11 SD; BMI 36 months after complete resection: −0.41 SD. **(c, d)** Patient with craniopharyngioma involving the anterior hypothalamus [I° = hypothalamic involvement (**c**)/surgical lesion of the anterior hypothalamic area (**d**)]. BMI at diagnosis: −1.75 SD; BMI 36 months after complete resection: −0.43 SD. **(e, f)** Patient with craniopharyngioma involving the anterior and posterior hypothalamus (II° = hypothalamic involvement (**e**)/surgical lesion of the anterior and posterior hypothalamic area (**f**)]. BMI at diagnosis: +6.08 SD; BMI 36 months after complete resection: +6.79 SD. *Arrows* indicate mammillary bodies, defining the border between anterior and posterior involvement/lesion. Reproduced from Muller et al. [43] with kind permission of Bioscientifica

clinical characteristics like gender, age, location or tumour size, and the presence of hydrocephalus.

All of the above-mentioned treatment strategies and algorithms recommend that (a) for childhood-onset adamantinomatous craniopharyngioma with hypothalamic involvement, limited surgical approaches and postoperative external irradiation are advisable, and (b) treatment of craniopharyngioma should be confined to experienced multidisciplinary teams.

A major step towards potential standardisation of preoperative staging in childhood-onset craniopharyngioma is the comparison of published grading systems for assessment of hypothalamic damage/involvement in regard to prediction value for severe hypothalamic obesity as the main sequelae impairing quality of survival. Mortini et al. [48] have published one of the first studies which identify radiological variables linked to hypothalamic involvement on preoperative MRI, and correlate them with clinical features, long-term outcome, and prognosis. Additionally, Mortini et al. [48] analyzed the sensitivity of three published grading systems [40, 42, 43, 45] for prediction of hypothalamic obesity in their single centre cohort. Hypothalamic hyperintensity in T2-weighted/FLAIR images, unidentifiable pituitary stalk, dislocated chiasm, mammillary body involvement, retrochiasmatic tumour extension, and either infundibular recess or unrecognisable supra-optic recess have proved to be useful to define the hypothalamus invasion. Variables identified as factors with high and comparable prediction value for postoperative hypothalamic syndrome were the degree of hypothalamic involvement according to the classification described by Sainte-Rose and Puget [40] ($p < 0.002$; grade 0 vs 2 $p < 0.001$), Van Gompel et al. [45] ($p < 0.002$; grade 0 vs 1, $p < 0.027$; and grade 0 vs 2, $p < 0.002$), and Muller et al. [42, 43] ($p < 0.006$; grade 0 vs 1, $p < 0.05$; and grade 0 vs 2, $p < 0.004$). These results support the hypothesis that disease or treatment-related hypothalamic alterations have relevant negative impact on quality of survival and prognosis in childhood-onset adamantinomatous craniopharyngioma [49, 50].

There are only a few studies analysing the prognosis of patients with childhood-onset adamantinomatous craniopharyngioma in relation to the neurosurgeons' experience [42, 43, 51–54]. Sanford [51] and Boop [52] reported clinically significant differences in outcome according to the neurosurgeons' experience with the condition. Degree of obesity and quality of life were analyzed in a recent report based on reference assessment of tumour location and post-surgical hypothalamic lesions [43]. Treatment was also analyzed regarding neurosurgical strategy and the neurosurgical centre sizes based on patient load. Surgical lesions of anterior and posterior hypothalamic areas were associated with post-surgical obesity, negatively impacting long-term quality of survival in patients with surgical posterior hypothalamic lesions (Fig. 3). Treatment strategies in large centres were less radical and the rates of complete resection and hypothalamic surgical lesions were lower than those of middle and small-sized centres [42, 43, 54]. However, in multivariable analysis preoperative hypothalamic involvement was the only independent risk factor for severe obesity [43].

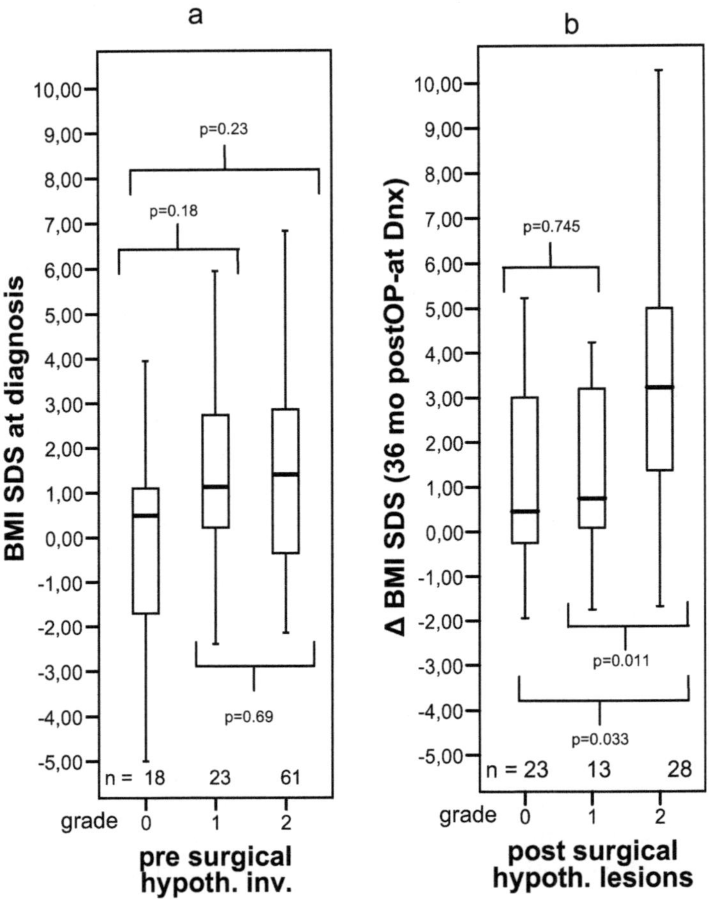

Fig. 3 Body mass index (BMI SDS) at diagnosis (**a**) and increases in BMI (ΔBMI SDS) during 36 months after surgery (**b**) in relation to presurgical hypothalamic involvement (**a**) and surgical hypothalamic lesions (**b**) of 117 childhood-onset adamantinomatous craniopharyngioma patients recruited in KRANIOPHARYNGEOM 2000. The *horizontal line* in the middle of the box depicts the median. *Edges* of box mark the 25th and 75th percentile. *Whiskers* indicate the range of values that fall within 1.5 box-lengths. Reproduced from Muller et al. [43] with kind permission of Bioscientifica

Based on the current literature, it is advisable to have a multidisciplinary team able to discuss diagnostic and treatment strategies, adopting the most sophisticated approaches feasible based on sufficient in-house surgical, radiooncological and psychosocial experience for treating patients with childhood-onset craniopharyngioma [3, 8, 49, 55].

Radiooncological Treatment

The site and rate of progression of childhood-onset adamantinomatous craniopharyngioma, as well as the patient's age, are important considerations when deciding whether reoperation and/or irradiation should be performed. Childhood-onset craniopharyngiomas are usually sharply bordered in the imaging. In contrast to primary brain tumours, craniopharyngiomas tend towards less infiltrative growth, permitting a small safety margin of 5 mm maximum [56, 57]. Biological characteristics usually allow the option of using high-precision, three-dimensional conformation technology. A conventional, fractionated irradiation total target volume dose of 54 Gray has been established worldwide [56] [58].

Fitzek et al. published the first series of 15 craniopharyngioma patients treated with combined proton-photon irradiation for residual or recurrent disease [59]. Five- and 10-year local control rates were 93% and 85%, respectively, with 10-year survival expectancy in 72% of patients. Functional status, academic skills, and professional abilities were unaltered after proton beam therapy; no treatment-related neurocognitive deficits were reported. Luu et al. published a preliminary report on 16 patients treated with proton beam therapy [60]. Local tumour control could be achieved in 87% of patients. Late sequelae included newly diagnosed panhypopituitarism, an out-of-proton-field posterior fossa meningioma (59 months following proton beam therapy administered to patient who previously received photon radiotherapy), and a cerebrovascular accident. One study of proton beam therapy in craniopharyngioma reported on cyst growth during the treatment course: 24% of patients demonstrated cyst enlargement and 5% cyst reduction requiring modification of the treatment plan, while one patient required cyst drainage during treatment [61].

Clinical outcome data are still limited in proton beam therapy compared to modern photon therapy. However, proton beam therapy has the advantages of better dose conformation to the target volume, sparing of critical structures, reduced integral dose, and lower dose of secondary neutrons, which should reduce the risk of secondary malignancies [62, 63]. Data should become available in due course with sufficient follow-up time and as more centres are able to deliver proton beam therapy.

Instillation of Sclerosing Substances for Cystic Tumours

A catheter insertion into a cystic craniopharyngioma may prevail over the transient success of a cyst fenestration by allowing repetitive drainage of the craniopharyngioma cyst and intracystic instillation of sclerosing substances. Different neurosurgical techniques are employed for catheter placement. Even though it often results in only transient relieve of pressure, it is a useful therapeutic method for cystic recurrent tumours whose anatomical location and configuration makes them difficult to resect [64]. Instillation of sclerosing substances into craniopharyngioma cysts such as bleomycin, using an intracystic catheter implanted by an open or stereotactic procedure, was used in such cases [65]. Severe neurotoxic side effects were observed in cases due to cystic leakage of bleomycin into cerebrospinal fluid [66]. Accordingly, a thorough neuroradiological imaging for exclusion of cystic leakage is mandatory before instillation of bleomycin.

Intracystic instillation of interferon alpha was first used by Cavalheiro et al., who published the largest series on cystic childhood-onset craniopharyngiomas [67]. Their recent publication [67] included 60 children, who were treated at three different institutions between 2000 and 2009. Twenty-nine of the 60 patients received intracystic interferon alpha after initial surgery or bleomycin treatment had failed and 31 were treated with interferon alpha as a first line treatment. While in 47 children (78%) more than 50% cyst shrinkage was achieved, 13 children (22%) presented with progression and required surgical intervention. Only one-third of the patients experienced side effects consisting of either headaches, fever, chronic fatigue, palpebral edema, or arthritis not necessitating discontinuation of treatment. There was no mortality. Based on these reports on the effect and tolerability, intracystic instillation of interferon alpha is a promising therapeutic option for predominantly monocystic craniopharyngiomas [67, 68].

Long-Term Outcome and Morbidities

Pituitary Deficiencies

Pituitary hormone deficiencies are common in craniopharyngioma. At the time of diagnosis, 40–87% of children [34, 69–71] present with at least one hormonal deficit. 17–27% [34, 70, 72] have been reported to present with diabetes insipidus neurohormonalis. The rate of post-surgical pituitary hormonal deficits increases due to the tumour's proximity or even involvement with the hypothalamic–pituitary axes [36, 38, 46, 69, 70, 72–74]. Transient post-surgical diabetes insipidus occurs in up to 80–100% of all cases [69, 75]. The rate of permanent post-surgical central diabetes insipidus ranges between 40 and 93% [38, 43, 69, 70, 72, 74–77].

In adult-onset craniopharyngioma patients anterior pituitary deficiencies and diabetes insipidus are most common and the majority of patients present with hypopituitarism [12, 71, 78, 79]. Endocrine deficiencies may worsen upon treatment. Mortini et al. [80] reported that 82%, 76%, 73%, and 67% of adult-onset craniopharyngioma patients with normal baseline levels for GH, ACTH, TSH, and gonadotropins developed a new deficits of the respective pituitary axes after surgery. Post-surgical onset of DI was observed in 70% of their patients. After transsphenoidal surgical approach the risk for new endocrine deficits appears to be lower [71, 80]. Recovery of preexisting pituitary dysfunction after surgery is rather rare. Most adult-onset craniopharyngioma patients suffer from partial or complete pituitary deficiencies as well as diabetes insipidus, with approximately 80% requiring the substitution of more than two pituitary hormones [12, 81].

Growth hormone deficiency has been reported at the time of diagnosis in 26–75% of childhood-onset craniopharyngioma patients [34, 77], and impaired growth, one of the early clinical manifestations of craniopharyngioma, often occurs years before diagnosis [36]. Growth hormone deficiency following tumour treatment for childhood-onset craniopharyngioma is found in about 70–92% of patients [36, 43, 82, 83] and a positive response to growth hormone treatment is seen in most cases [84]. Normal growth rates in childhood-onset craniopharyngioma patients with proven growth hormone deficiency are reported in the literature [85]. In fact, childhood-onset craniopharyngioma patients with hypothalamic involvement were found to achieve normal adult height more often than those without hypothalamic involvement [36]. Even though this phenomenon of "growth without growth hormone" was described in childhood-onset craniopharyngioma five decades ago [86], the physiology of growth in these cases is still not fully understood, although insulin and/or leptin are suspected to play a compensating role in this context. Both insulin and leptin have been hypothesised to induce growth in the fetus and in obese children [87–89], with leptin reported to function as a bone growth factor acting directly at the level of skeletal growth centres, independently of growth hormone action [87]. Mechanisms by which insulin stimulates growth include its known anabolic effects. At high serum concentrations, insulin may bind to the type 1 insulin-like growth factor (IGF) receptor and induce growth, mediated by its effect to decrease IGF-binding protein 1 levels, resulting in increased levels of free IGF-1 [87]. In support of this theory, obese childhood-onset craniopharyngioma patients were found to present with higher height SDS at the time of diagnosis and at last follow-up with no difference in endocrine substitution, including growth hormone treatment [31]. In contrast, another study found that childhood-onset craniopharyngioma patients who were growing despite growth hormone deficit had the same body composition, anthropometrical measures, and metabolic indexes, including insulin levels, as those patients requiring growth hormone substitution [85].

Visual and Neurological Outcomes

Due to frequent suprasellar tumour location, visual deficits (both visual acuity and visual fields) are relatively common in patients with craniopharyngioma: visual impairment as an initial clinical manifestation of craniopharyngioma is found in more than half of the affected patients [34], with some post-surgical improvement of vision in 41–48% of patients [69, 76]. Risk factors for post-surgical visual impairment include prechiasmatic tumour location and severe pre-surgical visual deficits [46, 69]. Improved ophthalmological outcome has been detected in surgical cases using the transsphenoidal approach [76], but such an approach is limited to resection of mainly intrasellar tumours. Because most childhood-onset adamantinomatous craniopharyngiomas typically extend to the suprasellar area, they are best removed through a transcranial or a combined transcranial and transsphenoidal approach.

Neurological sequelae include epilepsy, hemiparesis, cranial nerve deficits, and cerebrovascular disease manifestations [72, 74, 83]. Most of these sequelae are transient and the total prevalence of long-term neurological complications is reported to be 8% [69], but increases to 36% for large-sized tumours [72] and 30% when including both visual and neurological complications [75].

Morbidity Related to Treatment Strategies and Hypothalamic Injury

For favourably localised craniopharyngiomas, the preferred treatment of choice, especially at initial diagnosis, is an attempt at complete resection with preservation of hypothalamic and visual function [3, 4, 40, 53, 90–94]. For unfavourably localised tumours—those too close to or too entangled with the hypothalamus and/or the optic chiasm—a limited resection should be considered in order to preserve integrity of and/or to avoid further damage to optic and hypothalamic structures [5, 54, 95–101]. Endoscopic routes provide novel and less traumatic approaches for surgical resection [102–104].

The implantation of an intracystic catheter with a subcutaneous Ommaya reservoir enables the possibility of repeated decompression of the cyst and instillation of sclerosing agents [105, 106]. Bartels et al. [68] report high efficiency and good tolerability of interferon alpha as an intracystic sclerosing agent.

Irradiation is effective in preventing progression of residual tumour and relapses and is therefore a recommended treatment option in cases of limited surgical perspectives [107, 108]. First experiences with proton beam therapy applied to childhood-onset adamantinomatous craniopharyngioma are promising, offering a more protective radiooncological option than conventional photon irradiation, especially for tumours localised in the vicinity of hypothalamus or optic chiasm [109]. Stereotactic irradiation and gamma knife treatment are options in rare cases [110, 111].

Hypothalamic Obesity

Symptoms related to hypothalamic syndrome, such as obesity, behavioural changes, disturbed circadian rhythm, daytime sleepiness, and imbalances in regulation of thirst, heart rate, body temperature, and/or blood pressure, have been found at diagnosis in approximately 35% of childhood craniopharyngioma patients [3, 4, 10, 112–115]. The rate of hypothalamic dysfunction dramatically increases after treatment of craniopharyngioma; in some series up to 65–80% [5].

Suprachiasmatic lesions are associated with high morbidity. Surgical removal of sellar masses with suprasellar extension beyond the mammillary bodies risks hypothalamic structures and consequent hypothalamic syndrome with consecutive obesity [40, 42, 95, 116–118]. With the aid of imaging studies, recent reports have shown that the degree of obesity of affected patients is positively correlated with the degree of hypothalamic lesions [40, 42, 43, 119]. Based on this observation, novel classifications of pre-surgical involvement and post-surgical damage of hypothalamic structures based on MRI have been recently published [42, 43, 95] that might help to establish risk-adapted, i.e. hypothalamus-sparing surgical strategies.

Weight gain in childhood-onset adamantinomatous craniopharyngioma frequently occurs years before diagnosis [36], with 12–19% of patients reported to be obese at the time of diagnosis [4]. Weight gain and the development of obesity occur despite adequate endocrine replacement of pituitary deficiencies. The hypothalamic imbalances in energy management contribute to obesity and are exacerbated by factors limiting physical activity such as marked daytime sleepiness [120] and visual impairment. The degree of obesity frequently increases early after initial surgical treatment and rapid weight gain occurs the first 6–12 months after neurosurgical treatment [36]. Following initial treatment, the prevalence of severe obesity is higher, reaching up to 55% [49]. Severe obesity results in increased risks of cardiovascular disease and sequelae due to metabolic syndrome and [116].

The relation of severe obesity with hypothalamic lesions is obvious in childhood-onset adamantinomatous craniopharyngioma [7, 121, 122]. It is likely that in cases of craniopharyngioma with suprasellar extension, hypothalamic function will be compromised and will remain compromised after surgical and/or radio-oncological treatment. The hypothalamus contains several groups of nerve cell bodies forming distinct nuclei, which have highly diverse molecular, structural and functional organisations [121, 123]. The hypothalamus plays a dominant role in keeping the internal environment stable by synchronising circadian rhythms. Recent data indicate that appropriate balance of the autonomic nervous system has major impact on metabolism [124]. It is well known that adipose tissue is richly innervated by sympathetic nerve fibres controlling lipolysis. It now appears that lipogenesis is also controlled by parasympathetic innervation of adipose tissue originating from separate parasympathetic and sympathetic neurons located in the suprachiasmatic (SCN) and periventricular nucleus [122–124]. Such a high level of differentiation puts the SCN in a major position to balance circadian activity of both sympathetic and parasympathetic branches of the autonomous nervous system [122]. Considering the large proportion of craniopharyngioma patients with damage to suprasellar structures, it is most likely that craniopharyngiomas and/or the effects of treatment damage the SCN. This in turn influences the regulation of central clock mechanisms, which predisposes to metabolism alterations [120]. Accordingly, surgical strategies aiming at preservation of hypothalamic integrity are mandatory for the prevention of severe obesity owing to hypothalamic damage [3].

A study of Harz et al. [125] involving self-assessment by nutritional diaries showed that hypothalamic obesity occurs in patients with childhood-onset craniopharyngioma even though their caloric intake is similar to BMI-matched controls.

The analysis of physical activity using actimetry showed that childhood-onset craniopharyngioma patients presented with significantly lower levels of physical activity than BMI-matched healthy controls [125]. The aforementioned disturbances of circadian rhythms and marked daytime sleepiness have been demonstrated in patients with childhood-onset craniopharyngioma and hypothalamic obesity [120], which in turn were correlated with low, early morning and nocturnal melatonin concentrations in saliva [126, 127]. The proposed pathogenic mechanism involves impaired hypothalamic regulation of circadian melatonin secretion rhythms in patients with childhood-onset craniopharyngioma extending to suprasellar areas. Initial experiences with melatonin substitution in childhood-onset craniopharyngioma patients were promising: daytime sleepiness and physical activity improved and melatonin levels normalised [126].

Polysomnographic studies in patients with childhood-onset adamantinomatous craniopharyngioma and severe daytime sleepiness have revealed sleeping patterns typical for hypersomnia and secondary narcolepsy, i.e. frequent "sleep-onset REM phases" (SOREM) [120, 128, 129]. Medication with central stimulating agents (methylphenidate, modafinil) had a markedly beneficial effect on daytime sleepiness in these patients [129]. Secondary narcolepsy should be taken into consideration as a pathogenic factor in severely obese childhood-onset craniopharyngioma patients. Mason et al. [130] treated five patients with childhood-onset craniopharyngioma and severe hypothalamic obesity (age range: 6.0–9.8 years) with the central stimulating agent, dextroamphetamine, for the purpose of weight reduction. Dextroamphetamine therapy stabilised patient body mass index. The patients' parents reported noticeable improvements in their child's physical activity and alertness. Several reports [124, 131] hypothesised that decreased physical activity and severe obesity in patients with childhood craniopharyngioma are likely related to impaired central sympathetic output. For instance, Roth et al. [132] observed reduced urine concentrations of catecholamine metabolites correlating with the degree of obesity and the level of physical activity.

A decreased metabolic rate, in terms of both total and resting energy expenditure, contributes to the development of severe obesity in this population. Pediatric and adult patients with childhood-onset adamantinomatous craniopharyngioma were found to have a lower resting energy expenditure (REE) when compared to controls [133]. The energy intake/REE ratio was lower in patients with tumour involvement of the third ventricle [134]. Further factors that could potentially contribute to decreased physical activity are visual and neurological deficits [135, 136], psychosocial difficulties [49], and increased daytime sleepiness [120, 126].

Roemmler-Zehrer et al. [137] recently reported on the gastrointestinal hormones peptide YY and ghrelin and their effect on satiety in obese craniopharyngioma patients. Their findings support the hypothesis that reduced ghrelin secretion and reduced postprandial suppression of ghrelin and severe obesity lead to disturbances of appetite regulation in craniopharyngioma patients. However, peptide YY levels did not differ between normal weight, obese, and very obese patients. Further potential pathogenic roles of brain-derived neurotrophic factor and peripheral α-melanocyte-stimulating hormone in childhood-onset adamantinomatous craniopharyngioma obesity have been postulated [138, 139].

When elevated serum leptin levels relative to body mass index were found in childhood craniopharyngioma patients with a suprasellar tumour extension [140], researchers suggested that normal appetite inhibition failed to occur in these patients due to disruption of hypothalamic receptors that regulate the negative feedback loop in which leptin, formed in adipocytes, binds to hypothalamic leptin receptors.

Hoffmann et al. [141] recently reported on non-alcoholic fatty liver disease (NAFLD), a severe, previously underestimated sequelae in childhood-onset craniopharyngioma patients with hypothalamic obesity. NAFLD occurred in about 50% of childhood-onset craniopharyngioma patients with hypothalamic obesity and was associated with elevated liver enzymes and HOMA index. Over half of all patients (60%) with NAFLD were treated by stimulating agents for hypersomnia and increased daytime sleepiness as clinical manifestations of secondary narcolepsy, exerting considerable liver toxicity. The authors recommended that stimulating agents for treatment of secondary narcolepsy in childhood-onset adamantinomatous craniopharyngioma should be prescribed judiciously.

Challenges in Treating Hypothalamic Obesity

Due to the above-reported disturbances in energy expenditure, appetite regulation, and central sympathetic output, craniopharyngioma patients with hypothalamic involvement typically develop morbid obesity that is mainly unresponsive to conventional lifestyle modifications [3, 4, 14, 93, 123, 142, 143].

Recent studies on novel pharmaceutical treatment options for hypothalamic obesity in craniopharyngioma patients report mixed results [10].

Sibutramine is a neurotransmitter reuptake-inhibitor that reduces the reuptake of serotonin, norepinephrine, and dopamine, thereby increasing the levels of these substances in synaptic clefts and helping enhance satiety. Sibutramine has been widely used to treat obesity, leading to a weight loss of 7–10% when combined with a regulated diet. Sibutramine was tested in a randomised placebo-controlled crossover trial in patient cohorts with different obesity syndromes [144]. While the drug was well tolerated and safe, the weight loss response in patients with hypothalamic obesity was less pronounced in comparison to other participants (such as trisomy 21 and Prader–Willi syndrome). While the effect on body mass index was promising, sibutramine has been taken off the market due to adverse side effects and further clinical trials are not expected.

Based on impairment of epinephrine production and sympatho-adrenal activation manifesting as reduced hormonal response to hypoglycaemia, treating this disorder with amphetamine derivatives has been suggested [145]. In a prospective open-label 6-month pilot treatment trial in nine obese subjects with childhood-onset craniopharyngioma Hamilton et al. [146] assessed the effect of combined diazoxide-metformin therapy. Combined diazoxide-metformin therapy was associated with reduced weight gain in patients with hypothalamic obesity. Insulin levels at study commencement predicted effectiveness of the treatment. Kalina et al. [147] analysed the effect of fenofibrate and metformin treatment on metabolic status in 22 childhood-onset craniopharyngioma patients. The authors reported a positive effect on homeostatic model assessment (HOMA) and dyslipidemia. Zoicas et al. [148] treated eight adult patients with hypothalamic obesity (six craniopharyngioma) with GLP-1 analogues and observed a substantial and sustained weight loss associated with improvements in cardiovascular and metabolic risk profiles. Van Santen et al. [149] reported a case with hypothalamic obesity resulting from treatment of craniopharyngioma in which T3 mono-therapy was not effective in increasing REE or brown adipose tissue (BAT) activity. This may be explained by damage to the ventromedial hypothalamic region, which is a key area in the hypothalamus for T3-mediated BAT activation. Whereas substitution therapy with recombinant growth hormone is efficient in promoting normal growth, clinically relevant weight reducing effects are not observed in childhood-onset patients with hypothalamic obesity [150, 151].

Childhood-onset craniopharyngioma patients with hypothalamic obesity have a parasympathetic predominance of the autonomic nervous system induced by vagal activation, manifesting as reduced heart rate variability and daytime sleepiness [124]. Parasympathetic stimulation induces insulin secretion by direct β cells activation as well as adipogenesis. Insulin is an anabolic hormone and an important driver of weight gain in hypothalamic obesity. Octreotide as a somatostatin analogue reduces insulin secretion. Lustig et al. [152] tested octreotide in a double blind randomised controlled study in children with hypothalamic obesity, demonstrating moderate reductions in weight gain in which insulin levels during a proof-of-concept oral glucose tolerance test decreased without leading to relevant changes in glucose tolerance.

Initial experiences with laparoscopic gastric banding (LAGB) in severely obese childhood-onset adamantinomatous craniopharyngioma patients achieved sufficient tolerability and short-term weight reduction [153–155]. An instant improvement of binge-eating behaviour in craniopharyngioma patients was observed immediately after bariatric procedure (LAGB) [155], but failed in long-term weight reduction. Nevertheless, weight stabilisation could be achieved with regular follow-up monitoring [156]. In a systematic review and meta-analysis of the literature, Bretault et al. [153] analyzed the 12 months outcome after bariatric surgery for hypothalamic obesity due to craniopharyngioma treatment. At 1 year, among 18 cases with follow-up data, six had lost more than 20% of their initial weight; all had undergone either Roux Y gastric bypass ($n = 3$), sleeve gastrectomy, ($n = 2$), or biliopancreatic diversion ($n = 1$). All patients who had lost less than 5% of their initial weight had undergone LAGB except one Roux Y gastric bypass case. These findings indicate that Roux Y gastric bypass, sleeve gastrectomy, and biliopancreatic diversion are the most efficient bariatric procedures for weight reduction in hypothalamic obesity of childhood craniopharyngioma. However, treatment with invasive, non-reversible bariatric methods are controversial in the pediatric population because of medical, ethical, and legal considerations [156–158].

Reports on tolerability and efficacy of gastric pacemaking devices [159] and deep brain stimulation [160] in treatment of hypothalamic obesity have not been published in craniopharyngioma patients.

Despite the availability of promising therapeutic approaches [161], it must be emphasised that currently no generally accepted therapy for hypothalamic obesity in craniopharyngioma has been shown to be effective in randomised studies. Furthermore, structured rehabilitation programs for weight reduction in survivors of childhood-onset adamantinomatous craniopharyngioma have no proven persistent long-term effect on weight stabilisation or reduction [162].

However, an adequate communication of interdisciplinary decisions with patients' families [163] and purposeful care home environment [164] have been reported to exert beneficial effects on follow-up treatment of childhood-onset craniopharyngioma patients.

Eating Behaviour

Relevant associations between obesity and an obesogenic environment and consequent eating behaviour have been confirmed in children and adolescents [7, 10]. Due to the dearth of studies on eating behaviour in childhood-onset adamantinomatous craniopharyngioma patients, Hoffmann et al. [165] analysed eating disorders and eating behaviour in 101 survivors of childhood-onset craniopharyngioma and 85 BMI-matched healthy controls. Severely obese patients (BMI >+8 SD; $n = 9$) presented with pathological eating behaviours and more weight problems and eating disorders, as compared to obese (BMI +3 to +8 SD; $n = 44$) and normal or overweight patients (BMI <+3 SD; $n = 48$). However, childhood-onset craniopharyngioma patients with different degree of obesity showed similar or even less pathological findings as compared to BMI-matched normal controls. Hoffmann et al. conclude that the observed eating disorders are not disease-specific for childhood-onset craniopharyngioma.

Roth et al. [166] assessed pre- and post-meal responses to visual food cues in childhood-onset adamantinomatous craniopharyngioma patients' brain regions of interest using functional magnetic resonance imaging (fMRI). Following a test meal, BMI-matched controls showed suppression of activation by high-calorie food cues while childhood-onset craniopharyngioma patients showed trends towards higher activation. These findings support the hypothesis that perception of food cues especially after food intake may be altered in childhood-onset craniopharyngioma patients with hypothalamic obesity.

Roemmler-Zehrer et al. [167] compared eating behaviour in 26 craniopharyngioma patients (four childhood-onset cases) with 26 patients with non-functioning pituitary adenoma. Whereas craniopharyngioma patients scored higher in terms of conscious hunger perception, the rate of eating disorders was similar in both groups, supporting the hypothesis that eating disorders in patients with hypothalamic obesity are not disease specific.

Even though hypothalamic obesity is a frequent sequela in childhood-onset adamantinomatous craniopharyngioma [168], also diencephalic syndrome leading to severe weight loss and cachexia can occur as a rare hypothalamic disturbance of body composition in childhood-onset craniopharyngioma [169, 170]. Hoffmann et al. [170] analysed the incidence of diencephalic syndrome, its clinical manifestations before and after diagnosis of childhood-onset craniopharyngioma, and outcome in 485 patients recruited in the German childhood craniopharyngioma registry. Only 4.3% of all childhood-onset craniopharyngioma patients presented with low weight (BMI <−2 SD) at time of diagnosis. Initial significant differences between patients with low weight at the time of diagnosis and normal weight patients at diagnosis are usually observed at 5 years of age. Within the first 2 years after diagnosis of childhood-onset craniopharyngioma, the BMI of diencephalic syndrome patients and normal weight patients converge to a similar level. Hoffmann et al. concluded from their analysis of patients' histories that diencephalic syndrome at the time of diagnosis does not preclude subsequent weight gain caused by a childhood-onset adamantinomatous craniopharyngioma with hypothalamic involvement.

Sequelae, Prognosis, and Quality of Life

Quality of Life, Neurocognitive Outcome, and Psychosocial Functioning

Quality of life in childhood-onset craniopharyngioma patients can be affected by both the tumour itself and the treatment received [171]. Reports assessing physical and psychosocial functioning show variable results, ranging from excellent in a majority of subjects to impaired function in almost half of the patients [70, 75, 172–175]. The most common areas of impairments reported include emotional and social functioning, with patients rating their psychosocial status to be lower than their physical health [75]. Other reported challenges include somatic complaints such as pain, reduced mobility, and self-care [74, 75]. Behavioural studies indicate a high rate of psychopathological symptoms, including depression, anxiety, and withdrawal. The most frequent problems in children's everyday functional capacity include inability to control emotions, difficulties in learning, unsatisfactory peer relationships, and concerns regarding physical appearance and body image [83, 176].

Reported factors associated with impaired quality of survival outcomes as well as psychosocial and neurocognitive functioning include younger age at the time of diagnosis and preoperative functional impairment. Furthermore, tumour characteristics—including larger tumour volume, hypothalamic, and 3rd ventricle involvement at presentation—are reported in the literature to complicate both pre- and post-surgical conditions and therefore survival and quality of life outcomes in these patients [38, 42, 49, 177]. Furthermore, treatment strategies have been implicated with worse outcomes for surgery alone compared to limited surgery followed by irradiation and for multiple operations for tumour recurrence. Neurological, endocrine, and ophthalmological sequelae all adversely affect quality of life outcome [46, 70, 72, 74, 75, 172, 173]. Hypothalamic dysfunction has been found to have the most important negative impact on physical ability, body image, and social functioning [31, 75, 173].

Long-term neurocognitive complications following treatment for craniopharyngioma include cognitive problems, particularly those affecting attention, executive function, episodic memory, and working memory [75, 83, 176, 178–182]. Özyurt et al. [183] observed that childhood-onset craniopharyngioma patients had lower performance scores in tests of memory and executive functioning when compared with normal controls. Performance in functional capabilities and executive functions were negatively associated with the degree of hypothalamic involvement and damage.

Long-term survivors of childhood craniopharyngioma treated primarily with subtotal surgical resection followed by irradiation were also found to have significant psychological and educational deficits [83]. Neurocognitive deficits include slower cognitive speed, attention problems, memory disturbances, and behavioural instability [83, 176, 178, 180, 181, 184]. Whereas intact intellectual functioning has been reported in up to 82% of patients, visual memory is impaired despite normal visual-spatial abilities [83, 176]. The observed deficits in higher cognitive processing such as attention problems are considered precursors to poor academic achievement.

Despite over a quarter of century of literature documenting the neurocognitive challenges encountered by patients treated for craniopharyngioma, intervention efforts have lagged. Recent case studies have examined the efficacy of cognitive rehabilitation for dysexecutive problems and behavioural lability [185, 186]. In a case report of a 2-month intervention using a combination of goal management therapy and naturally occurring distractions within the patients' work environment, significant improvements in cognitive tests requiring organised behaviour were reported. Emotional, social, and/or behaviour problems, most notably aggressive behaviour, have been reported [75, 179, 187]. Despite their occurrence, the assumption of the biological underpinning of behavioural disturbances in these patients appears to result in limited attempts to effectively manage these disturbances with intervention. Behavioural treatment was used for severe aggressive behaviours [186]; the intervention included extinction of dysfunctional conditioned responses and functional behavioural analysis followed by differential reinforcement of alternative behaviours, with the goal of decreasing the frequency of aggressive behaviour. Aggressive behaviour subsequently decreased to below 88% of baseline levels and adaptive behaviours were found to increase significantly. These results suggest that patients' aggression is maintained by inadvertent social reinforcement. Taken together, these case studies suggest that cognitive rehabilitation efforts such as functional behavioural analysis and goal management therapy appear to be useful diagnostic and therapeutic options, compensating for psychosocial and cognitive challenges [188].

Cerebrovascular Morbidity

Radiation-induced vasculopathies are a rare consequence of radiation therapy for childhood-onset craniopharyngioma. In irradiated craniopharyngioma patients, moyamoya syndrome (a radiation-induced cerebrovascular condition predisposing to stroke) has been described [189, 190]. A retrospective estimate was that 27% of 22 patients treated with irradiation and some combination of surgery and intracystic chemotherapy with a median radiation dose of 52.2 Gray developed some type of vasculopathy, only half of which were symptomatic [190]. Although in this study no association was found between age, radiation dose, and maximum or mean dose to the internal carotid arteries with the presence of vascular abnormalities. Regine and colleagues reported a 13.7% rate of cerebrovascular events, all in irradiated cases receiving over 61 Gray [191]. No cerebrovascular clinical events have been reported in other series of conventionally fractionated radiotherapy for craniopharyngioma [58, 78, 192–196]. In 10–29% of childhood-onset craniopharyngioma patients fusiform dilatations of the carotid artery have been detected on postoperative magnetic resonance imaging after gross-total resection. The natural history appears to be benign. There have been no published reports of hemorrhage or stroke attributable to these lesions to date [197, 198]. However, studies on long-term prognosis in childhood-onset craniopharyngioma patients with fusiform dilatations of the carotid artery are still missing.

Other cardiac risk factors have also been observed in patients with craniopharyngioma. A small retrospective study of 12 craniopharyngioma patients found that half of them had at least one abnormality of cardiac function, structure, or rhythm, specifically prolonged QT interval in 25% of the cohort [199]. Besides cardio- and cerebrovascular mortality, other main causes of death in patients with childhood-onset craniopharyngioma are respiratory complications [200] and infections [79].

Second Malignant Neoplasms

In the largest reported series [192], no second malignancies in the irradiated field were observed in 173 irradiated patients with a median follow-up of 12 years after irradiation. Three long-term survivors (2%) died of malignancies with unspecified diagnoses. Overall, only four cases of second malignancies have been reported [61, 189, 201, 202], comprising two in-field glioblastomas [201, 202], one in-field glioma with unspecified grade of malignancy [189] and a posterior fossa meningioma [61].

Survival and Late Mortality

Craniopharyngiomas are associated with significant mortality, with reported overall mortality rates three to five times higher than those of the general population [79, 203]. When assessing mortality in patients with craniopharyngioma, the literature indicates that it is important to consider cases in adult- and childhood-onset craniopharyngioma separately. The overall survival rates (which reflect the effect of multiple treatments) described in exclusive children series range from 83 to 96% at 5 years [31, 34, 201, 202, 204–206] and 65 to 100% at 10 years [34, 202] [38, 70, 75, 189, 201, 205, 207–212] and average 62% at 20 years [191]. In adults or a mixed-age range population (adults and children) series, the overall survival rates range from 54 to 96% at 5 years [78, 79, 172, 192–194, 201, 213–215], from 40 to 93% at 10 years [72, 78, 79, 172, 192–194, 201, 203, 214, 215], and from 66 to 85% at 20 years [79, 192, 194]. The lower limits of survival rates usually reflected data from earlier series, before modern advances in microsurgery, neuroimaging, and radiooncological therapy. It is not clear whether the age at diagnosis represents a survival prognostic factor because some studies have shown that the youngest patients have better survival rates [201, 203, 213, 215]; others have found better outcome in older patients [177, 192], whereas still other studies report no difference between pediatric and adult populations [78, 193, 194, 216, 217]. The role of gender as a prognostic factor is not clear; some authors report a higher mortality among females [79, 203], but others have not found any gender differences [78, 193, 205, 217]. One of the two studies reporting higher mortality rates in females suggested a possible role of estrogen deficiency [79], but the other study did not consider that unsupplemented gonadal insufficiency had significant impact on mortality rates [203].

Sterkenburg et al. recently reported on an impaired 20-year overall survival in childhood-onset adamantinomatous craniopharyngioma patients with hypothalamic involvement [11, 218] (Fig. 4). Hypothalamic obesity has negative impact on long-term quality of survival [11]. Increased fatigue, daytime sleepiness, disturbances of circadian rhythms [120, 126, 219], memory deficits [183, 220, 221], pulmonary and gastrointestinal complaints (dyspnea, diarrhea) [11], and neuropsychological imbalances [112, 167, 222–224] are major long-term sequelae in childhood-onset craniopharyngioma patients with hypothalamic obesity.

Between 1998 and 2010, 567 patients with childhood-onset adamantinomatous craniopharyngioma have been recruited in the German craniopharyngioma registry. Only 5 of 567 recruited patients (<1%) presented without confirmed signs of relapse/progression, visual impairment, and neuroendocrine deficiencies during longitudinal follow-up of more than five years. The craniopharyngiomas in this small subgroup of patients without obvious symptoms indicating a chronic state of disease were characterised by a small non-recurrent tumour confined to the sellar area and without hypothalamic involvement, which could therefore be resected completely while preserving the integrity of the pituitary gland/stalk and visual and hypothalamic structures [50].

Disease-related mortality can occur even many years after treatment. Causes of late mortality include those directly related to the tumour or its treatment such as progressive disease with multiple recurrences, chronic hypothalamic insufficiency, hormonal deficiencies, cerebrovascular disease, and seizures [38, 46, 72, 75, 211]. Other causes have been described, including decreased mineral bone density and non-alcoholic steatohepatitis, leading to liver cirrhosis in some cases [32, 69, 75,

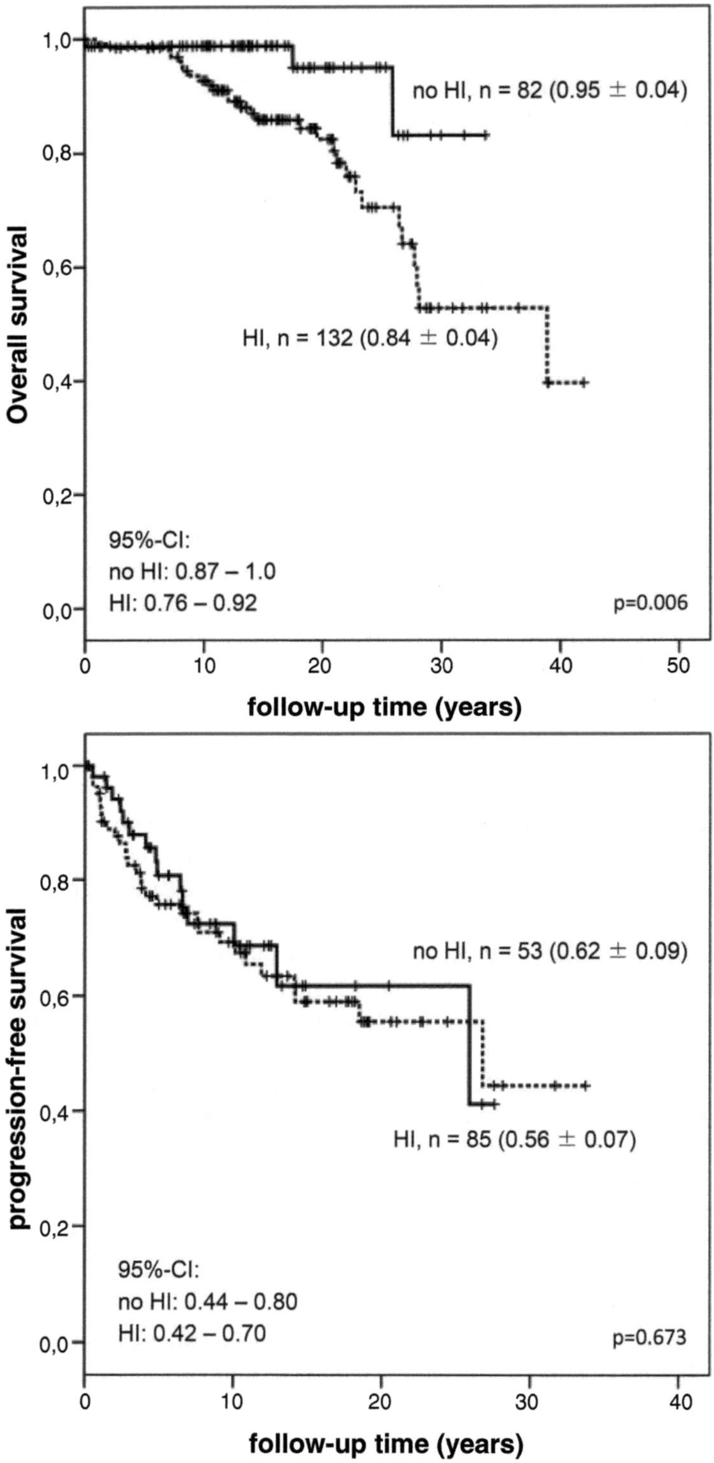

Fig. 4 Kaplan–Meier analyses of 20-year overall survival (**a**) and 20-year progression-free survival (PFS) (**b**) of patients with childhood-onset adamantinomatous craniopharyngioma recruited in HIT Endo related to hypothalamic involvement (HI). Reproduced from Sterkenburg et al. [11] with kind permission of Oxford University Press

141, 211, 225–228]. A recent review has reported substantial long-term morbidity with hypopituitarism, increased cardiovascular risk, hypothalamic damage, visual, and neurological deficits, reduced bone health, and reduction in quality of life and cognitive function. The standardised overall mortality rate varied from 2.88 to 9.28 in cohort studies covered in this review. According to the review, patients with adult-onset craniopharyngioma have a 3–19-fold higher cardiovascular mortality in comparison to the general population, and female adult-onset craniopharyngioma patients have an even higher risk [12].

Tumour size is likely to be a prognostic factor because increased survival rates have been shown in tumours with a diameter smaller than 3 cm [216]. The histological type as a prognostic factor is also controversial; better 5-year survival rates have been found in the squamous epithelial type vs. the adamantinomatous and combined histological types [229]. Higher perioperative deaths have also been reported in adult-onset adamantinomatous tumours [230], but other authors have not found significant differences between the two histological types [231, 232]. Several studies have described a more favourable prognosis when tumours lack calcification, especially in adult craniopharyngioma patients [216, 230], although no specific pathological feature predicted survival in childhood-onset craniopharyngioma patients [204]. In other studies, neither tumour histology [78, 216] nor tumour location [78, 205] had prognostic importance. In childhood-onset craniopharyngioma patients, the use of modern imaging as well as a good initial performance status (measured according to a functional classification that includes the presence of hypopituitarism, visual deficits, and neurological impairment) has been correlated with enhanced overall post-surgical survival at 10 years [202]. It is not clear whether the presence of hydrocephalus constitutes a prognostic factor because increased mortality [177] and lack of association with mortality have been reported [78, 193, 204, 205, 218]. Daubenbüchel et al. [218] recently published the results of the study on 163 long-term survivors of childhood-onset adamantinomatous craniopharyngioma showing that initial hydrocephalus had no impact on outcome. Overall survival rates and functional capacity were impaired in survivors presenting with initial hypothalamic involvement. Progression-free survival was not affected by hydrocephalus, hypothalamic involvement, or degree of initial resection. Accordingly, the authors conclude that gross-total resection is not recommended in adamantinomatous craniopharyngioma with initial hypothalamic involvement to prevent further hypothalamic damage [218].

Conclusions

Initial involvement and treatment-related hypothalamic lesions are major risk factors for impaired survival and reduced quality of survival due to hypothalamic obesity in craniopharyngioma [7]. Initial assessment of risk factors and execution of hypothalamus-sparing strategies by an experienced multidisciplinary team are recommended for prevention of further hypothalamic damage and consequent hypothalamic obesity. In the case of hypothalamic involvement, gross-total resection of craniopharyngioma should not be attempted. Based on the above-mentioned impact of initial involvement and treatment-related lesions of the hypothalamic–pituitary axis and optic structures on long-term morbidity and quality of survival, childhood-onset craniopharyngioma is a chronic, non-curable disease in most cases. Whereas 20-year overall survival rate is high (0.88 ± 0.03), low progression-free survival (0.58 ± 0.05) and low disease-free rate (<1%) in childhood-onset craniopharyngioma [50] support the concept of childhood-onset adamantinomatous craniopharyngioma as a chronic disease. Accordingly, treatment should be performed by multidisciplinary experienced teams in the context on national and international multicentre studies aiming at risk-adapted hypothalamus-sparing strategies and improvement of follow-up care. Due to the current lack of proven therapeutic options for hypothalamic syndrome and consecutive hypothalamic obesity, novel treatment approaches based on recent molecular findings are warranted [233].

Perspectives

As surgical expertise has been shown to have major impact on postoperative morbidity, medical societies should establish criteria of adequate professional expertise for diagnosis, initial grading, treatment, and care of adamantinomatous childhood-onset craniopharyngioma patients. Based on these criteria, health authorities should organise the certification of centres of excellence authorised for said treatment and care of patients with this disease.

Financial Disclosure This manuscript was composed in the absence of any commercial or financial relationships that could be construed as a potential conflict of interest. The author is supported by the German Childhood Cancer Foundation, Bonn, Germany.

References

1. Müller HL (2015) Craniopharyngioma – pediatric management. In: Evans JJ, Kenning TJ (eds) Craniopharyngiomas – comprehensive diagnosis, treatment and outcome, 1st edn. Elsevier, Munich, pp 429–458
2. Muller HL (2010) Childhood craniopharyngioma–current concepts in diagnosis, therapy and follow-up. Nat Rev Endocrinol 6(11):609–618. Epub 2010/09/30
3. Muller HL (2013) Paediatrics: surgical strategy and quality of life in craniopharyngioma. Nat Rev Endocrinol 9(8):447–449
4. Muller HL (2014) Craniopharyngioma. Endocr Rev 35(3):513–543
5. Daubenbuchel AM, Muller HL (2015) Neuroendocrine disorders in pediatric craniopharyngioma patients. J Clin Med 4(3):389–413
6. Muller HL (2016) Preoperative staging in childhood craniopharyngioma: standardization as a first step towards improved outcome. Endocrine 51(1):1–3
7. Muller HL (2016) Craniopharyngioma and hypothalamic injury: latest insights into consequent eating disorders and obesity. Curr Opin Endocrinol Diabetes Obes 23(1):81–89

8. Muller HL (2014) Craniopharyngioma. Handb Clin Neurol 124:235–253
9. Karavitaki N (2014) Management of craniopharyngiomas. J Endocrinol Invest 37(3):219–228
10. Bereket A, Kiess W, Lustig RH, Muller HL, Goldstone AP, Weiss R et al (2012) Hypothalamic obesity in children. Obes Rev 13(9):780–798
11. Sterkenburg AS, Hoffmann A, Gebhardt U, Warmuth-Metz M, Daubenbuchel AM, Muller HL (2015) Survival, hypothalamic obesity, and neuropsychological/psychosocial status after childhood-onset craniopharyngioma: newly reported long-term outcomes. Neuro Oncol 17(7):1029–1038
12. Erfurth EM, Holmer H, Fjalldal SB (2013) Mortality and morbidity in adult craniopharyngioma. Pituitary 16(1):46–55
13. Muller HL (2011) Diagnostics, treatment, and follow-up in craniopharyngioma. Front Endocrinol (Lausanne) 2:70
14. Muller HL (2013) Childhood craniopharyngioma. Pituitary 16(1):56–67
15. Muller HL, Albanese A, Calaminus G, Hargrave D, Garre ML, Gebhardt U et al (2006) Consensus and perspectives on treatment strategies in childhood craniopharyngioma: results of a meeting of the Craniopharyngioma Study Group (SIOP), Genova, 2004. J Pediatr Endocrinol Metab 19(Suppl 1):453–454
16. Nielsen EH, Feldt-Rasmussen U, Poulsgaard L, Kristensen LO, Astrup J, Jorgensen JO et al (2011) Incidence of craniopharyngioma in Denmark (n = 189) and estimated world incidence of craniopharyngioma in children and adults. J Neurooncol 104(3):755–763
17. Bunin GR, Surawicz TS, Witman PA, Preston-Martin S, Davis F, Bruner JM (1998) The descriptive epidemiology of craniopharyngioma. J Neurosurg 89(4):547–551
18. Muller-Scholden J, Lehrnbecher T, Muller HL, Bensch J, Hengen RH, Sorensen N et al (2000) Radical surgery in a neonate with craniopharyngioma. Report of a case. Pediatr Neurosurg 33(5):265–269
19. Larkin SJ, Ansorge O (2013) Pathology and pathogenesis of craniopharyngiomas. Pituitary 16(1):9–17
20. Rushing EJ, Giangaspero F, Paulus W, Burger PC (2007) Craniopharyngioma. In: Louis DN, Ohgaki H, Wiestler OD, Webster KC (eds) WHO classification of tumours of the central nervous system, 3rd edn. WHO, Geneva
21. Warmuth-Metz M, Gnekow AK, Muller H, Solymosi L (2004) Differential diagnosis of suprasellar tumors in children. Klin Padiatr 216(6):323–330
22. Boch AL, van Effenterre R, Kujas M (1997) Craniopharyngiomas in two consanguineous siblings: case report. Neurosurgery 41(5):1185–1187
23. Green AL, Yeh JS, Dias PS (2002) Craniopharyngioma in a mother and daughter. Acta Neurochir 144(4):403–404
24. Kordes U, Flitsch J, Hagel C, Goebell E, Schwarz R, Herberhold T et al (2011) Ectopic craniopharyngioma. Klin Padiatr 223(3):176–177
25. Muller HL, Kaatsch P, Warmuth-Metz M, Flentje M, Sorensen N (2003) Childhood craniopharyngioma - diagnostic and therapeutic strategies. Monatsschr Kinderheilkd 151:1056–1063
26. Sherlock M, Ayuk J, Tomlinson JW, Toogood AA, Aragon-Alonso A, Sheppard MC et al (2010) Mortality in patients with pituitary disease. Endocr Rev 31(3):301–342
27. Garre ML, Cama A (2007) Craniopharyngioma: modern concepts in pathogenesis and treatment. Curr Opin Pediatr 19(4):471–479
28. Kato K, Nakatani Y, Kanno H, Inayama Y, Ijiri R, Nagahara N et al (2004) Possible linkage between specific histological structures and aberrant reactivation of the Wnt pathway in adamantinomatous craniopharyngioma. J Pathol 203(3):814–821
29. Buslei R, Holsken A, Hofmann B, Kreutzer J, Siebzehnrubl F, Hans V et al (2007) Nuclear beta-catenin accumulation associates with epithelial morphogenesis in craniopharyngiomas. Acta Neuropathol 113(5):585–590
30. Martinez-Barbera JP (2015) 60 years of neuroendocrinology: biology of human craniopharyngioma: lessons from mouse models. J Endocrinol 226(2):T161–T172

31. Muller HL, Bueb K, Bartels U, Roth C, Harz K, Graf N et al (2001) Obesity after childhood craniopharyngioma–German multicenter study on pre-operative risk factors and quality of life. Klin Padiatr 213(4):244–249
32. Muller HL, Heinrich M, Bueb K, Etavard-Gorris N, Gebhardt U, Kolb R et al (2003) Perioperative dexamethasone treatment in childhood craniopharyngioma–influence on short-term and long-term weight gain. Exp Clin Endocrinol Diabetes 111(6):330–334
33. Sklar CA (1994) Craniopharyngioma: endocrine abnormalities at presentation. Pediatr Neurosurg 21(Suppl 1):18–20
34. Muller HL (2008) Childhood craniopharyngioma. Recent advances in diagnosis, treatment and follow-up. Horm Res 69(4):193–202
35. Elliott RE, Jane JA, Jr., Wisoff JH (2011) Surgical management of craniopharyngiomas in children: meta-analysis and comparison of transcranial and transsphenoidal approaches. Neurosurgery 69(3):630–643. discussion 43
36. Muller HL, Emser A, Faldum A, Bruhnken G, Etavard-Gorris N, Gebhardt U et al (2004) Longitudinal study on growth and body mass index before and after diagnosis of childhood craniopharyngioma. J Clin Endocrinol Metab 89(7):3298–3305
37. Hoffmann A, Boekhoff S, Gebhardt U, Sterkenburg AS, Daubenbuchel AM, Eveslage M et al (2015) History before diagnosis in childhood craniopharyngioma: associations with initial presentation and long-term prognosis. Eur J Endocrinol 173(6):853–862
38. De Vile CJ, Grant DB, Kendall BE, Neville BG, Stanhope R, Watkins KE et al (1996) Management of childhood craniopharyngioma: can the morbidity of radical surgery be predicted? J Neurosurg 85(1):73–81
39. de Vile CJ, Grant DB, Hayward RD, Kendall BE, Neville BG, Stanhope R (1996) Obesity in childhood craniopharyngioma: relation to post-operative hypothalamic damage shown by magnetic resonance imaging. J Clin Endocrinol Metab 81(7):2734–2737
40. Elowe-Gruau E, Beltrand J, Brauner R, Pinto G, Samara-Boustani D, Thalassinos C et al (2013) Childhood craniopharyngioma: hypothalamus-sparing surgery decreases the risk of obesity. J Clin Endocrinol Metab 98(6):2376–2382
41. Puget S, Garnett M, Wray A, Grill J, Habrand JL, Bodaert N et al (2007) Pediatric craniopharyngiomas: classification and treatment according to the degree of hypothalamic involvement. J Neurosurg 106(1 Suppl):3–12
42. Muller HL, Gebhardt U, Faldum A, Warmuth-Metz M, Pietsch T, Pohl F et al (2012) Xanthogranuloma, Rathke's cyst, and childhood craniopharyngioma: results of prospective multinational studies of children and adolescents with rare sellar malformations. J Clin Endocrinol Metab 97(11):3935–3943
43. Muller HL, Gebhardt U, Teske C, Faldum A, Zwiener I, Warmuth-Metz M et al (2011) Post-operative hypothalamic lesions and obesity in childhood craniopharyngioma: results of the multinational prospective trial KRANIOPHARYNGEOM 2000 after 3-year follow-up. Eur J Endocrinol 165(1):17–24
44. Mallucci C, Pizer B, Blair J, Didi M, Doss A, Upadrasta S et al (2012) Management of craniopharyngioma: the Liverpool experience following the introduction of the CCLG guidelines. Introducing a new risk assessment grading system. Childs Nerv Syst 28(8):1181–1192
45. Van Gompel JJ, Nippoldt TB, Higgins DM, Meyer FB (2010) Magnetic resonance imaging-graded hypothalamic compression in surgically treated adult craniopharyngiomas determining postoperative obesity. Neurosurg Focus 28(4):E3. Epub 2010/04/07
46. Steno J, Bizik I, Steno A, Matejcik V (2011) Craniopharyngiomas in children: how radical should the surgeon be? Childs Nerv Syst 27(1):41–54
47. Elliott RE, Sands SA, Strom RG, Wisoff JH (2010) Craniopharyngioma Clinical Status Scale: a standardized metric of preoperative function and posttreatment outcome. Neurosurg Focus 28(4):E2
48. Mortini P, Gagliardi F, Bailo M, Spina A, Parlangeli A, Falini A et al (2016) Magnetic resonance imaging as predictor of functional outcome in craniopharyngiomas. Endocrine 51(1):148–162

49. Muller HL (2011) Consequences of craniopharyngioma surgery in children. J Clin Endocrinol Metab 96(7):1981–1991
50. Muller HL (2015) Craniopharyngioma: long-term consequences of a chronic disease. Expert Rev Neurother 15(11):1241–1244
51. Sanford RA (1994) Craniopharyngioma: results of survey of the American Society of Pediatric Neurosurgery. Pediatr Neurosurg 21(Suppl 1):39–43
52. Boop FA 2007 Craniopharyngioma. J Neurosurg 106(1 Suppl):1–2. discussion
53. Hankinson TC, Palmeri NO, Williams SA, Torok MR, Serrano CA, Foreman NK et al (2013) Patterns of care for craniopharyngioma: survey of members of the american association of neurological surgeons. Pediatr Neurosurg 49(3):131–136
54. Hoffmann A, Warmth-Metz M, Gebhardt U, Pietsch T, Pohl F, Kortmann RD et al (2014) Childhood craniopharyngioma - changes of treatment strategies in the trials KRANIOPHARYNGEOM 2000/2007. Klin Padiatr 226(3):161–168
55. Muller HL (2006) More or less? Treatment strategies in childhood craniopharyngioma. Childs Nerv Syst 22(2):156–157
56. Kortmann RD (2011) Different approaches in radiation therapy of craniopharyngioma. Front Endocrinol (Lausanne) 2:100
57. Aggarwal A, Fersht N, Brada M (2013) Radiotherapy for craniopharyngioma. Pituitary 16(1):26–33
58. Merchant TE, Kiehna EN, Kun LE, Mulhern RK, Li C, Xiong X et al (2006) Phase II trial of conformal radiation therapy for pediatric patients with craniopharyngioma and correlation of surgical factors and radiation dosimetry with change in cognitive function. J Neurosurg 104(2 Suppl):94–102
59. Fitzek MM, Linggood RM, Adams J, Munzenrider JE (2006) Combined proton and photon irradiation for craniopharyngioma: long-term results of the early cohort of patients treated at Harvard Cyclotron Laboratory and Massachusetts General Hospital. Int J Radiat Oncol Biol Phys 64(5):1348–1354
60. Luu QT, Loredo LN, Archambeau JO, Yonemoto LT, Slater JM, Slater JD (2006) Fractionated proton radiation treatment for pediatric craniopharyngioma: preliminary report. Cancer J 12(2):155–159
61. Winkfield KM, Linsenmeier C, Yock TI, Grant PE, Yeap BY, Butler WE et al (2009) Surveillance of craniopharyngioma cyst growth in children treated with proton radiotherapy. Int J Radiat Oncol Biol Phys 73(3):716–721
62. Boehling NS, Grosshans DR, Bluett JB, Palmer MT, Song X, Amos RA et al (2012) Dosimetric comparison of three-dimensional conformal proton radiotherapy, intensity-modulated proton therapy, and intensity-modulated radiotherapy for treatment of pediatric craniopharyngiomas. Int J Radiat Oncol Biol Phys 82(2):643–652
63. Beltran C, Roca M, Merchant TE (2012) On the benefits and risks of proton therapy in pediatric craniopharyngioma. Int J Radiat Oncol Biol Phys 82(2):e281–e287
64. Cavalheiro S, Sparapani FV, Franco JO, da Silva MC, Braga FM (1996) Use of bleomycin in intratumoral chemotherapy for cystic craniopharyngioma. Case report. J Neurosurg 84(1):124–126
65. Schubert T, Trippel M, Tacke U, van Velthoven V, Gumpp V, Bartelt S et al (2009) Neurosurgical treatment strategies in childhood craniopharyngiomas: is less more? Childs Nerv Syst 25(11):1419–1427
66. Lafay-Cousin L, Bartels U, Raybaud C, Kulkarni AV, Guger S, Huang A et al (2007) Neuroradiological findings of bleomycin leakage in cystic craniopharyngioma. Report of three cases. J Neurosurg 107(4 Suppl):318–323
67. Cavalheiro S, Di Rocco C, Valenzuela S, Dastoli PA, Tamburrini G, Massimi L et al (2010) Craniopharyngiomas: intratumoral chemotherapy with interferon-alpha: a multicenter preliminary study with 60 cases. Neurosurg Focus 28(4):E12
68. Bartels U, Laperriere N, Bouffet E, Drake J (2012) Intracystic therapies for cystic craniopharyngioma in childhood. Front Endocrinol (Lausanne) 3:39

69. Caldarelli M, Massimi L, Tamburrini G, Cappa M, Di Rocco C (2005) Long-term results of the surgical treatment of craniopharyngioma: the experience at the Policlinico Gemelli, Catholic University, Rome. Childs Nerv Syst 21(8–9):747–757
70. Hoffman HJ, De Silva M, Humphreys RP, Drake JM, Smith ML, Blaser SI (1992) Aggressive surgical management of craniopharyngiomas in children. J Neurosurg 76(1):47–52
71. Honegger J, Buchfelder M, Fahlbusch R (1999) Surgical treatment of craniopharyngiomas: endocrinological results. J Neurosurg 90(2):251–257
72. Elliott RE, Wisoff JH (2010) Surgical management of giant pediatric craniopharyngiomas. J Neurosurg Pediatr 6(5):403–416
73. Jung TY, Jung S, Moon KS, Kim IY, Kang SS, Kim JH (2010) Endocrinological outcomes of pediatric craniopharyngiomas with anatomical pituitary stalk preservation: preliminary study. Pediatr Neurosurg 46(3):205–212
74. Merchant TE, Kiehna EN, Sanford RA, Mulhern RK, Thompson SJ, Wilson MW et al (2002) Craniopharyngioma: the St. Jude Children's Research Hospital experience 1984-2001. Int J Radiat Oncol Biol Phys 53(3):533–542
75. Poretti A, Grotzer MA, Ribi K, Schonle E, Boltshauser E (2004) Outcome of craniopharyngioma in children: long-term complications and quality of life. Dev Med Child Neurol 46(4):220–229
76. Elliott RE, Jane JA Jr, Wisoff JH (2011) Surgical management of craniopharyngiomas in children: meta-analysis and comparison of transcranial and transsphenoidal approaches. Neurosurgery 69(3):630–643
77. Ahmet A, Blaser S, Stephens D, Guger S, Rutkas JT, Hamilton J (2006) Weight gain in craniopharyngioma–a model for hypothalamic obesity. J Pediatr Endocrinol Metab 19(2):121–127
78. Karavitaki N, Brufani C, Warner JT, Adams CB, Richards P, Ansorge O et al (2005) Craniopharyngiomas in children and adults: systematic analysis of 121 cases with long-term follow-up. Clin Endocrinol (Oxf) 62(4):397–409
79. Pereira AM, Schmid EM, Schutte PJ, Voormolen JH, Biermasz NR, van Thiel SW et al (2005) High prevalence of long-term cardiovascular, neurological and psychosocial morbidity after treatment for craniopharyngioma. Clin Endocrinol (Oxf) 62(2):197–204
80. Mortini P, Losa M, Pozzobon G, Barzaghi R, Riva M, Acerno S et al (2011) Neurosurgical treatment of craniopharyngioma in adults and children: early and long-term results in a large case series. J Neurosurg 114(5):1350–1359
81. Kendall-Taylor P, Jonsson PJ, Abs R, Erfurth EM, Koltowska-Haggstrom M, Price DA et al (2005) The clinical, metabolic and endocrine features and the quality of life in adults with childhood-onset craniopharyngioma compared with adult-onset craniopharyngioma. Eur J Endocrinol 152(4):557–567
82. Halac I, Zimmerman D (2005) Endocrine manifestations of craniopharyngioma. Childs Nerv Syst 21(8–9):640–648
83. Crom D, Smith D, Xiong Z, Onar A, Hudson M, Merchant T et al (2010) Health status in long-term survivors of pediatric craniopharyngiomas. J Neurosci Nurs 42(6):323–328
84. Geffner M, Lundberg M, Koltowska-Haggstrom M, Abs R, Verhelst J, Erfurth EM et al (2004) Changes in height, weight, and body mass index in children with craniopharyngioma after three years of growth hormone therapy: analysis of KIGS (Pfizer International Growth Database). J Clin Endocrinol Metab 89(11):5435–5440
85. Srinivasan S, Ogle GD, Garnett SP, Briody JN, Lee JW, Cowell CT (2004) Features of the metabolic syndrome after childhood craniopharyngioma. J Clin Endocrinol Metab 89(1):81–86
86. Matson DD (1964) Craniopharyngioma. Clin Neurosurg 10:14
87. Phillip M, Moran O, Lazar L (2002) Growth without growth hormone. J Pediatr Endocrinol Metab 15(Suppl 5):1267–1272
88. Geffner ME (1996) The growth without growth hormone syndrome. Endocrinol Metab Clin North Am 25(3):649–663

89. Costin G, Kogut MD, Phillips LS, Daughaday WH (1976) Craniopharyngioma: the role of insulin in promoting postoperative growth. J Clin Endocrinol Metab 42(2):370–379
90. Rath SR, Lee S, Kotecha RS, Taylor M, Junckerstorff RC, Choong CS (2013) Childhood craniopharyngioma: 20-year institutional experience in Western Australia. J Paediatr Child Health 49(5):403–408
91. Lo AC, Howard AF, Nichol A, Sidhu K, Abdulsatar F, Hasan H et al (2014) Long-term outcomes and complications in patients with craniopharyngioma: the British Columbia Cancer Agency experience. Int J Radiat Oncol Biol Phys 88(5):1011–1018
92. Zaidi HA, Chapple K, Little AS (2014) National treatment trends, complications, and predictors of in-hospital charges for the surgical management of craniopharyngiomas in adults from 2007 to 2011. Neurosurg Focus 37(5):E6
93. Muller HL (2014) Childhood craniopharyngioma: treatment strategies and outcomes. Expert Rev Neurother 14(2):187–197
94. Solari D, Cavallo LM, Cappabianca P (2014) Surgical approach to pituitary tumors. Handb Clin Neurol 124:291–301
95. Flitsch J, Aberle J, Burkhardt T (2015) Surgery for pediatric craniopharyngiomas: is less more? J Pediatr Endocrinol Metab 28(1–2):27–33
96. Buchfelder M, Schlaffer SM, Lin F, Kleindienst A (2013) Surgery for craniopharyngioma. Pituitary 16(1):18–25
97. Mortini P, Gagliardi F, Boari N, Losa M (2013) Surgical strategies and modern therapeutic options in the treatment of craniopharyngiomas. Crit Rev Oncol Hematol 88(3):514–529
98. Ali ZS, Bailey RL, Daniels LB, Vakhshori V, Lewis DJ, Hossain AT et al (2014) Comparative effectiveness of treatment options for pediatric craniopharyngiomas. J Neurosurg Pediatr 13(2):178–188
99. Ogawa Y, Kawaguchi T, Tominaga T (2014) Outcome and mid-term prognosis after maximum and radical removal of craniopharyngiomas with the priority to the extended transsphenoidal approach–a single center experience. Clin Neurol Neurosurg 125:41–46
100. Raza SM, Schwartz TH (2014) How to achieve the best possible outcomes in the management of retroinfundibular craniopharyngiomas? World Neurosurg 82(5):614–616
101. Flitsch J, Muller HL, Burkhardt T (2011) Surgical strategies in childhood craniopharyngioma. Front Endocrinol (Lausanne) 2:96
102. Patel KS, Raza SM, McCoul ED, Patrona A, Greenfield JP, Souweidane MM et al (2015) Long-term quality of life after endonasal endoscopic resection of adult craniopharyngiomas. J Neurosurg 123:571–580
103. Cavallo LM, Solari D, Esposito F, Villa A, Minniti G, Cappabianca P (2014) The role of the endoscopic endonasal route in the management of craniopharyngiomas. World Neurosurg 82(6 Suppl):S32–S40
104. Gardner PA, Kassam AB, Snyderman CH, Carrau RL, Mintz AH, Grahovac S et al (2008) Outcomes following endoscopic, expanded endonasal resection of suprasellar craniopharyngiomas: a case series. J Neurosurg 109(1):6–16
105. Bailey S, Parkes J (2015) Intracystic interferon therapy in childhood craniopharyngioma: who, when and how? Clin Endocrinol (Oxf) 82(1):29–34
106. Zheng J, Fang Y, Cai BW, Zhang H, Liu W, Wu B et al (2014) Intracystic bleomycin for cystic craniopharyngiomas in children. Cochrane Database Syst Rev 9:CD008890
107. Klimo P Jr, Venable GT, Boop FA, Merchant TE (2015) Recurrent craniopharyngioma after conformal radiation in children and the burden of treatment. J Neurosurg Pediatr 15(5):499–505
108. Greenfield BJ, Okcu MF, Baxter PA, Chintagumpala M, Teh BS, Dauser RC et al (2015) Long-term disease control and toxicity outcomes following surgery and intensity modulated radiation therapy (IMRT) in pediatric craniopharyngioma. Radiother Oncol 114(2):224–229
109. Bishop AJ, Greenfield B, Mahajan A, Paulino AC, Okcu MF, Allen PK et al (2014) Proton beam therapy versus conformal photon radiation therapy for childhood craniopharyngioma:

multi-institutional analysis of outcomes, cyst dynamics, and toxicity. Int J Radiat Oncol Biol Phys 90(2):354–361
110. Harrabi SB, Adeberg S, Welzel T, Rieken S, Habermehl D, Debus J et al (2014) Long term results after fractionated stereotactic radiotherapy (FSRT) in patients with craniopharyngioma: maximal tumor control with minimal side effects. Radiat Oncol 9:203
111. Lee CC, Yang HC, Chen CJ, Hung YC, Wu HM, Shiau CY et al (2014) Gamma Knife surgery for craniopharyngioma: report on a 20-year experience. J Neurosurg 121(Suppl):167–178
112. Fjalldal S, Holmer H, Rylander L, Elfving M, Ekman B, Osterberg K et al (2013) Hypothalamic involvement predicts cognitive performance and psychosocial health in long-term survivors of childhood craniopharyngioma. J Clin Endocrinol Metab 98(8):3253–3262
113. Steno J, Bizik I, Steno A, Matejcik V (2013) Craniopharyngiomas and the hypothalamus. J Neurosurg 119(6):1646–1650
114. Rosenfeld A, Arrington D, Miller J, Olson M, Gieseking A, Etzl M et al (2014) A review of childhood and adolescent craniopharyngiomas with particular attention to hypothalamic obesity. Pediatr Neurol 50(1):4–10
115. Khan MJ, Humayun KN, Donaldson M, Ahmed SF, Shaikh MG (2014) Longitudinal changes in body mass index in children with craniopharyngioma. Horm Res Paediatr 82(6):372–379
116. Erfurth EM (2015) Endocrine aspects and sequel in patients with craniopharyngioma. J Pediatr Endocrinol Metab 28(1–2):19–26
117. Muller HL (2010) Childhood craniopharyngioma: current controversies on management in diagnostics, treatment and follow-up. Exp Rev Neurother 10(4):515–524
118. Muller HL, Gebhardt U, Schroder S, Pohl F, Kortmann RD, Faldum A et al (2010) Analyses of treatment variables for patients with childhood craniopharyngioma–results of the multicenter prospective trial KRANIOPHARYNGEOM 2000 after three years of follow-up. Horm Res Paediatr 73(3):175–180
119. Roth CL, Eslamy H, Werny D, Elfers C, Shaffer ML, Pihoker C et al (2015) Semiquantitative analysis of hypothalamic damage on MRI predicts risk for hypothalamic obesity. Obesity (Silver Spring) 23(6):1226–1233
120. Muller HL (2010) Increased daytime sleepiness in patients with childhood craniopharyngioma and hypothalamic tumor involvement: review of the literature and perspectives. Int J Endocrinol 2010:519607
121. Castro DC, Cole SL, Berridge KC (2015) Lateral hypothalamus, nucleus accumbens, and ventral pallidum roles in eating and hunger: interactions between homeostatic and reward circuitry. Front Syst Neurosci 9:90
122. Kim JH, Choi JH (2013) Pathophysiology and clinical characteristics of hypothalamic obesity in children and adolescents. Ann Pediatr Endocrinol Metab 18(4):161–167
123. Haliloglu B, Bereket A (2015) Hypothalamic obesity in children: pathophysiology to clinical management. J Pediatr Endocrinol Metab 28(5–6):503–513
124. Cohen M, Syme C, McCrindle BW, Hamilton J (2013) Autonomic nervous system balance in children and adolescents with craniopharyngioma and hypothalamic obesity. Eur J Endocrinol 168(6):845–852
125. Harz KJ, Muller HL, Waldeck E, Pudel V, Roth C (2003) Obesity in patients with craniopharyngioma: assessment of food intake and movement counts indicating physical activity. J Clin Endocrinol Metab 88(11):5227–5231
126. Muller HL, Handwerker G, Gebhardt U, Faldum A, Emser A, Kolb R et al (2006) Melatonin treatment in obese patients with childhood craniopharyngioma and increased daytime sleepiness. Cancer Causes Control 17(4):583–589
127. Muller HL, Handwerker G, Wollny B, Faldum A, Sorensen N (2002) Melatonin secretion and increased daytime sleepiness in childhood craniopharyngioma patients. J Clin Endocrinol Metab 87(8):3993–3996
128. O'Gorman CS, Simoneau-Roy J, Pencharz P, MacFarlane J, MacLusky I, Narang I et al (2010) Sleep-disordered breathing is increased in obese adolescents with craniopharyngioma compared with obese controls. J Clin Endocrinol Metab 95(5):2211–2218

129. Muller HL, Muller-Stover S, Gebhardt U, Kolb R, Sorensen N, Handwerker G (2006) Secondary narcolepsy may be a causative factor of increased daytime sleepiness in obese childhood craniopharyngioma patients. J Pediatr Endocrinol Metab 19(Suppl 1):423–429
130. Mason PW, Krawiecki N, Meacham LR (2002) The use of dextroamphetamine to treat obesity and hyperphagia in children treated for craniopharyngioma. Arch Pediatr Adolesc Med 156(9):887–892
131. Roth CL, Hunneman DH, Gebhardt U, Stoffel-Wagner B, Reinehr T, Muller HL (2007) Reduced sympathetic metabolites in urine of obese patients with craniopharyngioma. Pediatr Res 61(4):496–501
132. Roth CL (2011) Hypothalamic obesity in patients with craniopharyngioma: profound changes of several weight regulatory circuits. Front Endocrinol 2:49
133. Bomer I, Saure C, Caminiti C, Ramos JG, Zuccaro G, Brea M et al (2015) Comparison of energy expenditure, body composition, metabolic disorders, and energy intake between obese children with a history of craniopharyngioma and children with multifactorial obesity. J Pediatr Endocrinol Metab 28(11–12):1305–1312
134. Holmer H, Pozarek G, Wirfalt E, Popovic V, Ekman B, Bjork J et al (2010) Reduced energy expenditure and impaired feeding-related signals but not high energy intake reinforces hypothalamic obesity in adults with childhood onset craniopharyngioma. J Clin Endocrinol Metab 95(12):5395–5402
135. Prieto R, Pascual JM, Barrios L (2015) Optic chiasm distortions caused by craniopharyngiomas: clinical and magnetic resonance imaging correlation and influence on visual outcome. World Neurosurg 83(4):500–529
136. Drimtzias E, Falzon K, Picton S, Jeeva I, Guy D, Nelson O et al (2014) The ophthalmic natural history of paediatric craniopharyngioma: a long-term review. J Neurooncol 120(3):651–656
137. Roemmler-Zehrer J, Geigenberger V, Stormann S, Losa M, Crippa V, Otto B et al (2014) Food intake regulating hormones in adult craniopharyngioma patients. Eur J Endocrinol 170(4):627–635
138. Roth CL, Elfers C, Gebhardt U, Muller HL, Reinehr T (2013) Brain-derived neurotrophic factor and its relation to leptin in obese children before and after weight loss. Metabolism 62(2):226–234
139. Roth CL, Enriori PJ, Gebhardt U, Hinney A, Muller HL, Hebebrand J et al (2010) Changes of peripheral alpha-melanocyte-stimulating hormone in childhood obesity. Metabolism 59(2):186–194
140. Roth C, Wilken B, Hanefeld F, Schroter W, Leonhardt U (1998) Hyperphagia in children with craniopharyngioma is associated with hyperleptinaemia and a failure in the downregulation of appetite. Eur J Endocrinol 138(1):89–91
141. Hoffmann A, Bootsveld K, Gebhardt U, Daubenbuchel AM, Sterkenburg AS, Muller HL (2015) Nonalcoholic fatty liver disease and fatigue in long-term survivors of childhood-onset craniopharyngioma. Eur J Endocrinol 173(3):389–397
142. Muller HL (2015) Childhood craniopharyngioma - current status and recent perspectives in diagnostics and treatment. J Pediatr Endocrinol Metab 28(1–2):1–2
143. Lustig RH (2011) Hypothalamic obesity after craniopharyngioma: mechanisms, diagnosis, and treatment. Front Endocrinol (Lausanne) 2:60
144. Danielsson P, Janson A, Norgren S, Marcus C (2007) Impact sibutramine therapy in children with hypothalamic obesity or obesity with aggravating syndromes. J Clin Endocrinol Metab 92(11):4101–4106
145. Elfers CT, Roth CL (2011) Effects of methylphenidate on weight gain and food intake in hypothalamic obesity. Front Endocrinology (Lausanne) 2:78
146. Hamilton JK, Conwell LS, Syme C, Ahmet A, Jeffery A, Daneman D (2011) Hypothalamic obesity following craniopharyngioma surgery: results of a pilot trial of combined diazoxide and metformin therapy. J Pediatr Endocrinol Metab 2011:417949

147. Kalina MA, Wilczek M, Kalina-Faska B, Skala-Zamorowska E, Mandera M, Malecka TE (2015) Carbohydrate-lipid profile and use of metformin with micronized fenofibrate in reducing metabolic consequences of craniopharyngioma treatment in children: single institution experience. J Pediatr Endocrinol Metab 28(1–2):45–51
148. Zoicas F, Droste M, Mayr B, Buchfelder M, Schofl C (2013) GLP-1 analogues as a new treatment option for hypothalamic obesity in adults: report of nine cases. Eur J Endocrinol 168(5):699–706
149. van Santen HM, Schouten-Meeteren AY, Serlie M, Meijneke RW, van Trotsenburg AS, Verberne H et al (2015) Effects of T3 treatment on brown adipose tissue and energy expenditure in a patient with craniopharyngioma and hypothalamic obesity. J Pediatr Endocrinol Metab 28(1–2):53–57
150. Profka E, Giavoli C, Bergamaschi S, Ferrante E, Malchiodi E, Sala E et al (2015) Analysis of short- and long-term metabolic effects of growth hormone replacement therapy in adult patients with craniopharyngioma and non-functioning pituitary adenoma. J Endocrinol Invest 38(4):413–420
151. Yuen KC, Koltowska-Haggstrom M, Cook DM, Fox JL, Jonsson PJ, Geffner ME et al (2013) Clinical characteristics and effects of GH replacement therapy in adults with childhood-onset craniopharyngioma compared with those in adults with other causes of childhood-onset hypothalamic-pituitary dysfunction. Eur J Endocrinol 169(4):511–519
152. Lustig RH, Hinds PS, Ringwald-Smith K, Christensen RK, Kaste SC, Schreiber RE et al (2003) Octreotide therapy of pediatric hypothalamic obesity: a double-blind, placebo-controlled trial. J Clin Endocrinol Metab 88(6):2586–2592
153. Bretault M, Boillot A, Muzard L, Poitou C, Oppert JM, Barsamian C et al (2013) Clinical review: bariatric surgery following treatment for craniopharyngioma: a systematic review and individual-level data meta-analysis. J Clin Endocrinol Metab 98(6):2239–2246
154. Gatta B, Nunes ML, Bailacq-Auder C, Etchechoury L, Collet D, Tabarin A (2013) Is bariatric surgery really inefficient in hypothalamic obesity? Clin Endocrinol (Oxf) 78(4):636–638
155. Muller HL, Gebhardt U, Wessel V, Schroder S, Kolb R, Sorensen N et al (2007) First experiences with laparoscopic adjustable gastric banding (LAGB) in the treatment of patients with childhood craniopharyngioma and morbid obesity. Klin Padiatr 219(6):323–325
156. Muller HL, Gebhardt U, Maroske J, Hanisch E (2011) Long-term follow-up of morbidly obese patients with childhood craniopharyngioma after laparoscopic adjustable gastric banding (LAGB). Klin Padiatr 223(6):372–373
157. Schultes B, Ernst B, Schmid F, Thurnheer M (2009) Distal gastric bypass surgery for the treatment of hypothalamic obesity after childhood craniopharyngioma. Eur J Endocrinol 161(1):201–206
158. Rottembourg D, O'Gorman CS, Urbach S, Garneau PY, Langer JC, Van Vliet G et al (2009) Outcome after bariatric surgery in two adolescents with hypothalamic obesity following treatment of craniopharyngioma. J Pediatr Endocrinol Metab 22(9):867–872
159. Zonca P, Hoppe C, Cambal M, Jacobi CA (2014) Gastric stimulation in treatment in type 2 diabetes mellitus. Bratisl Lek Listy 115(1):34–37. Epub 2014/01/30
160. Ho AL, Sussman ES, Zhang M, Pendharkar AV, Azagury DE, Bohon C, et al (2015) Deep brain stimulation for obesity. Cureus 7(3):e259. Epub 2015/07/17
161. Gump JM, Donson AM, Birks DK, Amani VM, Rao KK, Griesinger AM et al (2015) Identification of targets for rational pharmacological therapy in childhood craniopharyngioma. Acta Neuropathol Commun 3:30
162. Sterkenburg AS, Hoffmann A, Gebhardt U, Waldeck E, Springer S, Muller HL (2014) Childhood craniopharyngioma with hypothalamic obesity - no long-term weight reduction due to rehabilitation programs. Klin Padiatr 226(6–7):344–350
163. Nemergut DR, Townsend AR (2015) The importance of interdisciplinary communication with patients about complex, chronic illnesses: our experiences as parents of a child with a craniopharyngioma. J Pediatr Endocrinol Metab 28(1–2):3–5
164. Meijneke RW, Schouten-van Meeteren AY, de Boer NY, van Zundert S, van Trotsenburg PA, Stoelinga F et al (2015) Hypothalamic obesity after treatment for craniopharyngioma: the importance of the home environment. J Pediatr Endocrinol Metab 28(1–2):59–63

165. Hoffmann A, Postma FP, Sterkenburg AS, Gebhardt U, Muller HL (2015) Eating behavior, weight problems and eating disorders in 101 long-term survivors of childhood-onset craniopharyngioma. J Pediatr Endocrinol Metab 28(1–2):35–43
166. Roth CL, Aylward E, Liang O, Kleinhans NM, Pauley G, Schur EA (2012) Functional neuroimaging in craniopharyngioma: a useful tool to better understand hypothalamic obesity? Obes Facts 5(2):243–253
167. Roemmler-Zehrer J, Geigenberger V, Stormann S, Ising M, Pfister H, Sievers C et al (2015) Specific behaviour, mood and personality traits may contribute to obesity in patients with craniopharyngioma. Clin Endocrinol (Oxf) 82(1):106–114
168. Roth CL, Gebhardt U, Muller HL (2011) Appetite-regulating hormone changes in patients with craniopharyngioma. Obesity (Silver Spring) 19(1):36–42
169. Kilday JP, Bartels U, Huang A, Barron M, Shago M, Mistry M et al (2014) Favorable survival and metabolic outcome for children with diencephalic syndrome using a radiation-sparing approach. J Neurooncol 116(1):195–204
170. Hoffmann A, Gebhardt U, Sterkenburg AS, Warmuth-Metz M, Muller HL (2014) Diencephalic syndrome in childhood craniopharyngioma–results of German multicenter studies on 485 long-term survivors of childhood craniopharyngioma. J Clin Endocrinol Metab 99(11):3972–3977
171. Muller HL, Gebhardt U, Etavard-Gorris N, Korenke E, Warmuth-Metz M, Kolb R et al (2004) Prognosis and sequela in patients with childhood craniopharyngioma – results of HIT-ENDO and update on KRANIOPHARYNGEOM 2000. Klin Padiatr 216(6):343–348
172. Van Effenterre R, Boch AL (2002) Craniopharyngioma in adults and children: a study of 122 surgical cases. J Neurosurg 97(1):3–11
173. Muller HL, Faldum A, Etavard-Gorris N, Gebhardt U, Oeverink R, Kolb R et al (2003) Functional capacity, obesity and hypothalamic involvement: cross-sectional study on 212 patients with childhood craniopharyngioma. Klin Padiatr 215(6):310–314
174. Muller HL, Bruhnken G, Emser A, Faldum A, Etavard-Gorris N, Gebhardt U et al (2005) Longitudinal study on quality of life in 102 survivors of childhood craniopharyngioma. Childs Nerv Syst 21(11):975–980
175. Muller HL, Gebhardt U, Faldum A, Emser A, Etavard-Gorris N, Kolb R et al (2005) Functional capacity and body mass index in patients with sellar masses–cross-sectional study on 403 patients diagnosed during childhood and adolescence. Childs Nerv Syst 21(7):539–545
176. Ondruch A, Maryniak A, Kropiwnicki T, Roszkowski M, Daszkiewicz P (2011) Cognitive and social functioning in children and adolescents after the removal of craniopharyngioma. Childs Nerv Syst 27(3):391–397
177. Yasargil MG, Curcic M, Kis M, Siegenthaler G, Teddy PJ, Roth P (1990) Total removal of craniopharyngiomas. Approaches and long-term results in 144 patients. J Neurosurg 73(1):3–11
178. Cavazzuti V, Fischer E, Welch K, Belli J, Winston K (1983) Neurological and psychophysiological sequelae following different treatments of craniopharyngioma in children. J Neurosurg 59(3):409–417
179. Riva D, Pantaleoni C, Devoti M, Saletti V, Nichelli F, Giorgi C (1998) Late neuropsychological and behavioral outcome of children surgically treated for craniopharyngioma. Childs Nerv Syst 14:179–184
180. Kiehna E, Mulhern R, Li C, Xiong X, Merchant T (2006) Changes in attentional performance of children and young adults iwth localized primary brain tumors after conformal radiation therapy. J Clin Oncol 24(33):5283–5290
181. Carpentieri S, Waber D, Scott R, Goumnerova L, Kieran M, Cohen L et al (2001) Memory deficits among children with craniopharyngioma. Neurosurgery 49(5):1053–1058
182. Sands S, Milner J, Goldberg J, Mukhi V, Moliterno J, Maxfield C et al (2005) Quality of life and behavioral follow-up study of pediatric survivors of cranipharyngioma. J Neurosurg Pediatr 103:302–311

183. Ozyurt J, Thiel CM, Lorenzen A, Gebhardt U, Calaminus G, Warmuth-Metz M, et al (2014) Neuropsychological outcome in patients with childhood craniopharyngioma and hypothalamic involvement. J Pediatr 164(4):876–881. e4
184. Colangelo M, Ambrosio A, Ambrosio C (1990) Neurological and behavioral sequelae following different approaches to craniopharyngioma. Long-term follow-up review and therapeutic guidelines. Childs Nerv Syst 6(7):379–382
185. Metzler-Baddeley C, Jones R (2010) Brief communication: cognitive rehabilitation of executive functioning in a case of craniopharyngioma. Appl Neuropsychol 17:299–304
186. Hammond J, Hall S (2011) Functional analysis and treatment of aggressive behavior following resection of a craniopharyngioma. Dev Med Child Neurol 53:369–374
187. Anderson C, Wilkening G, Filley C, Reardon M, Kleinschmidt-DeMasters B (1997) Neurobehavioral outcome in pediatric craniopharyngioma. Pediatr Neurosurg 26:255–260
188. Cohen M, Guger S, Hamilton J (2011) Long term sequelae of pediatric craniopharyngioma - literature review and 20 years of experience. Front Endocrinol (Lausanne) 2:81
189. Hetelekidis S, Barnes PD, Tao ML, Fischer EG, Schneider L, Scott RM et al (1993) 20-year experience in childhood craniopharyngioma. Int J Radiat Oncol Biol Phys 27(2):189–195
190. Liu AK, Bagrosky B, Fenton LZ, Gaspar LE, Handler MH, McNatt SA et al (2009) Vascular abnormalities in pediatric craniopharyngioma patients treated with radiation therapy. Pediatr Blood Cancer 52(2):227–230
191. Regine WF, Kramer S (1992) Pediatric craniopharyngiomas: long term results of combined treatment with surgery and radiation. Int J Radiat Oncol Biol Phys 24(4):611–617
192. Rajan B, Ashley S, Gorman C, Jose CC, Horwich A, Bloom HJ et al (1993) Craniopharyngioma–a long-term results following limited surgery and radiotherapy. Radiother Oncol 26(1):1–10
193. Stripp DC, Maity A, Janss AJ, Belasco JB, Tochner ZA, Goldwein JW et al (2004) Surgery with or without radiation therapy in the management of craniopharyngiomas in children and young adults. Int J Radiat Oncol Biol Phys 58(3):714–720
194. Pemberton LS, Dougal M, Magee B, Gattamaneni HR (2005) Experience of external beam radiotherapy given adjuvantly or at relapse following surgery for craniopharyngioma. Radiother Oncol 77(1):99–104
195. Moon SH, Kim IH, Park SW, Kim I, Hong S, Park CI et al (2005) Early adjuvant radiotherapy toward long-term survival and better quality of life for craniopharyngiomas–a study in single institute. Childs Nerv Syst 21(8–9):799–807
196. Combs SE, Thilmann C, Huber PE, Hoess A, Debus J, Schulz-Ertner D (2007) Achievement of long-term local control in patients with craniopharyngiomas using high precision stereotactic radiotherapy. Cancer 109(11):2308–2314
197. Nagata T, Goto T, Ichinose T, Mitsuhashi Y, Tsuyuguchi N, Ohata K (2010) Pathological findings of fusiform dilation of the internal carotid artery following radical dissection of a craniopharyngioma. J Neurosurg Pediatr 6(6):567–571
198. Bendszus M, Sorensen N, Hofmann E, Roll E, Solymosi L (1998) Fusiform dilatations of the internal carotid artery following surgery for pediatric suprasellar tumors. Pediatr Neurosurg 29(6):304–308
199. Mong S, Pomeroy SL, Cecchin F, Juraszek A, Alexander ME (2008) Cardiac risk after craniopharyngioma therapy. Pediatr Neurol 38(4):256–260
200. Tomlinson JW, Holden N, Hills RK, Wheatley K, Clayton RN, Bates AS et al (2001) Association between premature mortality and hypopituitarism. West Midlands Prospective Hypopituitary Study Group. Lancet 357(9254):425–431
201. Regine WF, Mohiuddin M, Kramer S (1993) Long-term results of pediatric and adult craniopharyngiomas treated with combined surgery and radiation. Radiother Oncol 27(1):13–21
202. Habrand JL, Ganry O, Couanet D, Rouxel V, Levy-Piedbois C, Pierre-Kahn A et al (1999) The role of radiation therapy in the management of craniopharyngioma: a 25-year experience and review of the literature. Int J Radiat Oncol Biol Phys 44(2):255–263
203. Bulow B, Attewell R, Hagmar L, Malmstrom P, Nordstrom CH, Erfurth EM (1998) Postoperative prognosis in craniopharyngioma with respect to cardiovascular mortality, survival, and tumor recurrence. J Clin Endocrinol Metab 83(11):3897–3904

204. Fisher PG, Jenab J, Gopldthwaite PT, Tihan T, Wharam MD, Foer DR et al (1998) Outcomes and failure patterns in childhood craniopharyngiomas. Childs Nerv Syst 14(10):558–563
205. Tomita T, Bowman RM (2005) Craniopharyngiomas in children: surgical experience at Children's Memorial Hospital. Childs Nerv Syst 21(8–9):729–746
206. Muller HL, Gebhardt U, Pohl F, Flentje M, Emser A, Warmuth-Metz M et al (2006) Relapse pattern after complete resection and early progression after incomplete resection of childhood craniopharyngioma. Klin Padiatr 218(6):315–320
207. Kalapurakal JA, Goldman S, Hsieh YC, Tomita T, Marymont MH (2003) Clinical outcome in children with craniopharyngioma treated with primary surgery and radiotherapy deferred until relapse. Med Pediatr Oncol 40(4):214–218
208. Khafaga Y, Jenkin D, Kanaan I, Hassounah M, Al Shabanah M, Gray A (1998) Craniopharyngioma in children. Int J Radiat Oncol Biol Phys 42(3):601–606
209. Lin LL, El Naqa I, Leonard JR, Park TS, Hollander AS, Michalski JM et al (2008) Long-term outcome in children treated for craniopharyngioma with and without radiotherapy. J Neurosurg Pediatr 1(2):126–130
210. Scott RM, Hetelekidis S, Barnes PD, Goumnerova L, Tarbell NJ (1994) Surgery, radiation, and combination therapy in the treatment of childhood craniopharyngioma–a 20-year experience. Pediatr Neurosurg 21(Suppl 1):75–81
211. Visser J, Hukin J, Sargent M, Steinbok P, Goddard K, Fryer C (2010) Late mortality in pediatric patients with craniopharyngioma. J Neurooncol 100(1):105–111
212. Cohen M, Bartels U, Branson H, Kulkarni AV, Hamilton J (2013) Trends in treatment and outcomes of pediatric craniopharyngioma, 1975–2011. Neuro Oncol 15(6):767–774
213. Bunin GR, Surawicz TS, Witman PA, Preston-Martin S, Davis F, Bruner JM (1997) The descriptive epidemiology of craniopharyngioma. Neurosurg Focus 3(6):e1
214. Bartlett JR (1971) Craniopharyngiomas–a summary of 85 cases. J Neurol Neurosurg Psychiatry 34(1):37–41
215. Fahlbusch R, Honegger J, Paulus W, Huk W, Buchfelder M (1999) Surgical treatment of craniopharyngiomas: experience with 168 patients. J Neurosurg 90(2):237–250
216. Petito CK, DeGirolami U, Earle KM (1976) Craniopharyngiomas: a clinical and pathological review. Cancer 37(4):1944–1952
217. Wara WM, Sneed PK, Larson DA (1994) The role of radiation therapy in the treatment of craniopharyngioma. Pediatr Neurosurg 21(Suppl 1):98–100
218. Daubenbuchel AM, Hoffmann A, Gebhardt U, Warmuth-Metz M, Sterkenburg AS, Muller HL (2015) Hydrocephalus and hypothalamic involvement in pediatric patients with craniopharyngioma or cysts of Rathke's pouch: impact on long-term prognosis. Eur J Endocrinol 172(5):561–569
219. Pickering L, Jennum P, Gammeltoft S, Poulsgaard L, Feldt-Rasmussen U, Klose M (2014) Sleep-wake and melatonin pattern in craniopharyngioma patients. Eur J Endocrinol 170(6):873–884
220. Ozyurt J, Lorenzen A, Gebhardt U, Warmuth-Metz M, Muller HL, Thiel CM (2014) Remote effects of hypothalamic lesions in the prefrontal cortex of craniopharyngioma patients. Neurobiol Learn Mem 111:71–80
221. Ozyurt J, Muller HL, Thiel CM (2015) A systematic review of cognitive performance in patients with childhood craniopharyngioma. J Neurooncol 125(1):9–21
222. Zada G, Kintz N, Pulido M, Amezcua L (2013) Prevalence of neurobehavioral, social, and emotional dysfunction in patients treated for childhood craniopharyngioma: a systematic literature review. PLoS One 8(11):e76562
223. Crespo I, Santos A, Webb SM (2015) Quality of life in patients with hypopituitarism. Curr Opin Endocrinol Diabetes Obes 22(4):306–312
224. Crespo I, Valassi E, Santos A, Webb SM (2015) Health-related quality of life in pituitary diseases. Endocrinol Metab Clin North Am 44(1):161–170
225. Altuntas B, Ozcakar B, Bideci A, Cinaz P (2002) Cirrhotic outcome in patients with craniopharyngioma. J Pediatr Endocrinol Metab 15(7):1057–1058
226. Basenau D, Stehphani U, Fischer G (1994) Development of complete liver cirrhosis in hyperphagia-induced fatty liver. Klin Padiatr 206(1):62–64

227. Holmer H, Popovic V, Ekman B, Follin C, Siversson AB, Erfurth EM (2011) Hypothalamic involvement and insufficient sex steroid supplementation are associated with low bone mineral density in women with childhood onset craniopharyngioma. Eur J Endocrinol 165(1):25–31
228. Muller HL, Schneider P, Bueb K, Etavard-Gorris N, Gebhardt U, Kolb R et al (2003) Volumetric bone mineral density in patients with childhood craniopharyngioma. Exp Clin Endocrinol Diabetes 111(3):168–173
229. Szeifert GT, Sipos L, Horvath M, Sarker MH, Major O, Salomvary B et al (1993) Pathological characteristics of surgically removed craniopharyngiomas: analysis of 131 cases. Acta Neurochir 124(2–4):139–143
230. Adamson TE, Wiestler OD, Kleihues P, Yasargil MG (1990) Correlation of clinical and pathological features in surgically treated craniopharyngiomas. J Neurosurg 73(1):12–17
231. Crotty TB, Scheithauer BW, Young WF Jr, Davis DH, Shaw EG, Miller GM et al (1995) Papillary craniopharyngioma: a clinicopathological study of 48 cases. J Neurosurg 83(2):206–214
232. Weiner HL, Wisoff JH, Rosenberg ME, Kupersmith MJ, Cohen H, Zagzag D, et al 1994 Craniopharyngiomas: a clinicopathological analysis of factors predictive of recurrence and functional outcome. Neurosurgery 35(6):1001–1010. discussion 10-1
233. Muller HL (2012) Craniopharyngioma - a childhood and adult disease with challenging characteristics. Front Endocrinol (Lausanne) 3:80

Index

J.P. Martinez-Barbera, C.L. Andoniadou (eds.), *Basic Research and Clinical Aspects of Adamantinomatous Craniopharyngioma*,
DOI 10.1007/978-3-319-51890-9

MIX
Papier aus verantwortungsvollen Quellen
Paper from responsible sources
FSC® C105338

If you have any concerns about our products,
you can contact us on
ProductSafety@springernature.com

In case Publisher is established outside the EU,
the EU authorized representative is:
Springer Nature Customer Service Center GmbH
Europaplatz 3, 69115 Heidelberg, Germany

Printed by Libri Plureos GmbH
in Hamburg, Germany

Springer Series in Transitional Justice

Series Editor:
Olivera Simić

For further volumes:
http://www.springer.com/series/11233

Olivera Simić · Zala Volčič
Editors

Transitional Justice and Civil Society in the Balkans

Springer

Editors
Olivera Simić
Griffith Law School
Griffith University
Brisbane, QLD, Australia

Zala Volčič
School of Journalism and Communication
Centre for Critical and Cultural Studies
University of Queensland
Brisbane, QLD, Australia

ISBN 978-1-4614-5421-2 ISBN 978-1-4614-5422-9 (eBook)
DOI 10.1007/978-1-4614-5422-9
Springer New York Heidelberg Dordrecht London

Library of Congress Control Number: 2012951310

Printed on acid-free paper

Springer is part of Springer Science+Business Media (www.springer.com)

Acknowledgments

Throughout the process of preparing our edited volume, we have enjoyed the support of many friends and colleagues who have given us an unconditional support and stimulated our thinking with their critical feedback on our own work. We want to especially thank Springer, our publisher, for their continuous support, advice, and guidance.

We are most grateful to Jelena Subotic, Jasmina Brankovic, Dinka Corkalo Biruski, Mirjana Beara, Huma Haider, Martine Hawkes, Melanie O'Brien, Trish Luker, Fayen d'Evie, Nesam Mcmillan, Sandina Begic, and Vasile Cernat for offering a number of helpful advices and feedback on the chapters in this collection. We are indebted to all our contributors for their patience throughout the endless rounds of revisions that this process of mutual learning has entailed.

Olivera Simić would like to thank Law School at the Griffith University and its staff who provided a wonderful and supportive environment that enabled her to finish this volume. Many thanks to Stojan Ignjatović for all the personal and professional support, love, and priceless companionship.

Zala Volčič would like to thank both, School of Journalism and Communication and Centre for Critical and Cultural studies at the University of Queensland, their staff and administrators, since they provided a supportive and creative environment during the development of this volume. Many thanks go to Mark Andrejevic, Karmen Erjavec, and Helga Tawil, who remain constant source of energy and strength.

Contents

Part II Transnational Civil Society Practices

Contributors

Jamie Rowen is presently a doctoral fellow at the American Bar Foundation in Chicago, IL. She received her PhD in 2012 from the Jurisprudence and Social Policy Program at the University of California, Berkeley School of Law, where she also received her JD and MA in 2009. Her doctoral research includes a comparative ethnographic analysis of mobilization around truth commissions in Bosnia and Herzegovina, Colombia, and the United States, and elucidates current beliefs and practices related to transitional justice. She first came to the Balkans in 2002 to work in a summer camp program for youth from throughout the region, and has since studied religion and post-conflict justice in Vietnam, developed life skills educational programs for orphans and vulnerable children in South Africa, examined the impact of the truth commission in Morocco, and conducted research on human rights protection in Latin America with the Inter-American Institute for Human Rights in Costa Rica. Jamie speaks Spanish and French fluently, conversational Bosnian/Serbian/Croatian, remnants of Vietnamese, Arabic and Hebrew, and is currently studying Russian.

Eunice Castro Seixas lives in Porto, Portugal, where she graduated in psychology and completed a Master in social psychology. She has worked as a school and clinical psychologist and also in the educational area (as an assistant at the University of Fernando Pessoa, Porto). Presently, she finishes her PhD studies at the Center for Social Studies, in the University of Coimbra and she receives a doctoral scholarship from the Portuguese Foundation of Science and Technology (FCT). The topic of her project concerns the international assistance to democratization in Bosnia–Herzegovina and in the Russian Federation. She has published several articles and given presentations at scientific events and her current research interests are postcolonial and decolonial approaches; civil society; social movements; democratization and international development; migration and contemporary expatriates; critical discourse analysis; critical ethnography.

Michael Potter Since graduating from Queen's University Belfast in 2002, has worked in the field of equality, human rights, and peacebuilding in Northern Ireland. As Policy and Research Officer for Training for Women Network in Belfast, he carried out comparative research on women and post-conflict transition, including on victims of the conflict, minority ethnic communities, women and entrepreneurship, women in nontraditional employment, and women in Loyalist communities. Projects investigating and sharing experiences of gender aspects of transition in the former Yugoslavia have been undertaken in the context of the EU-funded Northern Ireland Peace Programme. As Policy and Research Worker at the South Tyrone Empowerment Programme (STEP) in Dungannon, Michael carried out research into the experiences of migrant and minority ethnic communities in a transitional society, including a major project looking at forced labor in Northern Ireland, funded by the Joseph Rowntree Foundation. Academic publications include works on reconciliation in Northern Ireland, women and peacebuilding, women in nontraditional employment, and Irish Travellers. Michael is a doctoral candidate at Queen's University Belfast, carrying out research comparing the post-conflict institutions in Northern Ireland and Kosovo, and is a member of the Board of Directors of the Women's Tec in Belfast.

Orli Fridman is the Academic director of the SIT study abroad program in the Balkans (Peace and Conflict Studies in Serbia, Bosnia and Kosovo) and a lecturer at the Faculty of Media and Communications (FMK), Singidunum University. She is the director of the Center for Comparative Conflict Studies (CFCCS) at FMK. Her interdisciplinary research interests focus on the internal dynamics of societies in conflict, the role of Social Memory studies in teaching and researching post-conflict transformation, and critical approaches to encounters of groups in conflict. Contact: orli.fridman@sit.edu

Hedley Abernethy is a trainer with WAVE Trauma Centre an NGO that supports individuals who have been bereaved, injured, or traumatized as a result of the conflict in and about Northern Ireland. He has developed training courses that deal with a society transitioning from violent conflict and also lectures at Queen's University, Belfast. Prior to working with WAVE he was the Peacebuilding Education Advisor for Catholic Relief Services, the humanitarian aid agency of the Catholic Church in the United States. Hedley has also worked for the National Council of YMCAs, helping to promote and develop community relations and peacebuilding programs throughout Northern Ireland.

Despina Angelovska is a Lecturer at the University for audiovisual arts, ESRA and at Euro-Balkan Institute for Social and Humanities research, in Skopje, Macedonia. She is also an associate editor and member of the editorial board of *Identities: Journal for Politics, Gender and Culture*. Her fields of interest are cultural and gender studies, nationalism studies, post-colonial theory, and Balkanism. Her work seeks to connect literature, performing arts, and cinema with gender and nationalism studies. Her publications include "Écriture et déplacement: Les retrouvailles de soi comme autre dans le roman *La chaise d'Elie* d'Igor Štiks," in *Actes du 4ème congrès de REELC* (2012), "*OpenOption*: Roma, discrimination and peacebuilding in Šuto Orizari" in *Peace Psychology in the Balkans: Dealing with a Violent Past while Building Peace* (2012); "Intimité communiste: revendication de justice transitionnelle" in *Studia Europeae*, 3/2011; "Negotiating the name, renaming the nation-state," in *Renaming machine* (2010); "Identités translationnelles et perspectives postcoloniales" in *Identities: Journal for Politics, Gender and Culture,* 1/2006; "Le retour à la poussière" in *Identities: Journal for Politics, Gender and Culture* 3/2002; "Transition trouble" in *Capital and gender* (2001); "(Re)situating the self: nation, gender and multicultural perspectives" in *Komsi-kapicik, culture and politics* (2000). Her book *The dramaturgy of decentrement: Bernard-Marie Koltès, a voyage toward the edges* was published by Az-buki, Skopje, in 2010.

Briony Jones is a Research Fellow for the Dealing with the Past Program at swisspeace, an affiliated research institute of the University of Basel. Dr. Jones is an advisor on a Swiss National Science Foundation funded 3-year institutional partnership between the University of Zurich and the University of Sarajevo on *Memory, History and State-Building in Bosnia–Herzegovina: Critically Assessing and Re-Thinking the Teaching of Memory Politics in Bosnian University Curricula* and is a Co-Investigator for a 3-year Swiss National Science Foundation funded research project on *Resisting Transitional Justice? Alternative Visions of Peace and Justice.* In addition, she teaches a seminar course for Masters and PhD student at the University of Basel on *The Politics of Transitional Justice in Comparative Perspective.*

Prior to joining swisspeace, Dr. Jones worked as a Lecturer in International Development at the University of Manchester, as a Research Consultant for the Chronic Poverty Research Centre and as a Visiting Lecturer in Refugees, Migration and Development at the University of East London. She completed her PhD on reconciliation and citizenship in contemporary Bosnia–Herzegovina in 2009 and has published on topics including education reform and reconciliation in Bosnia–Herzegovina, contesting citizenship in Bosnia–Herzegovina, the use of interpreters for qualitative research fieldwork, and the role of emotions in research.

Alex Jeffrey is a University Lecturer in Human Geography in the Department of Geography at the University of Cambridge and a Fellow of Emmanuel College. Between 2006 and 2012 he was a Lecturer and then Senior Lecturer in Political Geography in the School of Geography, Politics, and Sociology at Newcastle University. His research has focused on the governance of post-conflict environments, particularly the former Yugoslavia, and the role of nongovernmental organizations in fostering democracy. He has published in journals across geography, political science, and development studies and has collaborated with Prof. Joe Painter on *Political Geography and Introduction to Space and Power* (Sage, 2009) and with Prof. Anoop Nayak on *Geographical Thought* (Pearson, 2011). In October 2012 he will publish *The Improvised State: Sovereignty, Performance and Agency in Dayton Bosnia* in the RGS-IBG Book Series (Oxford: Wiley-Blackwell). He is currently undertaking a 2-year research project funded by the UK Economic and Social Research Council (RES-061-25-0479) examining the nature of public outreach programs from the War Crimes Chamber of the Court of Bosnia and Herzegovina.

Michaelina Jakala received her PhD from the University of Bradford, UK in 2011. Her PhD research focused on the effects of post-conflict peacebuilding amongst women, more specifically survivors of wartime sexual violence, in Sarajevo, Bosnia–Herzegovina. She also has experience working with victims' associations in Sarajevo and cofounded the Wings of Hope, Center for Women and Youth Development in 2009. She currently works with Dr. Alex Jeffrey as a Research Associate at The School of Geography, Politics and Sociology at Newcastle University on the ESRC funded Localising International Law: Examining the Outreach Strategies of the War Crimes Chamber of the Court of Bosnia and Herzegovina project.

Arolda Elbasani is a visiting fellow at Columbia University. She has a PHD in Social and Political Sciences from the European University Institute, Florence; and an MA in International Relations from Central European University. She has held several academic positions among which Social Science Award Fellow at WZB, Berlin; Chair of International Relations Department at European University, Tirana; and Post-doc Fellow at Free University, Berlin. Her research interests lay at the intersection of post-communist politics, EU democracy promotion, and political Islam. She has published an edited book on Europeanization in Western Balkans forthcoming with Routledge, and a number of peer-reviewed articles on these topics.

Jill A. Irvine is President's Associates Presidential Professor and Director of Women's and Gender Studies at the University of Oklahoma. She is author of *The Croat Question, Partisan Politics in the Formation of the Yugoslav Socialist State* and coeditor of *State-Society Relations in Yugoslavia 1945–1991* and *Natalija: Life in the Balkan Powderkeg 1880–1956*. She has written numerous articles, book chapters, and government reports about democratic transformations in the Balkans. She is currently working on a book on women, war, and political transformation in the Balkans funded by the International Research and Exchanges Board and the National Council for Eurasian and East European Research.

Arnaud Kurze is a Postdoctoral 2012–2013 Visiting Fellow at the Center for Global Studies (CGS) at George Mason University (Mason). He received his Ph.D. in Political Science at Mason in May 2012. While at Mason, he was the CGS Publication Editor and the Coordinator of CGS' "Human Rights, Justice & Democracy Project," funded by the Open Society Institute. He holds a master's degree in Governance from University of Hagen, Germany, and a bachelor's degree in International Relations from Sciences Po, France. His dissertation "Justice Beyond Borders? The Politics to Democratize Human Rights in the Post-Conflict Balkans" analyzes civil society efforts to account for crimes against humanity. His research interests include Southeast European transitional justice issues; EU integration and identity politics; and critical security studies. He has published numerous articles in academic journals, reports, and policy briefs for government institutions and think tanks, as well as op-eds and commentaries. He has held visiting fellow positions at Sciences Po (Paris, France) and at the University of Zagreb (Croatia). He has also taught several undergraduate and graduate classes on globalization and identity politics; social movements in the Balkans; corporate practices and geopolitics; and transitional justice issues in post-conflict societies, among others.

Iva Vukusic is currently working at the Sense News Agency in The Hague where she manages the archive containing the footage and documents from war crimes trials held at the International Criminal Tribunal for the former Yugoslavia (ICTY). Before moving to The Netherlands Iva worked in Sarajevo, Bosnia and Herzegovina for 3 years, as an analyst and researcher at the Special War Crimes Department of the State Prosecutor's office and as a project manager at the Research and Documentation Centre. She holds a bachelors degree in political science from Zagreb University and a masters in human rights from the University of Bologna, Italy and the University of Sarajevo, BiH. Currently Iva is also working on a project called *Pictures of Peace and Justice: Documentation, Evidence and Impact of Visual Material in International War Crimes Prosecution* at King's College London, as well as participating in a research project at the University of East London, Centre on Human Rights in Conflict. In the past 5 years, she participated in numerous conferences and panel discussions around the world, speaking on war crimes prosecution in the former Yugoslavia, transitional justice and, more broadly, post-conflict human rights-related topics.

Elidor Mëhilli is a Postdoctoral Fellow at the Harriman Institute at Columbia University. He received his Ph.D. in modern European and Eurasian history at Princeton University in 2011. Based on extensive archival research in six countries, Mëhilli's work explores socialist transnational exchange in ideas, practices, and technologies during the Cold War. He was a co-winner of the 2011 Webb-Smith prize at the 46th Annual Walter Prescott Webb Memorial Lecture Series on "Transnational Perspectives on the Soviet Bloc, 1944–1991." He has also published on Cold War politics in the *Journal of Cold War Studies* and in *Kritika: Explorations in Russian and Eurasian History.* Previously, he held a Mellon fellowship in contemporary history at The George Washington University in Washington DC and a Whiting fellowship at Princeton.

Tamara Banjeglav works as coordinator of the culture of memory program in a civil society organization *Documenta*—Centre for Dealing with the Past from Zagreb, Croatia. She has earned her BA in English language and comparative literature and holds two MA degrees, in European Studies and in Nationalism Studies. She is currently working towards her PhD degree and is writing her doctoral dissertation on the topic of transitional justice mechanisms and their impact on politics of remembrance in post-war Croatia. She is a member of a regional research forum on transitional justice gathered around the Alumni Centre for Interdisciplinary Postgraduate Studies in Sarajevo. Moreover, she is a member of an organization team which organizes workshops on the issue of conflicting memories in post-conflict societies, as part of an informal trans-European platform for exchange and cooperation in the field of dealing with difficult pasts in Western Balkans and Western Europe.

Lavinia Stan is an Associate Professor of Political Science at St. Francis Xavier University, Canada, working mainly on issues related to post-communist democratization, especially transitional justice, and religion and politics. She has published over 40 articles in peer-reviewed journals and seven books, including *Transitional Justice in Eastern Europe and the Former Soviet Union: Reckoning with the Communist Past* (Routledge, 2009), *Church, State and Democracy in the Expanding Europe*, and *Religion and Politics in Post-Communist Romania* (both coauthored with Lucian Turcescu and published with Oxford University Press in 2011 and 2007, respectively). Stan is the Regional Editor for Europe of the internationally acclaimed *Women's Studies International Forum*, the Vice-President of the Society for Romanian Studies, a Scientific Council member of the Institute for the Investigation of Communist Crimes and the Memory of Romanian Exile in Bucharest, past chair of the Wildavsky Award Committee of the American Political Science Association, adjudicator for the Social Science and Humanities Research Council of Canada, and member of the editorial boards of 11 North American and European scholarly journals. Her two upcoming volumes are *Transitional Justice in Post-Communist Romania: The Politics of Memory* and the *Encyclopedia of Transitional Justice* (both in production with Cambridge University Press).

Artur Lipinski is an Assistant Professor at the Faculty of Political Science and Journalism, Adam Mickiewicz University, Poznan, Poland. He defended a PhD thesis on right wing anticommunism in Poland after 1989. Currently, he is interested in the role of discourse in the construction of political identities in Poland. Among his other research interests are Polish party politics and foreign policy, discourse analysis, social movements theory, and sociology of memory.

Editors

Olivera Simić is a Lecturer with the Griffith Law School, Griffith University, Australia. She holds an LLB from Nis University's Law School (Serbia), an LLM from Essex University (UK), an MA from UN University for Peace (Costa Rica), and a PHD from Melbourne Law School (Australia). Her research engages with transitional justice, international law, gender and crime from an interdisciplinary perspective. Olivera has published in journals such as *International Journal of Transitional Justice, Law Text Culture, Women' Studies International Forum, Journal of International Women Studies, International Peacekeeping* as well as in books and book chapters. Her latest book, *Regulation of Sexual Conduct in UN Peacekeeping Operations, has been published by Springer* in 2012.

Zala Volčič is a Postdoctoral Fellow at the Centre for Critical and Cultural Studies, and a Senior Lecturer at the School of Journalism and Communication, University of Queensland, Australia. She has published numerous articles and book chapters on topics including nationalisms in the Balkans, the cultural consequences of globalization, and media identities in former Yugoslavia. Her work has appeared in journals such as *Discourse and Society, Popular Communication, Journal of Children and Media, Journal of Global Mass Communication, Canadian Journal of Communication, Social Semiotics, Information Communication & Society, ICMR/ Islam and Christian–Muslim Relations, Critical Studies of Media Communication, and Harvard International Journal of Press/Politics.* Her latest book, *Serbian Spaces of identity, has been published by Hampton Press in 2011.*

Chapter 1
Localizing Transitional Justice: Civil Society Practices and Initiatives in the Balkans

Zala Volčič and Olivera Simić

Introduction

In the early twenty-first century, we are faced with processes of globalization, different forms of transitions, post-communist conditions, and increased violence along ethnic and class lines. We live in times of endless transitions, fragmentations, insecurities, and uncertainties, during which individuals and entire nations struggle to overcome the legacies of suffering "ranging from rape and domestic violence to collective atrocities of state-sponsored dirty wars, genocides, and ethnic conflicts" (Scheper-Hughes 2004: 374). In such times, a question of how to recover from episodes of mass violence demands to be addressed. Transitional justice is a growing field, which responds to dilemmas over how successor regimes should deal with past human rights abuses of their authoritarian predecessors. Transitional justice as such is a process, by which a state seeks to redress the violations of a former regime. It is clear that transitional justice scholars have much to learn from civil society literature, which argues that the notion of civil society encompasses various (political) entities, including institutions such as media, and associations based on voluntary agreements among citizens together with a public sphere, where citizens can engage in public activities and debate.

This book attempts to cover various transitional justice processes within civil society engagements in the Balkans. The Balkans are positioned in South-Eastern Europe and in most scholarly accounts they include the former Yugoslavia as well as Albania,

Z. Volčič (✉)
School of Journalism and Communication
Centre for Critical and Cultural Studies, The University of Queensland,
Level 4 Forgan Smith Tower, Brisbane, QLD 4072, Australia
e-mail: z.volcic@uq.edu.au

O. Simić
Griffith Law School, Griffith University,
Law Bldg (N61), Brisbane, QLD 4111, Australia

O. Simić and Z. Volčič (eds.), *Transitional Justice and Civil Society in the Balkans*, Springer Series in Transitional Justice, DOI 10.1007/978-1-4614-5422-9_1,

Greece, Romania, and Bulgaria. The Balkans, as a region, are marked by the post-communist and post-conflict transitional turmoil. Selected chapters here represent a series of empirically oriented and theoretically grounded reflections upon the transitional justice and civil society projects, written by local and international scholars.

The collapse of communism and the Yugoslav wars left the Balkan countries in an ambiguous state economically, politically, culturally, and psychologically. Economically, they are going through the process of transition from centralized, socialist, state-run economies to privatized, market-driven economies, while the region as a whole faces decline in economic security. The contradictions are rife: new economic opportunities and the emergence of economic elites coincide with rampant unemployment, the dismantling of a once taken-for-granted social safety net and the widening gap between the wealthy few and the struggling many. Importantly, we suggest, the question of how the region is addressing its past must be viewed in relation to other sociopolitical developments such as the rise of neoliberalism. Politically, the Balkan states are involved in building a specific type of (ethno) national identity, while, at the same time, seeking to address dilemmas of traumatic past, migration, and multicultural past and present. Culturally and psychologically, the Balkans are going through the painful process of reconciliation with the traumatic past. The nations involved are faced with the challenge of creating a shared cultural memory of the communist times and of fostering reconciliation between various ethnic groups, while attempting to build a strong civil society and a durable form of peaceful co-existence.

While the term "civil society" is a centuries-old product of European political thought, it has, especially over the last 20 years, enjoyed renewed popularity. The idea of civil society has gained a polemical power in Western societies (due to the decline of public institutions) and in Eastern European societies, including the Balkans, (due, in turn, to the collapse of communism). The disintegration of communist regimes in the Balkans paralleled increasing interest in the notion of civil society not only in the newly post-communist countries, but in the Western nations as well. The ideals of civil society have been praised by Western and non-Western scholars alike. But in spite of its almost universal appeal, the term "civil society" is replete with tensions and paradoxes. In general, civil society denotes spheres of social life "between household and state … which affords (the possibility) of concerted action and social self-organization" (Bryant 1995: 399). There is the implicit normative element in almost all analyses of civil society, which connects the term with some form of ideal and "authentic" democracy. It is argued, however, that "civil society exists where there are free associations that are not under the tutelage of state power" (Bryant: 402). A more forceful argument, however, suggests that civil society can be said to exist where those associations that constitute civil society can determine or redirect the course of state policy (Taylor 1994: 208). As Giner (1995: 304) argues, civil society is a historically evolved sphere of individual rights, freedoms, and voluntary associations whose politically undisturbed competition with each other in pursuit of their respective private concerns, interests, preferences, and intentions is guaranteed by a public institution, called the state. Any mature civil society exhibits at least five prominent dimensions: individualism, privacy, market, pluralism, and class.

Today, in the Balkans, it is precisely around these notions of individualism, privacy, market, pluralism, class, and, in addition, democracy, national interest, and civil society, that the past is being redefined and reshaped. In response to these developments, our collection foregrounds the relatively unexplored and under-researched role of civil society groups and social movements such as local women's groups, as well as the role of art, theater, and other grass-roots transitional justice mechanisms and initiatives. Grounded in specific case studies, the unique contribution of this volume is not only that it covers a part of the world that is not adequately represented in transitional justice and civil society fields, but also that the volume is the first project originally researched and written by experts from the region independently or in collaboration with international scholars. It brings together innovative scholarship that examines interdisciplinary aspects of transitional justice in Serbia, Bosnia and Herzegovina, Bulgaria, Romania, Macedonia, Croatia, Kosovo, and Albania. Our authors have sought to engage with a combination of theories that would allow for an in-depth examination of the specific cases of dealing with the past. To that end, theories of civil society, transitional justice, nationalism, and transnationalism guide their entry and discussion. The richness of the available literature both within theories of civil society *and* transitional justice studies provides room for the two to be brought together to produce an analysis that can effectively ground investigations into how communities deal with the past. The analysis relies on documenting, discussing, and inferring from the intersections between these different bodies of literature. By bridging these intersections and incorporating diverse theoretical and methodological approaches, the work here attempts to fill the research lacuna of transitional justice, while exploring issues of dealing with the past in a particular region.

Because the Balkans should be understood as a specific transnational space, we pay attention both to the concepts of transnationalism and to the processes that coalesce around nationalism. Our engagement with the Balkans is especially concerned with the region's unique positioning: it has been understood by some as marginal, but by others as a crossroads—all our authors see the Balkans as historically a region that represents the crossroads of different cultures and civilizations. Specifically, for example, the disintegration of Yugoslavia in the 1990s helped transform what had once been a national culture into a transnational one and significantly altered the broader cultural and social context of the region. After the wars and the subsequent fall of Serbian president Slobodan Milosevic in 2000, new transnational cultural relations and practices emerged between the seven resulting states (Slovenia, Croatia, Bosnia and Herzegovina, Serbia, Montenegro, Macedonia, and Kosovo). The authors in this volume are interested in how transitional justice is being "localized," and in that, they pay special attention to the formation and theories of civil society. All authors agree that survivors of both communism and the Yugoslav wars on the whole feel the need to share their stories and memories, and they want to learn from others about what actually happened during the times of violence and unrest. The authors suggest that inclusion of local communities increases the impact of transitional justice initiatives. Rowen (this volume), for example, argues that civil society initiatives in the former Yugoslavia that focus on investigating, documenting, and publicizing information provide crucial insights into how transitional justice initiatives develop and take shape. Similarly, Jones,

Jeffrey, and Jakala (this volume) focus on how transitional justice is practiced, since the fostering of civic virtues, trust, and civically minded behavior is vital for supporting transition toward democracy. Their research challenges a vision of the transitional citizen as a passive recipient of new legal or political programs and points to the emergence of an alternative understanding of justice and democracy through public outreach programs in the region. Some of the contributors to this volume employ more national lenses to explore specific case studies, while some adopt a transnational approach in their analysis.

In this introductory chapter, we attempt to engage with and provide theoretical arguments about civil society and transitional justice. Civil society can be a slippery, broad and overused term, which means that to consider the questions it raises for our examination of the Balkans requires that we first situate it in a historical and theoretical context. We will situate notions of civil society and transitional justice, within the Balkan context and identify their historical origins. Civil society is understood, firstly, in its modern Western variant, as the antithesis of "the state" (Wood 1990: 63). It is often defined in terms of simple oppositions: the state and the non-state or political vs. social power (Wood: 63). The modern conception of civil society differs from its definition in classical political philosophy, where civil society was seen as co-existing with the state. Before the mid-eighteenth century, civil society and the state were interchangeable notions (Keane 1988: 35–36). In that, the concept was used simply to draw the distinction between the "civilized" and "uncivilized" conditions of "humanity:" "whether in a hypothesized state of nature or, more particularly, under an "unnatural" system of government that rules by despotic decree rather than by laws" (Kumar 1993: 376).

For our discussion of transitional justice and ways of dealing with the past, conceptual distinction has to be made between the role of public sphere and the role of civil society in transitional justice. Two questions are crucial to this discussion: whether the elements of state and society are essential components of a strong civil society and which element—state or society—is to prevail. Calhoun (1994: 309) claims that the idea of civil society describes the possibility of a society to organize itself without being organized by a state: "If society had such capacity, then "the people" integrated in that society could better be seen as the source of political legitimacy rather than merely the object of rule." The arguments of this book are aligned with Calhoun's (1993: 276) argument that a strong public sphere where the past can be addressed depends upon a favorable organization of civil society. The most appropriate definition of civil society which fits former communist-European context can be found in Gellner's argument (1995: 32)—civil society is that set of diverse non-governmental institutions, which is strong enough to counterbalance the state, and which, while not preventing the state from fulfilling its role of keeper of the peace and arbitrator between major interests, can nevertheless prevent the state from dominating the rest of society. According to Keane (1988), the democratic changes in the Balkan countries emerged (at the beginning at least) on the border between the economy and the state and were motivated by ethical and cultural dimensions, not by profit or a pursuit of political power: the ultimate aim was to create a public sphere based on democratic principles. Similarly, transitional justice requires an existence of a public sphere, where different projects and initiatives—the ones, which converge around questions of judicial responses to human rights abuses, repressive regimes, and war, and the ones that focus on memorialization, public apology, and the work of forgiveness—connect the work of the state to a broader context of transforming social, economic, and political conditions. Furthermore, as transitional justice scholars argue, attention has to be paid to the unofficial local practices of remembering, commemorating, and responding to the violence of the past.

Mapping Civil Society

Since the nineteenth century, the idea of the separation of the state from civil society has become a basic conceptual framework in the social sciences. The notion of civil society has a complex history. There are two historically different lines of the discourse on civil society. One focuses upon its economic character: for example, Scottish moralists such as Ferguson and Smith, who call for free markets and private property as a condition for the citizen to be part of a public sphere. The other stresses social relations: for example, arguments found in the work of Tocqueville and Rousseau. Calhoun criticizes today's overlapping of the two characteristics of the civil society:

Capitalism did historically and can still play a special and crucial role in the growth of a civil society. The early growth of capitalist business relations provided essential support to the development of a sphere of political discourse outside the realm of state control. The development of a public discourse in which private persons addressed public issues was made possible, in part, by both the policy issues posed by the growth of the non-state dominated market-activity, and the creation of settings for such discourse in coffee-houses, journals, and other forums operated as businesses (Calhoun 1994: 311).

With the question "who may be members of a discursive public or a civil society (1994: 311)," Calhoun points to the heart of the problem of the public sphere—not all the people are citizens who participate. This can be fruitfully applied to today's Balkans—we need to consider at length who participates and why in current projects within civil society, and who is symbolically allowed or encouraged to debate and reflect upon the traumatic past. Some of the articles in this volume address this question of inclusion and exclusion within civil society: Simic, for example, writes about the project "Pillar of Shame" launched by "Mothers of Srebrenica" and the German "Centre for Political Beauty," which aimed to construct a permanent monument from the collected shoes as a reminder of United Nation's shame and responsibility for not preventing the Srebrenica genocide. While the project was successful in collecting the shoes transnationally, it failed to engage citizens of Serb origins in either Bosnia and Herzegovina or Serbia and in that, it reinforced the principle of ethnic exclusion of participation in civil society.

Importantly, the notion of civil society has a contested historical meaning. At the beginning of the modern era, civil society stood dichotomously against the state, as described in John Locke's contract theory. Locke did not see civil society as an autonomous sphere, but more as a sphere, where private rights and freedoms and individual ownership can be expressed. Similarly, the civil society posited by Hegel stands in the dichotomous opposition to the state. Hegel conceived of the civil society as the sphere of social relations not controlled by the state. It is, in Hegelian philosophy, an "anti-state," which contains both the economic labor and the exchange process as well as incorporating the free opinion-building process and the private sphere. From Hegel on, civil society came to mean the domain of civil rights and freedoms guaranteed and protected by the state. Further, Hegel's civil society is formed by individuals who had become independent of traditional loyalties, and were able to meet their own needs. His civil society is an essential condition for individual freedom. This civil society is equivalent to the whole of the public sphere in the case of Habermas's conception of civil society. Habermas (1992: 453) claims,

It (the concept "relations of association") also recalls the new current meaning of the term "civil society," which no longer includes a sphere of an economy regulated via labor, capital and commodity markets and thus differs from the modern translation, common since Hegel and Marx, of "societas civilis" as "bourgeois society."

In other words, the institutional base of civil society is constituted by voluntary unions outside the realm of state and the economy. Habermas provides examples ranging from churches and cultural associations to independent media, sports, and leisure clubs, debating societies, groups of concerned citizens, labor unions, and political parties. He sees the public sphere as a place of appearance, representative publicness, and publicity among private citizens, including the public sphere of politics. Habermas views civil society as *an organizational construct* within this public sphere that brings together people with common interests, and "as a corollary of a depersonalized state authority" (Habermas 1989: 19). He places public sphere between the state and civil society. Most of the authors in this volume address this historical tension—Stan, for example, writes in depth about the state, civil society and public sphere in Romania, and about how state actors have generally blocked—while civil society actors have supported—the quest for transitional justice.

As Gramsci was the first to adopt a three-part conceptual framework consisting of civil society, economic power, and state power, we will focus on the Gramscian definition of civil society. Gramsci's work clearly shifts civil society from the sphere of economic necessity to the sphere of social superstructure, and, as a result, his ideas allow us to reinvigorate discussions about the importance of both civil society and transitional justice.

Gramsci: Civil Society as a Hegemonic Project

Gramsci viewed the particular substance of civil society as the outcome and object of class struggle. For example, when the middle-class is hegemonic in this struggle, civil society takes on middle-class values and cultural norms. In spite of this rather schematized view of civil society as either bourgeois or socialist, Gramsci's tripartite scheme does avoid the economic and political reductionism that usually befalls simpler distinctions between the public and private sectors (Rowe 1997: 65).

The modern conception of civil society ignores Gramsci's understanding of the term, which was "unambiguously intended as a weapon against capitalism, not an accommodation to it" (Wood 1990: 63). What Gramsci recognized—and what makes his understanding more compelling than the modern liberal-democratic conception of civil society—is that the market is not a freely formed association like any other. When it comes to the market, participation is mandatory on the terms required by the system itself (Wood 1990). So the use of "civil society" that equates the term with the free market or freedom in the marketplace is misleading because it fails to understand the ability of the capitalist system to affect the exercise of personal freedom as significantly as that of state power.

Gramsci argued that both civil society and political society exercise the same functions: while political society organizes force, civil society organizes consensus (hegemony) by non-violent means in support of the groups that profit most by the maintenance of the liberal capitalist state (Buttigieg, 1995: 5). Thus civil society in the modern liberal state "is not some kind of benign or neutral zone where elements of society operate completely freely and openly" (Buttigieg 1995: 27). Unlike freely formed associations of civil society, argued Gramsci, "laissez-faire liberalism … must be introduced by law, through the intervention of political power: it is an act of will, not the spontaneous, automatic statement of economic facts or social will" (Buttigieg, 28). Some of the articles within this volume point to this very fact. Kurze and Vukusic, for example, focus on regional transitional justice activities of a number of NGOs across the former Yugoslavia. They write about how existing retributive justice mechanisms have been unsuccessful due to internal obstacles, such as controversial structures of regional truth committees or multiple and conflicting narratives of victimhood. In particular, they point to problematic retributive justice and fact-finding initiatives among specific war victims' groups in Bosnia and Herzegovina. Banjeglav (this volume) writes about civil society organization Documenta (Centre for Dealing with the Past from Zagreb), while, at the same time, considering the effect of a desperate economic situation on the attitudes of the Croatian public toward various ways of dealing with the region's past.

In their misuse of the notion of civil society, liberal democrats also tend to misuse Gramsci's insights so as to divert attention away from his critique of the liberal/capitalist state. What modern interpretations of Gramsci and modern uses of the idea of civil society ignore is the fact that civil society is such an important concept precisely because it acts as a critical force against the power exercised by both the state and the capitalist free market. In that, Angelovska's work (this volume) is of great import: in her contribution, Angelovska evaluates the civil society efforts in dismantling the totalitarian Communist heritage and in negotiating normative transitional justice practices. She writes about how Macedonian lustration, initially conceived as a response against the communist misdeeds and purges and as an attempt to overcome the totalitarian past, has, for now, failed to provide the necessary transparency for the confrontation with the past. Similarly, Elbasani (this volume) analyzes the reasons for the failure of the post-communist Albania to enforce the transitional justice for the abuses committed during communism. Elbasani points especially to the weak civil society and the very limited number of actors engaged in transitional justice projects as the factors, which have allowed economic exploitation to occur and have not, as a whole, led to the society-wide confrontation with the crimes of the past.

The market economy has been included, prior to Gramsci's definition, in the base of civil society, and, according to Walzer (1992), globalization and corporatization of the economy significantly change their relation to both the state and civil society and lead to the decline of civil society. Economic monopolies lead to political ones, and political ones turn into economic ones: democratic life can be heavily limited by concentrations of private ownership.

Alexis de Tocqueville provided the strongest case for a civil society made up of articulated associational groups that could rely on an informed leadership to shape public opinion in a positive manner. To bring the discussion of civil society back to the stated aims of this volume, it is important to stress that the contributing authors are careful to ground their analysis of civil society in the particular context of the Balkans. More specifically, the concept of civil society popularized during the 1980s—the concept, which denotes those forms of communal and associational life that are organized neither by the market nor by the state—introduced new arguments into both the theory and practice. For activists, the discourse around civil society still offers the language of freedom, while for social scientists and political theorists everywhere it is an important reminder that even in the modern world there is more to social life than political economy. In the non-Western world, the idea of civil society has gone from theoretical and academic conceptualizations to the rhetoric of politicians, investors, and "builders" of democracy.

Particular Uses of Civil Society in Contemporary Social and Political Thought

We focus here on the conceptual analysis of civil society in contemporary political thought, covering its meaning and its role in democracy. It is noted that a recent explosion of studies on the notion of civil society often draw on unclear conceptualizations. In this volume, civil society is understood as the realm of ordinary citizens, who join and participate in organizations based on their own interests and needs without directly seeking power or profit (even though they sometimes seek political influence and often need financial support).

The concept of civil society was revitalized in the 1980s and 1990s by the opposition movements in Eastern Europe and the Balkans and by the wave of democratization around the world. In the past decade, civil society as the object of study has grown, and today it is commonly applied to very different contexts throughout the world. In the new post-Cold War period, in which many countries have undergone various forms of transitions, civil society is seen as one of the main contributors of democratization. Almost every large foundation, association or institute has established some kind of "civil society initiative" with the express purpose of creating, monitoring, and strengthening civil society in countries across the globe, especially in the Balkans. Most scholars, who write on the subject or are civically active, argue that civil society exerts a positive influence.

The term refers to a peaceful, moral, and idealistic vision, which essentially stands in opposition to "un-civil society." In this framework, two fundamental "virtues" of civil society are generally emphasized: the first has to do with the ability of the organizations of civil society to protect citizens against a potentially aggressive state; the second posits the ability of civil society to serve as a "school for democratic values" (Levy 1999: 2). To put it simply, the existence of autonomous organizations allow citizens to interact with one another outside of the private sphere, thereby developing trust and tolerance, which is then transferred to the public sphere. Acting in civil society allows people to gain a greater sense of their own political role, thereby creating a more engaged citizenry. The more people participate in civil society, the more they internalize the norms of participatory democratic citizenry and the more they create their own political reality and are, as a result, able to deal with the traumatic pasts. Participation in civil society organizations is an important attribute of democratic performance. In this sense, we see it as essential for scholars to concentrate on the topic of civil society and transitional justice in contemporary Balkans.

Much of the literature has tended to focus on this problem of a sharp analytical dichotomy between state and civil society (Gellner 1995), but, especially in the case of the Balkans, this relationship was actually viewed as an opposition between two social forces.. In recent contributions, some scholars have introduced compelling counter-arguments to challenge the dichotomy of state and civil society in a democratic context. Another conceptual "gray area" has to do with the definition of less structured types of mass organization. Are demonstrations excluded from civil society? Unlike other forms of citizen activity and participation, civil society requires a degree of institutionalization that is usually absent in such forms of activism. Does civil society refer only to membership and participation in a formally organized group? This line of thinking leads to the necessary differentiation between civil society and social movements. Social movements generally consist of some combination of spontaneous and loose organizations. When they are formally organized, they are seen as belonging to civil society. However, what is distinctive about social movements is that their leaders often resist the institutionalization of their movement for a whole range of reasons, including a problematic transformation of their organization into a political party and, therefore, its incorporation into the establishment.

Another complex issue concerns those organizations where membership may not be completely voluntary, or where there is a lack of complete autonomy. For example, would a labor union, where all workers become members automatically, be defined as a part of civil society? What about labor unions in former communist countries, where they were controlled by the political party? Under communism, most people were members of multiple organizations, but their membership was often of a coercive nature.

A final and crucial problem for our analysis here is connected with those organizations and movements that fit the general definition of civil society—organizations, where ordinary citizens come together according to their common interests—but where those interests are undemocratic and intolerant, and, in some instances, call for a racial, ethnic, or gender-based exclusion. In this respect, Stan (this volume) writes about the vast diversity of Romanian civil society groups with a stake in the transitional justice process: these include groups that support, groups that oppose, and groups that are indifferent, to the politics of the past. Her chapter provides an overview of civil society efforts to promote Romanian post-communist transitional justice since 1989 by surveying methods such as lustration (the banning of communist decision-makers from post-communist public life), court trials against communist perpetrators, ordinary citizens' access to the files compiled by the secret political police, the Securitate, the presidential history commission, the citizens' opinion tribunal, property restitution, and memorialization. She also discusses state-independent groups that have opposed any meaningful reckoning with the recent past.

It is also important to point out that mainstream liberal assumptions about civil society, diversity, and multiculturalism tend to invoke stereotypical contrasts between the (peaceful, democratic, tolerant) "civic nationalism" of the West and the (aggressive, authoritarian) "ethnic nationalism" of the East (see Ignatieff 1993); or between the (ethnically inclusive) "constitutional patriotism" of the West and the (ethnically exclusive) "Balkanization" of the East. In so far as anyone starts with these dichotomies, they automatically interpret (ethnic) conflicts in the Balkans as a consequence of premodern tribalism that must be overcome if liberal democracy is to be achieved. These conflicts then become used as evidence that the Balkans are not suitable for liberal democracy and cannot sustain the functioning of civil society. However, some recent theorists of liberal pluralism have disputed these stereotypical dichotomies (Kymlicka 2001). The assumption that Western "civic" nations and Western form of civil society have surpassed all forms of ethnocultural particularism has been, we believe, strongly challenged by Kymlicka and some other liberal theorists (see Walzer 1992).

Transitional Justice

The field of transitional justice is relatively new, emerging in the wake of the Cold War (Hinton 2010: 2). Over the past 20 years, transitional justice has grown into a globalized form of intervention following civil war and political repression (Teitel 2003). There are various definitions of transitional justice, but, most commonly, transitional justice is defined as "a response to systematic or widespread violations of human rights that seeks recognition for victims and to promote possibilities for peace, reconciliation and democracy" (ICTJ 2011). Initially, transitional justice was narrowly focused on justice and retribution, but over the years "it has encompassed the study of how human rights abuses are confronted by societies emerging from violent conflicts or transitioning from authoritarian rule to democratic forms of government." (Crossley-Frolick 2011: 33)

In order to confront legacies of abuse, a variety of transitional justice mechanisms have been developed. They fall within two broad categories—judicial and non-judicial. The former focuses on civil or criminal trials either at local, national or international level, while the latter encompasses activities such as truth and reconciliation commissions, vetting, amnesty, lustration processes, reparation, institutional reform, reconciliation, memorialization, security sector reform as well as demobilization, disarmament, and reintegration.

Despite the plethora of transitional justice mechanisms, Weinstein et al. suggest that "it is time to reconsider whether the term *transitional justice* accurately captures the dynamic processes unfolding on the ground," and whether "we have limited our array of options, prematurely becoming closed to other interventions" (2010: 32). As Simic and Daly argue, "a sustainable peace will often require more than trials, prosecution and punishment of a relatively few high-placed officials; it will require more than state-sponsored or administered forms of truth-telling and reparations" (2011: 479). Such peace may require civil society justice mechanisms, which encourage citizens to reflect and remember the past in direct and meaningful ways and to participate in activities that can transform social relations in the places and societies they live in.

As Shaw and Waldorf argue, transitional justice has recently "undergone a shift toward the local" (2010: 4). As such, local justice mechanisms are an emerging area of research in the transitional justice literature (Lundy and McGovern 2008). One of these mechanisms is explored by Irvine and McMahon and by Kurze and Vukusic (this volume) who critically examined the establishment of the Regional Truth Commission for the Former Yugoslavia (RECOM) as a regional mechanism of truth seeking and truth telling. The authors conclude that, although the founders of RECOM support the formal judicial mechanisms for trying and punishing war criminals, these mechanisms could not fully satisfy the victims' need for information and truth telling; at the same time, any efforts to establish a comprehensive understanding of human rights abuses during the wars would necessitate a regional approach.

It has become widely acknowledged that it is critically important to engage survivors and those people, who live in conflict and post-conflict areas, in discussions about transitional justice. "Once size fits all" solution has become unacceptable, replaced by the recognition that local justice mechanisms need to correspond to people's priorities and local circumstances. The contributors in this volume explore ways, in which these mechanisms function in particular places and times and consider their interaction with local practices and experiences.

While initially the transitional justice field was the domain of primarily legal and human rights scholars and practitioners, it is now the concern of scholars, practitioners, and activists coming from a whole range of disciplines and contexts (Bell 2009). This interdisciplinarity is reflected in this volume: in it we bring together human rights scholars, sociologists, media scholars, social psychologists, criminologists, political scientists, practitioners, and activists.

The Balkan Region

The term "Balkan" has contested meanings. Todorova (1997) argues that the Balkans were and are seen as the dark side of Europe—an identity that the people of the Balkans themselves are eager to eschew. Bakic-Hayden and Hayden (1992) argue in a similar vein that the Balkans are identified with violence, incivility, and barbarism.

Historically, from a Western European viewpoint, the Balkans represented the cultural and religious "other." Furthermore, Bakic-Hayden and Hayden argue that during the Cold War the ideological "other"—communism—temporarily replaced the geographical or cultural "other"—the Balkans. However, at some level, the symbolic image of Balkan inferiority was preserved and reinforced throughout this period as well. Hammond (2004) echoes this argument by suggesting that during and after the 1990s wars in the region, the Balkans again became a symbol of intolerance and barbarity. Sandwiched between the end of the Cold War and 9/11, it was the Balkans that functioned as the uncivilized "other" in Western discourse.

The historical and present construction of the Balkan identity is discussed in depth by Todorova (1997). She argues that the negative connotations derived from the term are primarily linked to the Balkan wars (1912–1913) with little regard for the actual historical and cultural context of these wars or, in fact, for the level of violence elsewhere within the "civilized world." Todorova investigates the emergence of the Western image of the Balkans and queries its possible functions. She argues that at the beginning of the twentieth century "Balkanization" came to denote the parcelization of large political units into small ones, synonymous with the regression back to the tribal, the backward, and the barbarian. In the same manner as the Orient, the Balkans became "the other" of Europe. Even further, while the Orient was constructed by the West as "the other" located elsewhere, an anti-world, the Balkans had been continuously viewed as a dark side of Europe itself: in other words, it is the darkness within. The stigmatization of the Balkans could be conceivably interpreted, therefore, as an attempt to control the dark side of within Europe (Wolff 1994).

Nationalism and Transnationalism

The post-1989 enthusiasm for globalization has prompted many to deny the importance and endurance of nationalism, and to promulgate idyllic visions of a post-national, cosmopolitan society. Nation states and national belonging, we have been told, are merely anachronistic remains of the past, bound to disappear with the onset of a global modernity. On this account, if we are to truly understand the nature of the increasingly globalized industries, of transnational cultural flows and transitional justice, we need to dispense with our old vocabularies, steeped in methodological nationalism, and instead learn to think beyond the nation. Such account of the death of the nation-state and the forms of nationalism with which it is associated are premature. Research indicates that, for example, media continues to be one of the crucial sites of the reproduction of nationalism, despite the ongoing intensification of global media flows (Volcic 2011). Nation-states and national identities continue to function as the main building blocks of worldwide systems. The national lenses are the optics most of our authors deploy in their analysis of transitional justice developments in Bosnia and Herzegovina, Romania, Kosovo, Macedonia, Albania, Croatia, and Serbia.

At the same time, some of our chapters focus on transnational and regional approach in their exploration of civil society and transitional justice. The growing popularity of the concept of transnationalism stems, in part, from a general dissatisfaction with nation-state-centric understandings of society, identity, and politics. Scholars within traditionally state-centric disciplines, such as International Relations and political geography, have increasingly advocated new conceptualizations of justice, identities, political community, and territory that reflect the realities of globalization. These scholars argue that sociopolitical life often escapes the constraints of the territorial nation-state, and that new spaces of interaction are forged across and between many territories. The nation-state remains important for understanding forms and meanings of political identity, but in the case of transitional justice, the primary space of the political cannot be assumed to be circumscribed by the nation-state. Kurze and Vukusic, for example, explore the continuous struggle of human rights activists to create a transnational extra-legal space to deal with past atrocities across the former Yugoslavia. They survey recent fact-finding and documentation projects, in particular, the regional truth commission initiative (RECOM). In addition, they argue that the multiplicity of narratives of victimhood have emerged in response to the transnational character of restorative justice efforts in the post-conflict Balkans and explain why these recent efforts to establish a transnational fact-finding mechanism have been unsuccessful.

Transnational approach allows scholars to open up alternative ways of analyzing the implementation of transitional justice in post-conflict settings.

Throughout this volume, contributors strive to offer an insight into the complex realm of civil society engagement with transitional justice mechanisms in the context of Balkans. By taking a more critical look at national strategies, local practices and priorities, and by closely examining international transitional justice agendas, the authors in this volume explore the complex and unpredictable justice processes currently underway in the Balkans. They suggest lessons to be learned from those engagements and identify future directions that may be taken in order to bring a sustainable peace to the region.

References

Bakic-Hayden, Milica and Hayden, Robert M. 1992. Orientalist Variations on the Theme "Balkans": Symbolic Geography in Recent Yugoslav Cultural Politics. *In Slavic review* 51(1): 1–15.

Bryant, Christopher G.A. 1995. Civic nation, civil society, civil religion. In *Civil society. Theory, history, comparison*, ed. A.Hall John. Cambridge: Polity.

Bell, Christine. 2009. Transitional justice, interdisciplinarity and the state of the 'field' or 'non-field'. *In International Journal of Transitional Justice* 3: 5–27.

Buttigieg, Joseph 1995. Gramsci on Civil Society. *In boundary 2* 22(3): 1–32.

Calhoun, Craig. 1993. Civil society and the public sphere. *Public Culture* 5(2): 267–280.

Calhoun, Craig. 1994. Nationalism and civil society: Democracy, diversity and self-determination. In *Social theory and the politics of identity*, ed. David R. Dickens and Andrea Fontana, 304–335. Cambridge: Blackwell.

Crossley-Frolick, Katy. 2011. The European Union and transitional justice: Human rights and post-conflict reconciliation in Europe and beyond. *Contemporary Readings in Law and Social Justice* 3(1): 33–51.

Gellner, Ernest. 1995. The importance of being modular. In *Civil society*, ed. John Hall, 32–56. Cambridge: Polity.

Giner, Salvador. 1995. Civil society and its future. In *Civil society*, ed. John Hall, 301–326. Cambridge: Polity.

Habermas, Jurgen. 1989. *The structural transformation of the public sphere*. Cambridge: MIT.

Habermas, Jurgen. 1992. Further reflections on the public sphere. In *Habermas and the public sphere*, ed. Craig Calhoun. Cambridge: MIT.

Hammond, L.C. 2004. *This Place Will Become Home: Refugee Repatriation To Ethiopia*. Ithaca: Cornell University Press.

Hinton, Alexander, ed. 2010. Introduction: Towards and anthropology of transitional justice. In *Transitional justice: Global mechanisms and local realties after genocide and mass murder*. New Brunswick: Rutgers University Press.

Ignatieff, Michael. 1993. *Blood and belonging: Journeys into the new nationalism*. New York: Farrar, Straus and Giroux.

International Center for Transitional Justice website. http://ictj.org/our-work/transitional-justice-issues. Accessed 2 November 2011.

Keane, John. 1988. *Democracy and civil society*. London: Verso.

Kumar, Krishan. 1993. Civil society: an inquiry into the usefulness of an historical term. *British Journal of Sociology* 44(3): 375–395.

Kymlicka, Will. 2001. *Politics in the vernacular*. Oxford: Oxford University Press.

Levy, Jonah D. 1999. *Tocqueville's revenge: State, society, and economy in contemporary France*. Cambridge: HUP.

Lundy, Patricia, and Mark McGovern. 2008. The role of community in participatory transitional justice. In *Transitional justice from below: Grassroots activism and the struggle for change*, ed. Kieran McEvoy and Lorna McGregor. Oxford: Hart.

Michael Walzer. 1992. The idea of *civil society*: A path to social reconstruction. Dissent 39 (Spring): 293–304.

Rowe, Peter. 1997. *Civic realism*. Cambridge: MIT.

Scheper-Hughes, Nancy. 2004. *Violence in war and peace*. Oxford: Blackwell.

Shaw, Rosalind, and Lars Waldorf (eds.). 2010. *Localizing transitional justice, interventions and priorities after mass violence*. Stanford: Stanford University Press.

Simic, Olivera, and Kathleen Daly. 2011. "One pair of shoes, one life": Steps towards accountability for genocide in Srebrenica. *International Journal of Transitional Justice* 5(3): 477–491.

Taylor, Charles. 1994. *Multiculturalism: Examining the politics of recognition*. Princeton: Princeton University Press.

Teitel, Ruti. 2003. Transitional justice genealogy. *Harvard Human Rights Journal* 16: 69–94.

Todorova, Maria. 1997. *Imagining the Balkans*. New York: Oxford University Press.

Volcic, Zala. 2011. *Serbian spaces of identity*. New York: Hampton.

Walzer, Michael. 1992. The idea of *civil society*: A path to social reconstruction. Dissent 39 (Spring): 293–304.

Weinstein, Harvey, et al. 2010. Stay the hand of justice: Whose priorities take priority? In *Localizing transitional justice, interventions and priorities after mass violence*, ed. Rosalind Shaw and Lars Waldorf. Stanford: Stanford University Press.

Wolff, Larry. 1994. *Inventing Eastern Europe: The map of civilization on the mind of the enlightenment*. Stanford: Stanford University Press.

Wood, Ellen Meiskins. 1990. The uses and abuses of civil society. In *Socialist register*, ed. Ralph Miliband et al. London: Merlin.

Part I
Initiatives Within the Nation States

Chapter 2
Civil Society and Post-communist Transitional Justice in Romania

Lavinia Stan

Introduction

In the only text that examined the impact of state-free groups on the process of reckoning with the communist past in Romania, Grosescu (2007) argued that civil actors "were unable to build coherent methodologies and long-term projects" regarding transitional justice (p. 183). The statement rings true when applied to judicial methods. Nevertheless, Romanian civil society groups have been instrumental in maintaining the need to come to terms with the communist past in the public eye and on the policy agenda, and in proposing key methods designed to redress communist human rights violations. True, during the first two decades of post-communism, these groups faced a general public mainly concerned with its economic survival and political elites preoccupied with their narrow group interests (*Atitudini şi opinii despre regimul comunist din România. Sondaj de opinie publică*, 2012). In addition, as Grosescu (2007) noted, these groups were unable to come together and support common projects, and often criticized each other as much as they criticized state authorities.

This chapter provides an overview of civil society efforts to promote Romanian post-communist transitional justice since 1989 by surveying methods such as lustration (the banning of communist decision-makers from post-communist public life), court trials launched against communist perpetrators, ordinary citizens' access to the files compiled by the secret political police, the Securitate, the presidential history commission, the citizens' opinion tribunal, property restitution, and memorialization. Whereas most authors working on Eastern Europe have assumed that civil society generally promotes transitional justice, this chapter also discusses state-free groups that have opposed any meaningful reckoning with the recent past.

L. Stan (✉)
Department of Political Science, St. Francis Xavier University,
PO Box 5000, Antigonish, NS, Canada, B2G 2W5
e-mail: lstan@stfx.ca

O. Simić and Z. Volčič (eds.), *Transitional Justice and Civil Society in the Balkans*
Springer Series in Transitional Justice, DOI 10.1007/978-1-4614-5422-9_2,

Civil Society Groups Involved in Transitional Justice

Most of the crimes perpetrated by the Romanian communist regime occurred from 1945 to 1964, when many pre-communist political leaders and industrialists, anticommunist opponents, disgruntled army officers, students attempting to illegally cross the border, peasants resisting collectivization, and disgraced communist leaders were arrested and imprisoned often without observing minimal due process requirements (Deletant 2000; Judt 2005). The regime of Nicolae Ceauşescu (1965–1989) was restrictive even by Eastern European standards, although repression took the form of widespread surveillance by the Securitate more than outright killings, torture, and disappearances (Deletant 1996; Tismăneanu 2003). Decades of communist policies limiting freedom of speech, religion, and association, together with Ceauşescu's own brand of sultanism-cum-totalitarianism (Linz and Stepan 1996), decimated the civil society, and brought most churches, labor unions, and voluntary associations under the regime's firm control.

As such, when the communist regime collapsed in 1989, there were few state-free groups (Howard 2003; Ignatieff 1995). These groups lacked strong leadership and financial resources, and were mostly confined to the town where they originated, having weak representation in the country and constituting a "retarded civil society" facing an inefficient and incompetent state (Carothers 1999, p. 20). Groups created around pre-communist political and union leaders could draw on these leaders' expertise, but were also disadvantaged by their leaders' advanced age and serious health problems acquired in communist jails. By comparison, the most successful state-free groups have been constituted along the lines of communist-era associations and unions, but those were also the closest to the state and the least likely to confront it or propose policies that could hurt the interests of the new post-communist rulers. Transitional justice was among those policies because the new rulers were former second echelon Communist Party officials like Ion Iliescu, who got a new lease on political life in the 1989 revolution (Stan 2010).

A number of state-free groups constituted after 1989 have supported transitional justice. Some of them represent former political prisoners and owners of abusively confiscated property. Others include humanistic intellectuals concerned with the quality of the new democracy, the persistent legacies of communism, and the collaboration of post-communist politicians with the Ceauşescu regime. Both former victims and the intellectuals have called for resolute transitional justice in the form of property restitution, lustration, court trials against communist officials, the public identification of former Securitate part-time informers, access to secret archives, commemoration of victims, and research on and publications about communist repression. As public debates on the need to honor the victims and identify the perpetrators of communist crimes unfolded in the 1990s, civil society actors inimical to transitional justice started to organize.

Three main types of civil society groups are relevant in this context. The first includes former victims of communist human rights violations, who support resolute, prompt, and comprehensive transitional justice. The Association of Former

Political Prisoners in Romania (Asociația Foștilor Deținuți Politici din România), constituted in 1990 in Bucharest, has branches throughout the country and a membership that reportedly diminished from 98,700 in 1990 to 45,000 by 2009 (Asociația Foștilor Deținuți Politici din România 2009). It has been the most important group of communist-era political prisoners, deportees, and their relatives. In the mid-1990s, the smaller Association of Former Political Prisoners and Anticommunist Fighters (Asociația Foștilor Deținuți Politici și Luptători Anticomuniști) was created by communist-era political prisoners with ties to the inter-war fascist Iron Guard, which subscribed to nationalism, Orthodoxy, and anti-semitism. It commemorates mostly Iron Guard members imprisoned, tortured, or killed by the Securitate, keeping silent on the crimes in which the Guard engaged during pre-communist times. The Association of Owners of Property Abusively Confiscated by the State (Asociația Proprietarilor Deposedați Abuziv de Stat) was set up in 1999 in Bucharest by property owners, their descendants, and legal representatives to fight for property restitution and equitable compensation. With an initial membership of 11,000 and a current membership of 1,500, the Association is Romania's most important and vocal organization of initial homeowners (Oțoiu 2006; 2009). Constituted in 1991 in Bucharest, the ICAR Foundation provides medical rehabilitation for victims of communism. Besides the Bucharest clinic, it opened centers in Iași in 1995, and Craiova in 1998 (ICAR Foundation 2010a).

Other associations have local, not national, membership or represent smaller victims' groups. Created in 2009, the Association of Owners Abusively Deprived of Their Property, Former Deportees and Refugees (Asociația Persoanelor Deposedate Abuziv Foștilor Deportați Refugiați din România) represents property owners, deportees, and refugees seeking compensation and redress for the injustices they suffered. In mid-2009, it had 210 members, exclusively drawn from the town of Craiova (Asociația Persoanelor Deposedate Abuziv Foștilor Deportați Refugiați din România 2009). Former owners are also represented by the Association for Private Property (Asociația pentru Proprietate Privată) and the Community of Legitimate Owners and Descendants in Romania (Comunitatea Moștenitorilor și Proprietarilor Legitimi din România), both headquartered in Bucharest, and the more obscure Association of Victims of the Bolshevik Communist System and Its Legacy (Asociația Victimelor Sistemului Comunist Bolșevic și a Sechelelor Sale) in Râmnicu-Vâlcea. Since 1997, the Bucharest-based Association of the Victims of Magistrates (Asociația Victimelor Magistraților din România) represents those who believe they suffered as a result of court decisions handed down by corrupt communist and post-communist judges. Former owners residing in France and Germany are represented by the French Association for Defending Property Rights in Romania (Asociația Franceză pentru Apărarea Drepturilor de Proprietate în România) of Paris and the Restitution in Romania (Interessenvertretung Restitution in Rumanien) of Munich.

To these voluntary associations, add the Gheorghe Ursu Foundation (Fundația Gheorghe Ursu), set up by relatives of the engineer who was murdered in the Bucharest Militia jail in 1986. Ursu was arrested after being denounced by two workmates for keeping a journal critical of the communist leadership and policies. The Ion Gavrilă Ogoranu Foundation (Fundația Ion Gavrilă Ogoranu) was set up in

2008, 2 years after Ogoranu's death, to commemorate the leader of the Făgăraş anticommunist fighters group, active between 1947, when Romania became part of the communist bloc, and 1956, when the Securitate arrested most of the group's members.

The need to adopt transitional justice methods in view of addressing the legacy of communism has also been underscored by organizations uniting the victims of the 1989 revolution, which nonetheless remain preoccupied primarily with finding the truth about the revolution, publicly unmasking the involvement of post-communist leaders in those events, and securing financial benefits for their members. Best known among these groups has been the Association 21 December 1989 (Asociaţia 21 Decembrie 1989), set up in 1990 in Bucharest by some of those protesters who, during the revolution, took to the streets and defied the police and army forces deployed by the Ceauşescu regime. The Association has advocated in favor of resolute transitional justice, including radical lustration.

Last, this category further comprises the country's main religious denominations, especially the dominant Orthodox Church and the minority Greek Catholic Church. Compared to victims' groups, these denominations had been constituted well before 1989. Similarly to victims' groups, they faced persecution, confiscation of property, discrimination, and surveillance at the hands of the communist authorities. During the first decade of communist rule, many Orthodox leaders, priests, faithful were imprisoned, and the state requisitioned church property (Leuştean 2009). Afterward, the Orthodox Church was placed under extensive surveillance by the Securitate, which recruited some priests as secret agents (Stan and Turcescu 2005). In 1948, the communist state dismantled the Greek Catholic Church, imprisoned its bishops who refused to convert to Orthodoxy, and transferred the Greek Catholic churches, chapels, parish houses, and cemeteries to the Orthodox Church.

The second category includes the country's humanistic intellectuals and former anticommunist dissidents, who have lent support to some transitional justice methods (the truth commission, lustration, and access to secret files), but have mostly rejected efforts to unveil collaborators from among their members, the intelligentsia, and the political parties close to them. The Group for Social Dialogue (Grupul pentru Dialog Social) of Bucharest and the Timişoara Society (Societatea Timişoara), located in the town where the revolution first started, were both created immediately after December 1989 to "promote real democracy" by providing viewpoints alternative to those proposed by the former communists turned into the country's first post-communist rulers (Ştefănescu 1995, p. 47). While few in number and elitist in view, the intellectuals have been very vocal and able to propose some of the most coherent political programs advanced by the civil society. This category also includes the Civic Alliance (Alianţa Civică), created in 1990 by 216 respected writers, philosophers, and anticommunist dissidents in order "to channel social energies supporting faith, humanism and democracy" (Ştefănescu 1995, p. 104). The Alliance was most active during the early 1990s, when some of its members gained seats in Parliament, as representatives of the Civic Alliance Party. Most of its transitional justice programs have been run through the non-profit Civic Academy Foundation, led by two Alliance leaders, poet Ana Blandiana and writer Romulus Rusan.

The third type of civil society groups includes the tenants who occupied confiscated dwellings with the authorities' permission before and after 1989. While some tenants were poor workers who moved from village to town in order to work in the new industrial factories built throughout Romania, many others were privileged Communist Party leaders and nomenklatura members who had access to special food stores, medical clinics, and holiday resorts. Until 1989, all these tenants rented dwellings from the state, paying extremely low rents that did not reflect the real market value of the property. Most of these dwellings were large historical houses, located in leafy residential areas (Stan 2006). The Association of Tenants Living in Nationalized Dwellings (Asociația Chiriașilor din Casele Naționalizate) has branches in the country's most important towns. The Association of Tenants Who Acquired Ownership through Law 112/1995 (Asociația Proprietarilor pe Legea 112/1995) represents tenants who bought the confiscated dwelling in which they lived in virtue of Law 112/1995 (Benea 2008).

Tenants and secret agents were not the only ones to benefit from the communist-era infringement of other people's rights. Since 2003, the Motherland and Honor Solidarity Foundation (Fundația Solidaritatea Patrie și Onoare) has provided financial help to former Securitate officers and post-communist information service agents, and protected their interests (Andreescu 2003). Former Communist Party leaders, party activists, political officers, militia officers, and communist prosecutors and judges have not created civil society associations, since they could face public condemnation as a result and after 1989 they have retained enough political clout to advance their personal interests without feeling the need to create new political vehicles. In addition, former communist perpetrators have gained a public voice as members of formations successor to the Communist Party and the Communist Youth League: the Social Democratic Party (which formed the government in 1990–1996 and 2000–2004), the Democratic Party (which has ruled the country in 2004–2012), and the Greater Romania Party (represented in Parliament in 1990–2008). Both tenants and former perpetrators have consistently blocked transitional justice attempts.

Transitional Justice Methods

In Romania state actors have generally blocked, and civil society actors have supported, transitional justice. At the same time, after 1989 former perpetrators have retained considerable political influence, exerting far more leverage over the political process than former communist-era victims. This is why victims have promoted mostly non-judicial, and often local, reckoning processes of limited impact, Romania failed to enact radical lustration, and court trials have been few in number and deficient in procedure. The wavering and self-interested position of the intellectual groups and the dishonesty of some prominent civil society actors, who hid their own former collaboration (Grosescu 2007, p. 190), explain why the Romanian transitional justice program has been politicized and delegitimized in the eyes of the general public.

Lustration

The civil society has promoted lustration without convincing political elites to fully implement it. On 11 March 1990, the Timişoara Society made public its Declaration, whose Article 8 called for the adoption of electoral law amendments banning Communist Party leaders, state dignitaries, and Securitate agents from running in presidential elections and from being included on party lists for the first three consecutive legislative cycles, that is, 12 years (The Timişoara Declaration 1990). The Declaration was the first Romanian document to urge for the adoption of a method considered key to working through the communist past. It indelibly set the parameters of the lustration debate in Romania by promoting radical lustration. The German and Czech lustration programs, launched immediately after the collapse of the communist regime, were radical because they targeted many people and involved job loss (David 2011). By contrast, Poland and Hungary, which implemented lustration later, allowed former decision makers, party leaders, and secret agents to retain their post-communist public posts if they admitted to their past in written, signed declarations. While in the Czech Republic 10,000 people lost their positions because of lustration, in Hungary and Poland less than 500 were affected (Stan 2009). The Timişoara Society has always understood lustration as radical lustration, and has called for lustration even after Romanian public officials were repeatedly asked to disclose their tainted past, signaling that Hungarian-type lustration was not what the Society envisioned for Romania. Almost all the above-mentioned associations representing victims and intellectuals have shared this view.

Lustration came to the forefront on 7 December 1993 when senator Constantin Ticu Dumitrescu, the Association of Former Political Prisoners chairman, introduced in Parliament a simple motion on secret informers that amounted to a lustration proposal allowing "nobody to be blackmailed or smeared by false accusations" of collaboration with the Securitate (Şedinţa Senatului din 25 martie 1992, p. 2). Compared to the Timişoara Declaration, this proposal targeted only the Securitate part-time informers, but not the full-time officers who recruited them or the party leaders who masterminded repression and surveillance. While 74% of senators and 46% of deputies supported it, the motion had no effects, because the Senate leaders forbade the Romanian Information Service, heir to the Securitate domestic repression branch, to disclose the identity of former Securitate agents for fear of endangering national security (Şedinţa Senatului din 3 februarie 1994, p. 4).

Afterward, Dumitrescu promoted lustration through his Bill on Access to Files and Unveiling the Securitate as a Political Police which, in its original version, permitted citizens to read the secret files compiled on them by the Securitate and asked public officials and electoral candidates unveiled as former secret agents to give up their posts or renounce the electoral race. After bitter debates, Parliament stripped the bill of its lustration stipulations. When the bill was adopted as Law 187/1999, Dumitrescu was so dissatisfied with the changes operated by the house that he refused to accept the text's paternity (Stan 2000).

Civil society groups also brought lustration into the public eyes in 2005–2007, when the country was ruled by the Truth and Justice Alliance, which included the Liberals and the Democrats. Inspired by Bulgarian efforts, in 2006 Romanian journalists launched a Clean Voices campaign to identify secret agents from among television reporters, press contributors, and talk-show hosts. In response, Liberal legislators presented a lustration bill, the opposition lodged its own anti-nomenklatura legislative proposal, intellectuals, academics, and civil society representatives called on former and current spies to unveil their ties to communist and post-communist intelligence services, and many politicians admitted to past collaboration or were unmasked as former spies. More importantly, civil society groups convinced the Chamber of Deputies leaders to jointly organize a public debate on "Lustration: Principle or Instrument" on 25 May 2006. Besides legislators and ordinary citizens, representatives of 10 civil society groups—including the Timişoara Society, the Civic Alliance, and the Association of Former Political Prisoners—stated their position on this important transitional justice practice (Stan 2011).

Court Trials

Court proceedings related to communist human rights abuses have been few in Romania. Since 1989, civil society actors have repeatedly called for former Securitate officers, prison guards, and Communist Party leaders to be indicted, and for the "Trial of Communism" to be organized. Press campaigns, street demonstrations, and roundtable talks have been used to promote court trials. The civil society has also endeavored to collect the information needed to indict some of the most notorious communist crimes, to locate former perpetrators, and to bring them to justice. In 1991, the Association of Former Political Prisoners submitted to prosecutors a list of notorious communist perpetrators, including deputy head of the Securitate Alexandru Nicolschi and Minister of State Security Alexandru Drăghici. In 1998, the name of Gheorghe Crăciun, former head of the Aiud prison (1958–1964), was added to the list. All three accused died before the courts heard their cases (Muraru 2011). The Gheorghe Ursu Foundation collected evidence and applied pressure on the judiciary to indict Marian Clita, Ursu's assassin (Macovei 1999). In July 1999, Clita was sentenced to 20 years in jail, but his jail term was halved in virtue of a 1988 amnesty decree. When his sentence was handed down, Clita had already served 8 years in jail. As such, he was released in September 1999 (Artene 1999).

Since 1989, 20 different complaints have been submitted to military prosecutors by civil society representatives against former communist perpetrators, primarily Securitate officers, heads of detention centers, and prison guards. All these complaints were included in the so-called Case File 35 (titled the "Trial of Communism"), together with seven other complaints filed between 2006 and 2008 by the Bucharest-based Institute for the Investigation of Communist Crimes in Romania (Institutul pentru Investigarea Crimelor Comunismului în România), a transitional justice governmental agency. None of these formal complaints and the evidence they brought forth have been used to launch any court proceedings against human rights violators (Grosescu and Ursachi 2009; Muraru 2011). Unfortunately, few perpetrators are still alive today and most of the surviving ones are very old, so their future prosecution is highly improbable.

In Romania, the scarcity of court trials launched against communist perpetrators resulted not only from the constant opposition of the judiciary to hear such cases promptly and to recognize communist human rights violations as crimes for which the statute of limitations did not apply, but also from the support the civil society has provided to the generic "Trial of Communism" more than to specific court cases. The preference for an all-encompassing "Trial of Communism" was first voiced on 29 November 1990 by the Civic Alliance, which "in the name of the people, asks for a trial of the leftist ideology of communism, which inspired, generated and covered up crimes, protecting the perpetrators" (Ştefănescu 1995, p. 110). The call had little legal value, since only individuals, not abstract concepts such as an ideology or a political regime, can be put on trial. The Civic Alliance recognized this point in its March 1991 Declaration on National Reconciliation, which stated that "the Communist Party leaders, not all of the four million party members, should be considered morally responsible" for communist crimes. Whereas "any reference to a person should include his/her actions, not mere party membership," "former Central Committee members, party activists and Securitate agents should be banned from occupying public positions until year 2000" (Ştefănescu 1995, pp. 130–131). The preference for the generic "Trial of Communism" as opposed to specific cases stemmed from the civil society representatives' lack of legal expertise (Grosescu and Ursachi 2009, p. 182).

Access to Secret Files

Access to secret files was the brainchild of Dumitrescu, who almost singlehandedly convinced Parliament to legislate it. In his quest, Dumitrescu drew resources and support from the Association of Former Political Prisoners, the Group for Social Dialogue, and several other victims' groups. In 1997, the Senate amended the above-mentioned Bill on Access to Files and Unveiling the Securitate as a Political Police to include three changes not present in Dumitrescu's original proposal. First, personal files were made public only if their contents did not endanger national security, which was exclusively defined by the Romanian Information Service, recognized for its vested interest in keeping as many files as possible out of the public eye. Second, the leadership of the governmental agency in charge of file access (the National Council for the Study of Securitate Archives, Consiliul Național pentru Studierea Arhivelor Securității) included no longer prominent citizens, but nominees of the political parties represented in the Senate. Last, the archives remained with the institutions that produced them and wanted to keep them under lock to avoid exposing their links to the institutions and agencies of the communist regime (Stan 2000).

The Senate amendments provoked the ire of former victims and intellectuals. Before the Chamber of Deputies was scheduled to discuss the amended bill, in September 1998 the civic organizations and victims' groups collected signatures on a petition that called on deputies to consider Dumitrescu's original draft, as they believed that the amendments made it impossible for the truth about communism ever to be known. The petition was ignored, as the deputies decided to consider the Senate's modified version, not Dumitrescu's original proposal, and to ask two of their standing committees, not one, as it was customary, to review the bill. While unsuccessful, the petition clearly showed that civil society actors were able to come together in support of a transitional justice project, if and when they wanted.

The Presidential History Commission

Undoubtedly, the civil society scored its greatest success in the field of transitional justice in 2006, when it convinced Democrat-Liberal President Traian Băsescu to create the Presidential Commission for the Study of the Communist Dictatorship in Romania (Comisia Prezidențială pentru Analiza Dictaturii Comuniste din România), better known as the Tismăneanu Commission after its chair, the University of Maryland political science professor Vladimir Tismăneanu. In December 2006, weeks before the country's accession into the European Union on 1 January 2007, President Băsescu officially condemned the communist regime in front of the Romanian Parliament in a speech broadcast live by television stations. The condemnation was based on the Commission's 660-page-long final report, which detailed the mechanisms of repression in communist Romania, the ties between the Securitate and the Communist Party, as well as the continuity between the regimes of Ceaușescu and his predecessor, Gheorghe Gheorghiu Dej (Tănăşoiu 2007; Tismăneanu 2008). The Commission was created in response to civil society calls to renew the self-interested and corrupt political elites by legislating radical lustration. Under pressure from the Timişoara Society, the Association 21 December, and different victims' groups, President Băsescu declared that his ruling Democrat-Liberal Party could not support lustration without first condemning the communist regime as criminal and repressive. That was because those who had collaborated with the communist regime did nothing wrong, as they obeyed an internationally recognized regime and the laws in effect at the time. Only the exposure of the regime's repressive character could lay down the moral ground for a blanket policy like lustration. To condemn communism, President Băsescu needed a scientific expert report detailing the crimes of the 1945–1989 period in a systematic and dispassionate manner. As Parliament was divided between supporters and opponents of lustration, the civil society called on President Băsescu to create the commission as a presidential, not parliamentary, body. This is how the Tismăneanu Commission, which included representatives of the Association of Former Political Prisoners (Dumitrescu) and the Group for Social Dialogue (Radu Filipescu, Sorin Iliesiu, and Horia Patapievici), appeared.

The Citizen's Opinion Tribunal

Frustrated with the judiciary's unwillingness to organize a Nuremberg-type trial to condemn communist criminals, on 7 September 2006 civil society groups organized an opinion tribunal in Cluj-Napoca. This domestic tribunal, resembling opinion tribunals created in other parts of the world (Klinghoffer and Klinghoffer 2002), was composed of nine former victims of communist repression representing different Romanian counties, one councilor each for the prosecution and defense, and 150 audience members who acted as jurors. Charged with genocide and crimes against humanity (including premeditated murder, extermination, forced deportation, arrests, torture, disappearances, and ethnic and religious persecution), crimes not protected from prosecution by the statute of limitations, the communist regime was found guilty of all charges after the opinion tribunal discussed a summary of communist human rights abuses (Curtea Penală de Condamnare Juridico-Morală a Crimelor Regimurilor Comuniste 2006).

The opinion tribunal had little public echo inside and outside Romania, and was disregarded by the public, the political elite, the local press, and the main associations representing communist-era victims and intellectuals. Not only has the public remained reluctant to support a "Trial of Communism" that would indirectly implicate the four million ordinary party members, but former communist-era victims have argued that the post-communist state—the legal successor to the communist state that perpetrated those atrocities—should acknowledge responsibility for communist crimes through its regular courts of justice. The Romanian judiciary has ignored these demands, pointing to the statute of limitations applicable to those cases.

Victims' and intellectual groups challenged the legitimacy of the opinion tribunal, calling instead for a "Trial of Communism" in the courts of law, although the judiciary cannot indict an ideology or a political regime *in toto*, as already mentioned. In 2003, the ICAR Foundation filed with the Bucharest Court a complaint asking the government to acknowledge the "communist Holocaust" of 1945–1989, to officially apologize to the victims for the abuses they suffered, and to admit that the Securitate was a political police. According to the Foundation, the 'Trial of Communism' is "a duty to the victims who died with the hope that justice would be done and to those who survived and are still waiting for this justice to be done," "a necessary and expected acknowledgement of the state abuses of the past, and a sign of maturity for the Romanian democracy" (ICAR Foundation 2010b). President Iliescu and the Social Democrat government ignored the request. After the 2004 elections, President Băsescu received ICAR representatives at Cotroceni, the Bucharest-based presidential offices. During the meeting, the representatives convinced the president of the power of an official apology, coming from the country's most important state dignitary, for communist crimes addressed to the victims and the Romanian society at large. Băsescu agreed to deliver the apology, after an expert commission attested to the crimes of the communist regime, which the Tismăneanu Commission later did (see above). The apology never came, and the work of the presidential commission became a substitute for the "Trial of Communism" advocated by the civil society.

Property Restitution

The restitution of dwellings is an area where groups supporting the process have publicly clashed with groups opposing it. Since 1989, organizations representing the initial owners (most notably the Association of Owners of Property Abusively Confiscated by the State) have asked for the return of the property they lost to communist authorities or for fair compensation, when restitution in kind was not possible. They have been opposed by the Association of Tenants Living in Nationalized Dwellings and the Association of Tenants Who Acquired Ownership through Law 112/1995, which represents the tenants who rented the confiscated dwellings from communist authorities. All these groups have tried to influence public policy through street demonstrations, open letters addressed to the government, and press campaigns.

In 2000, Eugen Pleşa, the leader of the Association of Tenants Living in Nationalized Dwellings, joined the opposition Greater Romania Party in the hope of influencing legislation. That year, he entered Parliament, where he championed the interests of his Association, and convinced the house to accept as valid and legal the contracts through which tenants bought nationalized dwellings from the state. As a result, Law 10/2001 on the Legal Status of Property Abusively Taken Over by the Communist State during 6 March 1945–22 December 1989 allowed for the return of all dwellings except those bought by tenants in "good faith." The Association of Owners of Property Abusively Confiscated by the State warned that all tenants knew that the houses they occupied had been illegally confiscated by the communist state, in nationalization drives that ran counter even to the 1948 communist Constitution, which guaranteed the right to property. However, Law 10/2001 did not annul the provisions of Law 112/1995 which allowed tenants to buy nationalized property. In 1999, the Association addressed an open letter to all political parties represented in the Chamber of Deputies, but its plea for recognizing the property rights of initial homeowners remained unanswered (Oţoiu 2009, p. 70). The Association continues to monitor the activity of the Property Fund (Fondul Proprietatea), set up in 2005 to provide compensation to initial owners, and the cases lodged by Romanian owners with the European Court for Human Rights.

In Romania, property restitution also extended to ecclesiastical property, including the churches, chapels and cemeteries transferred in 1948 by the communist state from the minority Greek Catholic Church to the majority Orthodox Church. The Orthodox-Greek Catholic property dispute has been one of the bitterest disputes of post-communist times (Stan and Turcescu 2007). Since 1989, the two churches have behaved as latent interest groups that have mobilized to lobby the political establishment and to engage in high-profile public campaigns each time when their interests dictated such actions (Dickerson and Flanagan 2006). While their main institutional goals remain religious, both churches launched media campaigns against each other, wooed the public in support of their restitution demands, and lobbied the government for legislation and political action aimed at protecting their interests and hurting those of the other denomination.

At the pressure of the Orthodox Church, throughout the 1990s the post-communist authorities refused to get involved in the property dispute on grounds that it was a purely religious issue and denied the Greek Catholics their right to seek justice through the Romanian courts. In turn, the Greek Catholics complained to the Vatican and foreign governments sympathetic to their plight, and approached the European Court of Human Rights, which recognized the infringement of their rights. That decision obliged the Romanian state to allow the courts to hear cases of Greek Catholic church restitution starting in 2006. As it commands the loyalty of as much as 86% of the population and remains a key electoral ally for all political formations, the Orthodox Church has become the most important civil society actor shaping the politics of the past.

Memorialization

In the absence of a museum dedicated to the victims of the communist regime, the Sighet Memorial of the Victims of Communism and of the Resistance (Memorialul Victimelor Comunismului şi al Rezistenţei) remains the most significant memorialization project in post-communist Romania and the country's only museum opened in a communist-era political prison. Created in 1993 by the Civic Academy, the Memorial includes the Museum located in the Sighet prison (a small town in Northern Romania) and the International Center for the Study of Communism (Centrul Internaţional de Studii asupra Comunismului) located in Bucharest. Through them, the Civic Academy seeks "to promote civic education and revise the country's history falsified by the communist regime" (Ciobanu 2008, p. 60). Since 1997, the Memorial has received some funding from the Romanian government, but its most important donors have remained the Romanian diaspora, foundations such as the Konrad Adenauer Stiftung, and the Council of Europe. Besides a series of ongoing oral history programs, the Memorial organizes a yearly summer school for pre-university teachers and students, publishes a scholarly journal, and commemorates each year on the day of the Ascension a Day of Memory, dedicated to all those who opposed communism and suffered in communist prisons (Fundaţia Academia Civică 2009).

Victims' groups have also funded a large number of memorials. By 2004, for example, the Association of Former Political Prisoners had helped to erect hundreds of monuments, crosses, and commemorative plaques in every Romanian county (Asociaţia Foştilor Deţinuţi Politici din România 2004). Some of these memory projects mark the site of former political prisons or murders of anticommunist fighters, others remember the struggle of prominent local anticommunist heroes, while still others are dedicated to the Romanian people's opposition to the communist regime. Given the reluctance of successive post-communist governments to honor the victims and condemn the communist perpetrators, most important have been the Association's efforts to mark each and every political prison, regardless of whether it has remained a prison, was converted to some other use, or was demolished.

Conclusion

In the South African case, civil society efforts at redressing the past have been "undercut by the existence of a new democratically elected government that had high levels of legitimacy [and] served to discourage the need for active and unified action" by non-governmental organizations part of the civil society (Hamber et al. 1997, p. 1). By contrast, Romanian civil society efforts to ensure that victims have the right to reparation and redress (what the Germans call Wiedergutmachung, making right again) have been instrumentalized by successive post-communist governments interested in short-term electoral gains and largely indifferent to the victims' plight. Civil society groups have been listened to, and their projects have been supported, only when and for as long as political parties have anticipated possible gains from reevaluating the communist past. These groups have been promptly abandoned, ignored, or even silenced when their demands for justice threatened the careers of powerful political gate-keepers, the relative electoral advantage of the ruling party, its policy priorities, legitimacy, or popularity.

The Romanian civil society groups with a stake in the transitional justice process are quite diverse, spanning groups that support, groups that oppose, and groups that are indifferent to the politics of the past. The groups discussed in this chapter represent only a fraction of all civil society organizations in the country, most of which show no interest in decommunization. The numerous groups indifferent to transitional justice face similar difficulties related to organization, leadership, and resources as do groups interested in the politics of the past. Organizations that oppose de-communization, while much fewer, have benefited from convergence of goals with powerful post-communist politicians and parties. The challenge facing the pro-transitional justice civil society groups is to find common ground, a common voice, and a common platform to promote various methods, processes and practices of coming to terms with the past as new generations with no direct experience with the communist regime and its crimes come on the political stage. This implies more concerted action, a redesigned agenda and tool kit, and the political acumen needed to bring the public and the political elite behind the larger decommunization project.

References

Andreescu, G. 2003. Cazul Grigore Opriţa: A început vânătoarea zăpăciţilor şi a aiuriţilor. *Observator Cultural*, 180, August. http://www.observatorcultural.ro/Cazul-Grigore-Oprita-a-inceput-vinatoarea-zapacitilor-si-a-aiuritilor*articleID_8835-articles_details.html. Accessed 22 March 2012.

Artene, A. 1999. Asasinul inginerului Gheorghe Ursu a fost condamnat la 20 de ani de închisoare, *Ziua*, 1 July. http://www.halbjahresschrift.homepage.t-online.de/radu.htm. Accessed 19 July 2011.

Asociaţia Foştilor Deţinuţi Politici din România. 2009. *Memoriul Asociaţiei Foştilor Deţinuţi Politici din România.* http://www.bumbacari.net/news.php?extend.3. Accessed 23 June 2011.

Asociaţia Foştilor Deţinuţi Politici din România. 2004. *Album memorial. Monumente închinate jertfei, suferinţei şi luptei împotriva comunismului*. Bucharest: Ziua.

Asociația Persoanelor Deposedate Abuziv Foştilor Deportați Refugiați din România. 2009. *Ştiri, Legea 290/2003, Legea 10/2001, Legea 247/2005*, 7 August. http://caritas-craiova.piczo.com/stiri,legea2902003,legea102001,legea2472005,?cr=5&linkvar=000044. Accessed 23 June 2011.

Atitudini şi opinii despre regimul comunist din România. Sondaj de opinie publică. 2012. Bucharest: Institutul de Investigare a Crimelor Comunismului si Memoria Exilului Romanesc, 23 January. http://www.crimelecomunismului.ro/pdf/ro/sondaj/raport_iiccmer.pdf. Accessed 22 March 2012.

Benea, Daniela. 2008. Musceleni, la şedința pentru modificarea Legii Chiriaşilor. *Evenimentul Muscelean*, 28 October. http://www.evenimentulmuscelean.ro/index.php/Actualitate/789-musceleni-la-sedinta-pentru-modificarea-legii-chiriasilor.html. Accessed 19 July 2011.

Carothers, Thomas. 1999. Civil society. *Foreign Policy* 117: 18–29.

Ciobanu, Monica. 2008. Democracy and education—Teaching history and building a democratic future: Reflections from post-communist Romania. *Democracy and Education* 17(3): 58–62.

Curtea Penală de Condamnare Juridico-Morală a Crimelor Regimurilor Comuniste. 2006. *Procesul Comunismului*, October. http://redholocaust.org/procesul.html. Accessed 29 March 2011.

David, Roman. 2011. *Lustration and transitional justice. Personnel systems in the Czech Republic, Hungary and Poland*. Pennsylvania: University of Pennsylvania Press.

Deletant, Dennis. 1996. *Ceausescu and the securitate: Coercion and dissent in Romania, 1965–89*. London: C. Hurst.

Deletant, D. 2000. *Communist terror in Romania: Gheorghiu-Dej and the Police State, 1948–1965*. London: Palgrave MacMillan.

Dickerson, Mark, and Thomas Flanagan. 2006. *An introduction to government and politics. A conceptual approach*. Toronto: Nelson Thomson Learning.

Fundația Academia Civică. 2009. *Memorialul Victimelor Comunismului şi al Rezistenței*. http://www.memorialsighet.ro/index.php?option=com_content&view=frontpage&Itemid=1&lang=ro. Accessed 20 July 2011.

Grosescu, Raluca. 2007. The role of the civil society and anticommunist political actors in the Romanian transitional justice partial failure. In *Lustration and consolidation of democracy and the rule of law in Central and Eastern Europe*, ed. Vladimira Dvorakova and Andelko Milardovic, 183–195. Zagreb: Political Science Research Center.

Grosescu, Raluca, and Raluca Ursachi. 2009. *Justiția penală de tranziție: De la Nurenberg la postcomunismul românesc*. Iaşi: Editura Polirom.

Hamber, B., Mokofeng, T., and Simpson, G. 1997. Evaluating the role and the function of civil society in a changing South Africa: The truth and reconciliation commission as a case study. Presented at the seminar *The role of southern civil society organizations in the promotion of peace*. Catholic Institute of International Relations, London, 10 November.

Howard, Marc Morje. 2003. *The weakness of civil society in post-communist Europe*. New York: Cambridge University Press.

ICAR Foundation. 2010a. *Who is ICAR foundation*. http://www.icarfoundation.ro/. Accessed 1 September 2011.

ICAR Foundation. 2010b. *Communism trial*. http://www.icarfoundation.ro/index.php?option=com_content&task=view&id=29&Itemid=44. Accessed 1 September 2011.

Ignatieff, Michael. 1995. On civil society: Why Eastern Europe's revolutions could succeed? *Foreign Affairs* 74(2): 128–136.

Judt, Tony. 2005. *Postwar: A history of Europe since 1945*. London: Penguin.

Klinghoffer, Arthur, and Judith Klinghoffer. 2002. *International citizens' tribunals: Mobilizing public opinion to advance human rights*. London: Palgrave Macmillan.

Kligman, Gail. 1990. Reclaiming the public: A reflection on creating civil society in Romania. *East European Politics and Societies* 4(3): 393–438.

Leuştean, Lucian. 2009. *Orthodoxy and the Cold War. Religion and political power in Romania, 1947–1965*. New York, NY: Palgrave Macmillan.

Linz, Juan, and Alfred Stepan. 1996. *Problems of democratic transition and consolidation: Southern Europe, South America and post-communist Europe*. Baltimore: Johns Hopkins University Press.

Macovei, Monica. 1999. Procesul Gheorghe Ursu, *22*, no. 9. http://www.halbjahresschrift.homepage.t-online.de/radu.htm. Accessed 19 July 2011.

Muraru, A. 2011. Ne mor torţionarii, doamnă Kovesi! *Observatorul Cultural* 583, 15 July. http://www.observatorcultural.ro/2-comentarii.html*articleID_25608-articles_details. Accessed 19 July 2011.

Oţoiu, Damiana. 2009. Negocier la (re)constitution de la propriete privee en Roumanie postsocialiste. (Nouveaux) acteurs, (anciennes) strategies. *Options Mediterraneennes* A92: 65–75.

Oţoiu, D. 2006. La (re)construction de la propriete privee en Roumanie post-communiste. "Anciens" vs. "nouveaux" proprietaires?, paper presented at the At the Frontier of Land Issues: Social Embeddedness of Rights and Public Policy international workshop, 17–19 May 2006, Montpellier, p. 6. http://www.mpl.ird.fr/colloque_foncier/Communications/PDF/Otoiu.pdf. Accessed 26 July 2011.

Şedinţa Senatului din 3 februarie. 1994. *Monitorul Oficial al României, partea a II-a*, 4 February 1994.

Şedinţa Senatului din 25 martie 1992, *Monitorul Oficial al României, partea a II-a* pp. 2–10, 26 March 1992.

Stan, Lavinia. 2011. Witch-hunt or moral rebirth? Romanian parliamentary debates on lustration, *East European Politics and Societies*, April. http://eep.sagepub.com/content/early/2011/04/05/0888325411403922.abstract. Accessed 1 August 2011.

Stan, L. 2010. Romania: In the shadow of the past. In *Central and Southeastern European politics since 1989*, ed. Sabrina P. Ramet, 379–400. Cambridge: Cambridge University Press.

Stan, L. 2009. *Transitional Justice in Eastern Europe and the Former Soviet Union: Reckoning with the communist past*. London: Routledge.

Stan, L. 2006. The roof over our head: Property restitution in Romania. *Journal of Communist Studies and Transition Politics* 22(2): 180–205.

Stan, L. 2000. Access to securitate files: The trials and tribulations of a Romanian law. *East European Politics and Societies* 16(1): 55–90.

Stan, Lavinia, and Lucian Turcescu. 2007. *Religion and politics in post-communist Romania*. New York: Oxford University Press.

Stan, L., and Lucian Turcescu. 2005. The devil's confessors: Priests, communists, spies and informers. *East European Politics and Societies* 19(4): 655–685.

Ştefănescu, Domniţa. 1995. *Cinci ani din istoria României. O cronologie a evenimentelor decembrie 1989-decembrie 1994*. Bucharest: Editura Maşina de Scris.

Tănăşoiu, Cosmina. 2007. The Tismaneanu report: Romania revisits its past. *Problems of Post-Communism* 54(4): 60–69.

Timişoara Declaration. 1990. *Report on Eastern Europe*, 6 April 1990.

Tismăneanu, Vladimir. 2008. Democracy and memory: Romania confronts its communist past. *The Annals of the American Academy of Political and Social Science* 617: 166–180.

Tismăneanu, V. 2003. *Stalinism for all seasons: A political history of Romanian communism*. Berkeley: University of California Press.

Chapter 3
Dealing with the Past in Post-war Croatia: Perceptions, Problems, and Perspectives

Tamara Banjeglav

Introduction

Documenta—Centre for Dealing with the Past[1] is a civil society organization from Zagreb, which was established in 2004 in an attempt to encourage the processes of dealing with the past and the establishment of factual truth about the war. As part of its work, the organization conducted research concerning attitudes of the Croatian public about dealing with the past and the perception of victims of the 1991–1995 war. Multi-dimensional research of the public opinion in Croatia was conducted in the summer of 2006, and research results were collected, analyzed and published in October 2010[2] under the title "Dealing with the Past in Croatia: Attitudes and Opinions of Post-War Actors and Public". This research tried to cover some of the basic thematic units concerning dealing with the past in contemporary Croatia as a phenomenon which social scientists in Croatia have not found particularly relevant for researching so far.

---

[1] *Documenta*—Centre for Dealing with the Past is a civil society organization from Zagreb which aims to develop social and individual processes of dealing with the past in order to build sustainable peace in Croatia and the wider region, through deepening of public dialogue and initiating debate on public policies which stimulate dealing with the past, gathering and publishing documentation and research of war incidents, war crimes, and violations of human rights; as well as monitoring judicial processes at local and regional level as a contribution to the advancement of judicial standards and practices in the processing of war crimes. Results of the survey mentioned in this chapter have been analyzed, and the chapter written, during my employment with *Documenta*.

[2] Although conducted back in 2006, this research is still relevant for understanding dealing with the past processes in Croatia, since it usually takes longer time for changes in this field to take place and for people to change their attitudes and opinions as a result of certain public activities and processes. Moreover, no similar scientific research has been done in the meantime. The research was published as late as 2010 due to difficulties in securing financial funds for its publishing.

T. Banjeglav (✉)
Documenta—Centre for Dealing with the Past, Selska cesta 112c, 10000, Zagreb, Croatia
e-mail: tamara.banjeglav@documenta.hr

O. Simić and Z. Volčič (eds.), *Transitional Justice and Civil Society in the Balkans*,
Springer Series in Transitional Justice, DOI 10.1007/978-1-4614-5422-9_3,

Results of this first comprehensive research on this important subject in Croatia enabled a wide view on the relation of citizens and various public actors to many aspects of this issue. This book chapter offers a review and an analysis of these results, but also opens up a discussion on the meaning of these results for the process of dealing with the past in Croatia. Moreover, it gives an overview of the developments and improvements that have occurred since the time when the research was conducted.

Furthermore, the chapter discusses transitional justice mechanisms which foster reconciliation in post-war Croatia. It looks at mechanisms such as war crimes trials, regional commission for truth-telling (RECOM), but also at public apologies and memorialization practices, and gives a brief overview of the latest events in these fields, as well as an assessment of their role in post-conflict reconciliation and dealing with the past.

The hypothesis of this book chapter is that progress has been made in the field of transitional justice in Croatia during the past decade, particularly in war crimes trials, and that political will to deal with the past has also occurred since president Ivo Josipović was elected in 2010. However, much work still needs to be done. The chapter concludes with recommendations for future actions that need to take place in Croatia in the upcoming period, in order to foster reconciliation and speed up the process of dealing with the past.

Dealing with the Past and Transitional Justice: A Theoretical Overview

The term "dealing with the past" refers to coming to terms with violent history and violence committed against all victims, regardless of their ethnic, political, or any other background. This process usually needs to occur on a number of different levels, "from the micro-level of an individual in a small community, to the macro-level of national, regional and global political bodies" (Stubbs 2003). In this chapter, "dealing with the past" is understood as coming to terms with the events that took place during the war of the 1990s and with the consequences they have had for post-war reconstruction and reconciliation.

In former Yugoslavia, it has been often argued that failure to deal with the past after WWII left the people with a legacy of mistrust and with different and conflicting "truths" about past events, which were passed from one generation onto another. This legacy enabled political elites from different ethnic groups to use the power they had to change and influence the meaning of historical facts and to create different myths, producing, in this, way, their own "truths" at the beginning of the war of the 1990s. David Bruce Macdonald argues that "these 'stories' proved to be absolutely essential in creating and supporting war" (MacDonald 2002, p. 214). Such interpretations of history served to bring back into public memory past events which the public never dealt with and which were never publicly discussed or acknowledged. Each side portrayed themselves as the greatest victims, and managed, by manipulating public opinion within their own countries, to create an atmosphere of victimhood. Self-identification which is grounded in a collective victimization can end in dangerous results, particularly in terms of the group's relations with others and their reconciliation. Initiatives in the area of truth-telling, justice, and reparation contribute to the definition of who will be included in the category of victims. These initiatives, along with many others, are usually put under the common denominator of "transitional justice" and have the overall aim to serve society as a means to deal with past human rights abuses. A key goal of transitional justice is to contribute to sustainable peace and the rebuilding of a society based on the rule of law and respect for human rights (Teitel 2003).

According to Louis Bickford, "transitional justice refers to a field of activity and inquiry focused on how societies address legacies of past human rights abuses, mass atrocity, or other forms of severe social trauma, including genocide or civil war, in order to build a more democratic, just, or peaceful future" (Bickford 2004, p. 1045). Transitional justice seeks recognition for victims and promotion of possibilities for peace, reconciliation, and democracy. Since the Nuremberg trials, it has been possible to trace the genesis and history of politico-legal organization of international memory of victims and perpetrators of war crimes (Savić 2006).

Although the concept of transitional justice[3] has been constantly and dynamically developing, it usually includes a combination of different mechanisms such as war crimes trials, establishing of truth commissions, developing reparations for victims' families and survivors, memorialization initiatives for remembering the victims, etc. Possible combinations of the abovementioned mechanisms keep rising, but at least some of these mechanisms (not necessarily all of them) are applied in each particular case and they are in many respects complementary (Mobekk 2005, p. 280). Transitional justice has, however, often been criticized for being too normative and abstract, and lacking context and historical background. For this reason, the effectiveness of each of these measures depends on the broader context in which it unfolds, including its relationship with other dimensions of transitional justice (Orentlicher 2007). Thus, for example, public apologies made by high-ranking state officials do not make a great effect unless they are accompanied with other actions. These can vary from public campaigns, which would raise the public awareness about the suffering of all victims, to their inclusion in school text-books in order to make them part of educational materials.

Recently, the literature on transitional justice has focused less on perpetrators and more on the explicit goal of "healing" the victims (Andrieu 2010). Thus, some authors argue that the focus of transitional justice research has recently shifted more to the victims and their concerns, and has been less focused on political questions. "Victims and reparation for victims have become quintessential elements in the debate on transitional justice and how to deal with the past" (Rombouts 2002, p. 217).

Transitional justice initiatives, such as prosecutions and truth-telling, challenge distortions of the truth that allow groups only to see their own members as "victims".

[3] According to Louis Bickford, the term itself is misleading, as it more commonly refers to "justice during transition" than to any form of modified or altered justice. See (Bickford 2004, p. 1045).

They can also challenge claims of historical "victimhood" of entire groups and help to demonstrate the fact that members of other groups have been abused as well, which contributes to reconciliation among different groups and helps them deal with the difficult past.

Public Opinion About Dealing with the Past in Croatia

In order to put an end to elicitation and manipulation of war crimes for political purposes, human rights organizations and victim associations in Croatia have grown very insistent about dealing with the past. The key reason for this initiative emerged from the experience of the concealment and forgery of war crimes and other war events, which influenced the more recent past of the former Yugoslavia and post-Yugoslav societies.

In Croatia since the end of the war there has been a widespread public understanding of the nature of the 1991–1995 war as "just and legitimate, defensive and liberating, and neither aggressive nor conquering" (Declaration on the Homeland War 2000). A radical understanding of its nature was given by a Supreme Court judge and former president of the Supreme Court, Milan Vuković, who said in an interview that no war crime could be committed in a defensive war (Ivanišević 1995). Although much progress has been made since this statement, the opinion that there were no crimes in this war committed by "our side" is still present in the Croatian public. This shows that the problem of not understanding or having sympathy for "the other side" still remains present in Croatia. In order to answer the questions why this lack of sympathy and understanding for the victims from "the other side" occurs, why persons accused of war crimes still enjoy support among the Croatian public, even years after the war had ended, and what the public opinion on war crimes and violence committed during the 1991–1995 war in Croatia is, *Documenta* conducted, in 2006, a survey among the Croatian public.

The research consisted of three parts: qualitative research of the public opinion conducted with the use of focus groups method, qualitative research of the attitudes of public opinion creators conducted with the use of in-depth interviews method, and quantitative research of the citizens' attitudes conducted with the use of a survey. The research was conducted on a sample of the general population of Croatian citizens (regardless of their nationality), as well as two additional sub-samples. The first sub-sample consisted of citizens of Serbian nationality from war-affected areas, while the second sub-sample consisted of citizens of Croatian nationality from war-affected areas. The basic goals of the quantitative research included:

- Attitudes and level of awareness on dealing with the past, i.e., on the events and experience of the "Homeland War"[4]

[4] The "Homeland War" is the official name of the war fought for state sovereignty and territorial integrity of Croatia, which took place between 1991 and 1995.

- The general public's perception of victims and survivors
- Attitudes of the general public on the support to victims and survivors
- Perception of war crimes committed during the war
- Attitudes on processing of war crimes[5]

Survey results showed that, in general, the majority of Croatian citizens and actors supported dealing with the past in Croatia: they thought that dealing with the past was important for future stability of the Croatian society, as well as for sustainability of peace in Croatia and the region. Those who were directly affected by war and suffered its consequences more often felt that the term carried negative connotations. Thus, when analyzing survey results, it can be seen that, among the general population, 14% felt the term "dealing with the past" had an exclusively negative meaning, but for the majority, 41%, the term had a neither positive nor negative connotation (Kardov et al. 2010, p. 54).

Considering the perception of victims of the war in Croatia, more than half of the general population respondents (52%) thought victims of the war were exclusively Croats. Almost a third (31%) thought that victims from the Croatian side were far more numerous than Serbian victims, while none of the respondents from the general population said they thought there had been more victims on the Serbian side (Kardov et al. 2010, p. 69). There is also a significant difference in the perception of war victims between two target sub-samples, between Serbian respondents from war-affected areas and Croatian respondents from the same areas. The majority of respondents of Serbian nationality (84%) felt that Croats and Serbs were equally primary victims of the war, while only 12% of respondents of Croatian nationality felt the same (ibid.).

It is also interesting to look at the answers to the question which war crimes have the respondents heard about. Almost all of the general population of respondents (99%) have heard about the bombing and devastation of the town of Vukovar, while almost the same percentage (97%) have heard about killings of captives from the Vukovar hospital at Ovčara.[6] The same respondents have heard to a far lesser extent

[5] The research included a sample of the general Croatian public and two sub-samples. In the sample of general population, 700 people were interviewed. In each sub-sample (people of Serbian nationality from war-affected areas and people of Croatian nationality from war-affected areas) 150 people were interviewed. All respondents were older than 18.

[6] Ovčara is an agricultural property near the town of Vukovar where a war crime was committed by members of the Yugoslav National Army and Serbian paramilitary forces in the night between November 20, 1991 and November 21, 1991. More than 200 civilians and soldiers were killed, who were mostly patients at the Vukovar hospital from which they were taken and brought first to a camp and then killed at Ovčara. It is considered to be the largest slaughter of individuals committed during the war in Croatia. Thirteen people were found guilty in March 2009 before the War Crimes Council of the District Court in Belgrade for the war crime committed at Ovčara. They were sentenced to between five and twenty years in prison. Moreover, the ICTY sentenced, in the third non-appealable verdict made in December 2010, Yugoslav National Army major Veselin Šljivančanin to 10 years in prison for helping and supporting the crime at Ovčara. He was granted an early release on July 5, 2011. Mile Mrkšić was sentenced in 2009 to 20 years in prison for "having aided and abetted the murder and torture of prisoners" (ICTY 2007).

about killings of Serbian civilians during the operation "Storm"[7] (68%) and about seizure of property from Serbs in Croatia (58%), while the smallest percentage has heard about killings of Serbian civilians in the town of Sisak (44%) (Kardov et al. 2010, p. 86).

The survey also showed that only 22% of the general population was sure that Croatian citizens and army members committed war crimes against Serbs, while 64% of Serbian respondents from war-affected areas thought that the Croatian side had committed war crimes. However, it should also be emphasized that a quarter of respondents of Serbian nationality did not know or did not want to say whether they thought crimes had been committed against Croatian citizens of Serbian nationality by the Croatian army (Kardov et al. 2010, p. 83).

Support was expressed for punishing perpetrators of all war crimes and for prosecuting those responsible for crimes at war crimes trials. Thus, the majority of general population respondents (61%) thought that perpetrators of all crimes should be punished, while a third of Croatian citizens (32%) thought there were some crimes committed which could be justified and their perpetrators need not be punished. The majority of respondents of Serbian nationality (87%) believe all perpetrators should be punished, while only 8% feel that some crimes are justifiable (Kardov et al. 2010, p. 89).

The general support of the prosecution of all war crimes does not, however, mean that citizens supported every individual war crime prosecution. This could be best seen at the example of Croatian general Ante Gotovina, who was accused by the International Criminal Tribunal for the former Yugoslavia (ICTY) for committing war crimes against Serbian civilians after the operation "Storm" and for whom an indictment was issued after he had left Croatia and went into hiding in 2001. A poll conducted in early 2005, while Gotovina was still on the run,[8] during the broadcast of one of Croatia's most popular political TV shows *Latinica* showed how this war crimes suspect was considered to be a hero rather than a war criminal.[9] Viewers were asked if they were to encounter Gotovina, would they report him to the police or help him hide. Only 8% of the respondents said they would turn him in, while

[7] During the military operation "Storm," which happened on August 5, 1995, all of the occupied Croatian territory was brought back under the Croatian legal order, except for Eastern Slavonija which was peacefully re-integrated later. "Strom," next to the "Flash," was the crucial military operation, which led to the end of the war. During the operation, some 18.4% of Croatia's territory was liberated. In April 2011, the ICTY brought a first degree verdict against two Croatian army generals Ante Gotovina and Mladen Markač, who were sentenced to 24 and 18 years in prison respectively for war crimes committed against Serbian civilians during the operation "Storm." The verdict also directly identified President Franjo Tuđman as part of a joint criminal enterprise dedicated to expelling Serb residents of the country's Krajina region. The ICTY, however, did not rule that Operation "Storm" as a whole was a "joint criminal enterprise." Rather the judges ruled that some aspects of the military offensive violated international law (ICTY 2011).

[8] Ante Gotovina was arrested on the Canary Islands and extradited to the ICTY in The Hague in December 2005.

[9] For an extensive analysis of the "hero" and "martyr" symbolism of Ante Gotovina in the perception of the Croatian public see (Pavlaković 2010).

92% said they would help him avoid arrest. Of those who said they would help him, 76% said they believed he was innocent (Latinica 2005). Moreover, according to a poll conducted in 2006, after his arrest, 64.8% of respondents said they had had a "positive" or "mostly positive" opinion of Gotovina, while only 4.7% had a "negative" or "totally negative" view of him (Đula 2006). This testifies to the fact that socially constructed versions of Croatia's past and its role in the 1991–1995 war, which were firmly established in the collective public memory of the war through political propaganda and media reporting[10] and which were implemented in the public sphere during and after the war, are still relatively strong and powerful.

In accordance with such an interpretative framework, the survey showed that respondents mostly viewed crimes committed by "their" ethnic group as individual incidents which happened by accident and were directed at individuals, while they viewed crimes committed on the other side as hate crimes directed at the whole nation (Kardov et al. 2010, p. 74–91).

This problem's full capacity is understood when one takes into account that citizens primarily rely on their own experience and that collective memory, for many, comes before historical facts. A lack of information and of willingness to learn about victims "on the other side" and the reluctance of political structures to deal with the past contribute to slow pace of determining facts about the war. What also proved to be widespread among the Croatian citizens according to this survey is a lack of knowledge about already established facts, as well as low familiarity with the ongoing war crimes court proceedings.

It should be emphasized that, for these results to acquire full meaning, it is necessary to take into consideration that dealing with the past is a long-term and dynamic process which takes place at different levels in a society and that many activities that accompany it differ in their efficiency and extent. The dynamics of this process is not connected only to changes on the institutional and general support, but also to specific demands that dealing with the past puts before each member of a society. Thus, difficulties in meeting those demands result in different meanings which citizens attach to the term "dealing with the past."

Recent Developments and Changes Brought by Transitional Justice Mechanisms and Their Effect On Dealing with the Past in Croatia

War Crimes Trials

First war crimes trials in Croatia started already during the war in 1992, at the county courts in the towns of Karlovac, Šibenik, Sisak, and Varaždin. However, the

[10] For a more detailed discussion on the role of the media during the war in former Yugoslavia see (Đerić 2008), (Skopljanac Brunner et al. 2000), and (Thompson 1999).

trials were often characterized by ethnic partiality, lack of witness protection, and unprofessional court proceedings (Subotić 2009). Indictments which included a large number of persons were being issued exclusively against members of the Serb military and paramilitary formations. Those indictments were imprecise and insufficiently supported with evidence. Verdicts were pronounced, in most cases, in the absence of accused persons (464 persons were found guilty *in absentia* in 118 cases) (Stojanović and Kruhonja 2011).

A turning point in war crimes trials is related to the opposition coalition's coming to power in 2000,[11] when the Croatian Parliament, under the pressure from the EU, adopted the Declaration on Cooperation with the International Criminal Court in the Hague (2000).[12] The State Attorney's Office dropped charges against 1,403 members of the Serb military formations; the number of trials in the absence of accused persons was significantly reduced, and trials were also instituted for crimes committed by members of the Croatian military units (ibid.). Cooperation with the Prosecutor's Office of the Republic of Serbia and the Prosecutor's Office of Montenegro was established. Regional cooperation between Serbia and Croatia started officially on October 13, 2006 by signing an agreement which allowed transfer of war crimes trials to the country of the accused, which is not necessarily the country where the crime had been committed (ibid.).

Moreover, in order to fulfil the requirements for accession to the EU, concerning judicial reform, the Croatian parliament adopted, in October 2003, the Law on the Implementation of the Statute of the International Criminal Court and on the Prosecution of Crimes against the International Law of War and Humanitarian Law.[13] According to that law, four specialized war crimes chambers were set up at county courts in Zagreb, Osijek, Rijeka, and Split.[14]

Since 2001, war crimes committed by Croatian military forces members have also been investigated and prosecuted, while in 2010 the number of such cases has

[11] The so-called "coalition of six" won the parliamentary elections in 2000 against the ruling HDZ party (Croatian Democratic Union), which had been in power for 10 years. The coalition consisted of six left-oriented parties: Social-Democratic Party (SDP), Liberal Party (LS), Croatian Social-Liberal Party (HSLS), Istrian Democratic Party (IDS), Croatian Peasants' Party (HSS), and Croatian People's Party (HNS).

[12] See Narodne novine [the Official Gazette], 41/2000, April 18, 2000.

[13] See Narodne novine [the Official Gazette], 175/2003, November 4, 2003.

[14] However, war-crimes cases were not transferred exclusively to war-crimes chambers at these four courts, and such cases continued to be prosecuted at other county courts. Human rights organizations, however, continued insisting that exclusively these four courts needed to be authorized for prosecuting and trying war crimes, in order to make trials more professional, unbiased, and more effective. See (Stojanović and Kruhonja 2011) This was made legally possible due to the amendment to the Law on the Implementation of the Statute of the International Criminal Court and on the Prosecution of Crimes against the International Law of War and Humanitarian Law, which was adopted in October 2011 (Stojanović and Sjekavica 2012).

significantly risen.[15] Thus, for example, the trial of General Mirko Norac before the county court in Rijeka, which started in 2007, for war crimes committed against Serbian civilians in the town of Gospić and the renewed trial of officers of the military police for war crimes against prisoners of war at the military prison "Lora" in Split,[16] marked a break with the practice which existed in Croatia up to that moment to exclusively indict and try persons of Serbian nationality. In the course of the trial, Serbian victims testified for the first time.

However, trials still take place in an atmosphere of tolerance for "own" war criminals. Thus, for example, a county court judge in the town of Sisak, while reading a verdict to four Croatian army members tried for war crimes against a Serbian family, said she was sorry Croatian soldiers had to stand trial before Croatian courts for odious crimes which we were used to have been committed by "the other side", "especially in the days when we light candles for the victim of Vukovar" (Charges for War Crimes 2010). Moreover, the public space has been dominated by support for Branimir Glavaš, who was sentenced to 8 years in prison for war crimes against civilians committed in Osijek in 1991. The support came mostly from his own party (HDSSB), which nominated him in 2008, while he was still standing trial for war crimes, a member in the Parliamentary committee for human rights and rights of national minorities (Božić and Rešković 2008). Furthermore, in more recent events, Glavaš, although sentenced to prison by a non-appealable verdict, was named in October 2011, the head of HDSSB parliamentary elections list, which was possible due to a loophole in the Law on the Election of Representatives to the Croatian Parliament (2003).

The culture of non-prosecution of crimes also includes a lack of political responsibility for committed crimes, as for example former deputy Speaker of the Croatian Parliament, Vladimir Šeks, remained on duty despite the fact that the non-appealable verdict determined and proved the existence of a secret military troop, which was taking away, torturing and killing civilians, at the time when he was the head of the Regional Crisis Headquarters (Stojanović and Kruhonja 2011). Instead, the media speculated that he would be appointed coordinator for closing Croatia's EU negotiations in Chapter 23,[17] but this did not happen in the end. On the upside in 2010, an investigation was launched against Tomislav Merčep the Croatian Interior Ministry Adviser during the war, and commander of reserve police units in Pakračka Poljana and at the Zagreb Velesajam, for executing 43 civilians, while performing these duties (ibid.).

[15] According to the State Attorney's Office, quoted in their annual report on monitoring war-crimes trials by three human rights organizations, *Documenta*—Centre for Dealing with the Past, Civic Committee for Human Rights and Centre for Peace, Non-Violence and Human Rights Osijek, by the end of 2011, 104 members of Croatian army forces have been tried, while 29 have been convicted based on non-appealable verdicts. See (Stojanović and Sjekavica 2012).

[16] In March 2006, after the Croatian Supreme Court annulled the acquittals from August 2004, all eight accused were convicted and sentenced to 6–8 years in prison.

[17] Chapter 23 was a chapter in Croatia's negotiations on the accession to the European Union, which related to the state's judiciary and fundamental rights.

A serious setback happened in October 2011, when the Croatian Parliament adopted a law according to which all legal acts related to the 1991–1995 war, which find Croatian citizens accused or indicted of war crimes, became null. This was a direct consequence of Serbia sending Croatia a list of indictments against Croatian citizens charging them with war crimes. Croatian president Ivo Josipović publicly criticized this law as something which called into question Croatia's readiness to prosecute war criminals (President Josipović on the Occasion of Passing the Annulment Law 2011).

Further improvements in processing war crimes were stalled due to the lack of political will to strengthen the independence, expertise, and efficiency of judicial bodies through specialization of courts and state attorney's offices. The problem that was prevailing until the end of 2011 was the continuation of holding trials before local courts, in areas where crimes occurred, instead of before four specialized war crimes chambers (Centre for Peace, Non-Violence and Human Rights Osijek 2008). Legislations related to war crimes trials and trial proceedings mainly improved due to the EU accession process,[18] although the EU has made no serious conditions for progress in the accession process on domestic war crimes trials, as compared to the ones on cooperation with the ICTY (Human Rights Watch 2004). As Jelena Subotić argues, this allowed Croatian elites "to preserve almost intact the Croatian national understanding of the character of the war and Croatia's role in it" (Subotić 2009, p. 121).

However, in Croatia, no broader strategy in which war crimes prosecution is connected and supplemented with other key elements of transitional justice, such as truth-telling mechanisms, reparations to victims, public apologies, or institutional reform, is developed.

Truth-Telling Projects

The truth about past war crimes and human rights abuses has to be promoted by other mechanisms of transitional justice and peace-building and not just through war crimes trials. National legal instruments are not enough in order to achieve truth-telling and truth-seeking. What is needed is a regional level public agreement about the mechanisms for establishing and telling the facts about the past.

For this reason, civil society organizations from Croatia and other post-Yugoslav countries have been advocating joint strategies to address past human rights abuses: prosecution of perpetrators, addressing reparations and compensation issues, efforts to honor the memory of victims, analysis of institutional culpability, and efforts to reform institutions. A few years ago human rights and peace organizations have started discussing the possibility of establishing and eventual usefulness of a truth

[18] A discussion on the role of the EU accession process and EU conditionality on transitional justice processes in Croatia falls outside the scope of this chapter. For an extensive analysis see (Subotić 2009), (Batt and Obradovic-Wochnik 2009) and (Rangelov 2006).

commission. Recognizing the importance of systematic regional cooperation, on level of documenting war events and advocating justice for victims, a protocol on cooperation between three non-governmental organizations[19] was signed, followed with work on joint regional projects. At the First Regional Forum on transitional justice held in Sarajevo in May 2006, partners launched consultation process on the regional truth-telling mechanism leading to an initiative for the establishment of a Regional Commission mandated to establish and disclose the facts about war crimes committed in the former Yugoslavia (hereinafter RECOM) (Kostovica 2009). At the Fourth Regional Forum, held on October 28 and 29, 2008 in Prishtina, Kosovo, more than 100 organizations and individuals, including victims and victim associations from Bosnia and Herzegovina, Croatia, Kosovo, Montenegro and Serbia, as well as associations of citizens, human rights organizations, media associations, and other civil society groups from across the region, formed a Regional Coalition for RECOM (Consultation Process 2011).[20]

In April 2011, civil society activists across the region started collecting signatures for the establishment of RECOM, with the aim of handing in the signatures to governments and parliaments in the region, which are supposed to support it and adopt a decision on its establishment. The forming of RECOM has so far been publicly supported by Presidents of Croatia, Serbia and Montenegro (Ivo Josipović, Boris Tadić and Filip Vujanović), as well as President of BiH Federation Živko Budimir, Prime Minister of the Autonomous province of Vojvodina Bojan Pajtić, and Prime Minister of Montenegro Igor Lukšić (Political Support 2011).

In Croatia, the initiative received first strong public support by President Ivo Josipović at the Transitional Justice Forum held in Zagreb in October 2010. The President stressed the importance of "determining the faith of the missing and paying respects to victims, which is where I see the capacities of the future regional commission and it is necessary for the society as a whole to make additional efforts in order for families of victims to receive satisfaction" (Political Support 2011). A petition for the establishment of RECOM and the RECOM Statute proposal were submitted to the Croatian President on June 21, 2011, when he said he would promote the Initiative for RECOM as part of his regional activities and in communication with the Republic of Croatia government and political parties (Public Advocacy 2011).

However, a survey on the public opinion, conducted in March 2011 by an agency for public opinion research, showed that only 55% of Croatian citizens would sign a petition on the establishment of RECOM, as opposed to 95% of Kosovo citizens, 85% of BiH Federation citizens, and 63% of citizens of Serbia (Survey of Opinions 2011). Moreover, citizens of Croatia were the most pessimistic about the success of RECOM, compared to citizens of other post-Yugoslav states, while the largest

[19] These are: *Documenta*—Centre for Dealing with the Past from Zagreb, Humanitarian Law Centre from Belgrade, and Research and Documentation Centre from Sarajevo.

[20] For an extensive discussion about RECOM initiative see Jill Irvine and Patrice McMahon. *A Movement in the Making? The "REKOM" Coalition and Transitional Justice in the Balkans* in this volume.

percentage of citizens from Croatia never even heard about the initiative (only 6% have heard about it, compared to 25% of Montenegro's citizens or 11% of Serbia's citizens) (ibid.). Furthermore, a report issued by the Youth Initiative for Human Rights from Croatia, which coordinated the collection of signatures for the establishment of RECOM, showed that only some 20,000 signatures were collected in Croatia in the period April 26–June 30, 2011, as opposed to some 122,000 signatures collected in Bosnia and Herzegovina and some 100,000 collected in Kosovo (Youth Initiative for Human Rights Report 2011).

It is not surprising that Croatia's citizens did not show a high degree of interest in a regional initiative for dealing with the past, if we take into account that *Documenta*'s research on public attitudes on dealing with the past in Croatia showed that 41% of respondents felt the term had neither a positive nor a negative meaning, which points to low-level familiarity of respondents with the concept and low-level interest in it and its meaning (Kardov et al. 2010, p. 54).

Moreover, it has to be taken into account that low public support for RECOM in Croatia depends on a number of different factors, the most influential of them being the ICTY sentence to general Ante Gotovina from April 15, 2011,[21] when he was found guilty and sentenced to 24 years in prison for participation in a criminal enterprise the aim of which was to permanently remove the Serbian population from the Krajina region (ICTY 2011). Prime Minister Jadranka Kosor called the verdicts "unacceptable" and promised that the government would do everything it could to overturn them on appeal, and even President Ivo Josipović expressed his shock with the verdict (Kosor: The Verdict is Unacceptable 2011). The ruling provoked protests by, mainly, war veterans, as the indictment was perceived in the Croatian public as an indictment to the entire "Homeland War", and not just the accused generals. Most of the people in Croatia believe that the offensive was a legitimate military action taken to regain territory that had been seized by the Serb forces. What sparked the protests was a wrong interpretation of the indictment, which was understood as labelling the entire "Homeland War" as "a joint criminal enterprise". This was mostly due to the propaganda of the Croatian media, which did not ground its reporting in factual information and thus created a very biased and misleading picture of the indictment and its meaning.[22]

Public reactions after the reading of the indictment showed that Croatia was not yet ready to acknowledge and face the crimes committed by its army generals during the "Homeland War", mostly due to the existence of a strong public narrative about the war as a "just and legitimate defence…to defend its internationally recognized borders against Greater Serbia's aggression." (Subotić 2009, p. 91)

[21] However, many other factors which influenced low support to RECOM need to be taken into consideration, such as a hard economic crisis in the country and high level of unemployment of the population, which left people more concerned with problems of everyday survival, rather than with regional cooperation on fact-finding and reconciliation.

[22] For an extensive analysis of Croatian National Television's (HRT) reporting on the indictments to generals Ante Gotovina, Ivan Čermak, and Mladen Markač see (Preliminary Results 2011).

Public Apologies

Acknowledgment of crimes committed on every side during a conflict is crucial for reconciliation of all warring sides.

> Acknowledgment is an act, or, more precisely, a process of responding to the crime, whereby we publicly express and recognize our knowledge of the fact that killing and other forms of most brutal harming of the innocent people took place in the recent past, and that these atrocities was carried out in our name (Dimitrijević 2011)

Official apologies are expressions of regret issued by political elites in power, which represent a community or group in the name of which atrocities were committed.[23] The first official apology came from President Stjepan Mesić in September 2003, during his visit to Belgrade. Mesić said he apologized "to all those who have suffered pain or damage at any time from citizens of Croatia who misused or acted against the law" (Presidents apologize over Croatian war 2003). After his successor, Ivo Josipović, became president in 2010, more apologies followed. In mid-April 2010, Ivo Josipović visited Sarajevo and delivered a speech in which he expressed deep regret for Croatian politics toward Bosnia-Herzegovina in the early 1990s. He followed this apology with a visit to the Bosnian village of Ahmići, where Croat forces carried out a notorious massacre of Bosniak civilians in 1993. He was the first Croatian official to pay tribute to the Bosniak victims of the Ahmići massacre. However, the ruling party, HDZ, condemned President's speech and apology, saying they had a potential of "adding Croatia to the list of aggressors" in the wars of Yugoslav disintegration of the 1990s (Jović n.d.).

The next apology happened in the village of Paulin Dvor near Osijek, where members of the Croatian army killed 18 Serbian and one Hungarian civilian in December 1991. Josipović said that "the crime deserves to be condemned, the victims deserve to be respected, and those who were left behind the victims deserve our apology" (Presidents Josipović and Tadić Laid Wreaths at Paulin Dvor 2010). This apology followed Serbian president Boris Tadić's apology at Ovčara, who laid a wreath on the monument at Ovčara and expressed his regret for the crime committed there.[24] Part of the Croatian public saw the Serbian and Croatian presidents' visits as a new chapter in Croatian-Serbian relations and as a great contribution to the reconciliation process. However, the other part of the public condemned Josipović's apology seeing it as an attempt to belittle the tragedy that happened at

---

[23] Due to the number of official apologizes that can be heard, Nenad Dimitrijević remarks that "we seem to be living in an age of political apology" (Dimitrijević 2011).

[24] President of Serbia, Boris Tadić, is the first high-ranking Serbian official who apologized in June 2007 to citizens of Croatia and members of the Croatian nation for crimes committed in the past war by some of his co-citizens, in the name of the Serbian people. He apologized while appearing on the TV show "Nedjeljom u 2" (Sundays at 2) on the Croatian National Television. Moreover, some more informal apologies should also be mentioned, such as the one given by a non-governmental organization Women in Black from Belgrade. Members of the organization went to Vukovar in November 2006 to ask families of victims of crimes committed in Vukovar for forgiveness (Tadić ponovio ispriku Hrvatima zbog rata 2007).

Ovčara and as an attempt of relativizing the guilt for crimes committed in the war. Low-level support to public apologies by state representatives could already be discerned from *Documenta*'s survey, in which only 3% of the respondents said they thought an apology would help family members of war victims to deal with their losses (Kardov et al. 2010, p.73).

It could be, thus, concluded that, as much as they play an important role in facing the past and reconciling the sides in a conflict, apologies have to be understood and viewed in the context of other actions and efforts made by political elites. In case they are not accompanied with a real and strong political will for establishing dialogue and acknowledging all victims of a conflict, apologies face the danger of becoming empty gestures, void of any real importance and meaning in fostering reconciliation.

Monuments and Memorialization

Other actions which help recognizing the suffering of all victims during a conflict are monuments and memorials erected in the memory of victims. Public memorials after conflict can contribute to acknowledging the harm caused to victims and to restoring their dignity. However, they can also cement divisions between communities and strengthen barriers to communication across the conflict divide.

From the perspective of the state, the goals of public memorials are more often, in the words of Benedict Anderson, related to nation-building and defining an "imagined community" (Anderson 1983). Public monuments can play an important function in telling the story of a national group, which reinforces the group's chosen self-image. Through erecting monuments, the goal of states, therefore, may be to assert particular identities in the public space that support narratives needed for political legitimation, and these narratives may even be harmful for victims. It is, thus, interesting to note that the Croatian Law on Marking Sites of Mass Graves of the "Homeland War" Victims problematically defines victims of the "Homeland War" as "Croatian war veterans and civilians who died in mass executions during Serbian and Montenegrin chetnik aggression and aggression of the Yugoslav army on the Republic of Croatia" (Law on marking sites of mass graves of the Homeland War victims 1996).

Rare memorials to Serbian war victims in Croatia exist in two villages in the Dalmatian hinterland, Gošić and Varivode, where the Council of Serbian National Minority erected in August 2003 wooden crosses and plaques on which victims' names were written in Cyrillic. However, in April 2010 unidentified perpetrators damaged the monument to nine civilians killed in Varivode in August 1995, during the operation "Storm". The new, restructured monument was opened in October 2010 by President Josipović, who said that a terrible crime had been committed against them and that they were innocent victims of revenge which should not have taken place (President's Office Press Release 2010).

That the Croatian public is still not ready to face crimes committed by Croatian army members during the operation "Storm" can also be seen at the example of an attempt made by a civil society organization, Youth Initiative for Human Rights, which put up, in August 2010, a plaque in the memory of all civilians who had to flee their homes during operation "Storm", many of whom never returned. The plaque was put up at the exit from the town of Knin, through which refugees fled in August 1995. The plaque read that it was put up "at the 15th anniversary of the Croatian army action "Storm", (...) by Croatian citizens who offer their apology to victims, due to the lack of an apology from the responsible" (www.yihr.org). It also stated that its erection was financed by the Croatian government, but the government's human rights office immediately reacted denying financing the plaque. The plaque was removed by city authorities within 24 h, with the explanation that no permission had been issued for its erection.

Another controversial example happened in September 2011, when local population of the village Golubić near Knin put up a monument in the memory of all local Serbs from Golubić who died during the war. The controversy stemmed from the fact that some of the victims whose names were listed on the monument died in 1991 and 1992, which left the public speculating if those were members of rebel Serbs' army who died in fighting. Opening of the monument was banned by the interior ministry, saying it was put up unlawfully on state's property, but the monument was left standing, since names and dates of victims were removed and the new plaque read it was being erected in the name of all people from Golubić who died in wars, without mentioning when or in which war they died.

Thus, the scarcity of monuments to victims "of the other side" (and destruction of the few existing monuments) reflect the results of *Documenta*'s survey, in which more than a half of the respondents (52%) of the general population said they believed that victims of the war in Croatia were exclusively Croats (Kardov et al. 2010, p. 69), while 34% of the respondents felt that Serbian civilian victims and their families should not be given any reparations (ibid., 73). Moreover, only 5% of respondents from the general population felt that a memorial or a monument erected in the memory of victims would help victims' families to deal with their losses (ibid.).

The mentioned examples of erected monuments also point to how complicated, and potentially how dangerous, memorials are in terms of post-conflict reconstruction. New public memorials built after the war in Croatia are mainly ethnically exclusive and reflect the view that the role of victim belongs exclusively to the majority community. Although there are examples of allowing victims of "the other side" to enter commemorative practices in the memory of war victims, it seems that some time will still need to pass before official narratives start talking about and remembering all innocent victims of the war, regardless of their ethno-national and religious belonging.

Future Perspectives

Dealing with the past in post-war societies has an important role for reconciliation, because denial of violent past and refusing to take responsibility for past war crimes prevents full social, economic, and political development of a society. The prospect of further normalization, as well as insuring preconditions for social and economic development in Croatia, seems to be largely dependant on increasing the capacity of the civil society, political leaders, academic community, media, as well as of the judiciary, to deal with the past.

One of the mechanisms to achieve this is prosecution of war crimes at domestic courts. In Croatia, these trials have significantly improved in the last decade, although problems still exist. Trials are still conducted in the context of social tolerance toward "one's own" criminals, particularly in the context of the 2011 parliamentary elections. Moreover, in war crimes trials, the focus of public interest still lies on perpetrators instead on victims, while the general public still lacks compassion toward victims from other ethnic communities. Since the political will to prosecute criminal acts committed by members of Croatian army units was lacking, there was no awareness about the need to pay respects to the victims of those crimes either.

During the last several years, however, the situation has changed. Steps forward undertaken by the highest-ranking state officials during 2010 might significantly influence the creation of a political and public opinion which condemns crimes while at the same time supports reconciliation processes.

Thus, erecting monuments and paying respects to all victims by the highest state officials on the sites of detention and killings of innocent people not only represents an act of paying respects to all war crimes victims, but it is also necessary in order to make those sites become places of remembrance, with a clear message of condemning the crimes committed there.

However, a national perspective is not enough in order to achieve truth-telling and truth-seeking. A regional cooperation is necessary in all post-war societies of the former Yugoslavia, since a regional approach could have a better chance at dealing with the past than a national one. For this reason, it would be of great importance to the former Yugoslav states to form a regional commission, which would make efforts to establish facts about all crimes committed and all victims who suffered during these wars, so that these crimes could never again be manipulated for political purposes and would not lead into another conflict. A regional commission will certainly not be able to fulfil its mission without a clear and direct support from all states, which is why advocating for its establishment seems particularly important at the moment.

References

Anderson, B. 1983. *Imagined communities: Reflections on the origin and spread of nationalism*. London: Verso.

Andrieu, K. 2010. Transitional justice: A new discipline in human rights. *On-line Encyclopedia of Mass Violence*. http://www.massviolence.org/. Accessed 18 January 2010.

Batt, J., and J. Obradovic-Wochnik (eds.). 2009. *War crimes, conditionality and EU integration in the Western Balkans (Chaillot Paper, no. 116)*. Paris: Institute for Security Studies.

Bickford, L. 2004. The encyclopaedia of genocide and crimes against humanity. *Macmillan Reference USA* 3: 1045–1047.

Božić, N., and Rešković, V. 2008. Glavaš nominated to the human rights committee. *Jutarnji list*, p. 6, February 20.
Centre for Peace, Non-Violence and Human Rights Osijek. 2008. *Monitoring war crimes trials: Report for January–June 2008*, Osijek, September 26.
Charges for war crimes against four former 'Tigers' dropped. 2010. *Večernji list*. www.vecernji.hr. Accessed 19 November 2010.
Declaration on Cooperation with the International Criminal Court in the Hague. 2000. *Narodne novine* [*The Official Gazette*], no. 41/2000. www.nn.hr.
Declaration on the Homeland War. 2000. *Narodne novine* [*The Official Gazette*], no. 102/2000. www.nn.hr.
Dimitrijević, N. 2011. Apology instead of reconciliation: A view from Serbia. www.pescanik.net. Accessed 12 November 2011.
Đerić, G. (ed.). 2008. *Intima javnosti [Intimacy of the Public]*. Beograd: Fabrika knjiga.
Đula, N. 2006. Nine percent less of those convinced in general Gotovina's innocence. *Jutarnji list*, pp. 5, March 18.
Human Rights Watch. 2004. Croatia: EU must address domestic war crimes trials. www.hrw.org. Accessed 19 December 2004.
International Criminal Tribunal for Former Yugoslavia. 2007. Summary of Judgement for Mile Mrkšić, Miroslav Radić and Veselin Šljivančanin. www.icty.org. Accessed 27 September 2007
International Criminal Tribunal for Former Yugoslavia. 2011. Judgement Summary for Gotovina *et al.* www.icty.org. Accessed 15 April 2011.
Ivanišević, Đ. 1995. Defense is not a war crime. *Slobodna Dalmacija*, p. 5, March 28.
Jović, D. (n.d.). Croatia and Serbia: New initiatives for regional cooperation. www.boell.hr.
Kardov, K., D. Lalić, and V. Teršelič. 2010. *Suočavanje s prošlošću u Hrvatskoj, Stavovi i mišljenja aktera i javnosti u poraću [Dealing with the past in Croatia. Attitudes and opinions of actors and the public in a post-war context]*. Zagreb: Documenta—Centar za suočavanje s prošlošću.
Kosor: The verdict is unacceptable, an appeal is following. 2011. www.danas.hr. Accessed 15 April 2011.
Kostovica, D., ed. 2009. European Union and transitional justice: From retributive to restorative justice in the Western Balkans, *Forum for Transitional Justice*, Humanitarian Law Centre, No. 3.
Law on marking sites of mass graves of the Homeland War victims. (1996). Narodne novine [*The Official Gazette*], no. 100/96. www.nn.hr.
Law on the Implementation of the Statute of the International Criminal Court and on the Prosecution of Crimes against the International Law of War and Humanitarian Law. 2003. Narodne novine [*The Official Gazette*], no. 175/2003. www.nn.hr.
MacDonald, D.B. 2002. *Balkan holocausts? Serbian and Croatian victim-centred propaganda and the war in Yugoslavia*. Manchester: Manchester University Press.
Mobekk, E. 2005. Transitional justice in post-conflict societies—Approaches to reconciliation. In *After intervention: Public security management in post-conflict societies—From intervention to sustainable local ownership*, ed. A. Ebnother and P. Fluri, 280. Geneva Centre for the Democratic Control of Armed Forces (DCAF): Geneva.
Orentlicher, Diane F. 2007. 'Settling accounts' revisited: Reconciling global norms with local agency. *The International Journal of Transitional Justice* 1: 10–22.
Pavlaković, V. 2010. Croatia, the International Criminal Tribunal for the former Yugoslavia and general Gotovina as a political symbol. *Europe-Asia Studies* 62(10): 1707–1740.
Political support to the RECOM initiative in the region. 2011. www.documenta.hr. Accessed 20 May 2011.
Preliminary results of a research on the contents of the central news program *Dnevnik* of the Croatian. 2011. National Television after the indictments to generals Gotovina, Markač and Čermak in the period April 15 to 30, 2011. www.documenta.hr. Accessed 4 August 2011.
President Josipović on the Occasion of Passing the Annulment Law. 2011. www.predsjednik.hr. Accessed 21 October 2011.
Public advocacy of the RECOM initiative. Report. 2011. www.documenta.hr. Accessed 7 July 2011.

Presidents apologize over Croatian war. 2003. www.bbc.co.uk. Accessed 10 September 2003.
Presidents Josipović and Tadić laid wreaths at Paulin Dvor. 2010. www.predsjednik.hr. Accessed 4 November 2010.
President's Office Press Release. 2010. www.predsjednik.hr. Accessed 5 October 2010.
Rangelov, I. 2006. EU conditionality and transitional justice in the former Yugoslavia. *Croatian Yearbook of European Law and Policy* 2(2): 365–375.
Rombouts, H. 2002. Importance and difficulties of victim-based research in post-conflict societies. *European Journal of Crime, Criminal Law and Criminal Justice*, 10:2–3, 216–232.
Savić, O. 2006. Memory of war crimes: Can victims speak? *Belgrade Circle Journal,* 1–4
Skopljanac Brunner, N., et al. (eds.). 2000. *Media & war*. Belgrade: Agency Argument.
Stojanović, M., and K. Kruhonja (eds.). 2011. *Monitoring of war crimes trials. A Report for 2010.* Osijek: Centre for Peace, Non-violence and Human Rights.
Stojanović, M., and M. Sjekavica (eds.). 2012. *Praćenje suđenja za ratne zločine. Izvještaj za 2011. Godinu [Monitoring War Crimes Trials. Report for 2011]*. Osijek: Centre for Peace, Non-violence and Human Rights.
Stubbs, P. 2003. *Dealing with the past in Bosnia-Herzegovina, Croatia and Serbia & Montenegro*, Regional synthesis report, London: Quaker peace and social witness programme in post-Yugoslav countries, September 4.
Subotić, J. 2009. *Hijacked justice. Dealing with the past in the Balkans*. Ithaca: Cornell University Press.
Survey of opinions on the RECOM initiative. 2011. www.documenta.hr. Accessed March 2011
Tadić ponovio ispriku Hrvatima zbog rata [Tadić repeated his apology to Croats for the war]. 2007. www.index.hr. Accessed 24 June 2007.
Teitel, R. 2003. Transitional justice genealogy. *Harvard Human Rights Journal* 16: 70.
The consultation process on the establishment of the facts about war crimes and other gross violations of human rights committed on the territory of the former Yugoslavia. 2011. Belgrade: Humanitarian Law Centre
The Law on the Election of Representatives to the Croatian Parliament 2003. Narodne novine [*The official gazette*] 69/03. www.nn.hr.
Thompson, M. 1999. *Forging war .The media in Serbia, Croatia, Bosnia and Herzegovina*. Luton: University of Luton Press.
TV show *Latinica.* 2005. http://www.youtube.com/watch?v=YEvVcjLaDFw.
Youth Initiative for Human Rights report on the campaign. 2011. *1.000 000 Signatures for RECOM.* 26 April–30 June 2011. www.zarekom.org.

Chapter 4
The Failure of Macedonian Post-communist Transitional Justice: Lustration, Between Cleansing and Parody

Despina Angelovska

Introduction

Due to the wars in the former Yugoslavia (1991–1995), the fall of the Communist regime in Macedonia was not closely followed by a process of decommunization and a search for transitional justice, as was the case in other Communist countries such as Eastern Germany, the Czech Republic, Albania, Bulgaria, or Romania. Even though Macedonia did not take part in these wars, the sanguinary crimes in its neighborhood drew attention away from the misdeeds of the Communist regime. Furthermore, the war in Kosovo, NATO's intervention, as well as the large number of refugees that fled to Macedonia in 1999, were acute problems that had to be tackled without delay. In 2001, Macedonia confronted another important crisis, this time on its own territory, the armed conflict between the Albanian National Liberation Army and the Macedonian security forces that resulted in the Ohrid Framework Agreement adopted the same year. This permanence of crises contributed to the postponement of the transitional justice processes in Macedonia, similarly to those in other countries of the former Yugoslavia. Additionally, as highlighted by Lavinia Stan regarding Slovenian lustration (2006, p. 10), the Yugoslav "socialism with a human face" also explains to some extent why, at first, those countries were not very much interested in sanctioning their Communist officials. In this regard, the Law on Lustration, adopted by the Macedonian parliament only in 2008—17 years after the country's independence—was denounced by many political and civil actors as arriving too late.

D. Angelovska (✉)
University for Audio-Visual Arts ESRA Paris—Skopje-New York,
ul.Nikola Rusinski, n. 1, 1000, Skopje, Republic of Macedonia
e-mail: adespa@yahoo.com

O. Simić and Z. Volčič (eds.), *Transitional Justice and Civil Society in the Balkans*, Springer Series in Transitional Justice, DOI 10.1007/978-1-4614-5422-9_4,

This chapter analyzes how this late Macedonian lustration, aiming at screening public officials for their past involvement in the Communist regime, in particular collaboration with secret services (Sadurski 2003, p. 3), rapidly became not only an object of political revenge, but also of parody and distrust. Originally conceived as being only a part of larger processes of transitional justice, lustration, eventually, transformed itself into the complete opposite of its primary normative goals such as those defined by the guidelines of the Parliamentary Assembly of the Council of Europe (PACE) (1996a, b) based on the respect of human rights and the rule of law.

In this chapter, I consider how Macedonian lustration—intended as a response against Communist crimes and purges, and as an attempt to overcome the totalitarian past—only unveiled that the same dangerous authoritarian logic of surveillance and cleansing still operates in post-Communist Macedonia. In the first part of this study, I consider the introduction and turbulent evolution of the Macedonian law on lustration. In the second part, I address its controversial applications and designated deficiencies. Finally, in the last part, I focus on the contribution of civil society not only in the promulgation of lustration, but also in the critique and correction of this latter. The aim of the present chapter is to evaluate the civil society efforts in dismantling the totalitarian Communist heritage and in negotiating normative transitional justice practices. Furthermore, by placing at the center the reflection of some engaged citizens and intellectuals on lustration's processes and paradoxes, this study highlights the importance of their critical thought for the transitional justice scope. By focusing thus on the unexplored role of civil society engagements within transitional justice and lustration processes in Macedonia, and by also reflecting on the difficulties and ambiguities underlying their application on the ground, this chapter aims to contribute to the enrichment of transitional justice research.

Although there has been significant research done in the field of post-Communist transitional justice (Letki 2002; Dvořáková and Milardović 2007; Stan 2009; Nalepa 2010), there is a lack of academic studies on the controversial Macedonian lustration, both nationally and internationally, regardless of the fact that this precise subject had received close media and press coverage, while provoking huge debates in the public sphere. Because of this lack of academic and systemic studies of Macedonian transitional justice, wishing to provide a timely and in-depth analysis of the ongoing lustration, this study draws on field research and personal interviews with key local actors. At the same time, the analysis of the Macedonian case is situated within the larger scholarship of transitional justice.

The term "transitional justice," as analyzed by Christine Bell, emerged in late 1980s and early 1990s with reference to the transitions from authoritarian rule in Eastern Europe and Central America, while it was only consolidated as a field after 2000 (2009, p.7). Even if its definitions differ, Bell argues that transitional justice is generally viewed as "the attempt to deal with past violence in societies undergoing or attempting some form of political transition" (2009, p.7). Furthermore, in his anthropological approach of transitional justice, Alexander Laban Hinton defines it as the process of "redressing past wrongs committed in states shifting from a violent, authoritarian past toward a more liberal, democratic future" (2010, p.2).

Introduction of Lustration in the Context of Transitional Macedonia

Lustration in Post-communist Countries

Ruti Teitel interprets transitional justice more as the "conception of justice associated with periods of political change", stressing its "legal responses to the wrongdoings of repressive predecessor regimes" (2003, p.69). It is after the collapse of the Communist block between 1989 and 1991, that Teitel situates the beginning of the "post-Cold War phase of transitional justice" (2003, p.75). Conceived as a wide range of inter-related processes of coming to terms with the recent dictatorial past (Stan 2009, p.1), transitional justice, in post-Communist countries, ushered in lustration. Lustration received its name from the Greek and Roman ceremony of purification (Merryman and Tafel 2009), but even if this word seems to cover very effectively the problems of reckoning with the past, there has been no unanimity about its meaning and the form it should take (Letki 2002). Nevertheless, as defined by the guidelines of PACE (1996a, b), lustration has been conceived as the politics of disqualification and reduction of the participation of those associated with abuses of human rights under the Communist regime in state-appointed offices where they could represent a threat for the newly emerged democracies. The first countries to introduce lustration, in 1990 and 1991, were, respectively, East Germany and Czechoslovakia (Ellis 1996; Stan 2006), followed, with different delays and scopes, by most of the Central and East European countries. However, although lustration is the most common form of transitional justice in post-Communist Europe, used for "screening and 'prosecuting' former communist leaders, candidates for office, and selected public employees" (Ellis 1996, p.181), it remains controversial and often criticized by the national and international community (Letki 2004).

Macedonian Law on Lustration

Aiming at disclosing past misdeeds in order not to repeat the authoritarian abuses of the previous regime, lustration has also been linked to the opening of the Communist secret service archives. Preceding the Law on Lustration, the Law on Handling Personal Files kept by the State Security Service (Закон за Постапување со Досијеа за Лица Водени од Службата за Државна Безбедност 2000) was adopted in 2000. Later, after the parliamentary elections and the ascent to power of the coalition presided by the right wing VMRO-DPMNE party, and closely following PACE's Resolution 1481/2006, Need for International Condemnation of Crimes of Totalitarian Communist Regimes, the Macedonian Assembly, in 2006, approved the Declaration of Apology to the Victims of Repression by the regime between 1945 and 1990. By the end of the same year, Stojan Andov, the leader of the Liberal Party and a member of the coalition in power, proposed a draft-law on lustration, which took two more years to be approved. Finally, at the end of January 2008, the Assembly, with a large consensus among all political parties, adopted the new "benchmark", the Law on Lustration, officially called the Law on the Establishment of an Additional Condition for the Performance of Public Office (Закон за Определување Дополнителен Услов за Вршење Јавна Функција 2008).

This law forbids former secret collaborators, secret informants, those delivering orders, as well as the users of information obtained in such a way to hold public office. The covered period starts from August 2, 1944 and ends on the date of the adoption of the Law. Articles 1 and 2 of the Law stipulate that every candidate for public office or holder of public office should submit a Statement of Non-collaboration with the Secret Services for the period specified by the Law. The statement is to be processed by the Commission for the Verification of Facts, conceived as an autonomous and independent body, elected by the parliament. The additional condition for the performance of public office is fulfilled if the Commission has verified this statement of non-collaboration. The law was amended on May 22, 2009, with the amendments mainly concerning the status and working conditions of the Commission. The Commission started to implement the Law in September 2009.

Evolution and Revision of the Law

In a broader sense, transitional justice measures have often been criticized as allowing dangerous "legal exceptionalism" and thus creating legal controversies (Bell 2009, p. 12). In this respect, as argued by Alan Uzelac, it is the very principles of the rule of law that may be one of the important barriers to the success of lustration (2007, p.48). Indeed, on March 24, 2010, the Macedonian Constitutional Court contested four articles of the Law on Lustration as unconstitutional, among which was Article 34 stipulating that political parties, journalists associations, NGOs, as well as religious communities and association, could also demand the fulfillment of the additional criteria from its members and candidate members (Одлука на Уставен Суд на Република Македонија У. бр. 42/2008. У. бр. 77/2008, April 1, 2010). In its decision, the Court argued that this signified involvement of the State in the functioning of civil associations, violating the constitutional guaranties of citizens' freedom of association as well as the separation of the State from the Church and religious groups. The Court also abolished a paragraph of Article 2 of the Law, reducing the period covered by lustration, thereby limiting it to 1991, the year of Macedonian independence and of the adoption of the Constitution in power, when a democratic system was established based on the separation of powers and the protection of the rule of law and human rights. The Court concluded that the period after 1991 could not be subjected to lustration, since this process aims at reckoning with the abuses of Communist past and not with those of the democratic system, which has other correspondent legal mechanisms. The Court also abolished, because of violating the dignity as well as moral and personal integrity of citizens, the part of Article 8 that stipulated that the names of those who have not filed a Statement of Non-collaboration with the Secret Services would be published in the Official Gazette.

On November 10, 2010, a few months after this ruling, contested by the governing party VMRO-DPMNE, the first official to be disqualified by the Commission as an informant and to lose his position was precisely the presiding judge of the Constitutional Court, Trendafil Ivanovski, a choice denounced by some as being clearly politically motivated. On February 25, 2011, the Law on Lustration was amended a second time. The amended version—avoiding, among other things, to implement the Court's decision regarding the period covered by the law, and additionally introducing new pecuniary sanctions (Article 35-a) for those not filing a Statement of Non-collaboration—was described by some legal experts and civil society members as in contempt of the Constitutional Court and of the Constitution (Najčevska 2011b; FOSM 2011). Most recently, with its ruling from January 25, 2012, the Constitutional Court decided to review the anti-constitutionality of 12 articles of the latest Law of Lustration (Решение, У. Број: 76/2011-0-0).

The Failure of the Law on Lustration and its Application: Lustration, Between Cleansing and Parody

Lustration as Parody and Object of Distrust

As I have already argued, post-Communist lustration has provoked legal and moral controversy in many countries. Moreover, the question has been put forward whether this process supports or undermines democratic transition (Horne 2009). In a broader sense, this investigation joins an emerging body of research that reflects on whether "transitional justice works and whether it delivers the political goals to which it lays claim" (Bell 2009, p.10). In this respect, the "imperial dimension" of universal transitional justice goals and its "exceptional justice", with disturbing consequences for local justice and democracy more generally, have been stressed lately by academic research (p.12). Additionally, the application of Macedonian lustration seems to reinforce the dilemma about the "unintended consequences" often provoked by the translation of the normative goals of transitional justice into a particular context (Hinton 2010, p.17).

Indeed, lustration, rapidly after entering into force, became an object not only of political manipulation, but also of parody and distrust. From its very first implementation, it was followed by political scandals and mutual accusations between the principal political parties. On one occasion, secret files "fell from the sky", incriminating politicians and senior officials from one of the Albanian governing parties: in December 2010, the Lustration Commission received three secret files charging senior officials from the Democratic Union for Integration party (DUI), delivered by a professor from the city of Tetovo, claiming to have found them in his yard. Furthermore, the lustration procedure itself seems to have lost a great deal of its credibility, since some of the secret files, on which the allegations were based, have been denounced as falsified. More recently, and for the same case, on January 23, 2011, the criminal court sentenced to conditional discharge the former secret agent accused of having falsified the files compromising the senior officials from DUI. Even so, the accused declared himself a victim of an unprecedented politically mounted process.

Indeed, as pointed out by Stan (2009), an important element of lustration, linked to the reliability of the process, is its dependence on access to the archives of secret services. More generally, this reliance on information fabricated by the secret services has also been one of the main reasons for the doubts and criticism surrounding lustration. That is why legislation on file access is crucial for lustration, whose well-regulated implementation depends directly on "the quality of the basis for the screening process, i.e. the quality of the files of the secret service archives" (Letki 2004, p.8). In this regard, Stan also argues that the later a country launches lustration, "the more tampered the archives, the harder to identify the implicated and the more disputed the official findings" (2006, p.15). In this respect, the late Macedonian Law on Handling Personal Files, providing only very restricted access to the files, and the even later creation of the Lustration Commission, left great possibilities for their tampering and did not contribute to the successful implementation of lustration.

Lustration as a "Witch-Hunt"

As emphasized by Natalia Letki, lustration, in post-Communist countries, has often been used as a political tool for dealing with opponents, thus reinforcing non-democratic and manipulative political solutions (2004, p.6). In this sense, before it was even adopted, but especially afterward, the Macedonian Law on Lustration has been often qualified by civil society organizations, experts, and media as a "witch-hunt" (Milčin 2007, 2011; Najčevska 2011a). In this respect, the question that has often been asked is, whether lustration is at all possible in a society determined by party affiliation.

In this section, I will explore the way in which, in its first implementations, Macedonian lustration has appeared as used mainly for discrediting political opponents and civil society organizations. In this regard, the Lustration Commission's choice of the individuals to be disqualified could be easily interpreted as politically oriented. Indeed, the first to be designated as an informant, as we already saw, was presiding constitutional judge Ivanovski. The second to be disqualified was Tito Ugrinovski, a member of an opposition party and a counselor of the City council of Skopje. More recently, in Summer 2011, Vladimir Milčin, the executive director of the prominent civil society organization, the Foundation Open Society Macedonia

(FOSM), who was also one of the most virulent critics not only of the abuses of the central authorities, but also of those of lustration, was declared an informant. Denying the allegations, Milčin, at a press conference held on July 25, 2011 at the club of the association of Citizens for a European Macedonia (CEM), denounced his alleged lustration as an attempt to politically lynch and discredit FOSM as "one of the few persistent defenders of democracy, human rights, and the rule of law in Macedonia" (FOSM 2011b).

The Omissions of the Commission Procedure

As indicated in the Support for Improvement and Government Management (SIGMA) assessment on Macedonia (2009), unless lustration processes are truly professional and impartial, they may be easily misused. Therefore, the legitimacy of those who are supposed to be the implementers of the lustration practices on the ground is crucial. In this respect, the professionalism and objectivity of the Macedonian Lustration Commission has been often questioned, since the Commission was accused of violating the procedure defined by the Law.

In the case of judge Ivanovski, he claimed the Commission lustrated him without the necessary signed proofs of collaboration as stipulated in Article 4 of the Law (Dimitrioska 2011). The same criticism was formulated by Milčin. Furthermore, according to him, the Commission seriously violated the procedure since, although bound by the Law to protect the identity and dignity of persons subject to verification, in his case, as in the previous cases, this was not done (FOSM 2011b). Additionally, the Commission has been criticized for violating the procedure, since, instead of determining facts, it reached its verdicts by voting (CEM 2010, 2011a). Finally, the lack of competence of the Commission members, in regard to not only the functioning and mechanisms of the secret services, but also the nonexistence of forensic expertise for documents in Macedonia, was addressed as making the process more vulnerable to manipulation (CEM 2011b).

Lustration has thus provoked great controversy. Moreover, the process has not only failed in accomplishing its transitional justice goals, but also been transformed from "a mechanism of prosecution of abuse, to abuse of prosecution" (Naj evska 2011c).

Civil Society and Lustration

As argued in recent analysis, Macedonian civil society still needs to reinforce its capacities and independence. In its latest report on local civil society from 2008 to 2011, the Macedonian center for international cooperation, among other things, indicates the deficiencies of this sector and the need for its further development. According to this study, the civic participation in Macedonia remains low, while the influence of civil society on policy making is insufficient. Additionally, the non-sustainability of the civil sector diminishes its possibilities of action (Klekovski et al. 2011, p.3). This lack of financial resources and the strong dependence on foreign funding of the civil sector have also been quoted in the latest report of the European Commission on Macedonia (2011, p.17). Furthermore, as argued by the executive director of FOSM, Milčin, the deep polarization of Macedonian society by party, ethnic and religious affiliation also extends to the civil sector. While some civil society organizations seem to have been created directly by the current government in order to endorse its policy, other critically oriented associations have been publicly attacked and discredited by officials of political parties and Members of Parliament (Milčin 2010, European Commission 2011, p.17).

Despite those deficiencies and internal divisions of Macedonian civil society and despite the increasing political pressures suffered, it has played an important role in the process of lustration: firstly, in its initiation and promulgation, and secondly, in pointing out its dangers and abuses.

The Role of Civil Society in the Promulgation of Transitional Justice and Lustration

The debate on the taboo subject of lustration was initiated in Macedonia by civil actors, following the adoption of Resolutions 1096 and 1481 of PACE (1996, 2006). Effectively, one of the most active promoters of the adoption of a lustration law, and later its consultant, was the blogger and columnist Ivica Bocevski (2006), who later became a spokesman and a minister of the Government. Nikola Gelevski, the co-founder of the citizen association *Kontrapunkt,* published an article titled "The World of the Ghosts and the Sleeping" (2006), in which he addressed the essential need for Macedonian society to establish transitional justice mechanisms in order to stop the political manipulation of the issue of Communist crimes in the present. Furthermore, theorist and writer Jasna Koteska, the daughter of the poet Jovan Koteski, a victim of the Communist secret services, was one of the first to address this repressed subject. In 2005, after she began to speak publicly on her father's persecution, she even started receiving threats (Koteska 2008, p.41). Still, she persevered and published a text with extracts of her father's file in *Margina* (Koteska 2006). Her public interventions and her personal testimony had a great influence on raising awareness about the subject and have been often quoted in support of the promotion and application of lustration.

As outlined by Tchavdar Marinov (2009), the Law on Lustration itself was inspired by the NGO sector and the project entitled "Disclosing hidden history: Lustration in the Western Balkans," which was run from 2004 to 2006 by the Center for Democracy and Reconciliation in Southeast Europe (Hatschikjan et al. 2006). Yet, after the adoption of the Law and its first implementations, it was also the civil society voices that were the first to point out its dangers for democracy.

The Critique of Lustration: The Contribution of Civil Society as a Transitional Justice Watchdog

It is primarily in the denunciation of the abuses of lustration that Macedonian civil society has contributed the most, constituting itself as a defender of human rights and justice.

Indeed, in 2007, when the proposal of the Law was submitted by Ardov, it was civil society representatives that were the most critical towards it entering into force. In this regard, the Helsinki Committee for Human Rights of the Republic of Macedonia (HCHRRM), under the presidency of Mirjana Najčevska, in its analysis of lustration (2007) criticized the proposal for containing important deficiencies and for endangering human rights and the rule of law.

In a television debate on lustration in 2007, Najčevska also pointed out the risks of the proposed Law, since the Lustration Commission competences went beyond the law and allowed it to sentence a person without a trial (MS Television, 2007). In this respect, Najčevska's critique outlines existing critical examinations of transitional justice denounced often as promoting "legal exceptionism" and putting itself "beyond the law" (Bell 2009). The Law's proposal was also opposed by FOSM, while its chief executive, Milčin, labeled it as a "political and ideological purge" (2007).

As argued by Letki (2004, p.5), lustration should be proposed, discussed, and passed by the legislative body, and it should be approved or corrected by the Constitutional Court. On that point, it is on the basis of petitions submitted by engaged citizens and NGOs that the Constitutional Court, in 2010, abolished four contested articles. More recently, in 2011, petitions against the Law have been submitted again to the Court by FOSM representatives (2011) and by Filipov and Najčevska (Najčevska 2011b). It is in response to this civic initiatives that the Constitutional Court, on January 25, 2012, decided to review the anti-constitutionality of 12 articles of the Law on Lustration (Решение, У. Број: 76/2011-0-0). Those initiatives underline the importance of civil society and active citizenship for correcting transitional justice processes.

After the controversies provoked by the first applications of the Law, once again, civil society responded to the need for further discussion and consideration. Moreover, by critically examining the very concept and goals of lustration, the complexity of reflection on transitional justice has been broadened. In this respect, the cycle of public debates on lustration, organized by CEM, in which numerous experts, intellectuals, and citizens took part, enriched the democratic dialogue on lustration. A large number of civil society representatives and engaged intellectuals consider that the Law has been adopted too late and that it should be withdrawn or even annulled (CEM 2010a, b; Najčevska 2011b).

At the same time, different NGOs declared their public support for the applications of the law. Sometimes, in the case of lustration, the role of civil society could even be qualified as contradictory, as in the example of the HCHRRM. Joining other NGOs that affirmed their support for the Law on Lustration, the HCHRRM—previously very critical of the democratic deficiencies of the law

(2007)—adopted a contradictory decision that each of its members and candidate members should file a Statement of Non-collaboration with the Secret Services (2010a, b)—even after Article 34, concerning the lustration of NGOs and religious groups, was abolished by the Constitutional Court. As formulated by Najčevska, it seems sometimes that regarding lustration, Macedonian civil society has "lost its compass" (2010). Furthermore, regardless of the constitutional revision of the law, lustration has won some more battles within the civil society sphere: in 2011, for example, members of different religious organizations such as the Macedonian Orthodox Church and the Islamic Religious Community supported the lustration of its members (Islamic Center 2011). The support for this measure was also recently expressed by the Macedonian Association of Journalists (МПЦ Нема Против, ИВЗ Молчи 2011), which earlier, in 2007, had denounced it as means of political pressure on journalists critical of the government (Vojnovska 2007).

Finally, civil society has also been criticized for initiating the debate on lustration too late—thus allowing for manipulation of the secret services' archives—and for having missed the opportunity to prepare the Law well, and thus for contributing to its professionalism and success. Regarding this issue, Koteska points out the failure of Macedonian civil society to address the need of transitional justice in due time.

> As for transitional justice, I am also critical of FOSM and their contribution to the process. They have not opened the issue of transitional justice in time. (…) There were funds for transitional justice at disposal at that time and they did not use them. Also, by not tackling this issue, they left it in the hands of other people that have approached it more roughly. (…) And now, maybe also as a consequence of this failure to act timely, lustration is being used against them, as a boomerang effect (personal communication, September 23, 2011).

The Role of Individuals and Human Right Activists

Engaged intellectuals and human rights activists played a crucial role in the debate on lustration. In this regard, the former director of the HCHRRM, Najčevska, taking part in public discussions, writing in the press, but also regularly publishing on her blog, has been one of the most active ones in denouncing the law, although her father himself was a victim of the Communist regime. On her blog, Najčevska (2007) argued that, instead of lustration, Macedonian society should use established judicial procedures—that respect the presumption of innocence and the clear distinction between informants and perpetrators of crimes or human rights' violations—accompanied by a precise file expertise. Additionally, in the petition contesting the law that she, together with Filipov, submitted to the Constitutional Court, she has pointed out its unconstitutional and antidemocratic aspects and asked for its annulment (Najčevska 2011b). According to Najčevska (2011c), lustration has also been used as an instrument for intimidating citizens, allowing a prolonged period of control over them and depriving them of concrete civil rights.

Lustration and the Oblivion of the Victims

Reflecting and writing on lustration, another human rights activist, columnist, and member of CEM, Žarko Trajanovski, argues that post-Communist transitional justice is a far more complex process than its Macedonian translation into a simple Law on Lustration. Furthermore, this engaged intellectual claims that the Macedonian law, which does not conform to the recommendations of PACE (1996a, b; 2006), has, in fact, not much to do with transitional justice, as it makes the victims fall into oblivion (personal communication, September 30, 2011).

In his public argumentation, Trajanovski points out what he considers to be one of the principal failures of Macedonian lustration: by focusing on secret informants and collaborators, who are not the key figures in the process of violation of human rights, the Law on Lustration—that is, relying on the files of members of the secret services that have not yet been lustrated themselves—pursues only the "small fish" who, in some cases, were victims of the same secret services and were forced to collaborate. Consequently, according to him, the Law reproduces the same scheme: "Basically, today we are facing a lustration conducted by the non-lustrated services of state security, who are one of the principal violators of human rights, while the guilt is transferred to those whom they have sacrificed. In this respect, we could ask if this Law was in fact a law on amnesty of the true perpetrators of human rights violations" (personal communication, September 30, 2011).

However, the greatest failure of Macedonian lustration, in Trajanovski's analysis, is that it has completely put aside the victims. Instead of the misused lustration, which has muted the prosecuted, he considers that Macedonia needs a law on rehabilitation and financial compensation of the victims of the Communist regime. For him, lustration, without those two other crucial laws, represents a profound lack of understanding of the process of transitional justice. In this respect, Trajanovski formulates a general critique of the actions of civil society organizations in regard to transitional justice, as none of them really tried to hear the needs of the victims, to approach, to encourage, or to empower them (personal communication, September 30, 2011).

Furthermore, Trajanovski points to an important issue concerning the contradictions of Macedonian lustration, and that is the fabricated nature of the secret files. As also argued by some secret services employees within the debates on lustration, one of the principal tasks of these services is to produce collaborators and informants, even if sometimes the latter or the files are entirely invented (CEM 2011a). Consequently, Trajanovski stresses that this fictional dimension of the files, on which the lustration procedure relies, questions its objectivity (personal communication, September 30, 2011). Prolonging his reflection on the moral and legal controversies underlining transitional justice, Trajanovski analyzes the latter as intervening in transitional societies where confidence in the institutions of the system and in justice has been lost. According to him, that is exactly the current situation that Macedonian society faces today: it has lost that confidence, particularly because lustration has been transformed into an instrument of injustice (personal communication, September 30, 2011).

Communist Intimacy and Transitional Justice: Constructing more Complex Memories of the Past

Joining the reflection of Trajanovski on those who have been forgotten in the process of lustration, theorist and writer Koteska, daughter of a poet persecuted by the Communist secret services and imprisoned for 5 years, was one of the first to address publicly the issue of the unrecognized victims of the Communist regime. She was one of the first to affront the taboo of the totalitarian past and to speak publicly on her personal trauma caused by what she considers as the deeply provincial version of Macedonian Communism. In this respect, *Communist intimacy* (2008), by dismantling the sacrificial mechanisms of communism, represents an important contribution to transitional justice. With her book, Koteska primarily addresses the victims and tries to encourage them to cope with the past in search for recovery: "My basic premise was that if the post-communist states want healthy nations, they have to psychoanalyze themselves extensively, in order to know what happened to them and where they are now. So I invited the children of the former political prisoners to open the communist files they kept at home" (Koteska and Prueber 2011). Koteska also argues that the numerous public reactions to her book prove the need of people to speak about Communism, as this subject remained suppressed. This has created "traumatic pockets" within people, and, to this end, she wrote her book to resist the "cultural and political amnesia" where decades of history have been erased from the collective memory of people (Koteska and Galevski 2008).

Communist intimacy was published the same year the Law on Lustration was adopted. In that respect, the book was also criticized for putting accent on the informants as the principal villains, while denouncing some important personalities from the political and cultural Macedonian establishment for having cooperated with the Communist secret services. Yet, clearly distancing herself from the political and personal revenge that, according to her, was enacted by the Law on Lustration, Koteska argues that her intention was not to propose a "list of persons to kill," but instead to offer a critical analysis of the engineering that allowed the installation of Macedonian Communism: "It was not the informants that were my primary subject, it was the system in whose name they operated" (Koteska and Galevski 2008). Additionally putting Radomir Konstantinovic's reflection on provincialism and nationalism (1981) at the center of her work, Koteska focuses on the specific mechanisms of Macedonian Communism that she describes as a mixture of the tyranny of provincial minds and secret services. Furthermore, she identifies the latter as still moving contemporary Macedonian society and its lustration.

Positioning herself on the antipode of the Law on Lustration, perpetrating the spirit of authoritarian provincialism and provoking paranoia within people, Koteska calls for a lustration that she sees as "non-revolutionary and legal, allowing the opening of the Communist archives in order to understand the past" (Koteska and Stanković 2008). In this respect, criticizing the restricted access to the secret services archives that is still in force in Macedonia, Koteska outlines transparency as the essential condition of a successful lustration: "As a scientist, still for me, the unresolved question of the archives is crucial, and again with this Law on Lustration the principal problem remains–somebody still mediates the past, in this case, the Commission" (Koteska 2008, p.76). Reaffirming, more recently within a public debate, her criticism against the Law on Lustration, Koteska repeated her conviction that the non-selective publication of the files, in the spirit of Wikileaks, is crucial in order to fight against the secret services logic. On the same occasion, she stated that it was definitely too late for lustration in Macedonia (CEM 2010). Still, Koteska considers that it is not too late for a confrontation with the past, confrontation that should have historical and scientific value, not a political one (Koteska and Pureber 2011).

Communist intimacy was thus written searching for the answer to the question of what kind of reconciliation is still possible for those who were the children of the Communist system and the parents of this one (Koteska and Pureber 2011). In Macedonia, there are around 15000 persons who possess personal files as victims of the secret services (Koteska 2008, p.12). Yet, Koteska does not pretend to speak on the behalf of all the victims, since, as she says, every story is different, and mass crimes do not have a simple or unique solution (personal communication, September 23, 2011). It is in this spirit that *Communist Intimacy,* as argued by Koteska, was written as a part of the reconciliation process, "not in order to force people to forget their happy memories about the Communist past, but to help them render those more complex, and to place them next to the memories of those people amongst them, that were less happy" (Koteska and Galevski 2008).

In this respect, outlining the complexity of coping with mass crimes and trauma, Koteska's book represents a precious contribution to the field of transitional justice, and by also opening our memories of the Communist past to its unrecognized victims and thus acknowledging their right to grieve publicly, it indicates a possible path to recovery and collective reconciliation.

Conclusion

In this chapter, I have considered the difficulties and deficiencies of Macedonian lustration. Drawing on examples, I have explored the way in which it has failed to respond to the normative goals of transitional justice and rapidly become an object of political manipulation. As argued by Teitel on lustration "in fledgling democracies, where the administration of punishment can pose acute rule-of-law dilemmas, the contradictions to the uses of the law may become too great" (2003, p.77).

In this regard, Macedonian lustration has been qualified, entirely or partially, by different experts, members of the civil society, and by the Constitutional Court as violating the rule of law and human rights. The implementations of the Law provoked moral and legal controversies and debates on whether this process sustains or undermines democratic transition. Najčevska has outlined the dangers of the Law on Lustration as a measure of "exceptional justice" that allows for a verdict and a loss of concrete citizen rights without a trial. In this respect, Macedonian lustration, with its disturbing consequences for local justice and democracy, seems to reinforce the already existing dilemmas related to transitional justice initiatives. Lustration, in a broader sense, has been often criticized for creating divisiveness and excluding some people from the first category of citizenship (Sadurski 2003, p.6; Teitel 2000, p.98). This deficiency has shown as even more acute in a deeply politically divided society such as Macedonia.

Furthermore, in this study I have highlighted the particular difficulties of late lustration, which has shown serious failures in others post-Communist countries as well. As it has been argued, those difficulties are greatly linked to the crucial issue of access to archives of the secret services. The later file access has been granted, the greater the possibility of the archives being tampered with and manipulated, and thus the lustration process being discredited.

Additionally, in my analysis I pointed out that the crucial problem in the Macedonian case is the non-lustrated secret services on whose "corrupted" archives lustration relies, and which, consequently, continue to control the process. Lustration has lost its credibility since it has become clear that some of the secret files might have been entirely invented. This implies that, when rethinking lustration, we have to take into consideration the unreliability of the secret files and their constitutional fictional dimension, which seems to enlighten the inherent contradiction of this process more generally.

On the other hand, as noted by Hinton (2010, p.17) "transitional justice is a tool that, despite its limitations, can still potentially bring about positive social change, particularly if it more fully takes account of justice and locality". In this regard, even if Macedonia's late lustration seems to have failed in respect of universal transitional justice goals, it has still outlined the complexities of the local application of this tool. I argue that those revealed difficulties and contradictions, at the same time, could contribute to the enrichment of transitional justice field by bringing it closer to realities on the ground.

Furthermore, Macedonian lustration has shown the crucial role that civil society has in monitoring of these processes and in rethinking transitional justice permissible goals. As stated by Bell (2009, p.17), transitional justice ends require that "they be constantly argued and won throughout design and implementation". In this sense, local actors have played an essential part in correcting Macedonian lustration. Trajanovski denounced thus the Macedonian lustration as being wrongly targeted towards the informants, masking the ones principally responsible. Joining Trajanovski's argumentation, this chapter has also outlined the need of a law on rehabilitation and financial compensation of the victims of communism, who have been completely forgotten by lustration.

Finally, this study has noted the precious contribution to Macedonian transitional justice also coming from personal testimonies, as in the case of Koteska, the author of *Communist intimacy.* Resisting political, historical, and cultural amnesia, Koteska, with her book, copes with the repressed totalitarian past in search for recovery. In this respect, I consider her engaged efforts toward collective reconciliation as also enabling us, her readers and listeners, to enrich the memory of our common Communist past by the memories of its unrecognized victims, rendering it more complex, more inclusive, and thus more just.

Reflecting on the important challenges that transitional justice faces today, Bell writes (2009):

> The deep justice project is one in which all must participate jointly, with: awareness of the contingency of their own positions; a willingness to negotiate justice while adhering to a belief that justice has a normative core content capable of delivering meaningful change in people's lives; and a conscious sense of how much is at stake in the negotiation (p.27).

In that sense, finally, it is precisely with their shared contribution, resisting, and correcting the authoritarian abuses of lustration, but also rethinking and reshaping transitional justice scope, that truly committed civil society actors have helped Macedonian citizens not to forget what meaningful justice should and could be.

References

Bell, C. 2009. Transitional justice, interdisciplinarity and the state of the "field" or "non- Field". *International Journal of Transitional Justice* 3: 5–29.

Bocevski, I. 2006. Соочување со сеништа од минатото. *Utrinski Vesnik*. http://star.utrinski.com.mk/?pBroj=1985&stID=60712&pR=7.

Citizens for European Macedonia. 2010. Нашата лустрација. http://www.gem.org.mk/vesti/659-zborno-mesto-vo-gem-klubot.

Citizens for European Macedonia. 2010. Санитарна енигма 2 или нашата лустрација. http://www.gem.org.mk/vesti/667-predocna-za-lustracija-nam-ni-treba-otvoranje-na-istorijata.

Citizens for European Macedonia. 2011. Лустрацијата хајка против неистомислениците. http://www.gem.org.mk/vesti/1026-zborno-mesto-hajka-vrz-neistomislenicite.

Citizens for European Macedonia. 2011. Бројни манипулации во процесот на лустрација. http://www.gem.org.mk/vesti/1032-zborno-mesto-brojni-manipulacii-vo-procesot-na-lustracija.

Dimitrioska, P. 2011. Стразбур: Лустрација со цврсти докази и строги правила. http://vesti.alfa.mk/default.aspx?mId=36&eventId=36664.

Dvořáková, V., and A. Milardović (eds.). 2007. *Lustration and consolidation of democracy and the rule of law in Central and Eastern Europe*. Zagreb: Political Science Research Centre.

Ellis, M.S. 1996. Purging the past: The current state of lustration laws in the former communist bloc. *Law and Contemporary Problems* 59(4): 181–196.

European Commission. 2011. *Commission Working Staff Document: The former Yugoslav Republic of Macedonia 2011 progress report.* Brussels: European Commission. http://ec.europa.eu/enlargement/pdf/key_documents/2011/package/mk_rapport_2011_en.pdf.

Foundation Open Society—Macedonia. 2011. Правна програма: ФИООМ по вторпат поднесува иницијатива за утврдување неуставност на Законот за дополнителен услов за вршење јавнафункција(Законзалустрација).http://soros.org.mk/default.asp?lang=eng&menuid=1847. Accessed Архива 2011.

Gelevski, N. 2006. Светот на соништата и на успаните. http://www.kontrapunkt-mk.org/kontra/kolumna/2006/2006-01-31.htm.

Hatschikjan, M., Reljić, D., and Šebek, N., eds. 2006. *Disclosing hidden history: Lustration in the Western Balkans—A project documentation.* http://www.lustration.net/.

Helsinki Committee for Human Rights of the Republic of Macedonia. 2007. Лустрација—дали може да се гради систем во кој се почитуваат човековите права и слободи врз основа на досијеата кои ги водела тајната полиција на претходниот политички систем? http://www.mhc.org.mk/WBStorage/Files/Posebni_analizi_Lustracija.pdf. 2007, January 15

Helsinki Committee for Human Rights of the Republic of Macedonia. 2010. Monthly report on the situation of human rights in The Republic of Macedonia. http://www.mhc.org.mk/default-en.asp?ItemID=920A8485A0ECB1448750BD54DE3770FD.

Helsinki Committee for Human Rights of the Republic of Macedonia. 2010. Членовите на Комитетот на проверка пред Комисијата за лустрација. http://www.mhc.org.mk/default.asp?ItemID=C7F5D8AB7027884AABE406023272F567.

Hinton, A.L. 2010. Introduction: Toward an anthropology of transitional justice. In *Transitional justice: Global mechanisms and local realities after genocide and mass violence*, ed. A.L. Hinton, 1–22. New Brunswick: Rutgers University Press.

Horne, C. 2009. Late lustration programs in Romania and Poland: Supporting or undermining democratic transitions? *Democratization* 16: 344–376

Islamic Center. 2011. ИВЗ сака лустрација за свештени лица. http://www.islamski-centar.org/web/index.php?option=com_content&view=article&id=419:2011-01-27-18-07-35&catid=51:2009-11-21-03-37-29&Itemid=112.

Klekovski, S., D. Stojanova, G. Jakovlevska, and E. Nuredinoska. 2011. *ЦИВИКУС—Индекс на граѓанското општество за Македонија: Долг пат до поголем граѓански ангажман*. Скопје: МЦМС.

Konstantinović, R. 1981. *Filozofija palanke*. Belgrade: Nolit.

Koteska, J. 2006. Јован Котески: Досије 5622. *Margina* 71: 133–173.

Koteska, J. 2008. *Комунистичка интима*, 2008. Skopje: Templum.

Koteska, J., and Galevski. V. 2008. Досије Интимист. *Forum*. http://jasnakoteska.blogspot.com/2009/11/interview-for-communist-intimacy-forum.html.

Koteska, J., and Stanković, S. 2008. Македонија живее во тиранија на паланката. *Globus*. http://jasnakoteska.blogspot.com/2009/11/interview-for-communist-intimacy-globus.html.

Koteska, J., and Pureber, J. 2011. Democracy finds its continuity in totalitarianism: a conversation with Jasna Koteska. http://www.bifc-hub.eu/interview/democracy-finds-its-continuity-in-totalitarianism-a-conversation-with-jasna-koteska.

Letki, N. 2002. Lustration and democratisation in East-Central Europe. *Europe-Asia Studies* 54(4): 529–552.

Letki, N. 2004. The consequences of lustration for democratisation: The experience of East Central Europe. In *Past and present: Consequences for democratization*, ed. M. Hatschkijan and C. Noack-Aeotopoulus. CDRSEE: Thessaloniki.

Marinov, T. 2009. Anticommunist, but Macedonian: Politics of memory in the Republic of Macedonia. *Tokovi Istorije* 1–2: 65–83.

Merryman, D., and Tafel, H. 2009. History in the hands of the politicians: Lustration, civil society, and unfinished revolutions in Poland, Hungary, Slovakia, and the Czech Republic. *Student Summer Scholars*. Paper 27. http://scholarworks.gvsu.edu/sss/27.

Milčin, V. 2007. Фрустрација, лустрација, конфискација. *Utrinski Vesnik*. http://www.utrinski.com.mk/?ItemID=48F94F381686E648866702FEB9C76D76.

Milčin, V. 2010. Учеството на граѓанското општество во политичките процеси во Република Македонија. http://www.gem.org.mk/vesti/513-ucestvo-na-graganskoto-opstestvo-vo-politickite-procesi-vo-republika-makedonija.

Milčin, V. 2011. Комисијата за лустрација е света инквизиција. http://daily.mk/cluster3/3dff28b1f7eadc86e3860c7dc25876c3/860238/.

МПЦ нема ништо против, ИВЗ молчи. 2011. *Macedonia news*. http://www.macedonia-news.info/component/content/article/1999-mpc-nema-nishto-protiv-ivz-molchi.html.

MS Television (Producer). 2007. *Ова е Македонија*. Skopje: MS Television. [Video file]. http://www.ndm.org.mk/ndm/index.php?option=com_content&view=section&layout=blog&id=12&Itemid=114&limitstart=56.

Najčevska, M. 2007. Македонија не може да спроведе правична лустрација. *Utrinski vesnik*. http://www.utrinski.com.mk/?ItemID=63124C0E3D9A544F8C8A00B49E9E0141.

Najčevska, M. 2010. Кога една невладина организација ќе го изгуби компасот. [Web log comment]. http://kvadraturanakrugot.crnaovca.mk/2010/10/08/koga-edna-nevladina-organizacija-kje-go-izgubi-kompasot/.

Najčevska, M. 2011a. Препишано од Facebook: За прогонот на вештерки/вештери во македонското современие. [Web log comment]. http://kvadraturanakrugot.crnaovca.mk/2011/07/26/prepishano-od-facebook-za-progonot-na-veshterkiveshteri-vo-makedonskoto-sovremie/.

Najčevska, M. 2011b. Наш прилог во борба со неправото. [Web log comment]. http://kvadraturanakrugot.crnaovca.mk/2011/08/26/nash-prilog-kon-borbata-so-nepravoto/.

Najčevska, M. 2011c. Лустрацијата треба да биде злоупотреба на прогонот, а кај нас се претвора во злоупотреба на прогонот. [Web log comment] http://kvadraturanakrugot.crnaovca.mk/2011/01/18/lustracijata-treba-da-bide-progon-na-zloupotrebata-a-kaj-nas-se-pretvora-vo-zloupotreba-na-progonot/.

Nalepa, M. 2010. *Skeletons in the closet: Transitional justice in post-communist Europe*. Cambridge, UK and New York, NY: Cambridge University Press.

Одлука на Уставен Суд на Република Македонија. 2010. У. бр. 42/2008. У. бр. 77/2008 *Služben Vesnik na Republika Makedonija,* n.45, 3–8, April 1.

Parliamentary Assembly of the Council of Europe. 1996a. Resolution 1096 on measures to dismantle the heritage of former communist totalitarian systems. http://www.assembly.coe.int/Mainf.asp?link=/Documents/AdoptedText/ta96/ERES1096.htm.

Parliamentary Assembly of the Council of Europe. 1996b. Measures to dismantle the heritage of former communist totalitarian systems. Doc. 7568. http://www.assembly.coe.int/Mainf.asplink=/Documents/WorkingDocs/Doc96/EDOC7568.htm.

Parliamentary Assembly of the Council of Europe. 2006. Resolution 1481. Need for international condemnation of crimes of totalitarian communist regimes. http://assembly.coe.int/Main.asp?link=/Documents/AdoptedText/ta06/Eres1481.htm

Решение, У. Број: 76/2011-0-0. Constitutional Court of Republic of Macedonia. http://www.constitutionalcourt.mk/domino/WEBSUD.nsf.

Sadurski, W. 2003. *"Decommunization", "lustration", and constitutional continuity. Dilemmas of transitional justice in Eastern Europe.* European University Institute. [Working paper]. LAW. n. 2003/15. http://cadmus.eui.eu/bitstream/handle/1814/1869/law03-15.pdf?sequence=2.

Stan, L. 2006. The politics of memory in post-communist Europe: A comparative analysis. Conference *Expériences et mémoire : partager en français la diversité du monde,* Bucharest. (Conference paper). http://www.celat.ulaval.ca/histoire.memoire/b2006/Stan.pdf

Stan, L. (ed.). 2009. *Transitional justice in Eastern Europe and the former Soviet Union:Reckoning with the communist past*. London: Routledge.

Support for Improvement for Government and Management. 2009. Assessment, The former Yugoslav Republic of Macedonia, OECD and EU. http://www.sigmaweb.org/dataoecd/19/8/47078508.pdf.

Teitel, R.G. 2000. *Transitional justice*. New York: Oxford University Press.

Teitel, R.G. 2003. Transitional justice genealogy. *Harvard Human Rights Journal* 16: 69–94.

Uzelac, A. 2007. (In)surpassable barriers to lustration: quis custodiet ipsos custodes? In *Lustration and consolidation of democracy and the rule of law in Central and Eastern Europe*, ed. V. Dvořáková and A. Milardović, 47–68. Zagreb: Political Science Research Centre.

Vojnovska, O. 2007. МАНУ и МПЦ не се плашат од лустрацијата. *Utrinski Vesnik*. http://www.utrinski.com.mk/?ItemID=A22EF17843B964418C36A9A9A9D664F4.

Закон за определување дополнителен услов за вршење јавна функција. 2008. *Služben Vesnik na Republika Makedonija,* n.14, 2–7.

Закон за постапување со досиеја за лица водени од Службата за државна безбедност. 2000. Služben *Vesnik na Republika Makedonija,* n. 52, 3736–3737.

Chapter 5
How Activists See Civil Society and the Political Elite in Bosnia: Relevance to Prospects of Transitional Justice

Eunice Castro Seixas

Introduction

Between 1992 and 1995, Bosnia–Herzegovina (BiH) was devastated by a war in which serious crimes under international law were committed, resulting in mass killings and a large number of refugees. Following this 'bloody war', it was widely recognized that some kind of process for dealing with the past is vital to help communities recover from the wounds of the war and move towards a culture of peace and justice. This process has recently been labelled as 'Transitional Justice'. In this chapter, I argue for a bottom-up and contextualized conceptualization of transitional justice, which takes into consideration a multiplicity of local voices at the grassroots level. Non-Governmental Organizations (NGOs) in BiH are important actors in democratization and peacebuilding processes. They represent multiple voices whose discourses and practices can have great relevance for the building of a contextualized and bottom-up approach to transitional justice. The purposes of this chapter are: to critically explore the representations that NGO members hold regarding some of the main actors in BiH, civil society and the political elite; and to discuss the impacts for the prospects of justice and peace in the country. How are these actors perceived in the NGO world? Are they considered legitimate political or moral authorities to promote peace and justice? What are the consequences of these representations for Transitional Justice in BiH? These are the main questions considered in this chapter.

The structure of this chapter is as follows: I begin by presenting a review of literature regarding the conceptualization of transitional justice and its relevant actors; then I provide a description of BiH actual context; and subsequently I argue for the relevance of a view from the grassroots, which includes an analysis of how the main

E.C. Seixas (✉)
Center for Social Studies (CES), Coimbra University, Av. 25 de Abril, 86,
1° Dto. C. 4420-355 Gondomar, Coimbra, Portugal
e-mail: euniceseixas@gmail.com

O. Simić and Z. Volčič (eds.), *Transitional Justice and Civil Society in the Balkans*,
Springer Series in Transitional Justice, DOI 10.1007/978-1-4614-5422-9_5,

actors involved in transitional justice in BiH are discursively represented by NGO members. Afterwards I present my methodological approach and my data, followed by a discussion of these and final concluding remarks.

The Conceptualization of Transitional Justice and the Expansion of Relevant Actors

Transitional Justice has been defined as 'measures that are implemented in order to redress the legacies of massive serious crimes under international law' (De Grieff 2010:2) or as 'the full range of processes and mechanisms associated with a society's attempt to come to terms with a legacy of large-scale past abuses, in order to ensure accountability, serve justice and achieve reconciliation' (UN 2010: 2). These conceptualizations of transitional justice have several inherent problems. One is the tendency to identify transitional justice with a pre-defined list of measures, organized hierarchically into core and secondary elements, and supposedly linked to abstract and consensual notions of justice and peace. It is therefore necessary to critically examine the values and premises that support each and every approach to transitional justice, and to clarify the concepts of justice, peace and reconciliation in use, as well as assumptions about the nature of the social reality in which one is intervening, including how that reality may change. An important aspect of social reality that the preceding definitions do not take into account concerns the actors involved in transitional justice. Who designs and implements these measures? How are these actors perceived in the community? Are they granted legitimacy as political or moral authorities capable of promoting peace and justice? I suggest that these are fundamental questions for building a bottom-up, contextually grounded and emancipatory approach to transitional justice.

As the concept of transitional justice has evolved, it has become less focused on the role of states compared to its early phases. It has become more complex, integrating various levels of analysis (international, national and local) and different actors, including states, IGOs, NGOs, epistemic communities and global policy networks (Crossley-Frolick 2009). These important changes in the conceptualization of transitional justice are described and analysed by Teitel (2003) as part of a proposal for a genealogy of transitional justice. This proposal is structured along critical political cycles, divided into three phases. Phase I, in the post-war period after 1945, reflects the importance of international law in framing transitional justice, and has the Nuremberg trials as its most recognized symbol. This first phase of transitional justice was associated with interstate cooperation, war crime trials and sanctions, and ended soon after the war. With the post-Cold War transition began Phase II of transitional justice, which was associated with a period of growing democratization and political fragmentation. Because the form of transitional justice emerging in this period was linked with the rise of nation-building, it tended to rely on more local understandings of the rule-of-law, which stood in tension with a more transnational conception of justice.

According to Teitel (ibid.: 78), this post-Cold War model of transitional justice constituted a critical response to Phase I, but was not suitable for the globalization period and the interdependence of national and international politics. The contemporary and third phase of transitional justice, that emerged by the end of the twentieth century, has been associated with globalization, heightened instability and violence. Importantly, the relevant actors also changed 'from those with legal and political authority to those with moral authority in civil society,' including churches, NGOs and human rights groups (ibid: 83). The contemporary phase has encompassed a whole new range of actors, at local, national and global levels, and likely increased tensions amongst those.

> The current evolution of the transitional justice discourse, as it is associated with globalizing politics, implies a complex interaction of the international, national and local. In the present dynamic political context, multiple alternative forms of transitional justice emerge, involving a range of international, transnational, national and private settlements (Teitel 2002: 899).

The multi-level and multi-actor nature of transitional justice requires an examination of the power relations established in discourse and social practices, between these different actors and in the integration of these different levels. In this chapter I argue the importance of analysing the representations of relevant social actors of transitional justice, specifically civil society and political actors.

Contemporary BiH Context

Since the *General Framework Agreement for Peace* (known as the Dayton Peace Agreement) ended war in 1995, BiH has been under international regulation and the supervision of the 'Office of the High Representative' (OHR). Since 2002, OHR has combined this role with the Special Representative of the European Union. The present High Representative is Valentin Inzko, the seventh High Representative of the International Community to BiH. The complex political structure established by the Dayton Peace Agreement to accommodate the various warring factions of BiH (Serbs, Croats and Bosnian Muslims (Bosniaks)) has divided the society into two so-called 'entities': the Federation of Bosnia and Herzegovina (FBiH) and the Republika Srpska (RS). The Federation is predominantly Bosniak and Bosnian Croat, while the RS is predominantly Bosnian Serb. The Constitution (Annex 4 of the Dayton Agreement) established a central government with a bicameral legislature, a three-member presidency (consisting of a Bosnian Croat, a Bosniak and a Bosnian Serb), a council of ministers, a constitutional court and a central bank. The Dayton Peace Agreement is often criticized for having transformed BiH into a shared sovereignty, where supra-national organizations implement liberal peace and market economy, leading to an increased dependency on foreign aid and a lack of ownership and accountability, either in civil society, politics or economy (see for example Bieber 2002; Chandler 2000; Jansen 2006).

Important donors, like the European Commission or the USAID, have focused their efforts and investment on democratization through a strategy of building civil society. But the building of civil society in BiH has been associated with a top-down imposition of a liberal/pluralist idea of civil society, and has led to an exponential increase of associations, presumed independent from the State and from the market (Belloni 2001).

In terms of the present political context, the 2010 elections signalled the rise of the multiethnic party, the SDP, in the FBiH. However, Nationalist parties remain highly represented, primarily the SDA party in the FBiH and the SNSD in the RS. Milorad Dodik, from the Bosnian Serb nationalist party SNSD, was re-elected president of the RS with 50.52% of the votes. It takes 14 months for the new central government of BiH to be formed, and only at that moment will a national budget be able to be approved and loans from the IMF de-blocked.

For a Contextually Grounded Transitional Justice Approach in BiH: A View from Civil Society

The building of a civil society in post-war BiH has led to an exponential increase in NGOs that are presumed to be independent. The NGOs represent a multiplicity of voices including think tanks, media organizations, human rights and peacebuilding organizations and cultural associations. Although some of these NGOs have dubious links to the political elite and oligarchies, it is also true that a considerable (but unknown) number are working for democracy, human rights, peace and justice in the country. Considering the large number of NGOs in BiH, as well as their relevance to democratization and peacebuilding, it would seem natural that a transitional justice approach in BiH would focus on the multiplicity of voices at the grassroots. However, in BiH, the International Criminal Tribunal for the former Yugoslavia (ICTY) has been the key instrument for transitional justice (Armakolas and Vossou 2008; Zupan 2006). The credibility and perceived impact of the ICTY in the former Yugoslavia countries has been low. The ICTY is widely considered to have failed in promoting reconciliation in the region, and is not generally associated with justice within the former Yugoslavia (UNDP 2009: 17).

The lack of an integrative, holistic approach to transitional justice has contributed to transforming the ICTY into 'a world unto itself,' removed psychically, culturally, and politically from those who would live most intimately with its success or failure (Flechter and Weinstein 2004: 33; see also UNDP 2009).

These critiques point to the limitations of a focus on international law and retributive justice, and also, to the need for a more contextually grounded approach to peace and justice in BiH. One needs to take into account the diversity of local voices in BiH and move towards a more contextualized engagement with civil society, as is argued by Kappler and Richmond (2011), in relation to peacebuilding. Analysing the diversity of local voices from the grassroots is also important to guard against prevailing weaknesses of international community actions in BiH: romanticizing civil society; selectively reinforcing existing power structures; or failing to embed a social contract in cultural agency (ibid.).

Methodology

Design

To study perceptions and analyse discourses, the following qualitative methods were used: in-depth qualitative face-to-face interviews with members of human rights NGOs, 'ethnographic' observation[1] and document analysis (NGO reports and other documents, news related to human rights activities, etc.).

Setting and Sampling Strategy

The setting of the research is BiH, and specifically the cities of Sarajevo and Banja Luka. These were chosen because they are the capital cities of each of the entities of BiH, the FBiH and the RS respectively.

The sampling strategy was based on making use of several sources—enquiry within my host institution, research within academic documents (Masters thesis on the subject) and use of World Wide Web search engines—to develop a list of the most active/important human rights/peacebuilding NGOs. Due to the difficulties in reaching subjects through email and institutional contacts, I used the snowball sampling technique or 'respondent-driven sampling' (Salganik and Heckathorn 2004), asking interviewees for further referrals. I also engaged the help of a Sarajevo journalist to establish some of the contacts.

Data Collection Techniques

Data collection was primarily based on in-depth interviews with representatives of human rights NGOs. Open-ended questions covered the following topics: the history of each organization; its major goals and frameworks; major achievements and obstacles/difficulties; relationships with donors and the influence of international

[1] I spent 5 months in BiH and collected my thoughts and feelings resulting from that experience in a field diary, especially in relation to conversations and interactions with locals. Although this was inspired by the methodology of ethnographic observation, it was limited by my poor knowledge of the language.

community; civil society and human rights organizations in BiH; human rights discourses in BiH and perception of the future of human rights system in BiH; prospects of reconciliation and peace in the region and ways towards reconciliation. Following informed consent,[2] interviews were recorded where possible, and later transcribed and subjected to critical discourse analysis. Altogether 23 interviews were made, 14 of them with representatives of NGOs, the remainder with other relevant social actors such as politicians, academics and students.[3]

Additional sources of data consisted of a field diary, where I stated my impressions, feelings and analyses during my fieldwork, as well as other relevant documents (NGOs reports and other material, news, surveys, economic data from each country, etc.).

In this chapter, I focus on interviews with members of NGOs. However, I then refer to the data from the other interviews as well as from the ethnographic observation as a way to triangulate these findings during the discussion.

Relationship Between Researcher and Participants

The majority of interviewees did not have any objections to their interviews being recorded. The fact that neither I nor the interviewee was native speaker of English was taken into consideration in the analysis, by focusing more on the ideational meanings than on the formal linguistic aspects of the text. Also, as a way of dealing with possible language bias, during the interview I consistently reformulated the key ideas (including feelings) I had gleaned from my interviewee's responses in order to verify my understanding.

[2] Before each interview, I briefly explained the purpose of my research as part of my PhD studies at the University of Coimbra (Portugal), and assured interviewees that data would be used solely for the purposes of my research and in an anonymous way. Then, I would ask their permission to record the interview. Interviewees were encouraged to express their personal opinions and hence they may not represent the position of the NGO they are working for.

[3] Within the FBiH, I interviewed members of the following NGOs: the International Crisis Group (ICG), Transparency international BiH, Wings of Hope, Nansen Dialogue Center (NDC), Youth Information Agency (OIA), a member of both Scouts of BIH and Youth Press of BIH, a member of the banks' Syndicate; a Judge working for USAID Justice Sector Development Project II, Sarajevo; a local journalist working for the UNDP; a local journalist working for the Center for Media Development and Analysis; a member of the Association Pravnik and a former member of the Helsinki Committee for Human Rights in BiH. Within the Republika Srpska I only had time to conduct four interviews, all in Banja Luka: with representatives from the Helsinki Citizens' Assembly (hCa) Banja Luka, the Centre for Informative Decontamination of Youth, with one local journalist working for Radio Free Europe, and with one of the political advisors for the Office of the High Representative. Given the aims and framework of this study, variables such as age, sex and ethnicity were not considered as a priori categories. The urban focus of the research as well as the imbalance between the number of the interviews in the two entities is acknowledged as constraining the generalization of the results.

Analysis

The analysis followed a critical discourse analysis approach, using the framework and categories proposed by Fairclough (2003). In this chapter I focused on 'styles' or identities, specifically, the representations of civil society and the political elite and also gave particular attention to the way difference is represented as related to these identities.

Results

Representations Concerning People and Civil Society

Amongst the interviewees, 'people' and civil society in BiH were often represented as apathetic, lacking initiative to participate in the social and political life, and unable to take responsibility for their actions. A widely expressed explanation for this apathy was that people do not believe in the possibility of change in the country, and hence, their actions and activism cannot bring any change. In some discourses, this passivity was linked to the previous socialist mentality. In other discourses it was linked to a fear of consequences. Often these discourses were presented with a strong, angry, frustrated or resentful emotional tone. Within this representation of people and civil society, there were however important differences, contrasting urban and rural environments and also younger generations with older generations in BiH.

For instance, according to the representative of Transparency International BIH (TI BIH):

> And citizens are completely apathetic because after 16 years it is obvious that nothing significantly has changed in this country.
>
> That was the biggest challenge for civil society, how to convince the people that they are part of civil society too.

In this discourse, 'citizens' and 'people' are used interchangeably, and 'civil society' is presumed to be linked to a more active role in society, encompassing both NGOs and 'people'. The problem of convincing people that they also are part of civil society is seen as the biggest challenge for civil society (including for TI BIH). This is also increasingly difficult because, according to this interviewee, Bosnia has no tradition of civil society. The apathy of people is thus linked to both the lack of real change in the country in the last 16 years, which causes disbelief that changes are possible, and a lack of civil society tradition in BiH.

Another complaint of the apathy of people comes from a member of the 'Center for Decontamination of Youth Banja Luka' (CID Banja Luka):

> Now our target group is people between 25-40 years old (…) People are keeping in silence, not voting, not participating in civil society, there is a kind of apathy.

This evaluation of peoples' attitudes as apathetic is linked to the mission of the NGO. But the association between the lack of change in society and the mentality/attitude of the people appears also less explained by the passivity of people than by their nationalist attitudes, as expressed elsewhere by the same interviewee:

> Our society doesn't want any kind of changes, they want to be separated, that is the biggest problem (…) Everything is the same, we don't have reasons to celebration, maybe if you come to BiH again in 5-6 years and you can't find us, maybe something can change.

The interviewee expressed frustration and anger (noticeable also in their tone of voice), and again portrays BiH as a place of no change because 'society' *does not want any kind of changes* and prefers to be *separated.* 'Society' as used here is linked to the pronoun 'they', which suggests the interviewee's efforts to distance himself and the organization. In the last statement 'we' relates to the organization and thus affirms the value of its work against the prevalent nationalist mentality.

Another example of this discourse linking 'people' to a passive and even lazy mentality is one that argues that 'people are the problem' of the country. In the words of a local analyst working for the International Crisis Group (ICG):

> People are the problem because they don't understand it's really their problem and it's up to them, not just politicians.

This relates to the previous idea that people are not convinced that they are civil society too and that they also should take an active role in society alongside politicians and NGOs. This interviewee refers to this mentality as linked to the previous 'socialist mentality' that was held responsible for the perception that *many people don't feel to be in control in their own lives.* Instead of a feeling of agency, there is a recurring victimization of everyone. At the same time, the interviewee attempts to establish a distinction between himself and the majority of people (especially the younger generation of BiH), by stating that, for him, it was never an option to do nothing, while the majority of people in BiH live their life according to the Bosnian saying *You can never pay me as little as little as I can work.*[4] He also expressed resentment regarding the way that he feels that young people in BiH regard him.

> The only thing that for me was never an option was to sit down and do nothing; that was never an option.
>
> They (young people) see my salary, my car, but they don't see how much I work.

This idea that people in BiH complain but fail to take initiative, do not want to work to improve things, and resort to victimization, is common, and emerged also in my interviews with representatives of foreign NGOs. For example, an interviewee from the United States working at the NGO Center for Women and Youth Development c/o Wings of Hope BiH stated:

> They complain but don't want to do nothing.

[4] In the original Bosnian language this saying is: 'Ne možeš me platiti tako malo, koliko malo ja mogu raditi'. The translation to the English language was made by my interviewee.

In this statement 'They' refers broadly to 'Bosnians' in a general, homogenizing, and essentializing manner. This sits in contrast with the portrayal these activists give of their project participants, a group of teenagers described as an educated and a 'special' group who do not represent the majority of the Bosnian.

Similarly, an American judge working for USAID mentioned the reluctance of Bosnians to take responsibility for their actions. Although this statement is presented with some modulation (*seemed to be*), it is contradicted by the expression *a lot*.

> There seemed to be a lot of reluctance to take responsibility.

At the same time, my interviewee recognizes the structural obstacles for those who want and try to take responsibility for their actions, referring to the post-Dayton governmental multi-level structure:

> A structure that frustrates those who try.

At the end of the same interview, the respondant acknowledged the efforts of the Bosnians, but at the same time, took a moral standpoint, evaluating Bosnians as deserving of American help.

> They (Bosnians) are courageous and they are trying to make things better and they deserve our help.

In this evaluation, the opposition between 'they' and 'us' is noticeable, placing 'us' (which in this case is an ambiguous term that may represent the USAID, the United States or even the international community) with the power to morally judge the Bosnian 'other' and to decide if they do in fact deserve help.

Fear is another explanation given for the passivity of civil society in Bosnia, as one can see from the statement of a member of the NGO 'Center for Media Development and Analysis':

> If you are scared for your life you won't be interested in activism.

Insecurity and fear as explanations take the blame away from the citizens, focusing attention instead on those who contribute to promote fear and insecurity in society.

Although negative representations of people and civil society as 'lacking' (strength, initiative, responsibility, etc.) seemed to prevail, some interviewees also expressed more positive representations. There was, for instance, a more encouraging representation of 'the people' of BiH, one linked to a temporal relation of difference, contrasting the time immediately after the war to the present time, and focusing on the change of mentality for the better. For example, my interviewee from Helsinki Citizens' Assembly (hCa) Banja Luka, when asked about the major achievements of the NGO, stated:

> Maybe the change in the attitude of people. After the war it was really hard.

Her statements show a relationship of difference contrasting two moments: after the war and now. The change in the people's attitude is mostly noticeable in their perception of human rights NGOs. There is also a narrative which legitimizes this claim, when my interviewee proudly told me about a fundamental social event: the first inter-ethnic meeting of women after the war in 1997, which was organized by the Helsinki Citizens' Assembly (hCa) Banja Luka:

> We organized one of the first meetings between Serbian, Croats and Muslim women, inter-ethnic, in the early beginnings. That time it was quite dangerous, really.
>
> NGOs were seen as traitors, spies for IC, money laundry—after the war that was the perception. (…) It is now different.

This idea is reinforced by other activists' accounts of the perception of the human rights NGOs in the past. For instance, a former member of the Helsinki Committee for Human Rights in BiH gives the following description:

> We had all kinds of problems, all kinds of labels, during these 15 years, we were accused of working for Milosevic, for XXX for the international community (…) all kinds of labels that were possible.

The same interviewee states that:

> What is clear is that civil society is not at this point as strong as to be able to make the difference. But it is becoming stronger and stronger.

There is thus an accent on the positive developments of civil society, while acknowledging that civil society is still not capable of making a difference in BiH. The same interviewee also describes how civil society in BiH started as an anti-war movement, active both during and after the war, and has played a crucial role in social support and protection of women and victims' rights. But most important, civil society is seen as fundamental in promoting the idea of a multicultural 'fourth Bosnia'.

> Interviewee: Civil societies as well as multiethnic parties were kind of trouble makers. Because we were coming up with a new concept. They prefer the concept that there are 3 parts in Bosnia; there are Serbs, Croats and Bosniaks. But no, there are also citizens! We come with a more complicated idea, yes of course the respect for the 3 ethical groups, but it's not the only basis for a country, has to be also a country of citizens, of a real democracy and ethnical rights cannot block all the reforms.
>
> Interviewer: Actually maybe you're saying that international community in Dayton kind of neglected ideas from civil society?
>
> Interviewee: Exactly, exactly! They considered that Bosnia is only what they heard from nationalists. What we call the fourth Bosnia …
>
> Interviewer: The 'fourth Bosnia'?
>
> Interviewee: The fourth Bosnia, a Bosnia that is European that is not infected by nationalists, that is cherishing human right values, democratic values so on. We are the results of elections, these are the winners, and these are the losers. … And that was a suicidal idea.

This strong critique to the international community in BiH ('They') who, according to my interviewee, considered civil society a 'trouble maker' and neglected the multiethnic idea of a 'fourth Bosnia', clearly states the importance and trust in civil society as a force to promote democracy, multi-ethnicity and multi-culturalism. The international community in Dayton is accused of taking the simplest option, by negotiating with nationalists and ignoring the 'more complicated idea' coming from civil society. Also, there is a strong critique of the idea that democracy is only about elections, while human rights and democratic rights are not being cherished.

Relations of Difference

Civil society is not always represented in a homogenized or essentialized manner and some fractures are visible in the discourses of my interviewees. There are two main relations of difference regarding civil society: one contrasts urban vs. rural civil society; another establishes a difference between the younger generation and the older generations in BiH.

Urban Versus Rural Civil Society

Civil society in the urban area is often contrasted to civil society in the rural area, in which the latter is described as weak and showing a lack of initiative and support. There is the assumption that cosmopolitanism is necessarily urban. Here is one example from an interview with a former member from the Helsinki Committee for Human Rights BiH:

> Interviewer: What about CSO (Civil Society Organizations) or movements in the rural area?
>
> Interviewee: Civil society is very weak in the rural area. There is a lack of initiative, there is a lack of support, but there is also a very strong pressure on the part of the government.
>
> Interviewer: So, they don't have a chance. It's like CSO only have a chance in a multi-cultural environment.
>
> Interviewee: Exactly, Exactly. It doesn't mean we are better here no, but it means the atmosphere is much better.

Younger Generation Versus Older Generations

The accentuation of difference between the younger generation in BiH compared to the older generations is usually framed as 'the problem of the younger generation', who grew up in BiH after 1992 and hence in a divided society. A local journalist working for a media organization in Sarajevo ('Center for Media Development and Analysis') explains this in the following manner:

> From 1992 till today, everyone younger than 20 years is brainwashed. It's completely different system of values, of education. Those kids never lived together with other ethnical group. Those kids, they don't have firsthand experience in living together. Day by day in the family, in the school, they are just listening how the other are cheating them, how the other are the enemy, how the others are prepared to kill them. I am really scared what is in the minds of those kids.
>
> If you tomorrow have a very high tension, those people will start a fight. (Interviewer: The young people?) The young people. Who is going to war? Only young people go and fight to die. Who has ever sent pensioners to work? (laughs) You need the young people, Young people are easier to manipulate. They want to prove themselves. And if you create an atmosphere that they are heroes and so on they will be very happy to go to work.

This interviewee is clearly expressing his concerns and even fear regarding the future in BiH, when the older generation disappears and the country is left to this generation that grew up during or after the war, with a totally different value system.

Another journalist working for Radio Free Europe in Banja Luka makes similar claims although he adds, perhaps as an excuse or explanation, that the youth are depressed:

> Youth is depressed and divided by ethnical lines.

We have already seen how the younger generation is sometimes accused of being lazy and unwilling to work for the fulfilment of their objectives (cf. *supra*). Nevertheless, in one case, the younger generation was mentioned in a positive sense, by an interviewee from the NGO Pravnik:

> We them (young people), we still have hope.

The idea expressed here is that with the young there is still hope possibly because there is still time and enough opening/flexibility to change.

Representations of the Political Elite

The political elite were represented exclusively in a homogenized and negative way, as the main group responsible for the lack of progress in the country and for creating a divided society, unable to reach any consensus and interested only in remaining in power and keeping their own benefits. In these representations, difference is accentuated between the interests of the political elite vs. people's interests; the 'political life' vs. the 'real life'; the politicians vs. the NGOs.

For instance, a member of the Pravnik organization mentioned the existence of strong political differences and nationalist ideas that divide society and create an obsession with three-sided approaches.

> If you are watching like media or reading newspapers we still have strong, strong political differences and really strong nationalistic ideas that are running the country. And division doesn't really help that much and also to me to have everything divided in the country (…) You know those are all the signs that we are pretty much obsessed with, you know, with three-sided.

The idea that information in BiH is always influenced by a nationalist framework was also suggested by an interviewee from USAID:

> I don't think they (people in BiH) have a lot of information and the information they have is colored by their ethnical group (regarding public acceptance of the justice system).

The same interviewee also refers to the election year (2010) as the most difficult period to bring about change because politicians were afraid to take controversial decisions which might have harmed their campaigns.

> And then, this year has been really, really frustrating because it's an election year. They are afraid to do something that may be controversy. So, it's really a difficult time to make changes.

The representation of the political elite often elicits anger, as noticeable in the statement of the interviewee from the 'Youth Press of BiH', regarding the problem of visa restrictions in 2009.

> I am very angry about this, with our politicians.

The politicians' image seems to be linked to the blocking of any reforms or important changes in the country. Lack of political will is another common way to refer to this problem and this is linked to the manipulative strategy of accusing others to distract attention from the serious questions, as an interviewee from Transparency International suggested:

> It's obvious the lack of political will. And they are using these accusations and threats against TI in order to distract attention from other serious questions.

Relations of Difference

Relations of difference concerning politicians accentuate differences between these social actors and civil society actors. For instance, a member of the Helsinki Citizens' Assembly (hCa) Banja Luka suggests that the nationalist political discourses do not represent the 'real life' and social relations. The 'political life' is thus contrasted to the 'real life'.

> There is difference between real life and political life, in reality there is pretty good cooperation, not in every sphere—people that return to their houses they still have trouble finding a job; now you can travel everywhere (…)

Although it is acknowledged that cooperation is not generalized to all spheres, the activist gives examples from her work in the hCa Banja Luka, reporting, for instance that:

> From all this period and work with women I never noticed any kind of animosity (…) There are issues that are common to women.

On the other hand, she sees political discourses as almost unchanged since the end of the war and still influencing people:

> But the political national discourses are almost the same, haven't changed and people are influenced by these discourses, hate speeches.

As 'political life' is contrasted to the 'real life', so the political leaders are contrasted to the people of BiH. The political elite are represented as working for their own interests and against the people, as was stated by the interviewee from the International Crisis Group:

> Political leaders first and foremost represent their political parties' interest; political leaders are working against the BiH population.

There was one exception: the interviewee from the Radio Free Europe, at Banja Luka. Instead of accentuating the difference between politicians and the people of BiH, he diminished this difference, by putting these actors on the same level:

I hope people will have a better government that they deserve.

This is a strong statement, one that can be associated with the previously analysed discourse portraying people as 'the problem'.

Additionally, one can also find an accentuation of difference between politicians and NGOs in BiH, with the two groups perceived as working for opposite objectives. An example is when a member of the NGO Youth Information Agency (OIA) contrasted the objectives of OIA to those of BiH politicians. One could schematize this relation as the following:

Politicians	OIA
Nationality discourses	Unite people
Make differences	

That's what politicians say, you know. We want to change that and connect people again. We just want to change the political atmosphere; because what they are doing (…) doesn't work, you know? We still need visa to go from one place to another (…)

The idea of change as a goal appears here in a pronounced way: to change the nationalist discourse, change the political atmosphere and unite people. The activist identifies himself with the NGOs, and there is a clear opposition between 'we' (the NGOs) and 'they' (the politicians).[5]

Real threats to the security of NGO members are the most serious sign of this 'struggle' between politicians and NGOs fighting for human rights, as the interviewee from Transparency International BiH shows, in his reference to Milorad Dodik's accusations against the NGO and the resulting temporary suspension of their activities in Banja Luka.

That was a very problematic thing when we were forced to suspend our activities in Banja Luka. He officially admitted that his advisors and he personally collected evidence against us.

Yes, the question is resolved after IC reacted (…) And this stopped in such drastic way But for example last year Dodik adopted official document about conspiracy against RS, international conspiracy. (laughs) That was not very pleasant. We ask ourselves after that what is the next step? I mean to arrest us or? What? (laughs)

From these statements it is clear that the problem has not been resolved and the NGO still expects to be the target of further threats in the future.

Discussion

The findings of this research reveal that civil society actors hold negative views of the Bosnian political elite: politicians are distrusted, seen as working for their own interests and blocking the possibility of positive changes in the country. Furthermore,

[5] In 2009, the problem of visa restrictions impeding BiH citizens from travelling to the EU was a prevailing concern.

they continue to be linked with the promotion of nationalist discourses and even pose security threats to civil society actors. As transitional justice presupposes an agreement between the three ethnic political elites and a certain degree of trust in the politicians, this representation may be considered a serious obstacle to transitional justice in BiH. These findings were corroborated by data from ethnographic observation and interviews with other social actors. Popular jokes concerning local politicians also indicate the amount of distrust, and sometimes anger, attached to these actors.

Civil society has more complex and nuanced representation, although most of the associations are also negative. 'People' and 'civil society' in BiH are often characterized as passive, apathetic, weak, depressed or lazy. Nonetheless, there are relations of difference in terms of the generations and rural vs. urban area. There are also some positive representations: descriptions of progress in people's mentality compared to the period immediately after the war; showing that even during the war, civil society was linked to the defense of peace and multiculturalism, although these ideas were neglected by the international community at Dayton; and expressions that civil society is getting stronger and that the political discourses do not represent 'real life'. These discourses point to the idea that some positive changes have occurred in civil society since the end of the war, but these are limited improvements are not enough to build a culture of peace and justice in BiH. Interestingly, from ethnographic observation and interviews with other actors, I also heard the opposite idea: that civil society was more active during the war. This could mean that members of NGOs, in their efforts to promote their work and show their good results, give a more positive picture of the social reality. During my fieldwork I noticed a recurrent feeling of pessimism and depression associated with the lack of change in the country and little confidence in potential changes. This feeling also emerged from the interviews and is possibly, one of the major obstacles to transitional justice in BiH.

Concluding Remarks

The findings of this study reveal prevailing negative representations of the political elite and more nuanced representations of civil society in BiH. It seems that any hope of positive change and moving towards a culture of peace and justice in the country will happen only in spite of the political elite's lack of will and people's apathy. There is also a common belief that only in the urban area can there be any drive to change. The participation of the younger generation in any transitional justice project remains a controversial question, given the different discourses regarding the young. One interesting aspect of the results is how data from both local and international actors included in the sample were consistent. In fact, most of these representations resonate with international organizations' discourses on civil society and the political elite in BiH. For instance, in USAID Country Assistance Strategy for Bosnia and Herzegovina 2009–2013, it is stated that (author's italicization of key words):

> Civil society organizations, which should be best suited to put pressure on government to reform, are weak and inexperienced, and strong, articulated public opinion is lacking to support their pressure (....) Most organizations are small and inexperienced and lack a stable domestic financial base, making them dependent on donor funding.

And also,

> Only *when citizens are confident in their country's institutions* and feel that they effectively promote and protect their interests will the country reach long-term stability.

And regarding the political elite in BiH:

> The presence of an international overseer, the Office of the High Representative (OHR), has ensured stability, *but has not instilled in local politicians the ability and will to implement reforms on their own.*

Thus, images of 'lacking' characterize both civil society and citizens (lack experience, size, strength, confidence in their institutions) and the political elite (lack the ability and will to implement reforms on their own). This correspondence between international and local representations surely merits further research to clarify its significance.

However, one has to remember that these representations constitute discourses and discourses are only part of the reality. If it is true that the way these social actors are represented by the members of NGOs may influence relations and cooperation, as well as subsequent transitional justice initiatives, it is also true that NGOs necessarily have to work with these actors in the pursuit of their projects. And from this encounter better things may come. Further research should focus on the social practices of the work of NGOs, specifically, in transitional justice. How are power relations between these actors negotiated in practice? How do practices influence representations and vice versa? The comparison of these practices and relations of power in urban and rural spaces would also be an interesting option of research.

Regarding representations, for reasons of space, I did not include here my data regarding the representations of the media or of the international community's powers and presence in BiH. These are clearly other fundamental actors in transitional justice. It would also be important to study local and international discourses about peace and justice and relate these to identities of the relevant actors.

As a concluding remark, I hope this chapter contributes to broaden and stimulate research promoting a contextually grounded and bottom-up approach to transitional justice, by taking into consideration the multiple voices and experiences from the grassroots.

References

Armakolas, Ioannis, and Vossou, Eleni. 2008. Transitional justice in practice: The International Criminal Tribunal for the Former Yugoslavia and Beyond. *UNISCI Discussion Papers*, 21–58.

Belloni, Roberto. 2001. Civil society and peacebuilding in BIH. *Journal of Peace Research* 38(2): 163–180.

Bieber, Florian. 2002. Aid dependency in Bosnian politics and civil society: Failures and successes of post-war peacebuilding in Bosnia-Herzegovina. *Croatian International Relations Review,* January-June, 25–29.

Chandler, David. 2000. *Faking democracy after Dayton*, 2nd ed. Pluto Press: London.

Crossley-Frolick, K.A. 2009. The European Union and transitional justice: Human rights and post-conflict reconciliation in Europe and Beyond. Paper presented at the *Eleventh Biennial International Conference, European Union Studies Association.* Los Angeles, CA, 23–25April 2009.

Fairclough, Norman. 2003. *Analysing discourse: Textual analysis for social research.* London: Routledge.

Fletcher L.L. & Weinstein, Harvey M. (2004). A world unto itself? The application of the international justice in the former Youguslavia. In Eric Stover and Harvey M. Weinstein (Eds.), *My Neighbour, My Enemy: Justice and Community in the Aftermath of Mass Atrocity* (pp. 29–48). Cambridge: Cambridge University Press.

Jansen, Stef. 2006. The privatisation of home and hope: Return, reforms and the foreign intervention in Bosnia-Herzegovina. *Dialectical Anthropology* 30: 177–199.

Kappler, Stefanie, and Oliver Richmond. 2011. Peacebuilding and culture in Bosnia and Herzegovina: Resistance or emancipation? *Security Dialogue* 42(3): 261–278.

Salganik, M.J., and D.D. Heckathorn. 2004. Sampling and estimation in hidden populations using respondent-driven sampling. *Sociological Methodology* 34(1): 193–239.

Teitel, Ruti G. 2002. Transitional justice in a new era. *Fordham International Law Journal* 26(4): 893–906.

Teitel, Ruti G. 2003. Transitional justice genealogy. *Harvard Human Rights Journal* 16: 69–94.

UN. 2010. Guidance note of the Secretary-General. United Nations approach to transitional justice. http://www.unrol.org/files/TJ_Guidance_Note_March_2010FINAL.pdf. Accessed 27 April 2012.

UNDP. 2009. Transitional justices guidebook for Bosnia and Herzegovina: Executive summary. http://www.undp.ba/upload/publications/executive_ENG_WEB.pdf. Accessed 20 October 2010.

Weinstein, Harvey M., and Eric Stover (eds.). 2004. *My neighbour, my enemy: Justice and community in the aftermath of mass atrocity*. Cambridge: Cambridge University Press.

Zupan, Natascha. 2006. Facing the past and transitional justice in countries of former Yugoslavia. In *Peacebuilding and civil society in Bosnia-Herzegovina. Ten years after Dayton*, ed. Martina Fischer, 327–342. Münster: Lit.

Chapter 6
The "Transitional Citizen": Civil Society, Political Agency and Hopes for Transitional Justice in Bosnia–Herzegovina

Briony Jones, Alex Jeffrey, and Michaelina Jakala

Introduction

Current approaches to transitional justice, while still retaining a central focus on questions of judicial responses to human rights abuses, repressive regimes and war, often include additional elements within their conceptualisation and policies. "Thicker" understandings of transitional justice (McEvoy 2008) which may include memorialisation, public apology and forgiveness connect the work of courts to a broader context of far reaching social, economic and political transformations. Increasingly, transitional justice is seen as part of liberal peacebuilding and thus connected to rule of law, good governance and democratisation processes (Sriram 2009, p.112). It is here that we come to the citizen, present in transitional justice assumptions, practices and outcomes. This chapter is concerned with the nature of the "transitional citizen", a subject which is implied in many transitional justice interventions and in much of the scholarly literature on their framing, practice and impact. However, "the citizen" is under-theorised in the literature and remains in many ways implicit, assumed and in the background behind more instrumental discussions of the workings and relevance of particular mechanisms. It is the argument

B. Jones (✉)
Dealing with the Past Program—swisspeace, Sonnenbergstrasse,
17 PO Box 3000, Bern 7, Switzerland
e-mail: briony.jones@swisspeace.ch

A. Jeffrey
Department of Geography, The University of Cambridge, Downing Place,
Cambridge, CB2 3EN, UK

M. Jakala
The School of Geography, Politics, and Sociology, Newcastle University,
Daysh Building, Newcastle upon Tyne, Tyne and Wear, NE1 7RU, UK

O. Simić and Z. Volčič (eds.), *Transitional Justice and Civil Society in the Balkans*,
Springer Series in Transitional Justice, DOI 10.1007/978-1-4614-5422-9_6,

made here that we need to know more about the type of citizen which is implied either implicitly or explicitly within transitional justice and what this means for civil society and political agency in societies undergoing transition.

The wealth of scholarship on citizenship, which cannot be fully discussed within the confines of this chapter, addresses not only the question of which rights and obligations are bestowed on individuals but goes further into analysing the ways in which citizenship is lived, practiced and made more substantive in context by citizens themselves. Indeed, citizenship is more than a status, but is also an identity and practice, and thus something which can be claimed through struggle and contestation (Hoffman 2004, p.1; Kabeer 2005, p.1; Lister 2003, p.3; Faulks 2000, p.7). What becomes clear when reading the literature is that citizenship is in many ways a malleable concept, capable of taking on new shapes according to which voices are trying to be heard (Jones 2009, p.60). If we see citizenship as a contested space then the question of "who should be included?" (Gibney 2006) i.e. "who is the citizen?" is one of fundamental importance. The citizen can be shaped by socio-economic contexts, cultural identification, gender and sexual orientation, political structures and experiences of violence, such that it exists not only in practice (who has the status of citizen and who is able to practice their citizenship and claim the rights this entails) but also in aspiration and in imagination. In this sense the citizen can have more or less agency in shaping the substance of citizenship in context, but it can also be instrumentalised within political discourses and national myth.

In the case of conflict and transitional justice the citizen is often central but usually under-theorised. Whilst there is a developed body of literature analysing the importance of contestations over citizenship for dynamics of conflict (see for example Mamdani 2001 and Adejumobi 2001) the role of the citizen in transitional justice is not as well understood. During transition the contexts in which the citizen is defined, both in terms of the legal and political definitions of who is included and in terms of the possibilities for practicing and living such a citizenship, undergo transformations at the same time as a consensus is sought over who can and should participate in the "new" society and what form such participation should take. Looking at transitional justice interventions it is clear that there is often an underpinning assumption of social transformation through which violent individuals or groups should (and could) become willing to coexist peacefully, to observe the rule of the law, and to practice a civic virtue which connects them to each other and the state. This is a normative project of which a particular understanding of the citizen is a part.

For some, citizenship can be used during transitions as a uniting and equalising force. Porter writes of reconciliation in Northern Ireland that, "Despite our differences there is no good reason why commitment to a common citizen dignity should not translate into a common civic identity. And if we care about the society in which we live, there are compelling reasons why it should" (2003, p.172). Here the citizen is mobilised as an agent of unity and holds the promise of an identification which can transcend divisions on which violence has been based. Porter acknowledges that this is complicated in contested spaces of transitional societies but suggests that fostering civic virtues such as forgiveness, magnanimity and reasonableness can go towards it (Ibid, p.67 and p.99). A danger may come when citizenship is hollowed out by policies which restrict active practice, as some have claimed to be the case in Rwanda. According to Buckley-Zistal (2006) a particular citizenship discourse has been mobilised by the current government of Paul Kagame, in order to force a de-ethnicisation of the public space in the name of unity whilst at the same time it supports a particular narrative of the past and future. This creates a public space in which what it is to be a Rwandan citizen is to some extent strictly controlled from the top down.

Recognising that citizenship is not only a core part of transition from violence to peace, but that it can also be instrumentalised in particular ways to achieve certain normative goals, we are prompted to ask how "the citizen" is shaped by, and shapes, transitional justice processes. There may be official policies and discourses of citizenship, such as in the Rwandan case, which link questions of citizenship to transition. But there are also the practices of citizens who themselves shape their own social and political membership in a way which is meaningful for them, and which interacts with transitional justice processes.

Van de Merwe et al. point out that actors will hold different perspectives on transitional justice which may also include "deep dissatisfaction with, and alienation from, the process" (2009, p.3). As Newman has written, "dealing with past abuses of human rights is as much about politics and trade-offs as it is about justice and accountability, especially in post-conflict and transitional situations" (2002, p.32). There has been an increasing emphasis placed by transitional justice scholars on trying to understand the nature of these contestations and the agency of citizens themselves, for example the recent edited collections on localising transitional justice (Shaw et al. 2010) and transitional justice from below (McEvoy and McGregor 2008a, b). Such work highlights the importance of context, agency and the everyday. This is important on two counts. Firstly, there is an instrumental argument: in order for particular transitional justice interventions to achieve their goals we need to understand how they will be received and reacted to by those who are the subjects of those interventions. Secondly, there is a more normative argument: it is important that those people who have been affected and marginalised by violence have not only a stake in the transition, but that their influence and agency is understood and recognised. Scholars of transitional justice have written about how discourses of law and rights tend to create regimes of truth which produce silences and legitimise certain types of speech (Wilson and Mitchell 2003, cited in Akanji 2011), and how in this way law can be complicit in a renewed disenfranchisement (Campbell and Turner 2008, p.381). With this danger in mind it is important to ask who determines the form and content of transitional justice? For whom does transitional justice take place? And how do the acts of citizens shape, and in turn are shaped by transitional justice processes?

Questions regarding the placement of the citizen within understandings of transitional justice are cast in sharp relief in contemporary Bosnia and Herzegovina (hereafter BiH). After the signing of the Dayton General Framework Agreement for Peace (GFAP) in 1995 there has developed a tension between a liberal democratic understanding of the citizen with a civic identity which can override ethnic division, and the ethnicisation of the political space. Mujkić (2008) has described BiH as an ethnopolis in which the citizen is first and foremost a member of an ethnic group. Voting rights, standing for political office and many other aspects of citizenship are directly connected to ethnic group membership and residency in one of the two entities which now form the country: the Federation of Muslims and Croats, and Republika Srpska. However, actors such as the Office of the High Representative (OHR—leading international authority in BiH) have at the same time supported a nation and state building process which is designed to develop consensus over an understanding of a Bosnian citizen (Sejfija 2006). Secondly, opinion surveys suggest that Bosnians themselves do not necessarily support a civic integrationist formula, and in addition Bosnian-Croats and Bosnian-Serbs are less likely to identify with the current BiH state (Ó Tuathail and O' Loughlin 2009). The political space in post-war BiH is thus to some extent defined by dichotomised debates over the nature of the polity and the demos. This has implications for the agency of the people of BiH. Some observers have suggested that there is a need for a development of a civil society which can engage and identify with the state (Fischer 2006), others have identified knowledge about participation opportunities as a key challenge for democratic participation (Bajrovi 2005), and others have asserted the need for more grassroots civil society activity to challenge the dominance of internationally funded non-governmental organisations (Fagan 2005; Jeffrey 2007).

This chapter began with the assertion that much more needs to be known about the nature of the citizen during times of transition. Given the contemporary background sketched above it is important to ask how transitional justice discourses and practices in BiH intersect with the contested nature of political identity and citizenship practice in BiH. The formal legal response to the human rights abuses and war crimes that were committed during the fragmentation of Yugoslavia in the 1990s has been led by the International Criminal Tribunal for the former Yugoslavia (ICTY) in The Hague. However, since 2005 the establishment of the State Court of Bosnia and Herzegovina (CBiH) has provided a new avenue for legal redress for victims of crimes during the conflict in BiH (1992–1995). The instrumentalisation of mechanisms of transitional justice as a state building practice is encapsulated by the establishment of this new judicial institution. The CBiH is much more than simply "localised" international law, since its creation marked a rare centralised form of authority in a state where power has been principally devolved to the two sub-state "entities" (the Muslim Croat Federation and the Republika Srpska). The creation of the CBiH was dependent upon the imposition of a *Law on the State Court of Bosnia and Herzegovina* by the then High Representative Wolfgang Petrisch in 2000.

In this chapter we will assess how this new legal institution casts light on the transitional citizen, exploring in particular the contested role of civil society within the implementation of programmes of transitional justice. As argued elsewhere, the establishment of the CBiH required a process of "public outreach": deliberate policies aimed at fostering participation within, and understanding of, the aims and activities of the CBiH (see Jeffrey 2011). Focusing on practices of public outreach identifies the significant role played by civil society organisations in fulfilling some vital aspects of citizenship entitlement. For example, it is envisaged that such schemes provide the opportunities for individuals and groups to participate in legal proceedings, give voice to concerns over the nature of the judicial process and articulate alternative concepts of justice after violent conflict. But the analysis extends beyond this normative placement of civil society as agents of popular engagement. The chapter moves on to explore how certain practices of civil society may constrain or limit the possibilities for redress, through the establishment of bureaucratic or administrative barriers that limit individual rights. In a sense, we see in these later examples of attempts to support survivors of wartime sexual violence the tension that exists in practice between civil society and citizenship. We would argue that recourse to the potential of civil society carries with it a contradiction: that while the organisations of civil society offer opportunities for a wide range of individuals to observe and participate in legal processes of transitional justice, these informal mechanisms are not necessarily emancipatory or inclusive. We would argue that this illuminates a tension between civil society and the citizen, where the opportunities inherent in associative life may only be fully realised in a context in which citizenship is able to be practiced in a way which is meaningful for the citizens themselves, and where citizenship status is matched by a substantive realisation of rights.

The arguments made during this chapter are informed both directly and indirectly by the empirical work of each author in BiH: Briony Jones in 2007, 2008 and 2011; Alex Jeffrey in 2007 and 2009, and Michaelina Jakala in 2006, 2007 and 2008. The two specific projects which are analysed directly were conducted by between 2009 and 2012. The first, carried out by Alex Jeffrey and Michaelina Jakala, is funded by Economic and Social Research Council (grant number RES-061-25-0479) and explores establishment of war crimes trials at the CBiH. To date, this research has involved 22 semi-structured interviews with officials within the CBiH, members of local and international NGOs, and employees of donor and intergovernmental agencies such as the Organisation for Security and Cooperation in Europe (OSCE) and the United Nation Development Programme. This research has been predominantly conducted in Sarajevo, though interviews have also been undertaken in Mostar and Bijeljina. The interviews have been conducted in Bosnian and English and, where necessary, translated by the authors. In addition to the interviews, the research has involved monitoring key trials at the CBiH, participating in NGO-organised events and attending workshops and seminars directed at sharing information concerning the war crimes trial processes in BiH. The second project the chapter draws upon was a series of life history interviews conducted in 2006–2008 as part of doctoral research by Michaelina Jakala with survivors of wartime sexual violence in BiH. The qualitative approaches of semi-structured interviews and participant observation have been selected by the research team in order to convey the plural accounts of public outreach from the CBiH that circulate in contemporary BiH. Attentive to forms of practice, these approaches allow an account of the lived experience of policies of transitional justice and forms of choices and deliberation that shape their implementation.

The chapter is divided into three further sections. In the following section we explore the history of the establishment of the CBiH and in particular identify its normative placement as an agent of new concepts of legal citizenship in BiH. In the second section we examine how the CBiH has attempted to engage with the public in BiH and beyond through practices of public outreach: initiatives that have invested particular faith in the role civil society as a "bridge" between legal institutions and a wider public. This section draws on the empirical research to view these initiatives from the perspectives of the civil society groups involved, testimonies that reflect on both the diversity of civil society as a concept and the range of perhaps unforeseen outcomes that have emerged from public outreach initiatives. The third section explores another articulation of civil society in contemporary BiH: the associative support available to survivors of sexual war violence. This work illuminates the ways in which just as civil society organisations invent new opportunities for individual citizenship practice, their scale and capacity can sometimes limit the reach of their support. This allows us to explore a conclusion that advances new sightlines for research exploring "the citizen" within transitional justice.

The Court of Bosnia and Herzegovina

There are a number of different avenues available for the prosecution of war crimes which took place during the 1992–1995 conflict in BiH. High-level perpetrators are tried at the ICTY in the Hague, Netherlands while the majority of the mid-to lower-level cases are tried at the state-level institution of the War Crimes Chamber (WCC) of the CBiH which is located just outside the centre of Sarajevo. There is also the possibility for the lower-level cases to be tried at the entity and municipal levels within the ten Cantons of the Federation of BiH, five district courts of Republika Srpska and the Basic Court of the Brčko District (Ivanišević 2008, p.5).

The CBiH is the first institution at state-level to deal with criminal offences within the jurisdiction of BiH. In 2003, the CBiH established the capacity to deal with organised crime, economic crimes and corruption cases. That same year the Criminal Procedure Code of BiH was established under which crimes such as crimes against humanity, genocide and war crimes were defined[1] and the ICTY and Office of the High Representative (OHR) agreed that there was a need to establish a local (i.e. state-level) WCC. The establishment of the WCC played a role in the ICTY completion deadline goal of completing all high-level cases. Furthermore, there was a need to localise the war crimes trials by bringing mid- to low-level perpetrators back to BiH. The concept behind the creation of the WCC is "that accountability for the gross violations of human rights that took place during the conflict is of concern to all humanity but ultimately remains the responsibility of the people of Bosnia and Herzegovina themselves" (OHR 2004, p.4).

[1] It is interesting to note that in cases of war crimes the state and entity level sentencing vary greatly. For example, the CBiH uses the 2003 Criminal Code Procedure which allows for a maximum sentence in war crimes of 45 years, while in the entities the Criminal Code of the Socialist Republic of Yugoslavia is most often applied which brings with it a maximum sentence of 20 years (OSCE 2011).

The WCC began operating in March 2005 just after the completion of vetting of domestic judges and prosecutors. Initial funding for the WCC came from international donors, a source which has been gradually phased out with the state taking over funding from the beginning of 2013 (Transition Council 2011). The first cases to be heard in the WCC were transfer cases from the ICTY. But while the bulk of the WCC's current cases have been locally initiated, the WCC is considered a "hybrid" court as it is a "national judiciary with capacity to run war crimes trials according to international standards." (Ivanišević 2008, p.6). Another contributing factor to the hybrid nature of the CBiH is the presence of internationals in the mixed trial panels of judges. But what makes the WCC truly unique compared to any other hybrid court is that fact that "key participants in the proceedings—prosecutors, defence lawyers, witness and presiding judges—all share the same language"[2] (Ibid, p.11).

The creation of the WCC can be seen as an institutional response to transitional justice which addressed the growing need to localise and address issues of mechanisms of transitional justice which are "spatially removed from the context within which the crimes took place" (Jeffrey 2011, p.347). With this in mind, the WCC, to date has completed proceedings in 73 cases which have "overall [. .] delivered efficient, fair, and human rights compliant proceedings" (OSCE 2011, p.7). Whilst there have been numerous successes of the WCC, criticism exists. According to the Organisation for Security and Cooperation in Europe (OSCE), the WCC has not successfully succeeded in the implementation of the National Strategy which provides recommendations as to how the CBiH and the entities can efficiently and effectively address and process war crimes cases. The OSCE identified a continued need to distribute the lower-level cases to the entity, cantonal and municipal courts so more pertinent cases could be addressed at a state-level. Another problem which the OSCE has identified is the "lack of political support for war crimes processing from certain quarters, evidenced by […] interference in proceedings, attempts to undermine existing judicial and legal reforms, and denial of war crimes established […] through binding legal decisions" (Ibid, p. 8). Finally, there is a lack of understanding amongst the public in regards to the activities of the CBiH and with the prosecution and sentencing of those found guilty. This creates an environment of misunderstanding and mistrust which is used by political elites to further their agendas.

The Significance of Public Outreach

The concept of "public outreach" poses questions concerning the relationship between law and society. One of the criticisms of an overly "legalistic" interpretation of transitional justice is that it conveys the normative impulse that law and society should be considered separate (McEvoy 2008). Within this optic, the

[2] The language is Bosnian/Croatia/Serbian.

purity of legal judgement is preserved through the cordon of spatial separation from the messy particularities within which crimes are committed. Of course, socio-legal studies scholars amongst others have contested this myth of conceptual separation. While the concept of public outreach may convey the transmission of legal knowledge from a centre (the CBiH) to a periphery (BiH society) its practice challenges this simple geometry. Our research exploring the public outreach programmes of the CBiH have illuminated the ways in which a number of civil society initiatives are firstly creating the possibility for individuals to shape the activities of new judicial institutions while secondly also creating new informal spaces where allied and alternative concepts of transitional justice are mobilised. These activities stretch the signifier of "outreach" since they instead illuminate a circulation of ideas of justice as they are formulated, contested and extended within BiH civil society (see also Lambourne 2012). Returning to the key argument of this chapter, this could be seen as a symbiosis between civil society and the citizen, as associations create arenas within which new articulations of citizenship in BiH may be voiced.

From the outset in 2005 the CBiH included a Public Information and Outreach Service (PIOS) which was directed at communicating the activities of the CBiH to a wider BiH public. This commitment to public outreach differed from the establishment of the ICTY in 1993, which had to wait 6 years before a programme of public outreach was included in its work. According to one of the initial employees in PIOS the CBiH was attempting, in those early years, to learn from the perceived mistakes made by the ICTY which were seen to contribute to a sense of alienation amongst the BiH public to its activities and agendas. Nettlefield's (2010) account of public perceptions of the ICTY speaks of the improved attitudes within BiH towards the tribunal through increased attempts to engage the BiH public. In the CBiH the initial prioritisation of public outreach was typified by the project in 2006 to establish a Court Support Network (CSN), a group of human rights NGOs that would serve as link points between the CBiH and wider society. The CSN was established by PIOS in 2005 with an aim "to integrate the mission of the CBiH into the wider Bosnian community."[3] Reflecting their normative placement as "democratising" agencies within the international intervention in BiH since Dayton (see Jeffrey 2007), PIOS looked to non-governmental organisations (NGOs) as the main instruments through which to achieve this objective.

Over the course of 2006, PIOS enrolled five human rights NGOs from across the territory of BiH into the CSN: *Žene ženama* (Sarajevo), *Centri civilnih inicijativa* (or CCI, Mostar), *Izvor* (Prijedor), *Helsinški komitet* (Bijeljina) and *Forum građana* (Tuzla). But the network envisaged by PIOS extended beyond these five organisations: the purpose of the initial 6 month programme (funded by the CBiH) was for each organisation to develop a network of other civil society organisations which could circulate information both from and to the CBiH. This approach focused on

[3] ICTY Liaison Officer and co-instigator of the CSN, Sarajevo, 9 October 2009.

the role of civil society groups as socially embedded organisations that may have developed long-term links with a range of institutions, such as victim associations,[4] youth groups, sports clubs and *Mjesne zajednice* (local associations or MZs). This model of public outreach envisages a particular normative role for civil society as a locus of trust acting outside particularistic agendas of a specific social group. In addition there is the more straightforward issue of approachability, with one PIOS official remarking in 2009 that people feel "more comfortable approaching an NGO than the CBiH," a situation that she felt was a reflection of a legacy from the "Yugoslav system" (PIOS interview, 2009) but perhaps reflects a wider internalisation of conceptual separation of law and society (see McEvoy 2008).

On some criteria the CSN initiative failed, since the organisations did not coalesce into an integrated network, but rather continued to pursue individual agendas through activities often focused on supporting the work of the CBiH. However, this limited assessment fails to grasp the wider set of practices and opportunities facilitated by participation in the CSN. Taking the example of *Helsinški komitet* in Bijeljina, this organisation participated actively in the work of the CSN but used it as a starting point for a much wider set of activities relating to transitional justice in BiH. As an organisation they had been involved in public outreach activities with the ICTY since 1999, expertise that made them well-placed to assist the CBiH in similar tasks. Certainly, their participation in the CSN saw them undertake work that seemed to accord with the expectations of civil society support for the legal process, for example they undertook publicity campaigns regarding the work of the CBiH and organised seminars and workshops that explained specific trial processes and outcomes. As one of the members of *Helsinški komitet* explained in 2009, these events were usually targeted at specific groups:

> For instance in Zvornik we did it with former camp detainees, because they raised the question several times in public that they were having difficulty approaching the CBiH, the Prosecutor's Office, no one is coming and listening to them, and so on, so we organised some events for them, bringing people from the CBiH and Prosecutors Office to Zvornik. [...] Now it is a question of whether [camp detainees] want to use it or not use it.[5]

Through such events *Helsinški komitet* was providing the opportunities for individuals and groups (specifically victims' associations and associations of camp detainees) to attempt to access justice through formal legal mechanisms. In a sense this reflects the failure of the citizenship model enacted since GFAP in 1995, since it is the feeling of anomie and powerlessness that prompts the intervention by the *Helsinški komitet*. Crucially for the argument in this chapter, the workshops organised by *Helsinški komitet* are not designed to produce an outcome in itself, since they are structured around the possibility of future prosecutions through the CBiH. Rather, the workshops create the conditions where expressions of citizenship rights may be secured.

[4] Many of these associations are often run by the survivors themselves who lack formal training in programme development and management and often speak little English. See for example Delpla (2007), Rangelov and Theros (2009), and Zupan (2007).

[5] Interview with representative from *Helsinški komitet*, Bijeljina, 18 October 2009.

While this example reflects the CSN as an extension of CBiH practice, enrolment in this network has also provided the opportunity for *Helsinški komitet* to challenge and extend ideas of transitional justice emanating from the CBiH. This challenge was evidenced in a number of aspects of their activities and agendas, but it was encapsulated by the idea of embedding discussions of transitional justice into education programmes. The organisation was using its access to the CBiH to organise regular student visits, but it was its more ambitious attempt to establish a new university course that illustrated a desire to explore ideas of transitional justice beyond the retributive model practiced through the legal process:

> We are now starting a relatively big project which is advocating establishing at universities a subject of transitional justice. We are now starting at the Bjeljina Faculty of Pedagogy [...] We presented it as a subject that was going to be financed for three years by us. And we are going to give all the documents and literature they need, it is about intercultural understanding, non-violent communication. And one part of it is dealing with the ICTY, the CBiH, the transitional justice initiatives, formal non-formal, judicial, non judicial and so on.[6]

These comments open up the possibility that the participation of civil society organisations in the formal processes of transitional justice in BiH is simultaneously cultivating more informal ideas of the possibilities and practices of transitional justice. We again see a symbiosis between the practices of civil society and the ideal of an engaged citizen, as associations attempt to use their privileged knowledge and access to cultivate individual engagement with ideas of transitional justice. This reflects the possibility of retributive mechanisms of justice acting as a vehicle through which alternative ideas of justice may be debated and enacted. Of course, we must not overlook the fact that the existence of the CSN speaks of a wider public suspicion of the CBiH, where political attacks on the CBiH (particularly by Republika Srpska) challenge its function and public acceptance. But this brief example of civil society activity between 2005 and 2009 illustrates how civil society agencies may facilitate civic participation where formal practices of citizenship emanating from the state have failed.

Civil Society and Agency

The second example drawn upon in this chapter challenges this virtuous narrative. As stated above, the model of citizenship adopted following the GFAP has failed to provide an inclusive formulation of political or social membership. As scholars have long argued, this situation reflects the failure of the wider peacebuilding process in BiH to address discrimination on grounds of gender, minority status or social vulnerability (see for example Domi 2002, and Greback and Zillen 2003). In the period since the GFAP these failings have been addressed through internationally funded civil society initiatives including victims associations who deal strictly with issues associated with torture and missing persons. These initiatives reflect a

[6] Interview with representative from *Helsinški komitet*, Bijeljina, 18 October 2009.

parallel process to those discussed above, whereby civil society organisations are cultivated in the absence of the legal processes of transitional justice enacted at the ICTY and later at the CBiH. In this second section we will draw on interviews conducted in Sarajevo with survivors of wartime rape[7] (hereafter survivors) from the periods of 2006–2007 and 2008–2009 and focus on how one "unofficial" mechanism of transitional justice—social assistance[8]—impacts a group of survivors in Sarajevo.

In BiH, there is no form of official state-level reparation or recognition for survivors of wartime rape, rather a system of Entity-sponsored social assistance exists for eligible survivors. This fragmentation exists despite the fact that wartime rape has been ruled a crime against humanity by the ICTY in 2001 with the Kunarac, Kovač and Vuković case (i.e. the Foča verdict). The status of the survivors only became a "real" issue when BiH was formally required to fulfil the criteria of international agreements to which it was a signatory, such as the United Nations Security Council Resolution 1325 (2000) on Women and Peace and Security. Despite this obligation, to date BiH offers no reparations or other forms of compensation for survivors at the state-level.

To address this issue, in 2005, the European Union funded a research project entitled "Improvement of the position of women victims of war" which was conducted by two local civil society organisations. The project held a series of roundtable discussions with the purpose to develop recommendations for the amendment of the *Law on the Basic Social Protection, Protection of Civilian Victims of War and Families with Children in the Federation of BiH*[9] *(Law on CVW)* to include survivors of sexual violence and rape. The campaign "For the dignity of the survivors" was launched in March 2006 as a form of active citizenship and publicised the issues of survivors through the collection of 50,000 signatures from the Federation through a series of advertisements in an attempt to address this group's feelings of marginalisation and powerlessness. The campaign also enlisted the Bosnian film director Jasmila Žbanić whose film *Grbavica* released 1 March 2006, follows the story of Sarajevans Esma and her daughter Sara and their coming to terms with Esma's rape and forced pregnancy in a concentration camp. The film was purposefully released to coincide with the launch of the campaign (Initiative Group of 'For Dignity of the Survivors' 2006) and brought wartime rape in BiH back into domestic and international spheres with the film winning a Golden Bear at the Berlin International Film Festival. This lobbying and international attention

[7] Official statistics suggest 20,000 women (mainly Muslim) were raped in BiH (European Community 1993). Yet we must remember that these statistics fail to take into account those who choose not to come forward, who have not been able to come to terms with their experiences, those who were killed, and non-Muslims victims. The statistics also do not include men who suffered sexual violence and/or rape (see Zawati 2007, p.1).

[8] This focuses on monetary assistance even though the law includes entitlement to professional training, employment priority, etc. which was never experienced by the survivors I spoke with.

[9] There is a similar law present in Republika Srpska which is the *Law on the Protection of Civilian Victims of War in the Republika Srpska* (Official Gazette of the RS No. 25/93, 1/94—special edition, 32/94, 37/07 and 60/07) but no existing State law and/or harmonisation of laws between the entities.

pressurised the Federation to amend the *Law on CVW* in 2006 to include wartime rape survivors. This significant step was a victory for the lobbying throughout the Federation by victim groups and associations and a rare example of active citizenship. This amended law currently entitles those officially registered as civilian victims of war and survivors of wartime rape to[10]:

> a personal disability allowance or a personal monthly allowance; a supplement for aid and assistance by another person; an allowance for orthopaedic supports; a family disability allowance; financial support for the cost of medical treatment and purchase of orthopaedic supports; the right to professional training (skills and competencies training and professional development); the right to worker employment priority; the right to housing priority; and, the right to psychological assistance and legal aid (International Commission for Missing Persons, 2007a, p.9).

In theory this unofficial transitional justice mechanism "addresses" the socio-economic needs of those officially registered to receive social assistance under the law but it does nothing in terms of reparation,[11] rather for the survivors spoken to it exacerbates their situation.[12] During the research[13] the survivors spoke of numerous barriers, misunderstandings and resentment in the process of qualifying and receiving the social assistance. First, there was an initial confusion of *where* to access information on their rights and entitlements and *how* to register and claim the social assistance. In one case, a survivor spoke with a Federation minister who asked *her* who was in charge of implementing the law (Jakala 2010, p.259).

But the logistical barriers to receiving assistance extend beyond limits to individual capacity. There is only one organisation, the Survivors of Wartime Sexual Violence Association[14] in the whole of the Federation[15] which issues the certificates which

---

[10] In theory access should be granted to *all* entitlements under the law. The survivors I spoke with only focused on and indicated that they could access the personal monthly allowance (Jakala 2010).

[11] De Grieff (2008) defines reparations as "… measures that provide benefits to victims directly. This use contrasts with measure which may have reparative effects, and which may be very important (such as punishment of perpetrators, or institutional reforms), but do not distribute a direct benefit to victims themselves" (2008, p.436).

[12] This can be experienced in the retelling of experiences, in feeling they need to justify their experience and making public their experiences which can lead to retraumatisation. See Track Impunity Always (TRIAL 2010), and Jakala (2010).

[13] Life history interviews were conducted with ten female research participants from a survivor's organisation in Sarajevo, BiH along with 22 months of intensive participant observation at the organisation.

[14] This is a pseudonym for the organisation.

[15] Whilst this research focussed on wartime rape survivors in the Federation it is known that wartime rape survivors in Republika Srpska also have means, albeit much more limited than the Federation, of claiming social assistance (see note 7). The law in Republika Srpska, "… limits the recognition of the civilian victim of war entitlements with a 5 year deadline, from the day the application of the Law started, that is from the day the person suffered damage to the body, or was killed, or went missing" (ICMP 2007a, b, introduction). This law has been amended a few times with the deadline for application set at 31 December 2007 and then later extended with a 6 month deadline (Human Rights Ombudsman BiH 2010).

allow survivors eligibility for the social assistance.[16] The women's organisation is located in the Sarajevo suburb of Ilidža and has a procedure of proving one was raped either by providing a witness or through documentary evidence. The survivors must also tell their stories to the committee (with little or no access to psychologists), risk retraumatisation, and face the possibility of their experiences not meeting the requirements needed to attain the certificate. Survivors have told of instances of being made to sign a release form for their statement in order to receive the assistance and some even spoke of being coerced into publically discussing their experiences.[17] The processing times to receive the social assistance vary from between 6 to 8 months with some cases taking up to 4 years. It was also suggested by some of the survivors that the monetary payments were inconsistent which created problems for those survivors who depended on the payments as their sole source of income.[18]

The social assistance granted from the 2006 amended law is a deficient and unofficial attempt at a transitional justice mechanism. Besides creating barriers, misunderstandings, confusion and for some retraumatisation[19]—the right to the social assistance is limited to those who can physically access the organisation and among other things willing to tell their experiences. As transitional citizens, this group of survivors have been rendered, for the most part, passive actors in BiH's transitional justice process which has marginalised this group (and other groups of civilian victims of war) by failing to draw upon their experiences and needs in this collective process. This passivity indicates the need for a strong state-level transitional justice mechanism which officially acknowledges the survivors experiences be it through symbolic and/or monetary reparations or other form of state-sponsored recognition (see for example Ilif et al. 2011) so to establish the foundations for an inclusive form of citizenship in BiH.

Conclusion

This chapter began with an assertion that the citizen plays a central role in transitional justice, both in terms of the way it is framed through policies and institutions, and the way in which it is enacted in practice. The agency of citizens interacts with

[16] This is an issue as the organisation is known to be biased towards Muslims and not physically accessible to all survivors especially those who are unable to travel from rural areas. It also is an issue because they do not have proper psychological support available to the survivors when they retell their experiences and do not account for the fact that many already belong to victim associations and have already told these associations of their experiences (see for example Amnesty 2009; Jakala 2010).

[17] This is problematic as it appears to be used as leverage—sign the form and get social assistance and if you do not sign it, then you do not get social assistance. The survivors also do not know what their statements will be used for. See for example Amnesty (2009), Jakala (2010) and TRIAL (2010).

[18] The social assistance is funded by the Federation and the cantons with 70% contributed through the Federation budget and the remaining 30% contributed by the local cantons. See Jakala (2010) and Popić and Panjeta (2010).

[19] See for example Amnesty International (2009), Jakala (2010), TRIAL (2010).

transitional justice policies and mechanisms to construct what are sometimes unexpected outcomes, but which may challenge or reinforce the implicit and explicit assumptions of the nature and role of the citizen in transition. The purpose of this chapter has been to illuminate these outcomes through two empirical examples from post-Dayton BiH. The aim of this approach is not integrative; we do not maintain that we can trace a unifying dynamic over these disparate cases. Rather, it is to illustrate the diversity of interpretations and practices of justice and citizenship as they are unfolding in contemporary BiH. Perhaps the enduring conclusion of this article is the limits to theoretical reflection in encapsulating the lived experience of transitional justice in BiH, and allied to the significance of long-term empirical examinations to understanding how policies of transitional justice are implemented.

The empirical examples discussed here (the WCC and social assistance for survivors of wartime rape) highlight a key tension between how civil society has been targeted as part of transitional justice, and the citizen which, in its more substantive form, requires more stable and secure citizenship rights than are afforded by the political arrangements of post-GFAP BiH. As far as WCC outreach is concerned we see a circulation of ideas of justice and the formation of new spaces in which alternative ideas of citizenship and justice can be articulated. In the case of social assistance for survivors of wartime rape we see the limits of a transitional justice mechanism which is designed to engage civil society but in fact marginalises survivors of wartime rape in a context of social inequality and social vulnerability. From this we learn that it is not enough to target civil society as a necessarily emancipatory space in which, and through which, transitional justice can be enacted. Instead, we must understand this concept as a realm of variegated political and social practices, and one which is connected to and shaped by the broader context of the possibility of a more substantive citizenship.

In this relatively new field and taking a new approach to understanding the nature of the citizen in transition, further fruitful areas for research can be identified. By seeking to understand the way in which the citizen is present in transitional justice processes we can identify the assumptions implicit in certain policies which, for example, may target civil society as agents for outreach. This can help us to understand whether transitional justice processes are meaningful for citizens living in those societies, and how they are shaped by broader contexts of social and political membership. Building on this, we create space for alternatives to dominant discourses and practices which may be articulated in the contestations over transitional justice. Finally, and following McEvoy and McGregor (2008a, b), it allows us to avoid losing sight of the social and political struggles which put transitional justice on the agenda in the first place.

References

Adejumobi, S. 2001. Citizenship, rights and the problem of conflicts and civil wars in Africa. *Human Rights Quarterly* 23(1): 148–170.

Akanji, O.O. 2011. Researching conflict in Africa: A researcher's account of Ife-Modakeke, South-Western Nigeria. In *Researching violence in Africa: Ethical and methodological challenges*, ed. C. Cramer, L. Hammond, and J. Potter, 23–174. Leiden and Boston: Brill.

Amnesty International. 2009. *Whose justice? The Women of Bosnia and Herzegovina are Still Waiting*. London: Amnesty International. http://www.amnesty.org/en/library/info/EUR63/006/2009. Accessed 5 November 2010.

Bajrović, R. 2005. *BiH municipalities and the EU: Direct participation of citizens in policy-making at the local level*. Sarajevo: Open Society Fund.

Buckley-Zistal, S. 2006. Dividing and uniting: The use of citizenship discourse in conflict and reconciliation in Rwanda. *Global Society* 20: 101–113.

Campbell, C., and C. Turner. 2008. Utopia and the Doubters: Truth, transition and the law. *Legal Studies* 28(3): 374–381.

De Grieff, P. 2008. Justice and reparations. In *The handbook of reparations*, ed. P. De Grieff, 436. Oxford: Oxford University Press.

Delpla, I. 2007. In the midst of injustice: The ICTY from the perspective of some victims associations. In *The new Bosnian mosaic: Identities, memories and moral claims in a post-war society*, ed. X. Bougarel, E. Helms, and G. Duijzings, 2007. Aldershot: Ashgate.

Domi, T. 2002. Advancing women's political rights in Bosnia and Herzegovina. *The Harriman Review* 14(1–2): 36–46.

European Community. 1993. *European community investigative mission into the treatment of Muslim women in the former Yugoslavia: Report to European Community Foreign Ministers*, U.N. SCOR, Annex I, U.N. Doc. S/25240 (1993).

Fagan, A. 2005. Civil society in Bosnia ten years after Dayton. *International Peacekeeping* 12: 406–419.

Faulks, K. 2000. *Citizenship*. London: Routledge.

Fischer, M. 2006. Introduction: Moving out of the Dayton era into the era of Brussels? In *Peacebuilding and civil society in Bosnia-Herzegovina ten years after Dayton*, ed. M. Fischer. Berlin: Lit.

Gibney, M. 2006. Who should be included? Non-citizens, conflict and the constitution of the citizenry. CRISE Working Paper 17. www.crise.ox.ac.uk/pubs.shtml. Accessed 10 January 2012.

Greback, K., and E. Zillen. 2003. *Endangering the peace process: A study on the Dayton accords*. Ottawa: United Nations Division on the Advancement of Women. EMG/Peace/2003/EP.2.

Hoffman, J. 2004. *Citizenship beyond the state*. London: SAGE.

Human Rights Ombudsman BiH. 2010. *Special report on the rights of persons with disabilities*. Sarajevo: Human Rights Ombudsman BiH. http://www.ombudsmen.gov.ba/docs/Invaliditet_ENG.pdf. Accessed 10 January 2012.

Ilif, F. et al. 2011. *Adverse consequences of reparations*. Colchester: Essex Transitional Justice Network, University of Essex. http://www.essex.ac.uk/tjn/documents/Paper_6_Adverse_Consequences.pdf. Accessed 1 November 2011.

ICMP. 2007a. *Guide for civilian victims of war: How to enjoy the right to protection as a civilian victim of war in the federation of Bosnia and Herzegovina.* Sarajevo: ICMP. http://www.ic-mp.org/wp-content/uploads/2008/02/guidebook-wictim-of-war-fbih.pdf. Accessed 3 November 2011.

ICMP. 2007b. *Guide for civilian victims of war: How to enjoy the right to protection as a civilian victim of war in the Republika Srpska.* Sarajevo: ICMP. http://www.ic-mp.org/wp-content/uploads/2008/02/guidebook-wictim-of-war-rs.pdf. Accessed 3 November 2011.

Initiative Group of 'For Dignity of the Survivors'. 2006. *Proposal for amendments to draft law on amendments to law on basic social protection measures, protection of civilian victims of war and protection of families with children of march 2006.* Sarajevo: Initiative Group of 'For Dignity of the Survivors'. http://www.stopvaw.org/uploads/a_FD_letter_to_parliament_FBiH.doc. Accessed 10 January 2012.

Ivanišević, B. (2008). *The war crimes chamber in Bosnia and Herzegovina: From hybrid to domestic court.* New York: International Center for Transitional Justice. http://ictj.org/sites/default/files/ICTJ-FormerYugoslavia-Domestic-Court-2008-English.pdf. Accessed 20 October 2011.

Jakala, M. 2010. *Wartime rape: War's worst experience? a feminist grounded theory study of female non-raped and wartime rape survivors in Sarajevo, Bosnia and Herzegovina.* Unpublished doctoral dissertation, University of Bradford, Bradford.

Jeffrey, A. 2007. The geopolitical framing of localized struggles: NGOs in Bosnia and Herzegovina. *Development and change* 38: 251–274.
Jeffrey, A. 2011. The political geographies of transitional justice. *Transactions of the Institute of British Geographers* 36(3): 344–359.
Jones, B. 2009. *Examining reconciliation's citizen: Insights from the multi-ethic district of Brčko, Bosnia-Herzegovina.* Unpublished doctoral dissertation, University of Manchester, Manchester.
Kabeer, N. 2005. Introduction: The search for inclusive citizenship, meanings and expressions in an interconnected world. In *Inclusive citizenship, meanings and expressions*, ed. N. Kabeer, 1–27. London: ZED.
Lambourne, W. 2012. Outreach, inreach and civil society participation in transitional justice. In *Critical perspectives in transitional justice*, ed. Phil Clark et al. Intersentia: Antwerp.
Lister, R. 2003. *Citizenship: Feminist perspectives*. Hampshire: Palgrave Macmillan.
Mamdani, M. 2001. *When victims become killers: Colonialism, nativism and the genocide in Rwanda.* Princeton: Princeton University Press.
McEvoy, K., and L. McGregor (eds.). 2008a. *Transitional justice from below: Grassroots activism and the struggle for change*. Oxford: Hart.
McEvoy, K., and L. McGregor. 2008b. Transitional justice from below: An agenda for research, policy and praxis. In *Transitional justice from below: Grassroots activism and the struggle for change*, ed. K. McEvoy and L. McGregor. Oxford: Hart.
McEvoy, K. 2008. Letting go of legalism: Developing a 'thicker' version of transitional justice. In *Transitional justice from below: Grassroots activism and the struggle for change*, ed. K. McEvoy and L. McGregor, 15–45. Oxford: Hart.
Mujkić, A. 2008. *We the citizens of Ethnopolis*. Sarajevo: Centar za Ljudska Prava Universiteta Sarajevu.
Nettlefield, L. 2010. *Courting democracy in Bosnia and Herzegovina: The Hague Tribunal's impact on the Postwar State*. Cambridge: Cambridge University Press.
Newman, E. 2002. Transitional justice: The impact of transnational norms and the UN. In *Recovering from civil conflict: Reconciliation, peace and development*, ed. E. Newman and A. Schnabel, 31–50. Portland: Cass.
OHR (2004). *War crimes chamber project: Project implementation plan registry progress report.* Sarajevo: OHR. http://www.ohr.int/ohr-dept/rule-of-law-pillar/pdf/wcc-project-plan-201004-eng.pdf. Accessed 18 November 2011.
OSCE (2011). *Delivering justice in Bosnia and Herzegovina: An overview of war crimes processing from 2005 to 2010.* Sarajevo: OSCE. http://www.oscebih.org/documents/osce_bih_doc_2011051909500706eng.pdf. Accessed 20 October 2011.
Ó Tuathail, G., and J. O' Loughlin. 2009. After ethnic cleansing: Return outcomes in Bosnia-Herzegovina a decade beyond war. *Annals of the Association of American Geographers* 99(5): 1045–1053.
Popić, L., and Panjeta, B. 2010. *Compensation, transitional justice and conditional international credit in Bosnia and Herzegovina: Attempts to reform government payments to victims and veterans of the 1992–1995 War*. Sarajevo: Norwegian Embassy. http://www.norveska.ba/PageFiles/417487/Study_Linda_Popic_english.pdf. Accessed 13 November 2011.
Porter, N. 2003. *The elusive quest: Reconciliation in northern Ireland.* Belfast: Blackstaff.
Rangelov, I., and Theros, M. (2009). *Maintaining the process in Bosnia and Herzegovina: Coherence and complementarity of EU institutions and civil society in the field of transitional justice*. Bonn: Working Group on Development and Peace. http://www.frient.de/downloads/RangelovTheros_kompl_071127.pdf. Accessed 4 January 2012.
Sejfija, I. 2006. From the "civil sector" to civil society? Progress and prospects. In *Peacebuilding and civil society in Bosnia-Herzegovina ten years after Dayton*, ed. M. Fischer. Berlin: Lit.
Shaw, R., L. Waldorf, and P. Hazan. 2010. *Localizing transitional justice: Interventions and priorities after mass violence*. Stanford: Stanford University Press.
Sriram, C.L. 2009. Transitional Justice and the Liberal Peace. In *New perspectives on liberal peacebuilding*, ed. R. Newman, R. Paris, and O.P. Richmond, 112–130. Tokyo and New York: United Nations University Press.

Transition Council. 2011. *Medium-term institutional development plan for the court and prosecutors office of BiH*. BiH: Sarajevo.

TRIAL. 2010. *Written information for the examination of Bosnia and Herzegovina's combined second to fifth periodic reports*. Sarajevo: TRIAL. http://www2.ohchr.org/english/bodies/cat/docs/ngos/Trial_BosniaHerzegovina45.pdf. Accessed 10 November 2011.

Van De Merwe, H., V. Baxter, and A.R. Chapman. 2009. Introduction. In *Assessing the impact of transitional justice: Challenges for empirical research*, ed. H. van de Merwe, V. Baxter, and A.R. Chapman, 1–11. Washington D.C: United States Institute of Peace Press.

Zawati, H. 2007. Impunity or immunity: Wartime male rape and sexual torture as a crime against humanity. *Torture Journal* [online], Vol. 7, No. 1. http://www.irct.org/Files/Filer/TortureJournal/17_1_2007/impunity_or_immunity.pdf. Accessed 25 October 2011.

Zupan, N. 2007. Facing the past and transitional justice in the countries of the Former Yugoslavia. In *Peacebuilding and civil society in Bosnia-Herzegovina*, ed. M. Fischer, 2007. Berlin: LIT.

Chapter 7
Transitional Justice in Albania: Historical Burden, Weak Civil Society, and Conflicting Interests

Arolda Elbasani and Artur Lipinski

Introduction

Albania has suffered one of the harshest communist dictatorships in the communist world. Despite the many victims brutalized by the former regime and the anticommunist policies advocated by the political forces that won the first free elections, however, the country has done very little in terms of dealing with the crimes committed during that dictatorship. This chapter asks why Albania has failed to ensure justice after the collapse of communism. What has been the role of civil society in this process and how has this role changed throughout the different stages of transition from communism?

Our proposed framework builds on alternative explanations of transitional justice processes: historical legacies, the configuration of political interests, and idea-based factors. We attempt to link and supplement these existing explanations by incorporating the role of civil society as a mediator that can screen and shape elite-driven initiatives on the issue. We ask how and when civil society has influenced processes of transitional justice in Albania by examining two different stages, which coincide with similar politically sponsored campaigns but feature different levels of civil society participation. The first of these campaigns, led by the anticommunist forces that took over in the early 1990s, featured little input from other social actors and resulted in the adoption of initial legislation on

A. Elbasani (✉)
Robert Schuman Centre for Advanced Studies, San Domenico di Fiesole,
Villa la Pagliaiuola, Via delle Palazine, 17, I-50014 (FI), Italy
e-mail: Arolda.Elbasani@EUI.eu

A. Lipinski
Faculty of Political Science and Journalism,
Adam Mickiewicz University, Poznan, Poland

O. Simić and Z. Volčič (eds.), *Transitional Justice and Civil Society in the Balkans*, 105
Springer Series in Transitional Justice, DOI 10.1007/978-1-4614-5422-9_7,

communist crimes in 1995. The second initiative, led by the same political actors but contested by a number of civil society actors, resulted in a second legal package on transitional justice in 2008. None of these initiatives actually materialized into coherent policies. During the first campaign, the political sponsors of the bill sought to use the legal package against their opponents and failed to ensure a fair process. The second time around, however, social and political actors spoke against the shortcomings and the timing of the proposed legislation. The contrast between these two examples suggests that a more active civil society, which triggered public debates and also served to channel different actors' objections, played an important role in 2008. Our empirical analysis draws from an extensive literature review, personal observations and conversations, official press releases, and representative samples drawn from the daily press.

The chapter is organized in four parts. The first part maps out existing explanations of the processes of transitional justice across the postcommunist world and develops our own analytical approach to the case study of Albania. The second part provides a historical overview of Albanian democratization and establishes the background for analyzing postcommunist transitional justice in light of the historical legacy of communism. The third part explores how different factors, including a weak civil society and vested political interests, have interacted to produce a profoundly politicized process of transitional justice in the period 1995–1997. The fourth part examines the role of growing civil activism against politicized policy initiatives undertaken in the years 2008–2009. While arguing that strengthened civil society actors—in conjunction with other institutionalized actors—played an important role to discredit the transitional justice bill in 2008, we also recognize that Albanian civil society has yet to introduce an alternative model for dealing with the communist past.

Transitional Justice After Communism

Quest for Justice, Different Models, and Varying Results

Transitional justice has been an issue of concern in most societies emerging from communist regimes (Teitel 2000: 15–18). The principal ideas and policy options that animated public debates and pushed forward the process came mainly from those groups who directly suffered under communism: repressed individuals, intellectuals, and anticommunist political groups with dissident credentials. Media outlets and the broader public have also shown intense interest in the debates regarding the communist past. Across Eastern Europe, television programs and news feature focusing on communist-era repression abound.

And yet, different countries have opted for different alternatives to transitional justice, depending on the strength of players advocating for or against it. The array of models ranges between different measures of lustration and/or excommunication, a mixture that determines the degree of moderation or extremeness of the system

(Sadurski 2003). Although opinions may vary on what constitutes the first and the second, we opt for a clear differentiation of the two (Nalepa and Kaminski 2006). Accordingly, lustration consists of the process of screening of all persons seeking to occupy certain public positions for evidence of involvement in communist-era abuses of human rights. As such, lustration is typical of moderate models, which aim only at the public disclosure of the perpetrators of human rights abuses. Excommunication, instead, is explicitly aimed at the banning of certain categories of former officials from holding certain public positions for a defined period of time. Hence, excommunication is the hallmark of a "wild" model that aims to "cleanse" the new system from those found guilty instead of merely ensuring transparency in dealing with the past (Nalepa and Kaminski 2006: 384).

Explaining Transitional Justice

The wide range of policy options, goals, and achievements has generated a wealth of research on the factors that might help explain such cross-country differences. This spectrum of explanations can be summarized in three main strands: one focusing on the role of *history*; one on the role of *politics*; and one on the role of *ideas*.

The most typical explanation emphasizes the significance of *historical* factors, especially the type of the communist regime and the mode of extrication from the old system. According to this line of argument, the higher the degree of liberalization of the old regime, the higher is the tendency to tolerate former abuses and embrace a moderate model of transitional justice (Linz and Stepan 1996: 38). This historical strand of explanation also maintains that where the exit from communism took the form of revolutionary action, the new regimes would tend to opt for a harsher line of transitional justice. In turn, where exit from the old system involved negotiations, it frequently led to rather lenient policies toward perpetrators (Moran 1994: 96). The politics-centered approach tends to emphasize major actors' *political strategies*, conceived in a competitive environment where anti-communist and postcommunist groupings strive to gain and maintain power (Williams et al. 2005). The general precept of this approach is that, with the passage of time, transitional justice becomes increasingly politicized and instrumental to the strategic interests of those with sufficient power to negotiate new rules (Welsh 1996: 421; Calhoun 2004: 16). The third current of explanations underlines the significance of *ideas* in the choices involving transitional justice. Accordingly, even though sometimes helpful in explaining the timing of certain measures related to crimes of the past, power politics alone can often fail to explain why certain actors, who one would expect to promote or reject reckoning with the past, abstain from it (Calhoun 2004: 17). In the political field, this argument goes, the advancement of political interests necessitates the ability to provide explanations and/or justifications for any given policy.

Most of the above approaches tend to dismiss civil society as a factor in explaining the pace and scope of dealing with the past in postcommunist countries. Historical approaches are predominantly focused on the balance of power and strategic games between reformers and hardliners within the old regime on the one hand and anticommunist opposition on the other. These approaches imply that transition is an elite-centered game where both sides strive to control or demobilize their supporters (Letki 2002: 537). The same pertains to the political approach, which emphasizes the role of political interests and competition. Transitional justice, therefore, remains entirely an "elite affair" (Nalepa 2010: 101). The third approach, which emphasizes the role of ideas and their public exchange, similarly employs a top-down elite-driven model of explanation. From this angle, too, political actors are the major players directing and shaping civil society attitudes (Ask 2006: 2).

Civil Society, Public Debates and Control over Politics

Our proposed conceptual framework attempts to link and supplement these existing explanations by bringing into focus the role of civil society actors. Various groups organized at the civil society level can be instrumental in exposing the shortcomings of top-down elite proposals. We certainly do not suggest that *history, political interests,* and *ideas* are not important. Rather, we argue that a numerically sizeable and politically active *civil society* can mediate and complement the role of other factors. Following the work of Backer, we suggest that civil society can play two distinct roles in influencing, controlling, and directing the process (2003: 302–305). First, it can perform a number of activities and services such as data collection and monitoring, service delivery, and compensation. Secondly, it can act as an advocate of certain choices, work against others, and even advance its own models and initiatives. Such activity contributes to keep an eye on the role of political interests and check out the political input on transitional justice programs. Although both dimensions of civil society's activities combine to reinforce social pressure on political choices, our analysis focuses mainly on the role played by civil actors via participation in open public debates.

We posit public debates as the crucial mechanism for civil society actors to articulate their ideas and, when necessary, exert pressure on other actors over specific alternatives to transitional justice. The exchange of arguments during public debates forces respective actors to take into account new ideas and information, in addition to elucidating the interests of opponents (Ask 2006: 14; Los 1995: 120). Debates in general require the existence of a multitude of actors driven by different interests who argue for the solutions they put forward. Public debates, moreover, belong to a different category of debates to the extent that they involve public contestation of a specific subject among elites and civil society actors (Ask 2006: 27). Those actors involved, including political elites, journalists, judges, religious authorities, public intellectuals, and trade unionists, among others, have the power to set the agenda and prime debates, thus influencing a wide audience and affecting the distribution of opinions within society (Entman 1989). Needless to say, not all public discussions have the same effect. Based on Ask's analysis of public debates, we distinguish between ordinary exchanges of ideas and concentrated public contestation. The latter includes both a wide range of actors (not only politicians, but also civil society participants), considerable duration (longer than 1 year), and intensity (frequent contributions to the national discussion) (2006: 30). It is concentrated debates that can work to disclose, reconfigure, and balance the inclination of some political actors to manipulate transitional justice at the service of their personal political agendas. The following section provides a general overview of political and social developments in postcommunist Albania and broadly introduces the actors involved in transitional justice initiatives.

Challenging Democratization: The Burden of the Past

Postcommunist Albania has gone through different waves of democratization and de-democratization, featuring institutional progress but also periods of stagnation coupled with recurrent crises of order and legitimacy. The difficult, chaotic, and at times uncertain path of regime change is often related to an inherited democratic deficit, as illustrated in the 'capturing' of transition politics by groups connected to the former communist regime (Elbasani 2008). Albania's was among the harshest dictatorships in the communist world; "opening up" to reforms, therefore, proved immensely challenging (Prifti 1978). Under the leadership of party leader Enver Hoxha, the communist state created a particularly repressive security apparatus, which, according to some estimates, affected the lives of as much as one-quarter of the country's total population (Biberaj 2000). A special law adopted in 1956 banned all independent organizations and restricted the activities of all existing associations (Chiodi 2007). To be sure, other communist countries in Eastern Europe widely used repressive methods, but the Albanian regime distinguished itself as a particularly tenacious follower of Stalinist methods even after the so-called de-Stalinization was launched in 1956.

By the late 1980s, the ruling communists were still determined to preserve the one-party system and their highly rigid economic policies (Goldman 2000). As a result of repression, moreover, there were hardly any dissidents who could advance alternatives to the one-party rule or work toward regime change (Biberaj 2000: 98–106). Instead, it was disaffected students who took to the streets in 1990 to seek limited reforms. These protests, which escalated into massive demonstrations joined by ordinary inhabitants, did not start out as political protests per se, but they quickly emerged as a threat to the regime's stability. Hoxha's successor, Ramiz Alia, who had reluctantly allowed for some minor reforms starting in 1989, opted to give up very limited controls and only when it became clear that the protests would escalate further (Vickers and Pettifer 2000: 33–34). In other words, given the mummified social and political climate created under decades of political repression, the impulses of change came from unorganized but growing groups of people dissatisfied with the regime. By the end of 1990, Alia reluctantly agreed to political pluralism.

The first opposition party, the Democratic Party (DP), was created soon thereafter and encompassed various anticommunist groups and individuals who capitalized on the student protests. The establishment's control over state institutions, however, ensured a communist victory in the first elections held in March 1991. Nevertheless, the victory proved insufficient in containing the escalating anticommunist protests. In light of ongoing clashes and turmoil, the country was forced into another round of elections. Promising large-scale reforms, the DP won the first free elections held in March 1992 (Vickers and Pettifer 2000: 31–34). The sweeping victory led many people to believe that Albania was about to embark on a quick program of transitional justice (Austin and Elisson 2008: 382). Intense power struggles and narrow political interests, however, trumped over demands for accountability for past repressions. Despite DP's ambitious political program, Albania failed to make a clear break with the communist past (Duffy 2000: 83–84).

At the helm of the DP emerged Sali Berisha, a former communist party member and a leading cardiologist. He was brought in by Alia to mediate between the regime and the student protesters, but he quickly seized the opportunity to emerge as a spokesman for the anticommunist forces. In the following years he established himself as an all-powerful president (Vickers and Pettifer 2000). After a round of contested elections in 1996 and the collapse of widespread ponzi schemes later that year, massive crowds of protesters once again took to the streets but this time denouncing the self-proclaimed anticommunist government for mismanaging the political and economic transition of the country (Biberaj 2000: 483–485). By 1997, those protests and clashes escalated further and threatened to bring Albania into total collapse.

Worried about the prospect of a failed state bordering the European Union, international actors intervened (Duffy 2000). They assisted the bitterly divided domestic political groups to negotiate a way out of the crisis and ensure some degree of stability until another round of elections in 1997. Substantial international support, including aid and direct supervision of reforms, proved essential to Albania's recovery and also enabled the emergence of a rudimentary civil society sector. Thus, both the discourse and practice of Albanian civil society was deeply connected from the beginning with Western donors' aid policies and their 'world of projects' (Chiodi 2007). But while foreign assistance has contributed to opening and enriching some kind of public sphere, donor-driven civil society remains weakly embedded in the Albanian social environment. It also remains incapable of standing up to the major political actors.

The 1997 elections brought in power the regenerated Socialist Party (SP), a self-described reformed strand of the former communists, which proclaimed a new path of democratization and fresh institutional reforms. In reality, the new government also excluded the opposition and relied on key figures who personalized political power (Biberaj 2000). Fatos Nano, the SP leader and a former member of the Institute of Marxist-Leninist Studies, served as prime minister for significant periods during the years 1997–2005. He also dominated the country's politics and often aggravated conflicts with the opposition but also within his own party, while 'using' the state as 'his own clientelistic enterprise' (Elbasani 2008).

Emphasizing deep reforms and introducing a host of new political faces, the DP returned to power with the relatively unproblematic 2005 elections. The new Berisha cabinet promised to tackle some of the long-standing problems in the country, including the 'capturing' of the state by corrupt politicians and their allies and the marked weakness of independent institutions. It also promised a more collaborative tone with the opposition. Nevertheless, the DP government has largely conformed to old patterns: centralizing powers in the office of the premier; subordinating independent state institutions; and coercing critics of the government, including through penalization of important media outlets (BTI 2012). In light of this historical overview, the following sections analyze projects of transitional justice as they have been introduced at different times in postcommunist Albania.

The First Wave of Transitional Justice 1992–2000: Anticommunist Politics and Weak Civil Society

Nonexistent Civil Society and Limited Public Debates

Given the level of repression under communism, Albania lacked organized opposition groups, or independent intellectuals, who could steer the process of transitional justice, as it happened elsewhere in the former Eastern bloc, at least in the first stage of transition (Chiodi 2007). Alia's timid concessions in the cultural sphere in the late 1980s, though encouraging the first signs of alternative thinking, hardly amounted to any organized effort from below, let alone the organization of autonomous or semiautonomous spaces of social engagement. Given extensive elite purges, even the most adventurous of intellectuals were intensely afraid that vocal opposition to the party-state could be deadly for them.

The DP was a motley of individuals from different backgrounds and driven by different interests. While this inclusion of diverse groups seemed to broaden its popular appeal in the short term, it also threatened party's cohesion on important political issues in the long term. From the very start, the party proved strongly divided on the issue of transitional justice. Young students and those persecuted under the communist regime, both well represented within party ranks, preferred a decisive break with communism. The reformist intellectuals who joined the DP in the latter stages, on the other hand, opted for a more gradual transition. The latter informed the main line of debate about the model of justice to be pursued in the country. Thus, while those persecuted under the communist regime asked for radical measures against perpetrators of communist crimes, a number of intellectuals advocated for a milder approach (Biberaj 2000). In general, very few within the decision-making core of the DP proved willing to dig deep into the past. Even fewer had any reasons to call for the declassification of the controversial files of the security apparatus. On the issue of transitional justice, in fact, Berisha has arguably played an important role in moderating demands for a tough policy line (Austin and Elisson 2008: 382).

At the other end of the political spectrum was the Socialist Party. Its members were widely perceived as collaborators with the former regime and, as such, enjoyed little credibility when it came to the issue of transitional justice. Moreover, the immediate postcommunist years were marked by a host of other intense challenges—deep social polarization, a nascent but extremely weak civil society, fierce conflicts between political groups, lack of experience with free press and professional media, a weak parliament, lack of rule of law, and massive waves of immigrants. In this context, debates about the communist past or policy proposals regarding transitional justice became a casualty of the political interests of those who sought to capitalize on the issue (Vickers and Pettifer 2000). Indeed, most initiatives related to transitional justice evolved separately from any meaningful public discussion, let alone the involvement of nonpolitical actors in serious and multiparty deliberations (Hatschikjan 2010).

Late-coming and Politicized De-communization Acts

The course of events during the first transitional phase in Albania confirms some authors' claims about a strong relationship between specific political interests and the timing of lustration packages (Williams et al. 2005). Despite the adoption of a strong anticommunist rhetoric, the ruling democrats were slow and ambiguous in formulating the promised decommunisation policies. The initial campaign was limited to public assaults against communist-era symbols, including the removal of communist names and landmarks, the confiscation of SP accounts, the banning of the communist party, which had shrunk to a handful of militants, and the prohibition of all parties of a Stalinist or Hoxhaist nature (Duffy 2000: 75). The government also undertook a large-scale indiscriminate "cleansing" of the administration from communist-era personalities (Elbasani 2009). The first decommunisation law, which targeted judiciary officials under the previous regime, was adopted in 1993 but was annulled shortly thereafter by the Constitutional Court. The DP also targeted key leaders of the previous regime, who were brought to trial and condemned for petty financial abuses (East and Pontin 1997: 217–218). Those public but disorganized and selective initiatives in dealing with the communist past failed to bring attention to the real abuses of the communist regime and created confusion about the goals of transitional justice.

Just ahead of the 1996 elections, the DP majority decided to enact a series of concrete and more radical measures in dealing with former communist officials. This came in the aftermath of the loss of the constitutional referendum in 1994, which the party interpreted as a concerning sign of diminishing popular support (Biberaj 2000). The decommunisation package, which was approved unilaterally by the DP in 1995, included an "anti-genocide" and a "verification" law. The "anti-genocide" law provided for the prosecution of "crimes against humanity" committed during the communist era and prohibited the election of those convicted for such crimes to a listed categories of offices until 2002 (article 3). The law did not have any legal impact given the fact that genocide and crimes against humanity were already indictable offenses under the penal code. Nevertheless, it served as a political move to show that the DP intended to enact its anticommunist promises. The law on the "verification of officials", on the other hand, provided the legal instruments to screen and ban from office a wide range of former communist officials. The Verification Committee (the so-called Mezini Commission), which was allowed full access to security files, was tasked with screening a wide range of enlisted categories of officials. If the persons found guilty refused to step down voluntarily, government institutions could remove them until the expiration of the law in 2001.

At first glance, the verification law opted for "wild" measures to the extent that it regulated the screening of an exceptionally wide list of positions and employed radical measures of banning from public office of all the listed categories. Yet, a closer look reveals that the law included a number of provisions that enabled the ruling majority to maintain political control over the process. First, the law contained a provision that enabled the government to appoint six out of seven members of the Verification Committee (article 4), casting doubts on the political impartiality of its institutional decisions. Similarly, provisions of exception for persons who "worked against the official line or distanced themselves publically" (article 3) were vague enough to allow the committee to protect certain individuals. Other provisions asserting that already employed individuals could be investigated only if required by the respective institution, and those appointed would have the option of requesting a review, seemed all tailored to protect current employees (Austin and Elisson 2008: 388). Additionally, all party leaders were automatically excluded from the verification process, unless it was "requested by other members of the party leadership" (article 12). Finally, the article concerning personal files (Article 16) called for their closure to the public until 2025, while also ensuring that the whole verification process would take place without any public scrutiny. Indeed, according to the law, both Verification Committee meetings and its screening decisions were to be taken behind closed doors, and none of its decisions would become public, unless the person concerned asked for a public hearing (article 11).

Sham Transitional Justice

If there were any doubts about the intention of the political motivations behind the 1995 decommunisation package, the selective implementation of the procedures actually dispelled them. To begin with, the list of the individuals screened between 1996 and 1997 was never disclosed, but estimates suggest that around 140 people were banned from participating in the forthcoming elections. The ban included only three DP members, while the rest were members of opposition parties. Moreover, the ban included 7 out of 12 members of the SP board and a quarter of all Socialist MPs (Biberaj 2000: 290). Needless to say, both laws had a detrimental influence on SP's performance in the subsequent national elections. Meanwhile, none of the ruling party leaders had their credentials questioned, even when their affiliations with the communist regime had become public knowledge.

Despite the obvious problems with the legal package adopted in 1995, no political or civil society groups contested the initiative. Even the Socialists, who were essentially the primary target of the initiative, did not contest the law, at least not openly, because they feared being perceived as defenders of communist crimes. Nevertheless, the Socialists went into the offensive once the DP-ruled government came under attack for mismanaging the transition process by late 1996. Rising popular protests forced the government to concede to the opposition's demands for relaxing decommunisation provisions and allow SP members to participate in the new elections of 1997 (Biberaj 2000). A package of amendments, agreed between the two major forces in March 1997, trimmed the banned categories and permitted all members of the opposition to run for office. After the elections, the SP majority pushed for new amendments, which ultimately transformed the decommunisation package into a merely formal act. In January 1998, the Socialists also voted to restrict the categories of those banned from public office to include only "senior officers," "leading functionaries" of security services, and "collaborators" in political trials, while preserving the provision of total secrecy around the Committee's decisions and nondisclosure of files. In May 1998, a new Verification Committee appointed by the Socialists claimed that it had reviewed the credentials of 3,000 individuals employed in various state positions, and had disqualified 84 of them. Nevertheless, the Committee never disclosed who these individuals were and what measures were taken against them (Austin and Elisson 2008). Only four members of parliament were declared guilty of collaboration with the communist regime, but they did not have to give up their positions since the revised law excluded elected officials from the ban. Shortly thereafter, the Committee declared that the new administration was 'clean' of any communist collaborators while the law itself was not exercised again.

The Second Wave of Transitional Justice 2005–2009: Political Agendas and Civil Society

The Challenge of Transitional Justice

Two decades after the fall of communism, Albania still faces the same questions regarding investigation, disclosure, and condemnation of perpetrators of communist-era abuses. According to commentators like Fatos Lubonja, a well-known public intellectual and former political prisoner under communism, Albanian citizens feel disappointed by the politicization of the "verification" process in the 1990s and have grown increasingly suspicious of politically motivated campaigns (Lubonja 2008). Though civil society organizations have sharply increased in number, they have not proposed a concrete alternative model of transitional justice (Hatschikjan 2010). Moreover, the lack of rule of law and the precarious nature of independent institutions continue to be fundamental challenges to any serious engagement with the communist past.

There are also plausible concerns that archival files may have been destroyed during moments of social unrest, especially during extreme social disorder in 1997. An important aspect of dealing with communist repression, access to communist-era archives remains problematic in Albania. After assuming power in 1997, the Socialists failed to liberalize access to archival material. Even as state archives have been largely opened to researchers and the public, security police files remain off limits. A law on archives passed in 2003 permits researchers to access files created more than 25 years ago. But the most significant challenge to anyone researching the communist past is a 1999 law restricting access to state secrets. Neither of these laws makes a distinction between secrets classified under communist rule or after 1991, thus creating a paradoxical situation in which certain documents are accessible according to one provision (the 25-year rule) but off-limits according to the other (due to their still-classified status). Even archive personnel are often confused about the declassification status of certain materials and, afraid that they might be fired, they tend to deny access to researchers or individuals lacking security clearances (Mëhilli 2009). Even if these laws were fixed, dealing with the archival legacy of the communist regime would continue to be a massive ordeal. Declassification has stalled, whereas funding, personnel training, and resources are extremely scarce. For example, the archive of the Ministry of Interior, which houses the infamous files of the security police, is located in a rundown building resembling a state prison. Materials are piled inside damp rooms and documents are not cataloged. Despite the heated public debate on these files, virtually no resources have been allocated to this particular institution and its staff—yet another indication of the lack of seriousness with which Albanian actors in either side of the political spectrum have approached the issue of historical reckoning. Under these conditions, some authors have expressed doubt whether the country can manage at all to deal with intricate questions of the past (Austin and Elisson 2008: 397).

New Political Initiatives and Growing Social Activism

Albanian politicians have continued to capitalize on anticommunist campaigns. This issue became especially salient after the 2005 elections and the return of the DP in power. The latest initiative, much like the previous one in the mid-1990s, concurred with the run-up to elections in 2009. Crucially, it also coincided with the prosecution of one of the most serious corruption charges related to the so-called Gërdec affair, which exposed several government ministers.

The adoption of the Council of Europe resolution on the need to condemn communist crimes in 2006 had already pushed this issue forward in the Albanian political agenda. Different parties, including the Socialists, had proposed as many as four draft laws in the Albanian parliament. Nevertheless, the DP decided to advance its own draft without consulting with other political forces. It subsequently passed the bill by a simple majority in December 2008 (Perndoj 2008). As adopted, the controversial law required the lustration of members of government and the legislative and judicial branches. It also provided for the creation of an Authority responsible for screening personal and professional backgrounds (Perndoj 2009).

This time around, however, a new coalition of social and political forces openly criticized what they perceived as highly problematic legislation. In addition to opposition members, independent observers, domestic civil society activists, and public intellectuals, foreign diplomats also voiced serious concerns. (Embassy of the US in Tirana 2008; OSCE 2008; Voice of America 2008; Council of Europe; 2009). The opponents brought forth a number of arguments against the bill. First and foremost, they pointed out that the bill threatened to create serious conflicts with constitutional rules guiding the election of high-level legislative and other officials, since the lustration act was passed as a separate law rather than as a constitutional amendment. Opponents also took the law to task for presuming guilt, insofar as the bill contained very broad and ambiguous references to all persons formerly "employed" by the security police but failed to specify who was to be included and excluded. There were also vague references to incriminating participation in "political trials" and "political processes" which worried some observers that overseeing officials could fault candidates at will. Finally, the provisions guiding the creation and activity of the Authority tasked with screening political candidates came under serious attack. Opponents pointed out, for example, that given the pervasive culture of political appointments in Albania, these individuals could be handpicked according to political interests. The law contained only vague references to the solicitation of recommendations for members of civil society organizations (Balaš 2010). A number of Albanian opponents suggested that the lustration campaign was aimed at independent institutions, especially the office of the public persecutor involved in the investigation of government corruption cases (Lubonja 2008). Indeed, the prosecutor in charge of the Gërdec case was publicly mentioned during the Democrats' promotional campaign. Fearing repercussions, he resigned as soon as the law entered into force in January 2009. The judicial process related to Gërdec, as a result, lost momentum.

The opponents of the bill also mobilized domestic institutions in their campaign. Once the law went into force in January 2009, a number of groups and individuals filed complaints with the Constitutional Court on grounds that the law was unconstitutional. These included a group of over 30 Socialist MPs (including 2 members of the DP majority), the association of prosecutors, and the Albanian Helsinki Committee. (The latter filed the complaint as a representative of the civil society.) When the Constitutional Court suspended the law, while forwarding it for review to the Venice Commission, the ruling majority mounted an aggressive campaign against the institutions involved. The Speaker of the Parliament, a high-ranking DP official, issued memos to the Court "asking" it not to suspend the law. The head of the DP parliamentary group declared that they would not recognize the court's decision because some of the Court members had a "conflict of interest," implying that the judges themselves had been involved with the previous communist regime (hence their opinion on the matter would be compromised). Through its legal department, the Council of Ministers requested the court to bar four out of its seven members from the deliberations, with the motivation that they had served under the communist regime. Premier Berisha, finally, charged that the judges who voted to defer the law did so with the intention of detracting from their own wrongdoings under communism, although no evidence was ever produced to back up these claims.

The Role of Plural Public Contestation

Both the number of actors involved and the duration and quality of public discussion on lustration were significantly higher in 2008–2009 than in the 1990s. A number of new civil society actors encouraged an intense public debate around the issue. Numerous analysts and intellectuals wrote editorials and debated the law's shortcomings in the press and on television. The media was very active in scrutinizing the law and providing space for arguing its merits and faults. This unfolding debate also showed how repeatedly botched attempts to institute a lustration law in the past had actually served to discredit any efforts to conceive a credible transitional justice model. For Lubonja, for example, lustration in Albania was not only impeded by the political maneuvers and tactical calculations of the DP. The Socialists, too, he wrote at the time of the law's passing, retain a powerful electoral base among aging former communists so they are very reluctant to engage meaningfully with the question of historical reckoning and communist crimes. If Albanian governments had failed so miserably in punishing corruption among their ranks, or raising the level of accountability, the author asked, what guarantees were there that these same people were now genuinely interested in a lustration project? (Lubonja 2008)

To be sure, one cannot idealize the role of civil society actors in the Albanian context. Civil society actors were not able to establish the hierarchy of the issues debated in the public sphere or outlined in the political agenda of the parties (Backer 2003). Particularly noteworthy during the debate of the lustration law in 2008, moreover, was the relative silence of precisely those individuals whose suffering politicians incessantly talked about: former political prisoners and persecuted individuals. The Pan-National Organization for the Integration of Prisoners and Persecuted Persons of Albania, the formal organization fighting for the rights of this group, held its first national convention in May 2006. Despite the strong bond allegedly linking many of its members and the DP, their pleas for support and accountability were only weakly represented in the political and public debate.

Sadly, though perhaps unsurprisingly, precisely because these people were so brutalized under communism, their voices are still weak and their interests mostly articulated by other actors carrying political weight. In a country where alternative forms of organization and resistance were nonexistent under communism, in short, those formerly persecuted have found it virtually impossible to find the means to mobilize on accounts of historical reckoning.

Besides the increasing involvement of nonpolitical agents in the 2008–2009 debate, external actors were also crucial to the outcome. In fact, it was the foreign observers, including US and EU diplomats, who energetically brought up a number of concerns about the proposed bill. Even the Czechs, whom the ruling Democrats referred to as an example in advocating for lustration, and who, coincidentally, held the presidency of the European Union at the time, panned the initiative. The opinion of the Venice Commission, which deemed the law to have infringed on a range of constitutional rights and guarantees, was the decisive factor in enabling the Constitutional Court to annul the bill. The involvement of the US embassy as well as interventions by the European Union also served to create a field in which civil society mobilization could gain in strength and importance. The multidirectional pressure exerted by this coalition of social and political forces put the law's sponsors continuously on the defensive. Under mounting international and domestic pressure, the ruling eventually agreed to submit to the Constitutional Court's ruling. This marked an unexpected setback for the government. Civil society activists and foreign diplomats hailed the decision as an important step forward.

Conclusions

A mixture of political maneuverings, electoral calculations, but also incrementally strengthening civil society and international involvement enabled a concentrated public discussion on the subject of dealing with communist abuses and what model of transitional justice is appropriate for Albania. Still, it remains to be seen how these actors—especially in light of the involvement of the European Union through its funding initiatives—will impact the ongoing effort in devising a lustration project or, more broadly, programs concerning historical reckoning (Gledhill 2011).

More recently, a number of civil society representatives, including the Albanian human rights group, have indicated that they intend to propose an alternative lustration law (Ballauri 2011; ABC News 2012). Foreign money and resources, especially from the embassies of the United States and major West European countries, have been critical in sustaining the few public commemoration initiatives that have taken place and in encouraging public debate on the communist past. German foundations have often been at the forefront of these civil society initiatives. In fact, German experience has been increasingly deemed important, and various local actors have sought to claim that experience in building their own platforms for an Albanian lustration project. The Konrad-Adenauer-Stiftung, for example, has organized roundtables on the subject matter, including recent discussions with

Rainer Eppelmann, the well-known East German opposition activist (Baze 2011). Nevertheless, it remains far from clear how such experience can inform the complicated Albanian processes in the 2010s. And though civil society and public engagement has made a difference in terms of scaling back politically motivated initiatives, a clear alternative has yet to emerge.

Acknowledgments The authors are grateful to Elidor Mehilli for assistance in writing the third section of this chapter and for linguistic editing.

References

ABC News. 2012. Lustracioni, nismë e re nga shoqëria civile. http://abcnews.al/lajme/politike/1/16290. Accessed 27 March 2012.
Appel, H. 2005. Anti-Communist justice and founding the post-communist order: lustration and restitution in Central Europe. *East European Politics and Societies* 19(3): 379–405.
Ask, D. 2006. *The politics of the Nazi past in Germany and Austria.* Cambridge: Cambridge University Press.
Austin, R., and J. Elisson. 2008. Post-Communist Transitional Justice in Albania. *East European Politics and Societies* 22(2): 373–401.
Backer, D. 2003. Civil society and transitional justice: possibilities, patterns and prospects. *Journal of Human Rights* 2(3): 297–313.
Balaš, V. 2010. Albanian lustration act, its constitutional and international law pros and cons. *Czech Yearbook of International Law* 1: 170–190.
Ballauri, E. 2011. Lustracioni …. Domosdo. *Panorama*, 4. http://www.panorama.com.al/opinion/lustracioni%E2%80%A6-domosdo. Accessed 9 September 2011.
Baze, O. 2011. Dosjet u hapën … "lustrum." *Panorama.* http://www.panorama.com.al/opinion/dosjet-u-hapen%E2%80%A6-%E2%80%9Clustrum%E2%80%9D/. Accessed 30 October 2011.
Bertelsmann Transformation Index (2012). *Albania Country Report.* http://www.bti-project.org/country-reports/ese/alb/.
Biberaj, E. 2000. *Shqipëria në Tranzicion [Albania in Transition].* Tiranë, Ora botime.
Calhoun, N. 2004. *Dilemmas of justice in Eastern Europe's democratic transitions.* New York: Palgrave Macmilan.
Chilton, P. 1987. Metaphor, Euphemism, and the Militarization of Language. *Current Research on Peace and Violence* 10: 7–19.
Chiodi, L. 2007. The profession of civil society in post-communist Albania. *Albanian Journal of Politics* 3: 1–32.
Council of Europe. 2009. Albanian Lustration Law. Statement by Council of Europe Secretary General Terry Davis. http://www.coe.al/index.php?faqe=content/detail&kat=news&id=462&mnu=0. Accessed 16 January 2009.
David, R. 2003. Lustration laws in action: the motives and evaluation of lustration policy in the Czech Republic and Poland (1989–2001). *Law and Social Inquiry* 28(2): 387–439.
David, R. 2004. Transitional injustice? Criteria for conformity of lustration to the right to political expression. *Europe Asia Studies* 56(6): 789–812.
Duffy, T. 2000. Albania: beyond the Hoxha's legacy. In *Transformations of post-communist states*, ed. W. Kostecki, K. Zukrowska, and B. Goralczyk, 37–52. London: Macmillan.
East, R., and J. Pontin. 1997. *Revolution and change in central and Eastern Europe.* London: Creative Print.
Elbasani, A. 2008. *Political transformation and implementation of EU democratic requirements in Albania.* Budapest: Centre for EU Enlargement Studies.

Elbasani, A. 2009. Enlargement instruments and domestic constraints: public administration reform in post-communist Albania. *Südosteuropa* 57: 70–90.
Elbasani, A., and Lipinski, A. 2011. *Public contestation and politics of transitional justice: Poland and Albania compared*. EUI/SPS working papers 2011/2.
Embassy of the US in Tirana. 2008. U.S. Government welcomes postponement of draft lustration law vote (Press release.). http://tirana.usembassy.gov/08pr_1216.html. Accessed 16 December 2008.
Entman, R. 1989. *Democracy without citizens. media and the decay of American politics*. New York: Oxford University Press.
Gledhill, J. 2011. Integrating the past: regional integration and historical reckoning in Central and Eastern Europe. *Nationalities Papers* 39(4): 481–506.
Goldman, M.F. 2000. *Revolution and change in Central and Eastern Europe*. New York: ME Sharpe.
Hatschikjan, M., ed. 2010. *Manual on lustration, public access to files of the secret services and public debates on the past in the Western Balkans*. Thessaloniki: CDRSEE. www.lustration.net.
Horne, C. 2009. Late lustration programmes in Romania and Poland: supporting or undermining democratic transitions? *Democratization* 16(2): 344–376.
Howard, M.M. 2003. *The weakness of civil society in post-communist Europe*. Cambridge: Cambridge University Press.
Kiss, C. 2006. The misuses of manipulation: the failure of transitional justice in post-communist Hungary. *Europe Asia Studies* 58(6): 925–940.
Kubik, J. 2005. How to study civil society: the state of the art and what to do next. *East European Politics and Societies* 19(1): 105–120.
Letki, N. 2002. Lustration and democratisation in East Central Europe. *Europe Asia Studies* 54(4): 529–552.
Linz, J., and A. Stepan. 1996. *Problems of Democratic consolidation: Southern Europe, South America and post-communist Europe*. Baltimore: John Hopkins University Press.
Los, M. 1995. Lustration and truth claims: unfinished revolutions in Central Europe. *Law and Social Inquiry* 20(1): 117–161.
Los, M., and A. Zybertowicz. 2000. *Privatizing the police state. The case of Poland*. Basingstoke: Palgrave Macmillan.
Lubonja, F. 2008. Qëllimi i keq i një ligji. *Korrieri*, 4. http://perpjekja.blogspot.com/2008/12/qllimi-i-keq-i-nj-ligji.html. Accessed 28 December 2008.
McEvoy, K., and L. McGregor. 2008. Transitional justice from below: an agenda for research, policy and praxis. In *Transitional justice from below*, ed. K. McEvoy and L. McGregor, 1–14. Oxford: Hart.
Mëhilli, E. 2009. 25 vjet heshtje. Shekulli, 6, p. 1. 25 May 2009.
Moran, J.P. 1994. The communist torturers of Eastern Europe: prosecute and punish or forgive and forget? *Communist and Post-Communist Studies* 27(1): 95–109.
Nalepa, M. 2008. To punish the guilty and protect the innocent: comparing truth revelation procedures. *Journal of Theoretical Politics* 20(2): 221–245.
Nalepa, M. 2010. *Skeletons in the closet. Transitional justice in post-communist Europe*. Cambridge: Cambridge University Press.
Nalepa, M., and M. Kaminski. 2006. Judging transitional justice: a new criterion for evaluating truth revelation procedure. *Journal of Conflict Resolution* 50(3): 383–408.
OSCE. 2008. OSCE Presence calls on Albanian government to ensure lustration law meets international standards (Press release). http://www.osce.org/albania/50530. Accessed 23 December 2008.
Perndoj, N. 2008. Berisha kundër SHBA: Do miratoj dosjet. *Shekulli*. www.shekulli.com.al/news/45/ARTICLE/38322/2008-12-22.html. Accessed 22 December 2008.
Perndoj, N. 2009. Dosjet, Berisha sulmon Kushtetuesen. *Shekulli*. www.shekulli.com.al/2009/01/dosjet-berisha-sulmon-kushtetuesen.html. Accessed 17 January 2009.
Prifti, P. 1978. *Socialist Albania Since 1944: domestic and foreign developments*. Cambridge: MIT.
Sadurski, W. 2003. "Decommunization", "lustration" and constitutional continuity: dilemmas of transitional justice in Central Europe. *European University Institute Working Paper LAW* 15: 1–50.

Schwartz, H. 1995. Lustration in Eastern Europe. In *Transitional justice: how emerging democracies reckon with former regimes Washington*, ed. N. Kritz, 461–483. Washington, DC: United States Institute of Peace Press.

Siegelman, P. 1995. The Problems of Lustration: Prosecution of Wrongdoers by Democratic Successor Regimes. *Law and Social Inquiry* 20(1): 1–6.

Teitel, R. 2000. *Transitional Justice*. Oxford: Oxford University Press.

Venice Commission. 2009. On the law on the cleanliness of the figure of high functionaries of the public administration and elected persons of Albania.' *Opinion*, 524/2009. Strasbourg.

Vickers, M., and J. Pettifer. 2000. *Albania: from anarchy to a Balkan identity*. London: Hurst.

Voice of America. 2008. Saga e Dosjeve në Shqipëri, reagojnë SHBA, Këshilli i Evropës. OSBE. http://zeriamerikes.wordpress.com/2008/12/23/247. Accessed 23 December 2008.

Welsh, H. 1996. Dealing with the communist past: Central and East European experiences after 1990. *Europe Asia Studies* 48(3): 413–428.

Williams, K., B. Fowler, and A. Szczerbiak. 2005. Explaining lustration in Central Europe: a 'post-communist politics' approach. *Democratization* 12(1): 22–43.

Chapter 8
Truth in the Shadow of Justice

Jamie Rowen

In Bosnia and Herzegovina (BiH), divisive nationalist narratives about the causes and consequences of the war dominate the social and political landscape. In order to maintain power, leaders within the country's two entities—the Federation, which is now primarily Muslim and Croat, and the Republika Srpska (RS), which is now primarily Serb—continue to point out how their constituencies were victimized during and after the war (see Chandler 2000; Toal and Dahlman 2011). In the meantime, survivors find themselves with few economic opportunities and remain concerned that the country may divide further (see Bougarel et al. 2007). In light of these challenges, survivors looking for information about the violence are losing hope that either domestic or international policy makers can provide some semblance of truth in the shadow of justice.[1]

To address ongoing social and political tensions, several civil society organizations, defined as those that enjoy autonomy in relation to the state and are formed voluntarily by like-minded individuals to protect or extend their interests or values, are trying to establish a fact-based account of the violence (White 1994, p. 379). Since the end of the war, organizations working in BiH have pursued a variety of innovative truth-seeking initiatives, defined here as strategies to investigate, document, and publicize facts about the violence. Given their relative autonomy, these civil society actors have had some success in pursuing their goals. However, these organizations also struggle to dispel ongoing disputes over the number of dead and missing, and which national group is to blame for the violence.

Transitional justice scholars have much to learn from civil society organizations that are pursuing truth-seeking initiatives. The goals of transitional justice include

[1] Although truth is a contested concept, I use it to refer to a fact-based account of the violence, and in reference to scholarly debates about truth and justice that defined the early literature on transitional justice (see Roht-Arriaza and Mariezcurrena 2006; Zalaquett 1994).

J. Rowen (✉)
American Bar Foundation, 1401 Scenic Ave., Berkeley, CA 94708, USA
e-mail: jamie.rowen@gmail.com

O. Simić and Z. Volčič (eds.), *Transitional Justice and Civil Society in the Balkans*, Springer Series in Transitional Justice, DOI 10.1007/978-1-4614-5422-9_8,

ensuring accountability, improving survivor well-being, and preventing future violence (see Elster 2004; Teitel 2002; Vinjamuri and Snyder 2004; Olsen et al. 2010: Roht-Arriaza 2006). Given the recent war, many civil society organizations in the Balkans also set their agendas around these goals. While many focus on helping individuals of only one national group, particularly victims or veterans from the war, others have been working since the war to help bridge social and political divides in the region. As the focus of this chapter is to examine civil society organizations (CSOs) that are pursuing truth-seeking initiatives in order to meet these goals, I will focus on three prominent BiH organizations: the Coalition for RECOM (ZaREKOM, which translates to "for RECOM"), the Information and Documentation Center (IDC), and the Youth Initiative for Human Rights (YIHR). In addition, this discussion includes information on the International Commission of Missing Persons (ICMP). This organization, though more an international organization than a civil society organization, is relevant to the analysis of transitional justice in BiH due to its ongoing efforts to improve the quality and impact of truth-seeking initiatives in the region.

The analysis reveals that these organizations face numerous challenges and points to ongoing dilemmas for those seeking new approaches to redress mass violence through truth-seeking in the Balkans and beyond. In addition, the study also highlights the various approaches these organizations use to involve survivors in their truth-seeking initiatives. With this strategy, the organizations work to counter the prevalent belief that foreign actors, domestic and international courts have not adequately addressed the needs of those who suffered (see Delpla 2007; Stover 2007).

Transitional Justice and Civil Society in BiH

Given the ongoing social and political divides in BiH, transitional justice scholars have much to learn from a study on how civil society organizations are trying to ensure accountability, improve survivor well-being, and prevent future violence. In the beginning of the 1990s, the republics in the former Yugoslavia began to declare independence, and the Muslim population in Bosnia, as the territory was then called, voted to create an independent state. The political leaders in Belgrade did not recognize their declaration and newly independent Croatia to the west and Serbia to the east attempted to take control of different parts of the territory. The violence intensified as military leaders tried to "cleanse" entire towns and cities of individuals with a different nationality. With 80% of the casualties occurring in BiH, this country experienced most of the violence in the wars that culminated in the dissolution of the former Yugoslavia. Following the Dayton Peace Accords, BiH was divided into two entities, the Federation, which is now primarily Croat and Bosniak, and the Republika Srpska, which is now primarily Serb.

During the war, the United Nations created the ad-hoc International Criminal Tribunal for the Former Yugoslavia (ICTY), hoping to deter ongoing war crimes and crimes against humanity (see Hagan 2003). National groups continue to express divided opinions about courts, particularly the ICTY (Nettelfield 2010). Bosnian Serbs feel victimized by the court, which they see as biased against them, and Bosniaks feel victimized by what they view as lenient sentences for convicted war criminals. This divide is also salient for the Bosnian War Crimes Chamber, which has struggled to overcome challenges related to funding and local politics (Barria and Roper 2008).

In addition, youth throughout the country are educated in "Two Schools Under One Roof," the name used to describe the segregated education system developed by the Organization for Security and Cooperation in Europe in 2001. While this educational system was designed to encourage survivors from different national groups to return to their communities, it has resulted in a generation of youth learning disparate, nationalist accounts of the war (Bartulović 2006; Kreso 2008). The courts and education system are two of many examples that highlight how well-intentioned policies created additional challenges for the already polarized country.

Though scholars increasingly discuss the need to examine nongovernmental efforts to ensure accountability, improve survivor well-being, and prevent future violence, there is not yet a robust literature on civil society truth-seeking initiatives (Backer 2003; McEvoy and McGregor 2008). Empirical studies reveal that survivors want to share their stories and learn from others about what actually happened during the violence (Crosby and Lykes 2011; Riaño-Alcalá and Baines 2011). However, as Daly (2008) notes, policy makers often respond to this desire and create truth commissions in countries where leaders may ignore "alternative suitors that might bestow similar outcomes" (23). Scholars suggest that outreach to local communities might increase the impact of transitional justice initiatives, yet it is not clear how civil society organizations can effectively communicate and involve survivors who want to provide information or learn more about the violence (Backer 2004; Hodžić 2010; Van der Merwe and Chapman 2008; Vinck and Pham 2010).

Scholars who are interested in how transitional justice is being "localized" must pay particular attention to civil society organizations (Shaw et al. 2010). The challenges they face, and the opportunities they create, illustrate broader theoretical and policy questions related to the value of investigating, documenting, and publicizing information about violence. By looking at advances in nongovernmental efforts for justice, one learns effective and efficient ways to bypass policy makers. Given that CSOs in post-conflict settings usually get funding from transnational actors in order to serve local communities, an examination of CSO goals and strategies can serve as window into the beliefs and practices of both local and transnational actors interested in promoting accountability, improving survivor well-being, and preventing future violence.

In order to examine new developments along these lines, the rest of this chapter will focus on four organizations that are working to meet the goals of transitional justice. While there are a variety of organizations that are trying to investigate, document, and publicize information about the violence, these organizations are particularly well-known and influential in the country, in the region, and internationally. Moreover, this sample of organizations was selected on account of their innovative truth-seeking initiatives, and due to the fact that several of them have adopted and adapted the concept of transitional justice to explain their goals and strategies. As a result, they are a useful sample to explore truth-seeking, civil society, and transitional justice in the Balkans. The data for this study were collected from 2009 to 2011 during four field visits to the Balkans, and include open-ended, in-depth interviews with 56 interviewees involved in various civil society initiatives to assist survivors and the new generations. All but two of the interviews were conducted in English, one relied on a professional interpreter and the other was conducted in the local language and translated by the author. Interviewees who chose to remain anonymous are identified by the number under which their interview transcript was filed. The interview questions focused on the goals and strategies of their work, as well as their understandings of transitional justice.[2]

The findings and analysis of this study suggest that civil society initiatives focused on investigating, documenting, and publicizing information can provide important insights into how transitional justice initiatives may develop and take shape in the future. In particular, they reveal how these organizations are taking advantage of science and technology to improve communication and outreach to those transitional justice is supposed to serve—those who must live together after the violence.

Coalition for RECOM (Regional Commission Tasked With Establishing Facts About The War 1991–2001)

The Coalition for RECOM (ZaREKOM or "for" RECOM) originated from a partnership between three leading human rights organizations in the Balkans: the Information and Documentation Center (IDC) in BiH, the Humanitarian Law Center (HLC) in Serbia, and Dokumenta in Croatia. During internal discussions, the founders became interested in a new initiative to address ongoing concerns that the international and domestic courts were not contributing to an authoritative account of the war. These organizations began holding regional forums around the region, seeking input from survivors, scholars, youth, and others interested in new strategies to investigate, document, and publicize facts about the violence.

The founders knew that they would have to address the skepticism with which survivors view both governmental and civil society initiatives designed to help them, particularly when foreigners are involved (Coles 2002). As a former HLC staff member who participated in early discussions explained, the organization was particularly wary about calling their proposal a truth commission:

[2] The data also include detailed notes on 14 public events in Sarajevo where civil society leaders presented their work, and a content analysis of dozens of documents provided by the individuals and organizations analyzed here. For more information on these methods, please see Jamie Rowen, (forthcoming 2012), "Mobilizing truth: Agenda setting in a transnational social movement," *Law and Social Inquiry*.

> The founders of the RECOM Initiative would say it's not about a truth commission. They would say it's about a fact-finding body. They want to get away from this idea from using the term *truth*. In essence, it remains the same. It's essentially a truth commission, but they were talking about a body that would establish the facts about human rights violations rather than a body that would establish the truth, which seems more sort of esoteric and just—a bit out there, really (67, personal communication, 16 May 2011).

Part of this skepticism towards truth commissions came from the leaders' awareness that most people in BiH are unfamiliar with truth commissions. For example, A 2010 United Nations survey study of 1,600 individuals throughout BiH found that two-thirds of the population do not know what a truth commission is and, among those who do, 90% want one in the country. 70% of the respondents said that "facts about the war have not yet been fully established."[3] Moreover, ZaREKOM was not the first attempt to create a truth commission in BiH, though it was the first effort initiated by regional civil society leaders. At the turn of the millennium, the United States Institute of Peace lobbied the governments for a truth commission, but some local civil society organizations as well as the ICTY prosecutor were concerned that a commission might usurp the authority of the tribunal and survivors were not properly consulted about its functions (Kritz and Finci 2001). Serb delegates in the BiH parliament proposed a commission that would uncover the "scientific truth" about the violence in Sarajevo, but this initiative failed when the parliament would not vote on the proposal (Buljugic 2007). There was commission of inquiry into the events at Srebrenica, which had no public hearings but helped establish the number of victims killed in the massacre, and another commission on violence in Bijeljina that began but never concluded its work (Somun 2010).

Rather than promote a commission that would investigate the violence in one town or in one country, the founders decided to promote a regional commission. It would be a treaty body signed by the governments of BiH, Croatia, Serbia, Montenegro, Macedonia, Slovenia, and Kosovo, though neither BiH nor Serbia recognizes Kosovo as an independent country. This issue caused immediate friction within the growing coalition, and other challenges quickly arose.

First, the IDC, which was to lead the effort in BiH, stepped out before the Coalition was formalized. The director explained to me his concern that the initiative was doomed to fail from lack of popular and governmental support:

> Yes, I was involved, and then I withdrew. Why? Simply because three years of our very intensive work, there is no reaction from State…. Then I don't want to travel, like a circus around the region, and to have every two, or one, or three months consultation with somebody, and stop my practical work. And then what will happen finally? Nothing. Why? Because there is no political atmosphere. The States are not ready, Serbia doesn't recognize Kosovo, Bosnia doesn't recognize Kosovo…. So this is not time for commissions basically…. And I have to reiterate. It is my moral act. I don't want to spend money, my dollars, and then to say, "Look, we failed" (Mirsad Tokaca, personal communication, 6 November 2009).

[3] Pajic, Zoran, and Dragan Popović, "Dealing with the Past and Access to Justice from a Public Perspective," Presentation from UNDP Conference on Transitional Justice, 31 March 2010.

Despite the withdrawal of the BiH leader, the HLC and Dokumenta decided to forge ahead with an ambitious set of goals: (1) to create an accurate, official, and unbiased account of war crimes and other serious violations of human rights, (2) to increase recognition of victims' suffering, and encourage recognition of each nationalist groups' suffering, and (3) to prevent a recurrence of violence (Alma Masic, university lecture, 20 May 2011). To ensure public support, individuals and organizations could become members of the Coalition by signing its charter. During 2008–2010, coordinators in each country would lead consultations with victims associations, religious leaders, public intellectuals, youth groups, and others interested in learning about the new truth-seeking initiative. At the consultations, leaders would provide information about transitional justice and the benefits and limitations of truth commissions in other countries. In this way, they hoped to educate survivors as well as learn about their hopes and desires for this proposed commission.

To replace the IDC, the Serbian and Croatian leaders chose civil society organizations throughout BiH to coordinate consultations with those interested in a new fact-finding strategy. These organizations included Izvor, a victims association from Prijedor, the Citizens Forum from Tuzla, the Youth Initiative for Human Rights from Sarajevo, and the Center for Civic Cooperation in Livno. The Helsinki Committee for Human Rights-Republika Srpska was chosen as the primary leader for the BiH activities. In addition, a working group of individuals knowledgeable about war crimes, law, and transitional justice would draft a mandate for the proposed commission. At the end of the consultations, the Coalition would begin a media and signature campaign to rally public support throughout the region. The leaders envisioned that there would be enough public pressure to compel the governments to create the proposed commission.

Even with the new BiH leaders, challenges arose throughout the consultation process. In particular, though the International Court of Justice (ICJ) determined that Serbia is not liable for its collaboration with Bosnian Serb generals, the ICTY and the ICJ declared the massacre at Srebrenica part of genocide against the Bosniaks.[4] Some Bosnian Serbs hoped that a commission might be able to undermine this judicial finding, while most Bosniaks would not support an initiative that calls the genocide into question. Early on, ZaREKOM's founders decided that the commission would not revisit the question of genocide and would simply build from the tribunals' conclusion. Two important RS organizations, the Association for Prisoners in Republika Srpska and the Republika Srpska Team for War Crimes Research, eventually called on its supporters to boycott the Coalition. Moreover, many Bosniaks worried that the HLC is from Belgrade and that the proposed commission would be used to justify Serb aggression. A Croat civil society leader noted his concerns that Bosnian Croats groups were not included in early discussions and, thus, he declined to join the Coalition (79, personal communication, 7 May 2010).

[4] Krstic Decision. (ICJ Bosnia and Herzegovina v. Serbia and Montenegro), case 91, International Court of Justice (ICJ) Judgment returned on 26 February 2007.

Thus, despite the founders' intent to reach out to survivors, important representatives from all of the national groups were reluctant to join the Coalition.[5]

From the perspective of some Coalition leaders, these problems had more to do with BiH civil society and nationalism rather than this particular initiative:

> And as time elapsed, we see that there are no strong organizations who are capable to, for example, convince people from other entity to come on consultations. So there is no authority among these organizations.... I see a weakness of civil society really in this fact that civil society is not independent from present political elite or context (52, personal communication, 1 April 2010).

This statement, from a Coalition leader in Serbia, was echoed by BiH civil society leaders. These actors noted that the tenuous political situation leaves little opportunity for initiatives that require independence from the government and collaboration across nationalist groups. Moreover, victims associations continue to hold tremendous political power in the country, and efforts to investigate, document, and publicize facts about the violence require their support.[6]

Among those who did participate in the consultations, some felt that they benefitted from having the opportunity to speak publicly about their experiences and hear the perspectives of other national groups.[7] However, several leaders from the victims' associations expressed that they were silenced in the consultations and leadership meetings, and nationalist newspapers reported on how tensions grew between the HLC, particularly its director, and the other BiH Coalition members.[8] At the end of 2009, the HLC brought in a media organization, the BiH Association of Journalists, to help their initiative gain public support in the country. Disputes continued through the first half of 2010, and the five BiH organizations that had been leading the consultations stepped out of the Coalition.

The BiH Association of Journalists, along with the Center for Informative-Legal Support in Zvornik, remained in charge of the consultations and was able to publicize ZaREKOM through its connections with local media outlets. In the previous year, leaders created an online website for individuals to sign the petition and aired TV commercials to prepare for the signature campaign in the spring. Paid volunteers stood at major thoroughfares in towns and cities throughout the region and people throughout BiH quickly learned about the initiative. However, even with a deadline extension from June 6 to June 26, the Coalition was unable to meet its target of gathering one million signatures on a petition asking the governments to

[5] For a detailed overview of these challenges, including the responses of the various victims' associations, see Igor Mekina, Analysis of Public Criticism and Support for the Initiative for RECOM. www.zarekom.org/uploads/documents/2011/09/i_1647/f_1/f_2826_en. Accessed August 2011.

[6] One well publicized example of their power can be seen in the dispute over Angelina Jolie's movie on a Bosnian war camp. The plot involved a romance between a rape victim and her perpetrator. A BiH rape victims group was able to lobby the government to temporarily withdraw the filmmaker's production permit.

[7] See Regional Debate on the Mandate of RECOM: Review of opinions, suggestions, and RECOMmendations Report May–December 2009, Novi Sad, Vojvodina, Serbia, 20–21 March 2010.

[8] See, e.g., Kandic Asks that Donations to her Critics be Revoked!? *Dnevni Avaz*, 25 March 2010.

create RECOM. At the end of the campaign, 122,473 signatures were collected in BiH and 542,660 in the whole region.[9] Several prominent BiH politicians expressed support for RECOM in theory, but it remains to be seen whether their statements will lead to any action.[10]

While nationalism in BiH might undermine any collaborative truth-seeking initiative, ZaREKOM's challenges extend beyond existing social and political divides. By setting its goal to create a government sponsored commission, the Coalition may have created expectations that are impossible to fulfill. Following the campaign, the director of the Helsinki Committee for Human Rights-Republika Srpska publicly stated "the entire RECOM project is envisioned completely wrong, too ambitious and is completely unrealistic."[11] Although the Coalition faced criticism from a number of civil society organizations in BiH, it engaged thousands of people and sparked public debate about truth-seeking initiatives and the organization leaders state that they will continue their efforts through 2013.[12] Its successes and ongoing challenges highlight both the challenges and the opportunities for civil society organizations hoping to engage survivors in a participatory truth-seeking initiative.

Investigation and Documentation Center

Like ZaREKOM, the IDC has set its agenda around investigating, documenting, and publicizing facts about the violence. The Center originated in 1992 from the State Commission on War Crimes and Genocide, and was tasked with collecting data on victims of the ongoing war. From that original data, the IDC created the Human Losses Project in order to develop an accurate count of deaths caused by the violence. At the time, leaders from different nationalist groups presented various figures, from 30,000 to 300,000 dead and missing, depending on whether they wanted to minimize or exaggerate the consequences of the war.[13] Based on years of

[9] These figures are cited in Youth Initiative for Human Rights, "Newsletter, July/August 2011," YIHR Index: YIHR-08-13727 24 August 2011. In order to understand these overall numbers, it is also important to highlight how the shadow of justice extended to neighboring countries. The Coalition leaders in Croatia seemed to have more success bringing together various parties, but the indictment of popular war-time general Gotovina in May 2011 undermined their efforts to collect signatures. Many Croatians believe that Gotovina is a hero who defended the country from Serb aggression and resent the international community for condemning him as a war criminal. The Coalition was only able to collect 19,668 signatures. ZaREKOM seemed to have most success in Serbia and Kosovo, where many youth appear interested in understanding the recent wars.

[10] Notably, the chairman of the BiH Presidency, Zeljko Komšić, pledged support for the establishment of RECOM.

[11] L. Kovacević, Is RECOM Collapsing? *SETimes*, http://www.eurasiareview.com/03072011-is-recom-collapsing/. Accessed 3 July 2011.

[12] To Establish RECOM in 2013, *Nezavisne Novine*, 9 July 2011.

[13] A prominent Croatian human rights advocate and scholar relayed to me that the only time he has been verbally attacked for speaking about the war was when he arrived in 1996 and said that 100,000 people were killed (79, personal communication, 8 May 2010).

data collection and statistical analysis, the organization published its estimate that 97,207 individuals were killed during the war in BiH. Three independent international experts, including one from the ICTY, validated the findings and the IDC became known internationally for its compilation of statistics in the study of the dead and missing.[14]

Despite the international attention, the Center has struggled to gain support from the survivors in BiH. As a former staff member explained:

> I can't say that we were physically attacked in any of those municipalities or threatened in any concrete way. The biggest obstacle—the biggest resistance [had to do with] our impartiality, trying to make a connection between offices that we have in Sarajevo, Goražde, and Srebrenica—places—mostly communities or places, towns, mostly inhabited by Bosnian Muslims. And [the] head of the IDC was the former chief of the state commission for gathering facts on war crimes and genocide established in 1992 by the President of Bosnia and Herzegovina. In several ways, those nationalistic representatives tried to blame the IDC in that direction of bias (73, personal communication, 18 May 2010).

The former staff member's statement highlights the ongoing problems that civil society organizations face when trying to change entrenched narratives about the war. Given that the director of the Center is a Bosniak, many survivors assume that he will enhance the number of Muslim victims. Ironically, given that the IDC's estimates of the dead are far less than what Bosniak leaders had claimed, the leader has struggled to convince members of his own national group that his findings are unbiased.

Following their work on the Human Losses Project, the Center decided to develop new strategies to investigate, document, and publicize facts about the violence. One project is called the "Positive Stories" and includes volumes of information about those who risked their lives to save others during the war.[15] In addition, the organization wanted to reach more survivors and interested parties than is possible with published reports and town visits. Using Google Earth technology, the organization developed the War Crimes Atlas, launching the project in November 2009.[16] At the launch event, the director projected the website, found the geo-coordinate of the building where the audience was sitting, and clicked on an icon. The audience sat still as a video showed the building ablaze during the war.

As the brochure for the Atlas states, the goal was to make "research more accessible to a wider public in a simple and efficient way both in the country and abroad." Individuals can search the site for information on loved ones who were killed, post pictures of memorials, and find documents from domestic and international tribunals. Like the ZaREKOM initiative, this project is designed to provide survivors with the opportunity to inform others about their personal experiences. As the

[14] See, e.g., Bosnia War Dead Figure Announced, *BBC News*, Accessed 21 June 2007. http://news.bbc.co.uk/2/hi/europe/6228152.stm; Balkan Investigative Reporting Network, Justice Report: Bosnia's Book of the Dead. http://birn.eu.com/en/88/10/3377/. Accessed 21 June 2007.

[15] Another Sarajevo-based organization, Gardens of the Righteous Worldwide (GARIWO), has been engaged in similar work to gather and disseminate stories of survivors who saved others.

[16] See Information and Documentation Center, War Crimes Atlas. http://www.idc.org.ba/.

director of the IDC explained in our interview, he presents truth-seeking as an individualized process involving self-expression:

> I need individual history. I need to offer you my story. I need to offer you my personal or family history … to have confidence in the people, the grass roots organizations, victims' groups, to offer them something, to respect the individual rights of people to know (Mirsad Tokaca, personal communication, 6 November 2009).

In this statement, he implicitly contrasts his project with that of the tribunals. The idea that people have a "right to know" shifts the notion of rights away from judicial accountability and provides a legal foundation for truth-seeking initiatives.[17] In a similar vein, the director also explained that War Crimes Atlas is designed to enable individuals to make their own judgments about the facts. In a conference on the legacy of the ICTY, the IDC's director explained his reasoning behind this decentralized approach to truth-seeking:

> [E]ach of us has our own personal truths. Munira's truth (a female audience member who commented)—is her truth about the suffering of her kids, her children, and no one has the right to take that away from her. As well is it is true for Mr. Dukic (a male audience member who commented) that someone took him out, beat him up, and shattered his ribs or whatever else they did. So we keep ignoring things because of our narrative. And our narrative, our past narrative, has been that only my truth and only my tragedy is what is valuable (Mirsad Tokaca, personal communication, 6 November 2009).

In his comments, he reiterates the belief that the narrative needs to change but truth-seeking initiatives should not be seeking to establish a unitary "truth," which he views as a goal of courts. Following this reasoning, a web-based approach to investigating, documenting, and publicizing facts can be particularly useful. It has the potential to efficiently inform many survivors and others who are interested in learning about the violence, but are skeptical of the courts. However, while creating a platform on the web may disseminate facts more effectively, it requires the IDC to assess the accuracy of the information and to protect the rights of those accused of violence. Moreover, only those with access to the web and an interest in this information will actually seek it out. Despite these limitations, this organization is but one example of several civil society initiatives that use the web to make sure that information of the violence reaches as many people as possible.

Youth Initiative for Human Rights: Sarajevo

The Youth Initiative for Human Rights (YIHR) is well-known in the region for bringing together youth from different national groups to learn about each other and the war. In the past few years, the organization has also engaged in new strategies to

---

[17] The Human Rights Council has developed a doctrine on the "right to the truth" as critical to "end impunity and to promote and protect human rights." United Nations Human Rights Council, *Resolution on the Right to Truth*, A/HRC/9/11, 18 September 2008.

investigate, document, and publicize information about the violence. For one project, "Mapping Genocide," the organization created a website with information about legal definitions of genocide and the history of violence at Srebrenica. The site states that the organization engaged in "comprehensive research" to develop this strategy for outreach:

> In order to get a modern and user friendly tool, particularly intended for younger population[s], after comprehensive research, we decided to adopt documentary animation as a new model of presenting facts, whilst applying modern technology for its user-friendly application and dissemination.[18]

This description reveals how the organization's goal was to reach a broad audience, particularly the youth of the region. As younger generations are accustomed to seeing images on film or on the web, utilizing documentary animation and the web seemed an effective and efficient strategy to meet its goals. Despite its stated goal of educating others on the history of Srebrenica and genocide, referring to Srebrenica as genocide is very problematic for those Bosnian Serbs who believe the violence should not be classified in this way. Thus, while the information may be more accessible, it may reinforce contentious narratives about the causes and consequences of the war.

The organization is also involved in a project called "Survivor House," which includes an online repository of documents with information on how residents of Sarajevo survived the 1992–1995 siege of the city. This latter project is being designed to memorialize the resilience alongside the suffering of survivors. As the YIHR director Alma Masić noted, truth-seeking initiatives that investigate, document, and publicize information about how people survived is necessary to create a sense of solidarity:

> The idea is that it is dangerous when everybody behaves as a victim. This is to make something dignified. Even the people reading texts on the mapping genocide project, the voices are plain. We don't want to tell you to make conclusions. Here are facts, judge it, it's not about victimization.... We will see a sense of solidarity came out, sharing an apple or an onion. No one had it but after the war, the world collapsed, you forget and become greedy, looking for status, grabbing, all basic values behind. Your sign of success is when you show it off. Big house, big car, skiing in exotic places, but no underwear on you.... Where is solidarity? Does something like this have to happen again (Alma Masic, personal communication, 17 May 2011)?

Here, Masić expresses a common concern among civil society leaders that courts entrench existing feelings of victimization. Moreover, her comments highlight the challenges of being a civil society leader who has been working for decades in order to bring together the different national groups. In discussing both projects, Masić provided insights into a major source of this frustration:

> One main issue here is the cost.... I was always a grassroots worker and believe you can do fabulous things with little money. It doesn't spoil you. It doesn't become the purpose of your existence (Alma Masic, personal communication, 17 May 2011).

[18] Mapping Srebrenića, About Us. Retrieved from http://www.srebrenica-mappinggenocide.com/en-m/.

Her comment is representative of statements by other civil society leaders who are worried about funding, both how to get it as well as how it shapes organizational goals and strategies. As civil society organizations continue to develop new approaches to investigate, document, and publicize facts about the war, they necessarily think about their own survival as well as the long-term stability of BiH. With this in mind, they will look for cost-effective ways to conduct outreach. The web may be a particularly useful strategy to address this challenge. Although it requires technical expertise, it can be a more cost-efficient medium than print and enables civil society actors to present the information in more creative way.

International Commission on Missing Persons

Although it is not technically a civil society organization, the International Commission on Missing Persons (ICMP) deserves the attention of scholars investigating civil society and transitional justice because of its unique organizational structure and ongoing work with courts, policy makers and survivors to investigate, document, and publicize the violence. The ICMP originated in a political agreement between the countries involved in the Dayton Peace Accords; though autonomous from BiH, and the Commissioners and steering committee include government representatives from around the world (Wagner 2008). Working with the BiH Council of Ministers, the organization helped found the Missing Person Institute (MPI), a BIH government agency that was developed to take over the ICMPs work on investigating missing people, storing the data, and working with the tribunals.

Like the IDC, ICMP has developed several innovative, science-based strategies around truth-seeking. The Commission's primary goal is "to ensure the cooperation of governments in locating and identifying those who have disappeared during armed conflict or as a result of human rights violations."[19] It relied on DNA-based techniques to identify the remains of victims exhumed throughout the region. Since its inception, the ICMP has developed a database of nearly 90,000 relatives of over 29,000 missing people, gathered more than 36,000 bone samples. By matching DNA from survivors' blood, the Commission has been able to identify over 16,000 victims.[20]

Referring to the organization's goals, the chief operating officer explained to me:

> For some people it's a forensic science program. For some people it's a humanitarian action to be taken with people who've lost family members. But really what it comes down to is ICMP supports the rule of law and … it really tries to build the capacity of governments to address the human rights needs of their citizens (Adam Boys, personal communication, 9 November 2009).

---

[19] International Commission on Missing Persons, About ICMP. http://www.ic-mp.org/about-icmp/.

[20] These numbers are regularly updated on the organizations website. See International Commission on Missing Persons, About ICMP. http://www.ic-mp.org/about-icmp/.

In this statement, he relates truth-seeking to justice by suggesting that accurate information is essential to ensure fair trials, and trials are a fundamental part of a democratic state. He further explained that there is no unitary truth that a judicial or quasi-judicial body can uncover but, rather, there are "some truths" to establish:

> We can at least limit the narrative that sort of changes the truth about what happened and, therefore, if we can establish some truths, for example the number of killed and missing, the names of those people, feed those into court processes, then we're no longer dealing in an abstract (Adam Boys, personal communication, 9 November 2009).

His comment reveals a common belief that truth-seeking initiatives might bridge the social and political divides. The goal of establishing "some truths" suggests that the organization believes its work might "limit the narrative" that nationalist leaders manipulate. To this end, the Commission has tried to avoid claims that it is biased in favor of one national group or the other. However, various media outlets and political leaders continue to criticize the organization in the same way they criticize the ICTY, claiming that the Commission is funded by governments who are biased against Serbs and staff contaminate their data to prove that Serbs defendants are genocidaires.[21] Such claims have made their way to the ICTY, where defense lawyers for Karadźić have asked for verification of the ICMP's work.[22]

In an effort to address these concerns and increase its impact in the local community, the Commission developed a new program entitled "mutual understanding and transitional justice."[23] The name of the program suggests that the Commission views transitional justice as a long-term process to prevent future violence. As part of this program, the Commission is engaged in discussions with local civil society leaders on how to memorialize victims and how to lobby the government for benefits. Moreover, the Commission has created an online inquiry center for family members to access information from the organization's database.[24] From the site, individuals looking for missing persons can track the status of a case and provide information about family members who might be able to provide DNA.

As one staff member explained, these new strategies are necessary in light of what the ICMP has already accomplished, and what remains to be done:

> There were 29,000, 30,000 people who went missing in Bosnia. ICMP has helped assist local institutions to find and identify about 20,000 of them—two thirds, which is just unprecedented anywhere in the world. And it's probably going to become ever harder to find the remaining third. And so for us, we're not turning our back on the exhumations and

[21] The ICMP has been targeted largely because of its close relationship with the ICTY. Questions about the number of dead have come up in several cases at the ICTY. See, e.g., *Prosecutor v. Vujadin Popovic (Judgment)*, IT-05-88-T, International Criminal Tribunal for the former Yugoslavia (ICTY). http://www.unhcr.org/refworld/docid/4c1f69fe2.html. Accessed 10 June 2010.

[22] International Criminal Tribunal for Former Yugoslavia, *Prosecutor v. Karadžić*, *Order on Selection of Cases for DNA Analysis*, IT-95-5/18-T (19 March 2010).

[23] International Commission on Missing Persons, Mutual Understanding and Transitional Justice. http://www.ic-mp.org/activities/public-involvement-civil-society-initiatives/mutual-understanding-and-transitional-justice/.

[24] International Commission on Missing Persons, Online Inquiry Center. http://www.ic-mp.org/fdmsweb/index.php?w=intro&l=en&2956.

> the technical assistance we provide in terms of DNA identifications, but it's time to start thinking about how we commemorate the 20,000 that have been identified, and also how do we commemorate those that haven't? Ultimately, these efforts are designed to improve social relationships between the nationalist groups and prevent a recurrence of violence, the long term goal of many who advocate for transitional justice (76, personal communication, 20 May 2011).

His statement reveals a common desire to improve outreach around these truth-seeking initiatives in order to prevent a recurrence of violence. While the connection between truth and prevention is unclear, this organization wants to dispel skepticism about its work. Rather than focus on the courts or BiH policy makers, the ICMP is now working more closely with civil society organizations and survivors who might be able to use the Commission's information to bridge the ongoing divides between national groups.

Civil Society, Truth-seeking, and the Future of Transitional Justice

Through investigating, documenting, and publicizing facts about the war, these four organizations are trying to realize the goals of transitional justice. The findings suggest that, although they value investigation, documentation, and publicizing information, civil society organizations are focusing their efforts on new strategies to publicize information. Moreover, they are going directly to the people rather than relying on the courts or policy makers to conduct outreach. ZaREKOM's strategy of regional consultations, in particular, highlights this new emphasis on outreach. The IDC, which has been holding discussions in local communities for years, decided to invest in a web-based strategy to expand its impact. Similarly, the YIHR used the web with the hope that this medium would publicize effectively and efficiently. Like the IDC, this organization chose to examine how people survived, not only how they were victimized. Even the ICMP, which worked mostly with the government and the courts over the past decade, has decided to engage directly with survivors.

Moreover, the study also points to the ways in which these organizations are utilizing science and technology to increase their impact. For years, scholars and scientists have been working together in order to improve advocacy efforts, and these organizations reveal the fruits of their labor (Doretti and Fondebrider 2001; Hagan et al. 2006; Stover 1985). The IDC is known for its statistical analysis of the number of dead and missing and the ICMP has been involved in investigations of the dead and missing outside of the Balkans. For scholars, policy makers, and advocates interested in improving the impact of transitional justice, the work of these organizations reveals a need for further inquiry into the opportunities and challenges of using scientific and technological advances to investigate, document, and publicize facts about violence.

These findings also raise important questions. First, what is the value of civil society truth-seeking initiatives when survivors distrust both governmental as well as nongovernmental actors? Next, will scientific methods, statistical or forensic, have more legitimacy than other approaches to investigate, document, and publicize facts about violence? Finally, should civil society initiatives focus on allowing individuals to share their stories, or should they try to develop as close to a factually accurate representation of history as possible?

The first question has been addressed by many scholars who have surveyed and interviewed survivors of the recent war in BiH, as well as those studying truth-seeking more generally (Daly 2008; Bougarel et al. 2007; Ivković and Hagan 2006; Stover 2007). The highly publicized failure of the ZaREKOM signature campaign reveals the difficulties faced by civil society organizations trying to reach across national divides in BiH. It is unclear whether any truth-seeking initiative may ever affect the pervasive belief that one's own national group was most victimized. Moreover, given that scholars, policy makers, and advocates around the world are increasingly interested in civil society and transitional justice, these truth-seeking initiatives may receive unwarranted approval. For example, despite its struggles in BiH, international media promoted the ZaREKOM campaign throughout the world, with one outlet calling it "one of the most serious efforts to promote reconciliation in the region."[25] Those evaluating civil society truth-seeking initiatives from abroad must pay close attention to what is going on *inside* the societies these initiatives are supposed to help.

The second question speaks to the need for more scholarly attention on how to utilize new technologies in truth-seeking initiatives. The organizations reviewed in this chapter have been taking advantage of new media to investigate, document, and publicize facts about the violence. They are using the web to inform others about their work, solicit feedback, and expand the scope of their investigations. However, focusing on web-based outreach excludes those without access to the web, which may leave out some of the most vulnerable survivors. Moreover, those designing and monitoring websites will need to carefully monitor the information they publish and how they present it. For example, should information on Srebrenića be presented as a massacre or as genocide? Moreover, as scholars such as Davenport and Ball (2002) note, more information does not necessarily lead to a valid account of the violence. Rather, the available information may reflect the objectives of the observer, leading to more biased accounts of the violence.

This latter dilemma about biased information is also relevant to the last question on whether truth-seeking initiatives should focus on story-telling or a factual account of violence. In addition to the challenge of evaluating the validity of available information, psychologists have found that individuals are more willing to accept the outcomes of a judicial process if they are given the opportunity to speak, feel heard, and validated by a legitimate authority (Lind and Tyler 1988; MacCoun 2005;

[25] How to Think about the Balkans, *The Economist*, 17 November 2010. Vesna Terselić, executive director of Dokumenta, was recently awarded the Right Livelihood award for her efforts to promote peace in the region. Speaking with reporters, explained the value of ZaREKOM See, Right Livelihood Laureates from Thailand, Burma and Croatia, *Democracy Now*. http://www.democracynow.org/2010/9/16/right_livelihood_laureates_from_croatia_thailand. Accessed 16 September 2010.

Tyler 1975). Thus, survivors may want a venue for public speaking or participation, and civil society organizations will want to facilitate this desire. At the same time, initiatives that prioritize participation may avoid dealing with root-causes of conflict and survivors may sacrifice fair outcomes if they are distracted by the fairness of the process (Folger 1977; Nader and Grande 2002; Törnblom and Vermunt 2007). It is thus important to examine whether or to what extent scholars, policy makers, and advocates working under the banner of transitional justice are promoting an understanding of justice that prioritizes process over outcome, and how this might undermine their goals of ensuring accountability, improving survivor well-being, and preventing future violence (see Nader 1999).

In conclusion, this chapter points to ongoing challenges for scholars, policy makers, and advocates who hope that investigating, documenting, and publicizing facts might bridge the social and political divides in BiH, as well as other countries struggling in the wake of violence. Though these civil society organizations have developed innovative approaches to outreach, their efforts are limited both on account of their methods as well as the pervasive distrust between different national groups. The dilemmas they face reveal that civil society, like international and domestic policy makers, must continue to reflect on how to effectively and efficiently reach those they seek to serve. While researchers outside of the country may learn from the new websites and applaud local leaders for their innovation, all must continue to question whether they can meet their goals within the country, especially as local leaders as well as domestic and international courts continue to reinforce narratives of victimization. In this way, transitional justice scholars can continue to discover the opportunities and limitations of seeking some truths in the shadow of limited justice.

Acknowledgments Funding for this study was generously provided by the National Science Foundation, The University of California, Berkeley, Institute for International Studies, and the Berkeley Empirical Legal Studies Program. Special thanks to all of my interviewees, and Damir Omergacic for his incredible generosity during my fieldwork. I would also like to thank Professor Lauren Edelman, Professor Calvin Morrill, Alexa Koenig, Gail Bensiger, for their ongoing feedback and support.

References

Backer, D. 2003. Civil society and transitional justice: Possibilities, patterns, and prospects. *Journal of Human Rights* 2(3): 297–313.

Backer, D. 2004. *The human face of justice: Victims' responses to the truth and reconciliation Commission Process in South Africa*. Ann Arbor: University of Michigan Press.

Barria, L.A., and S.D. Roper. 2008. Judicial capacity building in Bosnia and Herzegovina: Understanding legal reform beyond the completion strategy of the ICTY. *Human Rights Review* 9(3): 317–330.

Bartulović, A. 2006. Nationalism in the classroom: Narratives of the war in Bosnia-Herzegovina (1992–1995) in the history textbooks of the Republic of Srpska. *Studies in Ethnicity and Nationalism* 6(3): 51–72.

Bougarel, X., E. Helms, and G. Duijzings. 2007. *The new Bosnian mosaic: Identities, memories and moral claims in a post-war society*. Aldershot: Ashgate.

Buljugic, M. 2007. No progress for Sarajevo truth commission, Balkan Investigative Reporting Network, http://www.bim.ba/en/51/10/2327/. Accessed 20 February 2007.

Chandler, D. 2000. *Bosnia: Faking democracy after Dayton*. London: Pluto.

Coles, K.A. 2002. Ambivalent builders: Europeanization, the production of difference, and internationals in Bosnia-Herzegovina. *PoLAR: Political and Legal Anthropology Review* 25(1): 1–18.

Crosby, A., and M.B. Lykes. 2011. Mayan women survivors speak: The gendered relations of truth telling in postwar Guatemala. *International Journal of Transitional Justice* 5(3): 456–476.

Daly, E. 2008. Truth skepticism: An inquiry into the value of truth in times of transition. *International Journal of Transitional Justice* 2(1): 23–41.

Davenport, C., and P. Ball. 2002. Views to a kill. *Journal of Conflict Resolution* 46(3): 427–450.

Delpla, I. 2007. In the midst of injustice: The ICTY from the perspective of some victim associations. In *The new Bosnian mosaic: Identities, memories and identity in a post-war society*, ed. X. Bougarel et al., 211–234. Aldershot: Ashgate.

Doretti, M., and Fondebrider, L. 2001. Science and human rights: Truth, justice, reparation and reconciliation: A long way in third world countries. *Archaeologies of the Contemporary Past*, 138–144.

Elster, J. 2004. *Closing the books: Transitional justice in historical perspective*. Cambridge: Cambridge University Press.

Folger, R. 1977. Distributive and procedural justice: Combined impact of voice and improvement on experienced inequity. *Journal of Personality and Social Psychology* 35(2): 108.

Hagan, J. 2003. *Justice in the Balkans: Prosecuting war crimes in the Hague tribunal*. Chicago: University of Chicago Press.

Hagan, J., H. Schoenfeld, and A. Palloni. 2006. The science of human rights, war crimes, and humanitarian emergencies. *Annual Review Socioliology* 32: 329–349.

Hodžić, R. 2010. Living the legacy of mass atrocities: Victims' perspectives on war crimes trials. *Journal of International Criminal Justice* 8(1): 113–136.

Ivković, S.K., and J. Hagan. 2006. The politics of punishment and the siege of Sarajevo: Toward a conflict theory of perceived international (in) justice. *Law and Society Review* 40: 369–931.

Kreso, A.P. 2008. The war and post-war impact on the educational system of Bosnia and Herzegovina. *International Review of Education* 54(3): 353–374.

Kritz, N.J., and J. Finci. 2001. A truth and reconciliation commission in Bosnia and Herzegovina: An idea whose time has come. *International Law FORUM du Droit International* 3(1): 50–58.

Lind, E.A., and T.R. Tyler. 1988. *The social psychology of procedural justice*. New York: Plenum Press.

MacCoun, R. 2005. Voice, control, and belonging: The double-edged sword of procedural fairness. *Annual Review of Law and Social Science* 1: 171–201.

McEvoy, K., and L. McGregor. 2008. *Transitional justice from below: Grassroots activism and the struggle for change*. New York: Hart Publishing.

Nader, L. 1999. Globalization of law: ADR as soft technology. *The American Society of International Law Proceedings* 93: 24–27.

Nader, L., and E. Grande. 2002. Current illusions and delusions about conflict management—in Africa and elsewhere. *Law and Social Inquiry* 27(3): 573–591.

Nettelfield, L.J. 2010. *Courting democracy in Bosnia and Herzegovina: The Hague tribunal's impact in a postwar state*. Cambridge: Cambridge University Press.

Olsen, T.D., L.A. Payne, and A.G. Reiter. 2010. *Transitional justice in balance: Comparing processes, weighing efficacy*. Washington DC: United States Institute of Peace Press.

Riaño-Alcalá, P., and E. Baines. 2011. The archive in the witness: Documentation in settings of chronic insecurity. *International Journal of Transitional Justice* 5(3): 412–433.

Roht-Arriaza, N. 2006. The new landscape of transitional justice. In *transitional justice in the twenty-first century: Beyond truth versus justice*, ed. N. Roht-Arriaza and J. Mariezcurrena, 1–16. Cambridge: Cambridge University Press.

Roht-Arriaza, N., and J. Mariezcurrena. 2006. *Transitional justice in the twenty-first century: Beyond truth versus justice*. Cambridge: Cambridge University Press.
Shaw, R., J. Waldorf, and P. Hazan. 2010. *Localizing transitional justice*. Palo Alto: Stanford University Press.
Somun, A. 2010. Reports on the transitional justice experience in Bosnia and Herzegovina. *International Journal of Rule of Law, Transitional Justice And Human Rights* 1: 59–65.
Stover, E. 1985. Scientists aid search for Argentina's "desaparecidos". *Science* 230(4721): 56–57.
Stover, E. 2007. *The witnesses: War crimes and the promise of justice in the Hague*. Philadelphia: University of Pennsylvania Press.
Teitel, R. 2005. The law and politics of contemporary transitional justice. *Cornell International Law Journal* 38(3): 837–862.
Teitel, R.G. 2002. *Transitional justice*. New York: Oxford University Press.
Toal, G., and C.T. Dahlman. 2011. *Bosnia remade: Ethnic cleansing and its reversal*. New York: Oxford University Press.
Törnblom, K.Y., and R. Vermunt. 2007. *Distributive and procedural justice: Research and social applications*. Burlington: Ashgate.
Tyler, T. 1975. *Procedural justice: A psychological analysis*. Hillsdale: Laurence Earlbaum.
Van der Merwe, H., and A.R. Chapman. 2008. *Truth and reconciliation in South Africa: Did the TRC deliver?* Philadelphia: University of Pennsylvania Press.
Vinck, P., and P.N. Pham. 2010. Outreach evaluation: The international criminal court in the Central African Republic. *International Journal of Transitional Justice* 4(3): 421–442.
Vinjamuri, L., and J. Snyder. 2004. Advocacy and scholarship in the study of international war crime tribunals and transitional justice. *Annual Review of Political Science* 7: 245–362.
Wagner, S.E. 2008. *To know where he lies: DNA technology and the search for Srebrenica's missing*. Berkeley: University of California Press.
White, G. 1994. Civil society, democratization and development (I): Clearing the analytical ground. *Democratization* 1(2): 375–390.
Zalaquett, J. 1994. Balancing ethical imperatives and political constraints: The dilemmas of new democracies facing past human rights violations. *Hastings Law Journal* 43: 6–16.

Part II
Transnational Civil Society Practices

Chapter 9
Structured Encounters in Post-conflict/Post-Yugoslav Days: Visiting Belgrade and Prishtina

Orli Fridman

Introduction

In the past decade, an extensive number of publications have analyzed the Kosovo conflict, the declaration of independence, as well as the international presence in Kosovo. Often, when analyzing the conflict in Kosovo, more attention is given to the territory itself than to the people living in Kosovo and Serbia. While political processes and politicians are in the media and policy analysis spotlight, everyday realities and peoples' lives in Kosovo are often overlooked or forgotten. In the same line, on going local civic initiatives as part of transitional justice processes aiming to transform the conflict relations between citizens of Kosovo and Serbia were not systematically studied and analyzed.

This chapter therefore looks at meetings and projects initiated and facilitated by local civil society groups with the purpose of generating exchange between young people from Serbia proper and from Kosovo. I here refer to such meetings as to *structured encounters* and focus my analysis on meetings taking place between Kosovar Albanians and Serbs from Serbia proper in the form of the *Visiting Program*, a project initiated and facilitated by the Youth Initiative for Human Rights (YIHR), a Belgrade and Prishtina-based NGO. My aim here is not to offer a description of the project, but rather to offer a glimpse into what it generates: exchange of perceptions and firsthand experiences between young people from Kosovo and Serbia in relation to the recent war in Kosovo, the past relations and the present state of affairs.

The chapter analyzes the possible contribution of such a program as the *Visiting Program*, to the transformation of relations between Serbs and Albanian, in the context of transitional justice processes enhancing civic engagement and civic identities of young people in Kosovo and Serbia. It acknowledges and highlights the

O. Fridman (✉)
Academic Director and Lecturer, Peace and Conflict Studies in the Balkans,
SIT Study Abroad, a program of World Learning, Karadjordjeva 65, 11000 Belgrade, Serbia
e-mail: orli.fridman@sit.edu

O. Simić and Z. Volčič (eds.), *Transitional Justice and Civil Society in the Balkans*, Springer Series in Transitional Justice, DOI 10.1007/978-1-4614-5422-9_9,

shift in power relations that has occurred between Serbia and Kosovo as consequence of the 1999 international intervention. I understand this shift as crucial in reshaping intergroup relations inside Kosovo and between Kosovo and Serbia.

It is often argued that war tends to polarize and reduce the space where people can debate freely. Yet paradoxically war also tends to promote civil society and groups formed in reaction to it (Kaldor, Kostovicova and Said 2006, p.94). A number of liberal civil society groups were indeed established in Serbia early on during the wars of the 1990s.[1] Other groups were only formed later, in the aftermath of the wars, as it became clear that the regime change in Belgrade in October 2000 did not lead to the abolishment of the legacy of the Milošević regime (Fridman 2011, p.518). The YIHR was formed in that context in 2003 in Belgrade.[2]

Since its foundation, YIHR has become a regional NGO with programs in Serbia, Kosovo, Montenegro, Bosnia-Herzegovina, and Croatia. It was formed by young people from these countries in order to enhance youth participation in the democratization of society and empowerment of the rule of law by driving the processes of facing the past and establishing new progressive connections in the post-conflict region of the former Yugoslavia. The *Visiting Program* started in 2004, as an exchange program for journalists between Serbia and Kosovo and later developed into a broader exchange program for young people of various profiles from Serbia and Kosovo (Visiting Program 2006, p.47).

The *Visiting Program* is here analyzed in the context of bottom up approaches not only to transitional justice but also to peacebuilding and social change as generated by civil society groups. The concept of transitional justice is commonly understood as a framework for confronting past abuse as a component of a major political transformation that involves complementary judicial and nonjudicial strategies (Bickford 2004, p.1045). In this chapter, my interest is primarily in transitional justice in the context of conflict transformation and peacebuilding after armed conflict and less on its initial use in the context of societies transitioning from undemocratic regimes (Lambourne 2009, p.29).[3] The literature on peacebuilding points at the importance of and the need to transform relationship between people in addition to ending the state of war and violence between them (Lederach 2007). For such transformation to take place and remain sustainable, it must be based on recognition of the particular culture and conflict context and the effective participation of civil society (ibid, p.20). Based on this understanding, the model of *transformative justice* was developed proposing a focus on civil society participation in the design and implementation of transitional justice mechanisms

[1] For a discussion distinguishing between anti-war and anti-Milošević activism in Serbia in the 1990s see (Fridman 2011). The distinction between civil and uncivil society groups is also important in this context (Kopecky and Mudde 2003; Kostovicova 2010, pp.289–290).

[2] The fall of Milošević paradoxically also led to the proliferation of uncivil society and ultra nationalist groups in Serbia promoting illiberal ideologies, including anti-Semitism, xenophobia and racism (Kostovicova 2010, p.290).

[3] For additional discussion regarding developments and scholarly trends in the field of transitional justice see for example (Teitel 2003; McEvoy 2007).

(Lambourne 2009, p.35). Others have used the term *transitional justice from below* to denote a "resistant" or "mobilizing" character to the actions of community, civil society, and other non-state actors in their opposition to powerful hegemonic political, social, or economic forces (McEvoy and McGregor 2008, p.3).[4]

In discussing structured encounters as the *Visiting Program* in the context of the current Kosovo–Serbia frozen relations (or frozen conflict), I here highlight the need for scholarly analysis of such projects "from below" and explore their potential to become transformative. From that point of view, this chapter contributes to literature already analyzing the Dealing with the Past processes in the successor states of the former Yugoslavia in an encompassing manner (Franović 2008; Subotić 2009; Dragović-Soso and Gordy 2011).

Structured and Unstructured Daily Encounters: The Broader Frame

This chapter is part of a larger and broader research project comparing structured encounters such as the *Visiting Program*, with daily *unstructured encounters* referring to encounters between Albanians and Serbs from Kosovo. Given the limited length of this chapter, I here present the analysis of the structured encounters portion of the research only. In the larger frame of this research project, in addition to analyzing the *Visiting Program*, I also analyze unstructured encounters between Serbs and Albanians from Kosovo, from the point of view of Serbs who remained in Kosovo after 1999 and attempt to interact and part take in public and social life, in what is now an Albanian space. Such interactions are almost nonexistent as Serb communities in Kosovo are segregated and disconnected from the newborn state. Yet they do take place on a small scale. The analysis of the unstructured daily encounters focuses on three chosen sites of daily encounters: (a) the American University in Kosovo (AUK) campus in Prishtina; (b) work spaces, mostly of international organizations in Prishtina; and (c) the Albanian side of the Merdare border crossing between Serbia and Kosovo, controlled by local Kosovar Albanian forces (Fridman 2009). By analyzing and contrasting these two types of encounters, structured and unstructured, I attempt to shed light on hidden practices and exchanges between ordinary citizens, at times away from the spotlight of the international actors and policy analyzers.

In focusing on Serb–Albanian intergroup relations and highlighting the issue of power relations, I do not claim that this is a region torn between Serbs and Albanians only, nor do I wish to oversimplify other existing divisions and issues in Kosovo's population and societies today or ignore the rich history of other communities in

[4] For additional literature analyzing such processes see (Shaw and Waldorf 2010; Hinton 2010). I thank Olivera Simić for introducing me to these texts. For a discussion about transitional justice mechanism and truth telling to ensure a real political engagement and agency to a population that has been subject to violent conflict see (Lundy and McGovern 2008).

Kosovo (Duizings, 2003).[5] I do, however, focus my study here on Serb–Albanian relations in the context of the new post-Yugoslav realities, and more specifically, the realities created after June 1999.

The literature about Serb–Albanian relations in Kosovo mostly captures relationships defined by conflict and victimization, competing narratives of the past, animosity, and ongoing state of crisis (Zdravković-Zonta 2009; Zirojević 2000). The aim of this research is to uncover additional forms of relations existing today, resulting on one hand from everyday needs (and from the change in the power relations) (*unstructured daily encounters*), and on the other hand from civil society initiatives (*structured encounters*) that may offer an additional perspective and shed light on possibilities for transformation of current and future models of relationships.

About the Research

My interest in studying these issues in Serbia and Kosovo stemmed from my former experiences in Israel during the 1990s, first as a participant and later as a facilitator of political encounters between Israelis and Palestinians. The impact this work had on my political and social awareness and therefore my approach to the field of peace and conflict studies was and still is invaluable as it shaped my understanding of the internal dynamics of societies in conflict, group dynamics, and work towards transitional justice and conflict transformation. Seeing the similarities and differences between the dynamics of groups from the Middle East and the Balkans has sparked my interest in learning and writing about encounters in the post-Yugoslav context.

While the topic of structured facilitated encounters between Israelis and Palestinians was well researched in the last decades and generated a solid body of literature (Bekerman 2007; Maoz 2000; Sonnenschein 2008) and pedagogical discussions and debates,[6] publications analyzing similar work and issues in the context

---

[5] In Religion and the Politics of Identity in Kosovo, Duijzings captures realities... that in many ways no longer exist in Kosovo after 1999. I particularly think of the author's postscript comment in the Preface, as he explains that his manuscript was finalized before the start of the NATO actions against Serbia and therefore were not included at his account: "although these developments have put my work in a completely different light, I could not include them in my account. It is sad that this book now bears testimony to a world that may have ceased to exist" (Duijzings 2000, p.12).

[6] By pedagogical discussion I refer for example to debates among practitioners and educators in encounters between Israelis and Palestinians who over the years have come to criticize the "contact hypotheses" model of encounters. This model assumes that the act of bringing together people who belong to groups that are in conflict, cutting them off from their group affiliations, and introducing participants on a personal basis, can reduce both their alleged hatred for one another and the stereotypes they have about each other. Such an approach was highly criticized by some practitioners as it became clear that such encounters, not only disregard the controversial issues that are in the heart of the conflict, but also generate the same inequalities existing in the realities from which participants are coming, and therefore depoliticize the encounter and the issues between them. Approaching such encounters as political education offers an approach that understands any of such encounters to be a mean for political/social change, and not an aim in of itself (Halabi and Sonnenschein 2004).

of the post-Yugoslav conflicts are scarce. In that sense, this chapter aims to broaden the literature and discussions on encounters in the context of today's post-Yugoslav states looking at Kosovo–Serbia in particular.

In the successor states of the former Yugoslavia, all educational and encounter projects, dealing with the recent wars of the 1990s and their legacies, are initiated and implemented solely by civil society actors/groups, absent from official curriculums of state institutions. Leading actors in prominent civil society groups frame their work in the pedagogical lenses of what is generally referred to as to the "Dealing with the Past Project" theoretically leaning on transitional justice models of peacebuilding and reconciliation (Rangelov and Teitel 2011).[7] Projects of encounters between Serbs and Albanians were not thoroughly researched by social scientists, nor were they analyzed through the lenses of understanding groups in conflict or the political aspect/impact of group encounters. While descriptive materials of projects implementation are available (Nansen Center for Dialogue 2007), analysis and discussion of the pedagogical approaches to work with groups in conflict or of the impact this work has in shaping participants' perspectives towards the past and the present as well as their civic engagement is limited. In that sense, this chapter may also offer a bridge between literature analyzing bottom up approaches to transitional justice processes and literature focusing on peacebuilding and education for peace in the context of conflict transformation. As Lambourne suggested "analyzing and evaluating transitional justice in terms of its contribution to peacebuilding enables a more holistic perspective that takes into account the expectations of conflict participants, as well as links between dealing with the past and building peace for the future" (2009, p.29).

Based on the materials and data collected for this project, I here discuss not only structured encounters such as the *Visiting Program* and their impact, but more so, I generate a discussion on what else can such encounters tell us about the post-Yugoslav conflicts, and the related experiences and perceptions they generate. Given the limited scope of this chapter, my discussion will not focus on analyzing the pedagogical aspects of this work, but rather, it will focus on its contribution to our understanding of the conflicts, their legacies today, and the way they shaped Serb–Albanian relations.

The data collected and analyzed for this project comprises general observations and informal conversation I began to have from my very first visit to Kosovo and more than a decade ago, and culminated in two main phases of field research that took place from 2009 to 2011 and included: (a) interviews conducted in Belgrade and Prishtina with young participants of the *Visiting Program* from Serbia and

[7] Most of the available work and the writings focus on the triangle of Serbia-Croatia and Bosnia-Herzegovina, not including Kosovo. See for example (Franović 2008). The work of the Center for Non-Violent Action (*Centar za nenasilnu akciju* (CNA)) with former combatants is framed in terms of Peace Education (Vukosavljević 2007). An exception to that is a text about the work of the Nansen Center (Steinar 2011).

Kosovo as well as with the organizers from the Youth Initiative offices in both cities[8]; (b) participant observation of the *Visiting Program* itself.[9]

Structured Encounters

I approach structured encounters as the *Visiting Program* as a matter of choice, rather than the result of everyday needs or the necessities of life (as in the case of the unstructured daily encounters). Such meetings and projects are initiated and facilitated by local as well as international groups and organizations to generate exchange and interactions between young participants from Kosovo and Serbia proper who otherwise would not have the opportunity to meet each other as equals or to interact directly.

Such encounters should be understood in light (or in the shadow) of the reality of a deepening disconnect between Belgrade and Prishtina. The two cities and their populations seem to be miles away from each other, with barriers and distance that only grow, as the conflict remains frozen. Even though an official dialogue between Belgrade and Prishtina started again in late 2010, it has not yet generated any sense of change. On the contrary, official statements of politicians continue to create an atmosphere of unwillingness to transform the conflict, or the relationships it has generated over the past decades.[10]

Even if occurring on a small scale, structured encounters between activists, educators, and artists who believe in the importance of generating open lines of communication and exchange between the two capitals despite the general atmosphere of stagnation and conflict—may shed light on some links still existing between Belgrade and Prishtina, its populations, their pasts and futures, in spite of lack of will for change among politicians.[11]

One platform of such exchange takes place on Radio-Free Europe, in its program *Most* (Bridge) hosted by Omer Karabeg. The program has initiated a series of dialogues between prominent individuals from Serbia and Kosovo on the question of "how to unfreeze Serbia-Kosovo relations." On January 5, 2009, Karabeg's interlocutors were two writers: Migjen Kelmendi in Prishtina and Vladimir Arsenijević

[8] While the visits and interviews took place in Belgrade and Prishtina, some participants were born and raised in other towns or cities in Kosovo and Serbia, but currently attend universities and reside in the capitals.

[9] The data collected for the unstructured daily encounters portion of the project also includes interviews conducted in Prishtina and Gračanica with young Serbs from Kosovo working or studying in Prishtina and daily crossing imaginary borders by entering the city.

[10] See for example in the interview with the chief negotiator of the Serbian team upon his return from Brussels from negotiations with the Kosovar team who referred to Albanians as to "people from another planet" in an interview on Blic daily newspaper (Spaić 2011).

[11] See for example transcripts available from a seminar that took place in Prishtina in 2008, titled: "Cultural Policies as Crises Management?" (Heta and Osmani 2008).

in Belgrade (Belgrade is a Far Away Place for Albanians from Kosovo 2009).[12] The exchange between the authors was far more political and direct than any of the structured or unstructured encounters I have observed thus far. They were free to talk about real issues with a high level of self-confidence and self-criticism. Additionally, they were not bound by the language of the international community, in the sense of using the paradigm of multiethnicity and multiethnic society—a discourse that does not leave much room for meaningful/open discussion on issues of the past as well as the present.[13] In this exchange, the two writers spoke as equals, as two statesmen from their capital cities of Belgrade and Prishtina and no longer as the oppressor and the oppressed.

While the recognition of Kosovo's independence is experienced as a direct threat reinforcing the fears among many Serbs in Kosovo, for Arsenijević from his perspective as a Belgradian the declaration of Kosovo's independence has created "the first serious basis for a complete re-definition of Serb-Albanian relations, and thus for a new future" (Belgrade is a Far Away Place for Albanians from Kosovo). Such redefinition takes place in the interactions between young Kosovar Albanians and Serbs from Serbia Proper, as part of YIHR's *Visiting Program*. In this case, the change in power relations that occurred in 1999 offers the possibility for an encounter between equal citizens from Serbia and Kosovo, unlike the oppressive relations that existed prior to 1999.

Sites of Structured Encounter: The *Visiting Program*

The *Visiting Program* began in 2004 as an exchange program for journalists between Serbia and Kosovo, later developing into a broader exchange program for young people of various profiles from Serbia and Kosovo. The aim of the program is to "tear down the 'wall' between the Kosovo and Serbian societies, in order to enable young people to see the actual situation and social changes in Serbia and Kosovo through the direct experience of life in Pristina/Belgrade" (YIHR website). The program also aims to "connect (young people) with their peers 'from the other side,' meeting people with similar interests and establishing new relations and

[12] Karabeg also published his conversations between Albanians and Serbs (Karabeg 2000).

[13] The shift in the asymmetry of power relations in Kosovo has created a new reality on the ground, which culminated in Kosovo's declaration of independence in February 2008. However, much of the discourse on intergroup relations in Kosovo, shaped by the massive presence of the international community, is frequently reduced to the term *inter-ethnic society*, a phrase which in my opinion is emptied of the political context of minority-majority relations and identities whose power relations have been reversed. Terms such as *inter-ethnic society* and *inter-ethnic cooperation* shape meetings hosted by international organizations, constitute structured and planned facilitated encounters between local politicians as they focus on an open discussion that may "encourage cooperation on non-status issues ... and provide assistance in building a stronger multi-ethnic society in Kosovo" (Strengthening Interethnic Political Dialogue in (Kosovo 2008)). My question here is can such encounters help address the issues stemming from the change in power relations between Albanians and Serbs in Kosovo?

communication between the two societies" (Visiting Program 2006).[14] Finally, the *Visiting Program* aims to establish a network of young people with various interests who will remain involved in civic activism (ibid).

Visits usually last for 3 days and the agenda of the visits is created according to the interests and wishes of the participants, with different areas of interest: from visits to institutions, political parties, universities, local non-governmental organizations, museums, media houses, etc. In the discussion that follows I will address and analyze some of the dynamics and experiences of this program. I focus on differences in the aims and motivations between participants from Serbia choosing to visit Prishtina and participants from Kosovo who choose to visit Belgrade (often for the very first time), as well as differences in the motivations and experiences of organizers and participants.

The Ride from/to Prishtina and Belgrade and the Border Control

The participants of the *Visiting Program* use the Belgrade-Prishtina-Gračanica route, operated by a privately owned Gračanica-based company. In many cases, the participants do not know each other prior to the start of the program, and meet for the first time at the departure meeting point.[15] For those traveling to Belgrade or Prishtina for the first time, much remains unknown, and hesitation or even fear may color the start of their journey. These fears however are *not* similar.

The Belgrade participants going to Kosovo may at times keep their participation a secret from their immediate family and friends, worried about a general disapproval. Often the trip to the unknown is accompanied with hesitation about what it will be like in Prishtina. Will it be safe? Will they be able to speak in Serbian or will they need to conceal their Serbian identity and speak in English only?

For the Prishtina participants, visiting Belgrade primarily means traveling to a big city. While these participants also express concerns about safety, their fear as related to crossing the border is more specific. It is related to the fear of facing the Serbian border police, and men in Serbian uniform again. As some interviewees put it, this was the first time, since the end of the war that they encountered again the Serbian uniform. Such an experience immediately recalls painful memories from the 1990s, flashbacks from the war and harsh feelings. They are all of an age to clearly remember the war, to remember how they were forced to leave their houses by Serbian armed men. This was a theme that arose in all the interviews I conducted

[14] According to the organizers, more than a thousand young people, students, activists, journalists and professionals from various fields have gone through the Visiting Program, and for all of them it was the first time they saw Belgrade, Pristina and later Sarajevo.

[15] The Prishtina meeting point for the departure to Belgrade is Hotel Victory, while the Belgrade meeting point of departure to Prishtina is the train station. This is set in accordance with the existing Belgrade-Prishtina-Gračanica line.

with youngsters from Prishtina who visited Belgrade through the *Visiting Program*. In the words of one of them:

> in the border we saw the police, they reminded us of when we last saw the police in our streets … at that time I was scared. When I saw them at the border I was not scared but I remembered that bad time … I have strong memories of the Serbian police in Prishtina … I was ten years old when the police came to our house and ordered us to go out .. we then had to leave to the train station and on our way [to Macedonia] they were maltreating us … the police, uniforms and flags of Serbia, all make me feel unsafe (Interview with the author, Prishtina, June 30, 2011).

Indeed for the young participants from Prishtina, the border is an important reminder of the past inequalities as well as of the former power relations when Kosovo was under the control of the Serbian police.

The bus rides I joined in both directions revealed some of the existing gaps between the participants from Belgrade and from Prishtina. From either direction, approaching the border generates some loud silence among the passengers; conversations now take place in lower voices or grind to a complete halt. The distance of Belgrade from Kosovo reveals and emphasizes the lack of knowledge of the situation in Kosovo by ordinary people in Serbia. The daily news coverage from Kosovo in Serbia mostly emphasizes contested issues and violence, with barely any reference to Kosovo as a place where life has continued to develop since 1999, or as a place of normality. Even though Belgrade is covered with graffiti stating that "Kosovo je srce Srbije" ("Kosovo is the heart of Serbia"), most people (especially younger generation) have never been to Kosovo and in fact have very little interest in going there. As one of the visiting program organizers explained: "the majority of young people in Serbia are completely ignorant towards Kosovo and Prishtina, they truly do not care about it, do not want to think about it … " (Interview with the author, Belgrade, July 15, 2009).

Though program participants are clearly more curious than their peers, they too lack current knowledge of what they can expect. On the road to Prishtina, I noticed that some of the participants were not quite sure who will check their document at the border (internationals or local Kosovo police) or what the procedure will be like. Is there a border at the administrative line? If so, who is in control of it? This lack of knowledge in my understanding contributes to a sense of fear, nervousness or simply lack of interest to go. As a result, Kosovo's independence, and the recognition or lack of recognition of its independence, seems like a distant idea in the hands of politicians, detached from the lives of ordinary people.

For the young Albanian participants on their way to Belgrade, much more is known about Serbia and unlike for Serbs going to Prishtina, recognizing Serbia's independence and its sovereignty is not in question. The encounter with the Serbian border police after crossing the Albanian side of the border generates some tension and discomfort. On the ride that I joined, the policemen instructed all passengers with other documents than Serbian IDs to step off the bus. This included the four participants, the YIHR coordinator and myself. We were asked to get off the bus, show our bags, and answer to: "where are you going and why" type of questions. The participants were taken aback by the fact that the policeman addressed them in Serbian even though it was clear that they do not speak the language.

But interestingly enough, in spite of the discomfort and the imposed interaction in Serbian, there may have been something empowering about that border encounter for them: as some mentioned in interviews conducted few months later, being treated as foreigners meant at that moment, they are not considered to be coming from Serbia, but crossing the border, with what it meant, politically.[16]

Structured Encounters: Issues and Realities

Motivations

Civic Engagement and the "Dealing with the Past" Project

The difference in how the war in Kosovo was experienced by young Albanians from Kosovo and young Serbs from Serbia proper is important here in understanding the motivation of the organizers. While for young people from Serbia proper, the war was happening somewhere far away or seen on the TV screens only, with the NATO bombing being the event that shaped their war experiences, young Kosovar Albanians en mass experienced 1999 as a catastrophe, as they escaped their homes and became refugees. The levels of awareness are therefore different and generate somewhat different political agendas as well as understanding of the "dealing with the past" project.

From the point of view of the organizers at the Belgrade YIHR office, the aim of the *Visiting Program* is to inform people, raise awareness, and normalize relations between Belgrade and Prishtina. Additionally, they aim to expand the circles of young members of their society who are "dealing with the past"; increasing the number of young people interested in learning more about the events that took place during the wars of the 1990s and consequently enlarging the antinationalist circles among young people in Serbia. In that sense, this project has to do with enforcing the issues of the recent wars on a society that wishes to forget. It offers to break the silence, allowing young people to move away from discourses of victimization in order to enhance the engagement of young people in civic action.[17]

I therefore understand the *Visiting Program* to be an educational project about the recent past, offered to a generation that came of age during the 1990s, who did

[16] Later on, the visits from Prishtina to Belgrade were put on hold, as participants who after the declaration of independence carried Kosovar travel documents (while before they carried UNMIK passports or still the Yugoslav passports), could not enter Serbia who refused to recognize Kosovo's independence and all symbols generated from that unilateral move, including the new passports. Serbs are still able to enter Kosovo with their travel documents (ID is in fact sufficient). Following the February 2012 round of negotiations, it seemed like Serbia will allow Kosovar Albanians to enter Serbia with their Kosovo IDs. At the mean time YIHR were able to receive special permits for the visits, and the visitors had to leave their travel document at the border, and receiving them back upon their return to Kosovo.

[17] About Silence breakers in Serbia see (Fridman 2011).

not know about what was happening, at times in their name in the neighboring countries. As mentioned above, this agenda of raising political and social awareness in Serbia today is still completely in the hands of civil society organizations, absent from the official education system.[18]

The initiators of the program define their aim to raise political and social awareness, by highlighting the experiential component of the program: visiting Bosnia-Herzegovina (including visits to sites of war crimes) and visiting Kosovo.[19] In that sense such projects are at the heart of their social and political agenda, as stated by the organizers: "the visiting program aims to establish a network of young people with various interests who will remain involved in civic activism" (Visiting Program, p.47). While this statement may be read as a vague one, in essence this work is extremely political and should be framed and understood as political work and political education. However, the organizers do not often use terms as such and prefer to think of politics as something that they bring, but only from the back door.

Visiting Prishtina: Reclaiming the Yugoslav Space?

Rebuilding bridges between the citizens of the former Yugoslav republics, especially between the young generations who came of age during the post-Yugoslav wars, is a theme of numerous current civil society run educational projects in the post-Yugoslav states and particularly in Serbia. The participants of the *Visiting Program* are all young people in their 20s who were born during the very last years of Yugoslavia or as it was already disintegrating. They have never experienced Yugoslavia as adults, and barely as kids. Does the *Visiting Program* have elements of reclaiming the Yugoslav space?

The differences between the Serb and Albanian participants here is most apparent, as well as the generational differences with their parents' generation. While their parents' generation who grew up and came of age in Yugoslavia, may have met each other studying on faculties in larger cities, or (fathers) in military service, their generation already had no opportunities to meet each other.

---

[18] I heard reference from activists, referring to themselves as people who "professionally are dealing with the past" referring to that as to more than a job, more as a way of life a matter of values. Images of civil society in the Serbia vary among ordinary people but are mostly negative: as if it is common knowledge outside of these circles, that NGOs have a lot of money, but since the results of their work are not seen, they must be doing nothing, or as earlier addressed to, still in the 1990s, they are anti-war profiteers (who should be therefore seen just as culprits as the war profiteers). In a conversation with one of the YIHR coordinators, she explained how prior to joining YIHR she had a very bad image of NGOS in a very abstract way: "yes I was aware of their anti-war action, but even that was not enough for me; during my studies, in the late 1990's and even after 2000, I was not in these circles, I didn't know much about what they were doing, but I knew that a priori they were bad. I think it had to do with my ignorance and lack of knowledge. And this is still the connotation people have with NGOs" (Interview with the author, Belgrade, July 19, 2009).

[19] The importance of seeing in one own eyes is crucial in the approach to raising awareness and generating change in the process of knowing and breaking the silence. See also projects as the Helsinki committee for Human Rights School.

The difference between visiting Prishtina and visiting Sarajevo or other cities in Bosnia–Herzegovina from Serbia is interesting in understanding the motivation to organize and participate in the *Visiting Program* to Prishtina. While visiting Bosnia–Herzegovina especially Sarajevo was already normalized in the eyes of many people in Serbia, a similar process did not take place in regard to Kosovo and Prishtina. As one of the organizers explained: "most people in the streets of Belgrade will be happy to talk to someone from Sarajevo, someone who speaks with a Sarajevo accent … it is more familiar and acceptable. But at the same time, people from Belgrade would not be happy to know that next to their table in at café or at the restaurant someone who is from Kosovo sitting, someone who is Albanian" (Interview with the author, Belgrade, June 3, 2011). From that point of view, going to Prishtina is something that is still not acceptable nor is it normalized in Belgrade. It raises questions such as "why would you want to go there? What is in Prishtina for you?"

I see this as part of the motivation for young people from Serbia to go to Prishtina, searching for more cutting edge activism, for the adrenaline to flow: while going to Sarajevo is now normalized again and may be perceived by many as an ordinary or touristic thing to do: visiting Baščaršija, the summer film festival, enjoying the best Burek and Čevapi though not disregarding still the politics of that, there is no tourism to Prishtina from Belgrade, and going there may now be seen by some people as a betrayal, especially if it is in order to meet with Albanians and take part in such projects of structured encounters.

The generational difference among activists here is again insightful: while for the older generation of antiwar activists going to Sarajevo right after the war was emotional and painful, given that for them, Sarajevo was their city too, and was attacked brutally in their name; for the younger generation of activists it was already different. In the words of one of them who explained: "people my age, we do not have any memories or emotions regarding Sarajevo, it was never part of our country" (ibid). However, even for the older generation of activists, Prishtina never seem to hold a special or emotional role. For the younger activists it did become meaningful, as offering a real possibility to transform their relations with Albanians, and as an opportunity to allow them to form their own thoughts, independently, based on their own experiences, and based on access to knowledge and information, enjoying the kind of activism with a flavor of engaging with the forbidden.[20]

Visiting Belgrade

Some curiosity or attraction to the forbidden is also involved in the choice to visit Belgrade for youngsters from Prishtina. As one of them put it: "everything that is prohibited to do is attractive" (Interview with the author, Prishtina, July 1, 2010). Most of all, Belgrade is seen as a large, cosmopolitan city that has a lot to offer, both

[20] This as a motivation I especially noticed among the organizers of the program. Some of them joined activism in YIHR through first participating in the *Visiting Program*.

culturally and socially. But connections between Belgrade and Prishtina were completely broken, and while for some it is too early to reestablish them again, the *Visiting Program* is doing just that, allowing young people to meet each other and reestablish connections.

Discussing these broken ties on the radio program Most, Omer Karabeg (the host) asked Migjen Kelmendi: "do Albanian artists from Kosovo want to work with Serbia, to have their books translated there, to exhibit their paintings, to have a theatre performance there, or do they want artists from Serbia to come to Kosovo, to Prishtina?" Kelemendi replied:

> No. They are definitely no longer interested in Belgrade and Serbia. They are fully oriented towards Tirana and Albania. It is there that they wish their works to appear. I would say that the Kosovo Albanians have turned their backs on Serbia for good. There is not the smallest wish to know what is going on there. It seems at times that Belgrade for them is a faraway city, that Serbia is a faraway land … apart from occasional individual contacts, communication has practically ended (Belgrade is a Far Away Place for Albanians from Kosovo).

As an alternative to this perspective, it seems that most young people who choose to visit Belgrade, do see it as a city and that has something to offer them in spite of the recent past, the wars and the way Albanians were and are treated there. Visiting Belgrade means that one has to struggle with what the city, and Serbia in general, represent for many Kosovar Albanians: the home of the Milošević regime and of Serbia as a crime machine in the 1990s. As one of the organizers from the Prishtina YIHR office reflected on the first time he visited Belgrade: "in 2007 I went to Belgrade for the first time, it was April 1, 2007, exactly 8 years after I was forced to leave my home on April 1, 1999 by Serbian police" (Interview with the author, Prishtina, June 30, 2010). Going to Belgrade is a challenge, it means taking the time and space to think for oneself, and think critically, learning that in fact the way Serbia and Belgrade are represented in local media in Kosovo is also one-dimensional. Wanting to form their own opinion and to decide for themselves seems to be an important factor in the decision to go. Among friends and family, reactions may be discouraging. For many, Belgrade is the city of the enemy, especially for families who lost their loved ones during the war. For some, this is exactly the reason why they choose to go: "the *Visiting Program* is a small step for a big change in the relations between Kosovo and Serbia; this is the best way for us to get to know each other" (Interview with the author, Prishtina, July 1, 2010).

The wish to travel freely, to have options and opportunities, also has to do with the decision to go to Belgrade, as young women and men in Kosovo still feel isolated. Currently for those carrying a Kosovo passport, the only destinations they can travel freely to, with no need for a visa or a special permit, are Macedonia, Montenegro, and Albania. Traveling to Belgrade helps to break this sense of isolation.

Understanding this choice through generational differences may shed light on the communications that were broken and on the current sense of isolation. For young participants from Prishtina, visiting Belgrade is something their parents' generation can relate to. While for them there is almost nothing left from Yugoslav times or Yugoslav memories, their parents did come of age and live in Yugoslavia;

they speak fluent Serbian and in some cases are even nostalgic about the days of socialism as I heard in some of the interviews and conversations in Prishtina.[21] Many of their memories about Yugoslavia that they shared with their kids have to do with traveling. Thus, some of the parents do encourage their kids to take this opportunity and travel to Belgrade, talking about the days when they could travel freely, everywhere.[22]

One participant spoke about her father, who used to work in Belgrade: "when I went to Belgrade he wanted to come with me, for both my parents had jobs, security and stability during Tito's Yugoslavia" (Interview with the author, Prishtina, July 1, 2010). In that sense, reference to Socialist Yugoslavia was often mentioned, mostly in juxtaposition with what came after, the Milošević regime. For some, this shared past still holds meaning as something that should not be erased or ignored completely. The *Visiting Program* may allow these shared pasts to resurface again or at least to be explored.

While in Serbia, some participants from Kosovo see it is as an important opportunity to change the image people have about Kosovo and about Albanians, and even more so, about the right of Kosovars for an independent state. As one participant who returned to Belgrade several more times after her participation in the *Visiting Program* put it:

> During my very first visit, I was more in a position of observing and absorbing, trying to see what Serbian society is like, what do they have to offer me as an Albanian from Kosovo? Can they treat me equally? Did they change their approach towards Albanians [after the war]; the next time I visited the city, I was already louder, expressing my beliefs and also making a point of going there from a better positions, visiting a neighboring country Serbia, saying "I have a state now too" (ibid).

Connecting with like-minded young people from Serbia, those who are not nationalists or radicals, allows such statements to be made, and therefore may offer an empowering experience.

Revisiting What Used to be Home

A few of the participants of the *Visiting Program* choose to participate in the program and visit Prishtina because they were born and raised in Kosovo, members of families who left Kosovo when the war ended the way it did in 1999. For them, this is the first opportunity to go back and visit their sites of childhood, the city where they were born and grew up. This may be an emotional journey, mixed with personal memories and heavy feelings. I asked Jana, a young participant of the program

[21] More on memories of Socialist Yugoslavia in post-war Kosovo see (Schwander-Sievers 2010). The biography of Adem Demaçi is illuminating in portraying the images and of Yugoslavia in Kosovo (Gashi 2010).

[22] Similar sentiments towards the Yugoslav days were heard in Serbia prior to the abolishment of the visa requirement to enter the EU in December 2009.

who left Prishtina as a refugee to Belgrade when she was only 10 years old, if she had found what she was looking for. She replied: "I went there to search for feelings, I wanted to see the city and the people there and to compare it with my memories … I also wanted to talk to people from Prishtina to see what were their emotions towards the situation that Serbs are no longer there" (Interview with the author, Belgrade, December 29, 2010). Through her eyes, while walking in the streets of Prishtina, the other participants could learn more about the way the city was before; the layers of the city that changed completely, as the names of the streets and the buildings are no longer the same.[23]

The meetings held in local organizations and with local people while visiting Prishtina usually begin with the participants introducing themselves and where they are from. With younger presenters these meetings take place in English with people who are a bit older (in their late 30s and older) the exchange takes place in Serbian, as they are all bilingual. In a meeting at one Prishtina-based NGO, Jana introduced herself as born in Prishtina and now living in Belgrade. In a powerful gesture, the presenter welcomed her back to *HER* hometown, emphasizing how he would love to see all people who were forced to leave the city in 1999 return. This was a powerful way to enter the presentation with the group, as the presenter had positioned himself to make a point about where he stood politically.

Possible Transformations/Movements/Changes?

Unlike participants born in Kosovo who came back through the program to see their childhood city, most of the participants from Serbia who visited Prishtina through the program did so for the very first time. They had an image of Prishtina as a small town, even village-like (often using the term seljački, meaning peasant). After spending a few days in Prishtina, meeting young, energized people, and being exposed to fascinating information generates a whole new image of a city they realize they knew so little about. As one participant put it: "going to Prishtina was as if someone showed me I was in the darkness [in my thoughts about Kosovo and about Albanians] … it made me understand that my thoughts are based on pure ignorance" (Interview with the author, Belgrade, July 19, 2009). The concept of the *Visiting Program* exactly builds on this recognition that the dominant discourses in Serbia make use of such ignorance for manipulation and anti-Albanian propaganda. The organizers' way to resist that is to offer people an experience through which they can define their own thoughts for themselves, through their own eyes, to take a stand that is based on knowledge, information, and firsthand experience.

[23] For example she kept referring to names of the streets and buildings that have changed. The *Lepa Brena Building*, which is now called the KEKS building, Kosovo's electric distribution company seemed as a marker for her to find her way around. For a discussion and analysis of changes in space and landscape modifications as related to citizenship and identity in Kosovo see (Krasniqi 2012).

It is important to mention that stereotypes in Serbia towards Albanians are not new or so recent. In some ways Albanians were always seen as "the other," which goes back to relations during Socialist and the kingdom of Yugoslavia. The language was always a divider between these communities; Albanians were not Slavs and earlier in the second Yugoslavia were subject to oppressive mechanisms of control (Malcolm 1999, pp.314–334).[24] As one participant from Belgrade put it: "for me change only came when I realized I was a racist, that this is racism and that it was present in Yugoslavia towards Albanians, and not among Serbs only. But I needed to go there [to Kosovo] have that experience, so I could become aware of that" (Interview with the author, Belgrade, July 19, 2009).

Listening to firsthand stories from young Albanians who lived through the war seems to be a very powerful and important experience of the program. Such exchanges do not occur in the format of a lecture, but in arranged meetings by the organizers, taking place in offices and cafes, in quite an informal atmosphere. For some, the impressions left by such an encounter are so strong that they could not stop talking about it upon their return home. For some, this was their entry point to activism with YIHR, leading to volunteering and even taking full-time positions within the organization. What they encounter while visiting Prishtina is a celebration of Albanian identity, a sense belonging and pride, that many of them sense nothing of in Serbia. While in Serbia such sentiments would be connected with Serbian nationalism, from their point of view, in Kosovo this nationalism bears no such negative connotation yet, but a sense of pride and of freedom that people were denied of, for a long time.

While a change in opinion may occur for some, for others this encounter may generate more hesitation. Here as well as in other cases of such encounters, some participants end up engaged in a competition over victimization. This is a common phenomenon amongst other groups in conflict and participants in educational projects that aim at raising awareness (Fridman and Abu-Rish 2008). In this case, it may also be attributed to the way the war in Kosovo ended and how the events after 1999 have played out: after all it was quite clear that Serbia and Serbs in Kosovo emerged as the losers of this war and that the power relations shifted. As one participant put it: "visiting Kosovo made it clearer that they [Albanians] are the winners in this historic battle of Serbia and Kosovo; they got what they wanted, they got their independence and their state … in general, the public in Serbia has a feeling that we lost in this conflict, but it is hard to say it openly and publicly" (Interview with the author, Belgrade, June 16, 2010).

As a young man with political ambitions in Serbia, he was reluctant to openly acknowledge the independence of Kosovo, and chose not to post on his Facebook wall any of the many pictures he took, while visiting Prishtina with

[24] For further discussion on Serb stereotypes towards Albanians see (Marković 2003). Slobodan Naumović's analysis on popular narratives on Serbian disunity also sheds light on those dynamics (Naumović 2007).

the *Visiting Program*. And yet many others were surprised by the warm and positive welcoming attitude towards them, what helped break many of the images they had prior to the visit (such as that Albanians are less educated, hate Serbs, etc.), shaped by TV and newspapers images in Serbia. Being exposed to the Albanian point of view and narrative is almost impossible in Serbia, and yet is a key to any change or movement that may occur in the perceptions and world-views of young people.

Closing Remarks

Through an investigation of encounters, this chapter sought to offer a more complex approach to the study of conflict transformation in Kosovo and Serbia and more insights to dealing with the past processes in those societies. By analyzing structured encounters in the context of the change in power relations that took place in Kosovo, this study highlights the need for an in-depth analysis that approaches Kosovo not as a disputed territory only. A program such as the *Visiting Program* and the processes and dynamics it generates as analyzed above, shed light on the difference between Serbs from Serbia proper and Serbs from Kosovo in relation to the challenges and dilemmas they face in Kosovo today.

Both among young participants and activists from Serbia and from Kosovo, I did not encounter any individuals who discuss their everyday experiences or work in terms of peacebuilding or reconciliation; the use of these terms are usually made use of only by members of the international community. Nevertheless, the analysis of the *Visiting Program* offers an insight into its potential to become transformative in the awareness and understanding that it raises among its participants, towards the conflict. It does suggest though that relations between Prishtina and Belgrade may be easier to restore or transform than those between Serbs and Albanians within Kosovo itself.

Additionally, analysis of the dynamics and processes of the *Visiting Program* suggests that more work needs to be done in Serbia in order to transform the official and public approaches towards Kosovo. Public discourse in Serbia generally refers to Kosovo as a territory only, with no reference to the population living there, or with reference to the Serb population only. Stereotypes in Serbia that are tainted with racism towards Kosovo Albanians are commonly present, especially among those who have never been to Kosovo. Such racism is not a new phenomenon; however, the events of the past decades seem to have further entrenched such views as the result of a lack of direct knowledge of or contact with the population of Kosovo. In that sense, the *Visiting Program* interrupts and even combats such ignorance, offering a more engaged form of citizenship based on access to firsthand experience rather than propaganda. Even though such stereotypes are not completely absent among members of the Serbian communities in Kosovo, they too are often looked down at by urban population in Serbia proper.

These structured encounters therefore should be understood as encounters between young urban people. As such, they do not explicitly combat segregation but rather help to generate a sense of freedom as well as more critical thinking and the critical consumption of knowledge. As many young participants in the *Visiting Program* have explained, they wanted to have the opportunity to meet young people who are the same like them. This may raise the question as to what extent this program addresses otherness and differences in a systematic or complex way. More broadly, it sheds light on the contribution of such a project of structure encounters to the ongoing processes of transitional justice in the region; more particularly it sheds light on perspectives "from below" and their affect on young people. It raises yet another question as for how far below even civil society can reach. It points at the challenges of civil society groups to broaden the circles of participation in their projects that are often perceived as elitist; as well as at the complexities in enhancing more engaged forms of citizenship as part of transitional justice processes.[25]

Pedagogically speaking, while the program at first may seem to be leaning on the contact hypothesis model,[26] it may in fact offer a new and locally developed model for conflict transformation work, education for peace, and/or mechanisms for transitional justice from below in the post-Yugoslav successor states. While the program is not defined in pedagogical terms and is not even framed as political (or as political education) by its organizers, it clearly has the intent of including more young people in civic engagement, in particular of enlarging the circles of those involved in the Dealing with the Past project, which is seen as the main mechanism for peacebuilding and transitional justice in the region.

In such a model, the *Visiting Program* does offer elements of reclaiming the past Yugoslav space. But even more so, by offering young people from Serbia and Kosovo the opportunity to visit Belgrade and Prishtina, it accepts and acknowledges the change that has occurred in the power relations between Serbs and Albanians, as well as between Belgrade and Prishtina. This is the entry point of the project, which rebuilds bridges between the citizens of the former Yugoslav republics, especially between the young generations who came of age during the post-Yugoslav wars or after. As a consequence of their participation in the program, it may be possible for some of these young people who did not experience Yugoslavia as adults to find and redefine these lost connections as a platform for change in future relations, between equal citizens of now various independent states.

Acknowledgments The fieldwork and research on which this chapter is based was supported by a grant provided by the *Berghof Foundation*. I would like to thank Lea David, Jelena Tošić, and Nenad Porobić for their comments and helpful conversations with me on earlier drafts of this chapter. I would also like to thank Dana Johnson for her help in editing the first draft of this chapter and for her helpful comments and Dunja Resanović for her work and help as a research assistant at earlier stages of this project.

[25] My thinking about engaged forms of citizenship is shaped significantly by the work of Ariella Azoulay (Azoulay 2008).

[26] See endnote 5.

References

Azoulay, A. 2008. *The civil contract of photography*. New York: Zone Books.

Bekerman, Z. 2007. Rethinking intergroup encounters: Rescuing praxis from theory, activity from education, and peace/co-existence from identity and culture. *Journal of Peace Education* 4(1): 21–37.

Belgrade is a Far Away Place for Albanians from Kosovo. 2009. Omer Karabeg in conversation with Migjen Kelmendi and Vladimir Arsenijević. http://www.bosnia.org.uk/news/news_body.cfm?newsid=2533. Accessed 6 September 2011.

Bickford, L. 2004. Transitional justice. In *The encyclopedia of genocide and crimes against humanity*, vol. 3, ed. D. Shelton, 1045–1047. New York: Macmillan Reference USA.

Duijzings, G. 2000. *Religion and the Politics of Identity in Kosovo*. London: Hurst & Company.

Dragović-Soso, J., and E. Gordy. 2011. Coming to Terms with the Past: Transitional justice and reconciliation in the Post-Yugoslav lands. In *New Perspectives on Yugoslavia: Key Issues and Controversies*, ed. D. Djokić and J. Ker-Lindsay, 193–212. London: Routledge.

Franović, I. 2008. *Dealing with the past in the context of ethnonationalism: The case of Bosnia-Herzegovina*. Croatia and Serbia: Berghof Occasional Paper. 29.

Fridman, O. 2011. It was like fighting a war with our own people': Anti-war activism in Serbia during the 1990's. *Nationalities Papers* 39(4): 507–522.

Fridman, O. 2009. Structured and unstructured daily encounters in Kosovo. A paper presented at the annual meeting of the *Association of the Study of Nationalities (ASN)*, New York: Columbia University.

Fridman, O., and Abu-Rish, Z. 2008. (Re)Centring Europe: Competing Israeli and Palestinian narratives in the shadow of Europe. Paper presented at the *Inclusion and Exclusion in and on the Borders of Europe* Conference. Slovenia: Portoroz.

Gashi, S. 2010. *Adem Demaçi Biography*. Prishtina: Prokullia.

Halabi, R., and N. Sonnenschein. 2004. The Jewish-Palestinian encounter in a time of crisis. *Journal of Social Issues* 60(2): 373–387.

Heta, A., and V. Osmani (eds.). 2008. *The Way between Belgrade and Prishtina has 28000 unproper build objects. So, never it will be an autobahn!* Prishtina: Stacion Center for Contemporary Art.

Hinton, A. (ed.). 2010. *Transitional justice: Global mechanisms and local realities after genocide and mass violence*. Piscataway: Rutgers University Press.

Kaldor, M., D. Kostovicova, and Y. Said. 2006. War and peace: The role of global civil society. In *Global civil society 2006/7*, ed. M. Glasius, M. Kaldor, and H. Anheier, 94–121. London: Sage.

Karabeg, O. 2000. *Dijalog na buretu baruta: srpsko-albanski dijalog 1994–2000*, 2000. Belgrade: Medija Centar.

Kostovicova, D. 2010. Civil society in the Western Balkans: vehicle for or obstacle to transitional justice. In *Conflict and memory: bridging past and future in (South East) Europe*, ed. W. Petritsch and V. Dzihic, 287–294. Baden-Baden: Nomos.

Kopecky, P., and C. Mudde. 2003. Rethinking civil society. *Democratization* 10(3): 1–14.

Krasniqi, G. 2012. State borders, symbolic boundaries and contested geographical space: Citizenship struggles in Kosovo. Paper presented at the annual meeting of the *Association of the Study of Nationalities (ASN)*. New York: Columbia University.

Lambourne, W. 2009. Transitional justice and peacebuilding after mass violence. *The International Journal of Transitional Justice* 3(1): 28–48.

Lederach, J.P. 2007. *Building Peace: Sastainable Recinciliation in Divided Societies*. Washington DC: United States Institute for Peace.

Lundy, P., and M. McGovern. 2008. Whose justice? Rethinking justice from the bottom up. *Journal of Law and Society* 35(2): 265–292.

Malcolm, N. 1999. *Kosovo: A short history*. New York: Harper Collins.

Maoz, I. 2000. An experiment in peace: Reconciliation-aimed workshops of Jewish-Israeli and Palestinian youth. *Journal of Peace Research* 37(6): 721–736.

Marković, P. 2003. *Ethnic Stereotypes: ubiquitous, local or migrating phenomena? The Serb-Albanian case*. Bonn: Michael-Zikić-Stiftung, Southeast European Minorities Network.

McEvoy, K. 2007. Beyond legalism: Towards a thicker understanding of transitional justice. *Journal of Law and Society* 34(4): 411–440.

McEvoy, K., and L. McGregor. 2008. Transitional justice from below: An agenda for research policy and praxis. In *Transitional Justice from Below: Grassroots Activism and the Struggle for Change*, ed. K. McEvoy and L. McGregor, 1–13. Portland: Hart Publishing.

Nansen Center for Dialogue. 2007. Balkan regional conference on peace education and conflict resolution in schools. http://www.nansen-dialogue.net/ndcserbia/index.php?option=com_content&view=article&id=455&Itemid=541&lang=en. Accessed 7 September 2011.

Naumović, S. 2007. On 'us' and 'them': Understanding the historical bases and political uses of popular narratives on Serbian Disunity. *CAS Working Paper Series*. Center for Advanced Study Sofia. http://www.cas.bg/uploads/files/Sofia-Academic-Nexus-WP/Slobodan%20Naumovic.pdf. Accessed 6 September 2010.

Rangelov, I., and R. Teitel. 2011. Global civil society and transitional justice. In *Global civil society 2011: Globality and the absence of justice*, ed. M. Albrow and H. Seckinelgin, 162–178. New York: Palgrave MacMillan.

Schwander-Sievers, S. 2010. Invisible-inaudible: Albanian memories of socialism after the war in Kosovo. In *Post-Communist Nostalgia*, ed. M. Todorova and Z. Gille, 96–111. New Yrok: Berghahn Books.

Shaw, R., L. Waldorf, and P. Hazan (eds.). 2010. *Localizing transitional justice interventions and priorities after mass violence*. Palo-Alto: Stanford University Press.

Spaić, T. 2011, July 5. It looks like talking to people from another planet. *Blic*. http://english.blic.rs/News/7806/It-looks-like-talking-to-people-from-another-planet. Accessed 25 September 2011.

Sonnenschein, N. 2008. *Dialogue Challenging Identity: Jews Constructing their Identity through Encounter with Palestinians*. Haifa: Pardes.

Steinar, B. 2011. Inter-ethnic Dialogue between Serbs and Albanians in Serbia/Kosovo, 1996–2008. In *Civic and Uncivic Values: Serbia in the Post-Milošević Era*, ed. O. Listhaug, S. Ramet, and D. Dulić, 369–400. Budapest: CEU Press.

Strengthening Interethnic Political Dialogue in Kosovo. 2008. *Project on ethnic relations*. Meeting Report. Prishtina, January 31–February 1. http://www.per-usa.org/Reports/Political_dialogue_January_2008.pdf. Accessed 25 September 2011.

Subotić, J. 2009. *Hijacked justice: Dealing with the past in the Balkans*. Ithaca: Cornell University Press.

Teitel, R.G. 2003. Transitional justice geneology. *Harvard Human Rights Journal* 19: 69–94.

Visiting Program. (2006). Belgrade: Youth initiative for human rights.

Vukosavljević, N. 2007. Training for peacebuilding and conflict transformation: Experiences of 'the center for nonviolent action' in the Western Balkans. *Berghof Institute*. http://www.nenasilje.org/publikacije/pdf/articles/a-Training-BerghofHandbook.pdf. Accessed 7 September 2011.

YIHR website. http://rs.yihr.org/en/programmes. Accessed 10 August 2011.

Zdravković-Zonta, H. 2009. Narratives of victims and villians in Kosovo. *Nationalities Papers* 37(5): 665–692.

Zirojević, O. 2000. Kosovo in the Collective Memory. In *The road to war in Serbia*, ed. N. Popov, 189–211. Budapest: Central European University Press.

Chapter 10
What About the Women? Transitional Justice and Gender in Bosnia and Herzegovina and Northern Ireland

Michael Potter and Hedley Abernethy

Introduction

In the context of challenges to the traditional transitional justice framework, this chapter takes two transitions that featured peace agreements in the 1990s which brought overt armed conflict to an end and established new post-conflict political institutions—in Bosnia and Herzegovina (BiH) and Northern Ireland—and examines the experiences of women in these contexts. The chapter is written from the perspective of practitioners working with individuals affected by armed conflict and reflects on the nature of justice in transition in a wider sense in a way that is meaningful to a broad range of individuals affected by conflict.

Processes which bring violent conflicts based on disputed space between ethnonational identities to a conclusion—or at least a pause—create opportunities for shaping new political and social landscapes in the transition from conflict. Commonly referred to as "transitional justice" processes, these opportunities for change tend to focus on areas such as individual prosecutions, reparations, truth-seeking, institutional reform and vetting and dismissals (United Nations Security Council 2004: 4).

Received wisdom would suggest that all matters pertaining to the restructuring of society after violent conflict should be addressed. This must include issues of marginalisation, which, although not necessarily the root of the conflict is none the less exacerbated during or ignored by the conflict. The concepts underpinning transitional justice should result in widespread social and structural change. Yet civil and political rights often appear to trump other rights, including social and economic

M. Potter (✉)
School of Politics, International Studies and Philosophy, Queen's University Belfast,
97 Hilden Court, Lisburn, Co Antrim, BT27 4YL, Northern Ireland, UK
e-mail: Mpotter02@qub.ac.uk

H. Abernethy
WAVE Trauma Centre, 4 Clanroy Parade, Belfast, BT4 3BZ, Northern Ireland, UK
e-mail: Hedley.abernethy@gmail.com

O. Simić and Z. Volčič (eds.), *Transitional Justice and Civil Society in the Balkans*, Springer Series in Transitional Justice, DOI 10.1007/978-1-4614-5422-9_10,

rights. This has led to traditionally marginalised groups like the poor and women still unable to enjoy the benefits of the new order (Roht-Arriaza 2006).

The UN Security Council recognises that there is not a "one-size-fits-all" approach to transitional justice. However, the Security Council was clear to state explicitly that any transitional justice efforts are based upon meaningful public participation of a number of groups including women (UN Security Council 2004: 7). While the report goes on to state that reform must also be "cultivated" amongst elites and ex-combatants, it appears that a tacit hierarchy has developed with the needs of elites and ex-combatants taking precedence under transitional arrangements over marginalised groups such as women (ibid.). Women, being largely absent from the perpetration of violence in conflict, are often under-represented in the formation of post-conflict arrangements, although increasingly international involvement in transitions has led to the insistence on measures to increase the participation of women, at least outwardly.

It is not intended to imply that the conflicts in BiH and Northern Ireland are directly comparable: they are not, whether in terms of geography, economic situation, political influences, conflict structure, intensity or outcome. However, it is suggested that there are conceptual similarities in the way individuals respond to the impacts of conflict upon them and how they see their place in the post-conflict environment.

Women and Ethno-National Conflict

The notion of "ethnic conflict" can have the effect of simplifying complex phenomena to the extent that the terminology can be counter-productive and misleading for the purposes of analysis (McGrattan 2010). However, ethno-national identity can become a rallying point in times of tension or fear, leading to a retreat into a basic, secure group identity (Smith 1999: 276). This response is harnessed by "ethnic entrepreneurs" (Rothschild 1981), who formulate symbiotic relationships with fearful populations in conflict (Voutat 2000: 286), deriving power from both voicing the fears of a community under threat and also exacerbating concerns, real or imagined. In this context, "cohesion and identity in contemporary conflict tend to form within increasingly narrow lines" (Lederach 1999: 12).

This distilling of identity in simplistic terms with ethno-national reference points becomes such a strong bond that the process continues well after violent conflict has abated. Indeed, democratisation can lead to these essentialist ethnic markers being the dominant force in post-conflict elections (Snyder 2000). Evidence from BiH and from Northern Ireland demonstrates that political parties with more distinct ethno-national identities are those which have gained increasing electoral success (Mujkić and Hulsey 2010; Mitchell et al. 2009). Rather than ethno-national identity also being an area for negotiation after violence, the formation of identity during conflict has a significant influence on who is present and who is absent in post-conflict power relations.

Some authors have commented that the formation of national identity tends to take place in the public arena according to male norms and male reference points in the absence of women (Yuval-Davis 1997: 2). The idea of "nation" as a large, extended family consigns women to subordinate roles as carers for and reproducers of the community, but women are also appropriated to define the limits of the identity of the nation as "cultural markers" (Moghadam 1994: 2–3; Beasley and Bacchi 2000: 338; Timmerman 2000: 15). This has an impact on how the nation conducts its business in the spheres of governance and the application of justice with the criminal justice system tending to be retributive in nature and lacking any sense of gender-sensitivity. In its 2011–2012 report, the UN Women organisation asks if "the rule of law rules women out", continuing that "… laws tend to reflect and reinforce the privilege and the interests of the powerful whether on the basis of economic class, ethnicity, race, religion or gender" (UN Women 2011: 11). The difficulty in reconciling notions of the "nation" and social justice in gender terms, particularly in the context of nation-building, post-conflict or otherwise, has led to an observation that feminism and nationalism are incompatible ideologies for the manner in which this process has been acted out in reality (Kaplan 1997: 3).

Likewise, where ethno-national identity is the reference point for conflict, it is men who are predominantly active in the process. Northern Ireland during the conflict has been referred to as an "armed patriarchy" and the Stormont government as an "ethno-gender regime" (Monica McWilliams, quoted in Miller et al. 1996: 217; Racioppi and O'Sullivan See 2001: 93–5). Violent conflict has been viewed as a response to male needs in a male value system (Brock-Utne 1989: 15; Enloe 2000; Skjelsbaek and Smith 2001: 165) and gender relations even being seen as a root cause of war (Cockburn 2010). Whatever the explanation, there appears to be a distinct lack of women in decision-making roles during conflict.

While some have argued a biological pre-disposition to peace for women and to violence for men (Galtung 1996: 40–3), or that many women have developed a more protective stance towards the community due to learned mothering roles (Ruddick 1990: 23, 80), this stereotypical essentialising of women is seen as unhelpful (Moser and Clark 2001: 4; Reardon 1993: 15). Certainly, the impacts of war have a differential effect on women, who "suffer most, both in the perpetual violence against women during the occupation and through the deaths of sons, husbands, lovers, brothers and fathers" (Alonso 1993: 57).

The positioning of women as passive victims of conflict and men as dominant actors, it has been observed, plays into the hands of patriarchal norms (Karam 2001: 22). Yet, while there are women who play active roles in conflict, they are in a minority, and even then, the expectation has often been that once they have played their part, women would simply revert to an assumed pre-conflict role of domesticity and motherhood (Cherifati-Merabtine 1994: 40). In both Northern Ireland and BiH, women have been active in both fighting and peacebuilding, with the former in a minority and the latter largely unheard and celebrated. A stark contrast can be seen in BiH where the former President of the Republika Srpska, Biljana Plavšić, was indicted for genocide, crimes against humanity and violations of the laws or customs of war, while a figure such as Munira Subašić, head of the association

Mothers of Srebrenica, has been a significant figure in campaigning for justice for victims of the massacre in July 1995.

Whatever the circumstances of the involvement of women, it is male voices that are dominant during conflict, and if it is the conflict-related identities and reference points that exclude women that are carried over into the transition process, it is unsurprising that women's voices are absent there too.

The Nature of Post-conflict Transition

Attempts at ending conflicts perceived in terms of ethno-national identity over contested space have been approached by various methods of political accommodation. One to have gained significant currency in recent years is that of consociationalism, which, depending on the variant in use, comprises a "grand coalition" of ethnic groups, mutual veto in decision-making, ethnic proportionality in the allocation of offices and a certain degree of ethnic autonomy (Lijphart 1975: 166; Lijphart 1977; Horowitz 2001: 569–70). Essentially, elites of groups in conflict agree to manage their separate populations, divide up the executive roles and cannot make significant decisions without the consent of the other groups in the arrangement.

Such power-sharing systems have the potential to bring conflict elites together in a co-operative political structure in order to end violent conflict, but the approach has been criticised for privileging the identities that are associated with the conflict to the detriment of other identities and viewpoints, having the effect of perpetuating lines of division rather than fostering mixing or reconciliation between groups (Manning 2002: Taylor 2001; Wilson 2010). Political parties therefore develop along ethnic lines, having a centrifugal effect on social and political forces, rather than the cross-cutting, centripetal effect of non-ethnic parties (Horowitz 2001: 347). Conflict elites re-invent themselves as political elites while society remains divided along the lines of conflict in a form of suspended animation and the identities which lacked power and prominence during the conflict remain on the periphery.

However, this is not the full story. In the bargaining process that is a feature of post-conflict transition, Ní Aoláin (2009: 1056) notes that the "experiences and needs of women are markedly absent or silenced by the general discourse of accounting for the past." That said, actors in peacebuilding have sought to develop civil society in parallel to the political structures to foster the organic transition to a more peaceful, participative society, which has been referred to as "complementarity" (Bloomfield 1997). Non-governmental organisations (NGOs) are encouraged to become involved in the process of post-conflict transition through contributing to the political process while building capacity at community level to raise confidence, develop trust and provide opportunities for middle-range and grassroots actors to shape post-conflict relations (Lundy and McGovern 2008; Kumar 1997: 2; Lederach 1999: 39, 51).

If conflict elites are present at the negotiating table, then women are generally absent. It is assumed that if men are the ones with the guns, then women need not be involved in discussing decommissioning (Potter, A 2008). Yet the predominance of women in civil society organisations and the co-operative methods used by women active at community level makes women's presence in making decisions about the post-conflict society essential (Potter, M 2008). This is not to say that women conform to the "natural peacemakers" stereotype discussed above, but reflects the fact that women tend to predominate in civil society organisations and it is these organisations which have been at the forefront of reaching out to opposing communities.

Transitional justice requires a reassessment of violence and security to include re-examining repressive policies that impact upon women and not merely subsuming violence against women into existing and usually inadequate legislation (Ní Aoláin 2009: 1066; Oosterveld 2008: 75). The Nairobi Declaration goes some way to codifying this in that reparations "must address structural inequalities that negatively shape women's and girls' lives" (cited by Couillard 2007: 445). While conflict can impact upon women differentially to men, the process of change, violent or otherwise and the formation of a new society, can provide significant opportunities for women that may not have been available before (Timmerman 2000: 15; Karam 2001: 22; Hughes 2008). In its series of recommendations, the UN Women organisation says there is an imperative to "recognise the differential impact of conflict and rule of law deficits on women and children and the need to ensure gender sensitivity in restoration of rule of law and transitional justice, as well as the need to ensure the full participation of women" (2011: 21). These opportunity spaces can be brief and the participation of women is reliant on mechanisms that favour inclusion.

On 31 October 2000, the UN Security Council passed Resolution 1325 (2000), which called for the increased participation of women in decision-making roles, in all levels of negotiations, peacebuilding and UN field operations, and a gender perspective to be integral to all transitions from conflict. This was echoed the following month with a similar commitment by the European Union with European Parliament Resolution 2000/2025(INI) and 4 years later by the Council of Europe with Resolution 1385 (2004). These commitments amount to a structural application of gender equality provisions in post-conflict transition and the acknowledgement of the particular needs of women emerging from conflict and of the need for women to be present at every level and at every stage of the process.

While the key transition moments of the conflicts in BiH and Northern Ireland predate these resolutions, they are still cited as a rationale for the full participation of women in both contexts.

Challenges to the Transitional Justice Framework

Transitional justice can be broadly or narrowly defined. In its narrow sense, it refers to issues such as legal redress, truth recovery, reparations, prosecution of perpetrators and the establishment of international human rights standards. In its broader sense, "it involves anything that a society devises to deal with a legacy of conflict and/or widespread human rights violations" (Roht-Arriaza 2006: 2). The transitional justice framework has been increasingly challenged, with arguments that a "universal toolkit" is limited in its application to different contexts, for example, in a discarding of international mechanisms in favour of local processes (Shaw and Waldorf 2010: 3–4).

An expanding body of literature challenges the traditional transitional justice frameworks as being too narrowly focussed on legal and retributive processes, truth recovery and addressing human rights violations. While these issues are important in the transition from conflict, transitional justice, by these critiques, requires a wider application of inclusion and participation in post-conflict social and political life, particularly for women (Bell and Ní Aoláin 2004; Bell and O'Rourke 2007; Ní Aoláin and Rooney 2007; Rooney 2007).

It is contended in this chapter that meanings of justice in transition differ significantly between, on the one hand, those involved in judicial processes bringing to trial individuals who have committed human rights violations during the conflict, and on the other, people in wider society for whom the impact of the conflict varies considerably and whose understanding of justice has a broader meaning of participation and living in a fair society. Both BiH and Northern Ireland saw transitions that were subject to agreement between warring communities, where, although some recourse to justice processes for past wrongs are possible, the majority of victims will not see perpetrators come to trial.

Transitions from Conflict in Bosnia And Herzegovina and Northern Ireland

The armed conflict in BiH was halted with the Dayton Agreement of 1995, where new power-sharing structures were agreed in a short, sharp negotiating process (Holbrooke 1999). These new structures comprised three levels which assumed a need to accommodate ethnic differences through separation at community level. An overarching Parliamentary Assembly brings together the three identities in a grand coalition, below which are the two entities of the Parliament of the Federation of BiH and the Parliament of the Republika Srpska, the former of which also has a power-sharing arrangement. Then there are ten cantons within the Federation based on assumed ethnic identities. As a mechanism for ending the conflict, the arrangement has been successful, but divisions are maintained by the system and critics have suggested that it fails as a long-term solution, as elites are comfortable with a status quo that does not require inter-communal reconciliation (Glenny 1999: 651–2).

Women emerged from a conflict where there had been a significant level of gender-related violence and abuse (Benderley 1997: 66; Boeschoten 2003). In addition, women were the primary sustainers of community and the key elements in holding families together during and after the conflict (Robertson and Duckett 2007). While it has been claimed that transitional justice is merely a perpetuation of patriarchy (Ní Aoláin 2009), that is, a continuation of power systems that privilege male dominance, rather than a transformation to social justice, the experience of women is significantly different in that it has been a transition from a more liberal, egalitarian social context under the former Yugoslavia to what has been described as a "backlash" or "re-patriarchalisation" in the post-conflict experience (Skjelsbaek 2009; Majstorović 2011).

A Law on Gender Equality in BiH was passed in 2003, guaranteeing equality in a range of areas and a commitment to equal representation in government, including the judiciary, legislature and executive. At the time of writing, 9 of the 42 members of the House of Representatives are women and 2 of the 15 delegates to the House of Peoples. At the entity level, 23 of the 98 members of the House of Representatives of the Federation of BiH are women and 18 of the 83 members of the Parliament of Republika Srpska.

To complement the gender legislation, there is a range of structures dedicated to gender equality in BiH. The Gender Programme of the UN Development Programme oversees gender issues at the regional level and a Gender Equality Agency promotes gender equality at the national level. Both of the entities have Gender Centres with scrutiny roles in the area of advancing equality between women and men and there are Commissions for Gender Equality at cantonal and municipal levels.

Despite the efforts of the international community to develop sustainable gender equality norms and early optimism about the opportunities for women to take the initiative in the transition process (Draculić 1993: 129–30), it has been claimed that the majority of women have not benefited from the new structures, except for a small number of middle class women from an urban elite who have had the opportunity to do so due to the nature of top-down international development approaches (Pupavać 2005). Civil society organisations have been developed through international investment of which, on the one hand, it has been claimed that international models not suited to the context have been imposed (Belloni 2001), but on the other, while such critiques are valid to a certain extent, indigenous NGOs had existed from before and during the war and have had the opportunity to develop (Simmons 2007).

Generally speaking, the situation has been described in BiH as a model of men inside the political institutions in positions of power and women active in the community. Notions of women as "peacemakers" outside the formal political setting risk falling into essentialist stereotypes of male and female roles in the post-conflict society, but it has also been observed that politics is regarded as male and corrupt and women's activism in informal settings serves to maintain moral authority (Helms 2003).

Northern Ireland's conflict reached a settlement after protracted all-party negotiations (Mitchell 1999), culminating in the Belfast Agreement of 1998. After a number of false starts, primarily over the decommissioning of paramilitary weapons, the Northern Ireland Assembly was established, which has been described as consociational (O'Leary 1999). Members of the Legislative Assembly (MLAs) have to register as "nationalist", "unionist" or "other" on taking their seats and on important issues or where a petition of concern has been raised, a majority of nationalists and a majority of unionists are needed for decisions to be made. Therefore, the value of being an "other", i.e. not aligning oneself with the reference points of the conflict, is undermined when it comes to deciding key issues. In terms of gender balance in the institutions, at the time of writing 20 of the 108 MLAs are women.

While there were women who became directly involved in the conflict through participation in paramilitary organisations or security forces, women in Northern Ireland, as in BiH, have been regarded as sustainers of community, maintainers of families, tenders of the wounded and carers for those affected by the conflict (Potter 2004: 35). Women have been largely invisible in analyses of the conflict and the intersection of gender and other identities has been generally ignored (Rooney 2006). However, a sense of being left out of the transition from conflict led some women to organise to earn a place at the negotiating table, forming a political party for women's issues, the Women's Coalition (Fearon 1999). Fitzduff (1999: 91) describes how, initially, party members were treated with hostility including blatant sexism, although they provided a vital role in enabling the more extreme political parties to negotiate their way through the issues of releasing prisoners and decommissioning weapons. However, rather than benefitting from the tenets of consociational politics it was to be their demise as the electorate again voted along traditional political lines with the more extreme Unionist and Irish Nationalist parties seeing a rise in the number of seats gained to the Northern Ireland Assembly at the cost of moderates, including the Women's Coalition.

The Belfast Agreement and subsequent Northern Ireland Act 1998 contained commitments to equality and human rights, but rather than singling out gender in the post-conflict dispensation, equality between men and women was one of nine grounds about which equality impacts have to be taken into account and gender equality law proceeded in line with UK law in general. At the institutional level, there were no government agencies dealing specifically with gender. The Equal Opportunities Commission, the independent institution working in the field of sex discrimination, was absorbed into the more general Equality Commission and, although there is a Gender Equality Strategy, it is one of many across different equality grounds. Hence, there are no post-conflict mechanisms to promote gender equality in particular, and the UK Government does not intend applying UN Security Council resolution 1325 to Northern Ireland.

In some senses, to compare Northern Ireland and BiH as contexts of conflict fails to recognise specific differences in the nature and course of those conflicts. Deaths in Northern Ireland, at around 3,000 over 30 years, could not begin to equate with the around 100,000 of BiH over 3 years,[1] and the missing from Northern Ireland number fewer than 10, compared with the 13,000 still unaccounted for in BiH (Popović 2009: 32). Likewise, Northern Ireland was a low-intensity conflict in an affluent, resource-rich region of Europe, where casualties could expect high-quality

[1] The numbers of people killed in BiH have remained a matter of dispute. These figures are taken from the *Bosanska Knjiga Mrtvih*, a project by the Research and Documentation Centre in Sarajevo.

medical treatment within hours, if not minutes, of injury, and conflict-related damage to property was compensated for by the government. No such support or provision existed in BiH. Furthermore, the Northern Ireland conflict operated under a legal system which, although criticised as partial, did give some recourse to justice or appeal against state or third-party action was at least possible. BiH was in a state of war where legal and governmental structures were suspended.

Yet, the mechanisms and legacies of conflict fall into certain recognisable patterns and the differences are largely a matter of scale and severity. Loss or trauma can be as significant for an individual whether it takes place in isolation or in a context where they are commonplace. The experience of women can also be compared. While there were not the incidences of massed crimes against women in Northern Ireland that took place in BiH, patterns of essentialisation, appropriation and marginalisation of women are common patterns in the literature of both conflicts.

The Experience of Women

The evidence for this chapter is derived initially from consultations in 2005 as part of the development of the gender dimensions of the European Union Peace Programme for Peace and Reconciliation in Northern Ireland and from discussions at a workshop at the "Pathways to Reconciliation and Human Rights" conference in Sarajevo on gender and transition, which informed a formal research project in Northern Ireland and BiH with key individuals working with women at community level in 2007/2008. The themes of the research were then further expanded at the "Peace by Piece" conference in Belfast in 2008, involving women from conflict areas internationally, including BiH.

In Northern Ireland, the research was carried out with groups involved in women's training projects through the EU Peace Programme supported through the intermediary funding body Training for Women Network (TWN), including project officers at the funding body and project promoters and participants at Ballynafeigh Community Development Association, East Belfast Community Development Association, Short Strand Partnership and Women's Tec in Belfast, Positive Steps Learning Centre in Cookstown and the victims group Homes United by Recurring Threat (HURT) in Lurgan. In all, 20 individuals were interviewed, one of which was male. Information was also gleaned from project evaluations across 20 women's training projects. The Northern Ireland fieldwork was carried out at a time when the impact of the Peace Programme on women was being evaluated.

Interviews in BiH were carried out in Sarajevo with international agencies with gender programmes in the country (British Council, UNDP Gender Programme and Catholic Relief Services), the government Gender Equality Agency, the Centre for Interdisciplinary Studies at the University of Sarajevo and community-based organisations (Žene Ženama, Cure Foundation and Organisation Q), and an additional interview was carried out with an unaffiliated community activist. In all, 20 women were interviewed. The fieldwork in BiH was carried out in the context of a joint

Government and UN Development Programme project to promote and consult upon how to progress with transitional justice in BiH.

The war in BiH had a huge impact on women from many perspectives. Thousands of women were directly affected through being targeted for gender-based war crimes, menfolk were killed or disabled through the war, community infrastructures were destroyed, and, according to representatives of international agencies interviewed, facilities used by women, such as childcare providers, were appropriated for the purposes of the conflict. Justice for women in this instance lies in the restoration of community facilities that were formerly relied upon to facilitate social and economic participation as well as personal autonomy. In Northern Ireland, a range of community-based facilities for women developed as a result of the EU Peace Programme and other conflict-related programmes, but these facilities are heavily reliant on external funding, which has been diminishing with the increasing distance from the Belfast Agreement and the development of a more stable society.

One significant impact of the war in BiH has been the increase of women as heads of households. As in Northern Ireland, BiH is mostly rural, where, according to interviewees, there are stronger gender-based traditions of women in the home. These traditions also exist in rural areas of Northern Ireland, but there was more emphasis on this factor among interviewees in BiH. With the loss of the male head of household, many women have found themselves having to take the initiative with no credit history, no formal employment experience and the house remaining in the name of a husband who is dead or missing. A representative of the Catholic Relief Services interviewed stated: "now they are the man," indicating the change of gender roles still presents attitudinal and procedural barriers based on assumed positions in society of women and men. Women have had to organise out of necessity and try to make ends meet in a context of high unemployment and poor infrastructure.

Some international NGOs in BiH, such as Catholic Relief Services, have established micro-credit arrangements and provided training in business start-up and management for women. While successful businesses have been established by women, there are still attitudes among many in society who see entrepreneurship as a male preserve. In Northern Ireland, training and support projects for women's entrepreneurship have been funded through the EU Peace Programme, such as the Weave project in North Belfast, which trains women to develop traditional handicraft skills into viable businesses, but equivalent micro-credit programmes for women are not available. While born in adversity, transitions in both of these contexts provide opportunities for the building of more just gender relations in terms of economic independence for women and finding a place in a male-dominated economy.

Local NGOs such as Žene Ženama in Sarajevo have provided programmes for education, psychosocial support, campaigns against gender-based violence, health, gender awareness, promotion of rights and equality, civic participation, political participation, social care and peacebuilding. These activities represent the broad scope of issues that are affecting women's lives in the post-conflict context and, using the broader understanding of the term, cover many of the dimensions of transitional justice. For example, at the time of the interview, a course funded by the Spanish government was being developed focussing on women in transition from conflict, with supported discussion sessions around how conflict differentially affects women, understanding the processes of conflict and exploring how women can contribute to post-conflict reconstruction. Multi-faceted approaches to empowerment in settings in which women feel supported have a major role to play in promoting participation in both BiH and Northern Ireland.

However, while there have been some tentative cross-community initiatives among women's groups in Northern Ireland that endeavour to engage with divisive issues around the conflict itself and to challenge elements within their own communities over their role, it has been commented that this process has been more difficult in BiH, where there is a tendency not to challenge the dominant community perspective with regard to the conflict.

One community-based interviewee in Sarajevo commented that, while some NGOs have managed to secure funding from international NGOs for activities, the majority are small groups with little or no external support. Indeed, interviews with representatives of international agencies bemoaned the turning of UN attention to Kosovo, which has meant diminishing sources of funding for all NGOs and a need to become more self-sufficient. This sense of diminishing resources to fund empowerment activities was also a feature of interviews in Northern Ireland, where all project promoters interviewed held concerns regarding the finite nature of project-based funding in the community.

In theory, governmental structures in BiH, particularly the Gender Equality Agency, are to work co-operatively with community-based NGOs to promote gender equality, but a community group representative in the research complained that there is a sense of a division of labour, where some NGOs have felt that the Gender Agency takes the lead on policy, leaving NGOs to continue the work on the ground. This limits the potential of NGOs to convert civic activism into political activism and have an influence on the development of gender equality policy, leaving a feeling of exclusion from the political and policymaking arena.

Evidence from some international NGOs suggested that exclusion is also felt across BiH due to geography and the nature of the political structures, for example, co-operation between agencies is difficult. Also, it was claimed by a community-based interviewee that some of the smaller NGOs feel that an elite of strong personalities within the women's sector has developed their own work and organisations with international support, but there is a disconnection with other groups elsewhere in the country. This was also felt by a number of community-based project promoters in Northern Ireland, where groups with more resources were considered to have more access to decision-making processes. Indeed, a representative from a rural project in Northern Ireland highlighted how exclusion is compounded by rural isolation, the lower capacity and confidence of rural women and distance from the political centre, and rural groups receive fewer resources and less support than their urban counterparts. The distance from the centre of power, therefore, whether physically or in terms of access, has significant implications for a just transition.

An academic interviewee stated that gender issues are not popular in BiH from a range of perspectives. It was explained that the more liberal aspects of pre-war Yugoslav society did not penetrate far from urban centres in BiH, and traditional structures were disrupted by the conflict. By this, the interviewee meant that, while women's lifestyles and clothing, for example, were less prescribed in pre-war urban settings, there remained more restrictions on women's lives in rural areas, but the processes of conflict created opportunities and often compulsions to move out of traditional gender roles in order to adapt to new circumstances. Currently, there is a belief that the post-war social and political context is trying to place women back into the domestic sphere. This is complicated by the ethno-national nature of the political divide, where identities are defined in terms of mutually exclusive traditions. Women are caught in the middle of this process, where feminism and women's emancipation is seen not just as a challenge to a male social and political order, but a betrayal of cultural identity and a diluting of fundamental values on which conflict-related markers are based. Such processes have also taken place in Northern Ireland, for example, with the return of released prisoners to households where women had taken on new roles in the absence of male partners, but this was less stressed by interviewees in Northern Ireland compared with those in BiH.

This resistance to women's emancipation is complicated by a segregated education system, where there are different values taught from different perspectives, which in the context of transition and the formulation of new ethno-national identities does not have room for challenging social norms. Indeed, the reproduction of the national community takes place in the classroom and stereotypical images of women are markers of cultural identity. Justice in this sense is served by ensuring that individuals have equal access to participation and a fundamental mechanism for addressing stereotypes and prejudice is through education.

While there are extensive structural and legislative provisions for gender equality and the organisation of women emerging from the domestic sphere in the post-conflict context, women still struggle to be heard or be present at decision-making levels. As with many societies emerging from violent conflict, new systems of justice tend to be retributive in nature, more focused on individual perpetrators, less on victims and rarely cognisant of the gender consideration. Despite increased awareness of cases of sexual and reproductive violence, the place of women is still marginalised (Duggan et al. 2008: 193).

The interviews in Northern Ireland highlighted the extent to which women have been organising at community level, engaging in informal reconciliation processes and supporting people affected by the conflict. Thus, both bridging and bonding social capital have been developed to a high degree in some areas, largely through the efforts of small groups of women active in the community.

The conflict has produced a legacy of distrust and polarisation, particularly in areas with high levels of social housing. Representatives of rural groups noted that the conflict has been less overt in the countryside, but the impact of violence has led to fear and isolation, extending well beyond the cessation of direct hostilities and difficulty in addressing the conflict that has been possible in some urban areas, where populations live closer and contact is more regular. As one interviewee put it, there is a "rural politeness" which belies a simmering animosity. At the same time, there are many women for whom disconnection with the community has led to a considerable need for social, emotional and psychological support to re-engage with society. Re-connection with society is a central theme in work with victims and survivors of the conflict and also an important factor in processes connected with transitional justice in its broader, societal sense.

The violent conflict, according to one project promoter in Northern Ireland, led to equality being "put on hold." To a certain extent, women's empowerment projects have been trying to regain the ground that has been lost. The conflict created a division of labour while it was going on and also after the ceasefires. While it was mainly men engaged in violence and politics, the interviewee continued, "the women just kept things going: educating families and getting the children to school." After the violence, it was men at the negotiating table, but women are still dealing with the legacy, that while young people are no longer rioting, there are problems with drink, drugs and violence against older people. Both transitions from conflict recognise the need to realise equality goals, but if women have been marginalised during the conflicts and conflict elites take priority in the post-conflict arrangements, women are starting from a position of disadvantage that needs to be bridged before genuine equality can be realised.

Reflecting a general feeling among many interviewees in both contexts, a project promoter in Northern Ireland suggested women are more suited to reconciliation, a rationale that is seen as gender-specific; as another interviewee stated: "We need to do something for our children and grandchildren. We do not want a repeat of this." Acknowledging the gender stereotype this viewpoint perpetuates, women are seen as an important resource for reconciliation, but if women have been the main sustainers of families and communities, there is also an argument that it is they who are more in touch with the social impacts of violence and therefore are best situated to understand it.

The projects in Northern Ireland offered a space where women from different backgrounds could come together, form relationships across communities, develop personally, receive support and learn new skills. As one project promoter put it: "Women are more open, more willing to talk. There is a thing in men to fight, isn't there?" With practical measures to facilitate access, such as childcare provision, transport and individual support, spaces are provided for dialogue and to share experiences. For example, separate projects in a Loyalist area of East Belfast and the Irish Republican Short Strand area formed a cross-community women's group where it was found that, between women of different backgrounds, according to one group member: "the issues are the same; the only thing that divides us is politics." Justice, for these women, is working towards a society where conflict between them has been addressed and they and their families can live peacefully together.

Women working with women in the community is only part of the story. The lack of women in political, social and economic leadership creates a need to bridge the gap between community-based empowerment and taking an active part in decision making where power lies. Female role models in positions of power help to encourage women to follow in their footsteps and the development of skills in sectors where women are under-represented contributes to a more balanced, equal society, for example, in business or in information technology. For example, the Women's TEC in North Belfast combines cross-community and capacity-building activities with training in non-traditional skills, such as plumbing, electrical engineering, technology or business start-up and development, to place women in previously male-dominated industries. Other projects try to encourage more women into politics.

The dominant message from the interviews in both Northern Ireland and BiH is that projects developing women in the community have the potential to have wider impacts in cross-community engagement and in the building of community competences to deal with the legacy of the conflict and engage in the process of peace-building between communities. However, there are problems with this picture. Reinforcing the stereotype of community-based women activists and men in politics has to be addressed using the capacity-building process to bring more women into decision-making positions in economic and political arenas.

Bringing Justice into Transition

In terms of formal transitional justice processes, neither Northern Ireland nor BiH have established truth commissions, although there have been inquiries into individual events in Northern Ireland, such as the 1972 Bloody Sunday events, and commissions in BiH in individual areas, such as Srebrenica. However, the position of the UK Government is that there are to be no more open-ended inquiries into incidents relating to the conflict in Northern Ireland. The Historical Enquiries Team (HET) of the Police Service for Northern Ireland is not a satisfactory replacement for a formal truth process, due to a lack of resources, an extensive backlog, having been criticised for a lack of independence, being a police team, and lacking any wider involvement beyond the investigation of unsolved cases.

The International Criminal Tribunal for former Yugoslavia (ICTY) was established by UN Security Council Resolutions 808 and 827 in 1993, and departments for war crimes were set up in the Prosecutor's Office for BiH following UN Security Council Resolution 1503 in 2003. Significantly for women (and for some men), the ICTY defined rape as a war crime and almost half of all accused at the time of writing had charges of sexual violence included in their indictments. In terms of gender justice, this was a major step forward in comparison with other transition processes. In contrast, there have been no significant studies of conflict-related sexual violence in Northern Ireland. These judicial processes bring a degree of truth and closure to many and they are a symbolic demonstration of the re-establishment of justice and human rights norms following conflict. However, the majority of the population remains untouched by their deliberations.

Transitional justice encompasses all these things, which remain essential to any post-conflict process, but also to a far greater extent the embedding of equality, human rights and reconciliation throughout the communities in conflict.

Without a transformed society, punishment and restitution are transient, more or less clearing up the mess of one conflict, without laying the foundations of preventing the next one.

It is true that there are victims' groups in Northern Ireland and in BiH that do not see the possibilities of reconciliation without justice, that is, the punishment of those who have wronged their loved ones or the return of property that they have lost in the conflict. The needs of people directly affected by conflict cannot be overlooked or ignored. Yet the participants in the research referred to here, which included individuals who had lost family members in the conflict and, in the case of Northern Ireland, a victims' group, considered justice to mean equal participation in and contribution to the post-conflict society. With all their cultural, geographical and historical differences, this theme remained constant with interviewees in both Northern Ireland and BiH.

The representation of women in decision making is one area that, in international terms, is promoted as a means for promoting justice in the political arena. However, it is in civil society organisations in both Northern Ireland and BiH that women are most abundant and in significant leadership roles. Mechanisms for giving civil society access to the processes of deliberation, such as meaningful consultation, and the promotion of civic activism are just as important in empowering women in post-conflict settings.

Transitional justice for the women interviewed meant gender justice and a just and peaceful future. To some, this means equal and meaningful representation at political and decision-making levels. For others, this means being able to run a business on an equal footing to men. For others again, this means access to facilities where they and their families can develop personally, professionally and socially. However, the focus for all involved was that creating a just society must be undertaken together, rather than apart.

Acknowledgments Thanks are due to NI-CO in Belfast, which funded the research in Bosnia and Herzegovina, and the EU Building Sustainable Prosperity programme in Northern Ireland, under which the research in Northern Ireland was funded. Much gratitude is also owed to all those who took part in interviews and focus groups for the purposes of the research.

References

Aiken, N. 2010. Learning to live together: Transitional justice and intergroup reconciliation in Northern Ireland. *International Journal of Transitional Justice* 4.

Alonso, H. 1993. *Peace as a women's issue: A history of the us movement for world peace and women's rights*. New York: Syracuse University Press.

Beasley, C., and C. Bacchi. 2000. Citizen bodies: Embodying citizens - a feminist analysis. *International Feminist Journal of Politics* 2(3): 337–358.

Bell, C., C. Campbell, and F. Ní Aoláin. 2004. Justice discourses in transition. *Social and Legal Studies* 13(3): 305–328.

Bell, C., and C. O'Rourke. 2007. Does feminism need a theory of transitional justice? *International Journal of Transitional Justice* 1: 23–44.

Belloni, R. 2001. Civil society and peacebuilding in Bosnia and Herzegovina. *Journal of Peace Research* 38(2): 163–180.

Benderley, J. 1997. Rape, feminism and nationalism in the war in the Yugoslav successor states. In *Feminist nationalism*, ed. L. West, 59–74 New York: Routledge.

Bloomfield, D. 1997. *Peacemaking strategies in Northern Ireland: Building complementarity in conflict management*. Houndmills: Macmillan.

Boeschoten, R. 2003. The trauma of war rape: A comparative view on the Bosnian conflict and the Greek Civil War. *History and Anthropology*. 14(1): 41–44.

Brock-Utne, B. 1989. *Educating for peace: A feminist perspective*. New York: Pergamon.

Cherifati-Merabtine, D. 1994. Algeria at a crossroads: National liberation, Islamicisation and women. In *Gender and national identity: Women and politics in Muslim societies*, ed. V. Moghadam, 40–62 London: Zed.

Cockburn, C. 2010. Gender relations as causal in militarization and war. *International Feminist Journal of Politics* 12(2): 139–157.

Couillard, V. 2007. The Nairobi Declaration: Redefining reparations for women victims of sexual violence. *The International Journal of Transitional Justice* 1: 444–453.

Draculić, S. 1993. Women and the New Democracy in the Former Yugoslavia. In *Gender, politics and post-communism: Nationalism, war and the great powers*, ed. N. Funk and M. Mueller, 123–130 London: Routledge.

Duggan, C., Bailey, C., and Guillerot, J. 2008. Reparations for sexual and reproductive violence: Prospects for achieving gender justice in Guatemala and Peru. *The International Journal of Transitional Justice* 2: 192–213.

Fitzduff, M. 1999. Changing history—Peace building in northern Ireland. In European Centre for Conflict Prevention, *People building peace: 35 Inspiring stories from around the world*, 87–103 Utrecht: ECCP.

Enloe, C. 2000. *Bananas, beaches and bases: Making feminist sense of international politics*. Berkley: University of California Press.

Fearon, K. 1999. *Women's work: The story of the Northern Ireland women's coalition*. Belfast: Blackstaff.

Galtung, J. 1996. *Peace by peaceful means: Peace, conflict development and civilisation*. London: Sage.

Glenny, M. 1999. *The Balkans 1804–1999*. London: Granta.

Helms, E. 2003. Women as agents of ethnic reconciliation? Women's NGOs and international intervention in postwar Bosnia-Herzegovina. *Women's Studies International Forum* 26(1): 15–33.

Holbrooke, R. 1999. *To end a war*. New York: Modern Library.

Horowitz, D. 2001. *Ethnic groups in conflict*. Berkley: University of California Press.

Hughes, M. 2008. Armed conflict, international linkages and women's parliamentary representation in developing nations. *Social Problems* 56(1): 174–204.

Kaplan, G. 1997. Feminism and Nationalism: The European Case. In *Feminism and nationalism*, ed. L. West, 3–40 New York: Routledge.

Karam, A. 2001. Women in war and peacebuilding. *International Feminist Journal of Politics* 3(1): 2–25.

Kerr, R., and E. Mobekk. 2007. *Seeking accountability after war*. Cambridge: Polity.

Kumar, K. (ed.). 1997. *Rebuilding societies after civil war: Critical roles for international assistance*. Boulder: Lynne Rienner.

Lederach, J.P. 1999. *Building peace: Sustainable reconciliation in divided societies*. Washington DC: US Institute of Peace Press.

Lijphart, A. 1975. The Northern Ireland problem: Cases, theories and solutions. *British Journal of Political Science* 5(1): 83–106.

Lijphart, A. 1977. *Democracy in plural societies: A comparative exploration*. New Haven: Yale University Press.

Lundy, P., and McGovern, M. 2008. Whose justice? Rethinking transitional justice from the bottom up. *Journal of Law and Society* 35(2): 265–292.

Majstorović, D. 2011. Femininity, patriarchy and resistance in postwar Bosnia and Herzegovina. *International Review of Sociology* 21(2): 277–299.

Manning, C. 2002. Conflict management and elite habituation in postwar democracy. *Comparative Politics* 35(1): 63–84.

McGrattan, C. 2010. Explaining Northern Ireland? The limitations of the ethnic conflict model. *National Identities*12(2): 181–197.

Miller, R., R. Wilford, and F. Donoghue. 1996. *Women and political participation in Northern Ireland*. Aldershot: Avebury.

Mitchell, G. 1999. *Making peace*. London: Heinemann.

Mitchell, P., Evans, G., and O'Leary, B. 2009. Extremist outbidding in ethnic party systems is not inevitable: Tribune parties in Northern Ireland. *Political Studies* (57): 397–421

Moghadam, V. 1994. *Gender and national identity: Women and politics in Muslim societies*. London: Zed.

Moser, C., and F. Clark (eds.). 2001. *Victims, perpetrators or actors? Gender, armed conflict and political violence*. London: Zed.

Mujkić, A., and Hulsey, J. 2010. Explaining the success of nationalist parties in Bosnia and Herzegovina. *Politička Misao* 47(2): 143–158.

Ní Aoláin, F. 2009. Women, security, and the patriarchy of internationalized transitional justice. *Human Rights Quarterly* 31(4): 1055–1085.

Ní Aoláin, F., and Rooney, E. 2007. Underenforcement and intersectionality: Gendered aspects of transition for women. *International Journal of Transitional Justice* 1: 338–354.

O'Leary, B. 1999. The nature of the agreement. *Fordham International Law Review* 22(4): 1628–1667.

Oosterveld, V. 2008. The special court for Sierra Leone's consideration of gender-based violence: Contributing to transitional justice. *Human Rights Review* 10: 73–98.

Popović, D. 2009. *Transitional justice guidebook for Bosnia and Herzegovina: Executive summary*. Sarajevo: UNDP.

Potter, A. 2008. Women, gender and peacemaking in civil wars. In *Contemporary peacemaking: Conflict, peace processes and post-war reconstruction*, ed. John Darby and Roger MacGinty, 105–119 Houndmills: Palgrave Macmillan.

Potter, M. 2004. *In their own words: A research report into the victims sector in Northern Ireland*. Belfast: Training for Women Network.

Potter, M. 2008. Women, civil society and peacebuilding. In *Global change, civil society and the Northern Ireland peace process*, ed. C. Farrington. Houndmills: Palgrave Macmillan.

Pupavać, V. 2005. Empowering women? An assessment of international gender policies in Bosnia. *International Peacekeeping* 12(3).

Racioppi, L. and O'Sullivan See, K. 2001. "This we will maintain": gender, ethno-nationalism and the politics of unionism in Northern Ireland. *Nations and Nationalism* 7(1): 93–112.

Reardon, B. 1993. *Women and peace: Feminist visions of global society*. Albany: State University of New York Press.

Robertson, C., and Duckett, L. 2007. Mothering during war and Postwar Bosnia. *Journal of Family Nursing* 13(4): 461–483.

Roeder, P., and D. Rothchild. 2005. *Sustainable peace: Power and democracy after civil wars*. Ithaca: Cornell University Press.

Roht-Arriaza, N. 2006. The new landscape of transitional justice. In *Transitional justice in the justice-first century: Beyond truth and justice*, ed. N. Roht-Arriaza and J. Mariezcurrena, 1–16 Cambridge: Cambridge University Press.

Rooney, E. 2006. Women's equality in Northern Ireland's transition: Intersectionality in theory and place. *Feminist Legal Studies* (14): 353–375.

Rooney, E. 2007. Engendering transitional justice: Questions of absence and silence. *International Journal of Law in Context* 3(2): 173–187.

Rothschild, J. 1981. *Ethnopolitics: A conceptual framework*. New York: Columbia University Press.

Ruddick, S. 1990. *Maternal thinking: Towards a politics of peace*. London: Women's Press.
Shaw, R., and L. Waldorff (eds.). 2010. *Localising transitional justice: Intervention and priorities after mass violence*. Stanford: Stanford University Press.
Simmons, C. 2007. Women's work and the growth of civil society in post-War Bosnia. *Nationalities Papers* 35(1): 171–185.
Skjelsbaek, I., and D. Smith (eds.). 2001. *Gender, peace and conflict*. London: Sage.
Skjelsbaek, I. 2009. Perceptions of 'good womanhood' among twenty Bosnian focus group participants. *International Feminist Journal of Politics* 11(3): 392–411.
Smith, A. 1999. *Myths and memories of the nation*. Oxford: Oxford University Press.
Snyder, J. 2000. *From voting to violence: Democratisation and nationalist conflict*. New York: Norton.
Taylor, R. 2001. Northern Ireland: Consociation or social transformation? In *Northern Ireland and The Northern Ireland conflict and the Good Friday Agreement in comparative perspective*, ed. J. McGarry, 36–52 Oxford: University Press.
Timmermann, C. 2000. Muslim Women and Nationalism: The power of the image. *Current Sociology* 48(4): 15–27.
UN Security Council. 2004. *The rule of law and transitional justice in conflict and post-conflict societies*. New York: United Nations.
Women, U.N. 2011. *Progress of the world's women 2011–2012: In pursuit of justice*. New York: United Nations.
Voutat, B. 2000. Territorial identity in Europe: The political processes of the construction of identities in Corsica, the Basque country, Italy, Macedonia and the Swiss Jura. *Contemporary European History* 9(2): 285–294.
Wilson, R. 2010. *The Northern Ireland experience of conflict and agreement: A model for export?* Manchester: Manchester University Press.
Yuval-Davis, M. 1997. *Gender and nation*. London: Sage.

Chapter 11
'Pillar of Shame': Civil Society, UN Accountability and Genocide in Srebrenica

Olivera Simić

Introduction

Our wish is to create a world without genocide and war. This can only be achieved by learning from mistakes and by assuming responsibility for them[1] (Fig. 11.1).

The genocide in Srebrenica was the worst massacre in Europe since WWII. In July 1995, more than 8,000 Bosnian Muslim men and boys were killed while their female relatives were deported, killed, or raped by Bosnian Serb forces.[2] The killings took place in a UN 'safe area',[3] where Bosnian women, men and children had sought refuge after being displaced by the ethnic conflict that followed the dissolution

[1] Interview with Hatidza Mehmedovic, President of the Mothers of Srebrenica Association, http://stubsrama.com/index.php?option=com_content&view=article&id=13%3Aintervju-sa-hatidom-mehmedovi-predsjednicom-udruenja-srebrenike-majke&catid=3&Itemid=11&lang=en. Accessed 11 November 2011.

[2] The number of women who were raped or killed during the genocide has not yet been established.

[3] UN Safe Areas were established in 1993 in the territory of BH by the UN Security Council. The territories under the protection of the UN Protection Force (UNPROFOR) included Srebrenica, Sarajevo, Zepa, Gorazde, Tuzla and Bihac. UN Resolution 819 (S/RES/819, 16 April 1993) and 836 had established Srebrenica as a 'safe area' to be protected by all necessary means including the use of force.

Dr Olivera Simić is a lecturer with the Griffith University Law School, o.simic@griffith.edu.au.
I would like to thank Dr Fayen d'Evie, Dr Zala Volcic, Dr Nesam Mcmillan, Dr Trish Luker, Dr and Melanie O'Brien for their careful reading and useful comments on an earlier draft of this paper. I am indebted to Philipp Ruch, Belma Zulcic and Aldijana Mustafic for their participation in the research. I also wish to thank Professor Kathleen Daly for invaluable conversations on the symbolism of shoes and civil society justice mechanisms.

O. Simić (✉)
Griffith Law School, Griffith University,
Law Bldg (N61), Brisbane, QLD, 4111, Australia
e-mail: o.simic@griffith.edu.au

O. Simić and Z. Volčič (eds.), *Transitional Justice and Civil Society in the Balkans*,
Springer Series in Transitional Justice, DOI 10.1007/978-1-4614-5422-9_11,

Fig. 11.1 The image from the campaign 'Pillar of Shame'. Photo is courtesy of Phillip Ruch

of Yugoslavia. Each year, a commemoration for victims of the genocide is held in Srebrenica–Potocari, and thousands of people from BH and abroad come to pay their respects to those whose lives were taken so abruptly. Over the years, the Srebrenica annual commemoration has become not only a place of mourning but also an event of 'international consumption' where people from all walks of life gather to pay their respects (Duijzings 2007, p.161; Simic 2009). The Srebrenica anniversary is organized annually on July 11 and attracts and mobilizes local and international human rights activists, relatives, survivors and governmental representatives from around the world. Tens of thousands of people gather to pay their respects to the dead and to bury those who have been found and identified after years of being missing.[4] However, although a number of heads of state have made their way to Srebrenica–Potocari to express their condolences and remorse,[5] no UN Secretary-General has ever visited the graveyard.

The responsibility for not preventing the Srebrenica genocide has been apportioned largely to the UN and the Dutch peacekeepers that were stationed in Srebrenica at the time the killings took place. The release of a 2002 Dutch report on Srebrenica led to the resignation of the Dutch Government, even though the report claimed that the Dutch troops were put in an 'impossible situation' (BBC News

---

[4] An estimated 40,000 people gathered in Srebrenica on July 2011 to mark the 16th anniversary of the genocide. See, 'Thousands Gather to Remember Srebrenica Massacre', *VOANews*, 12 July 2011, http://sofiaecho.com/2011/07/12/1121775_thousands-gather-to-remember-srebrenica-massacre; 'Tens of Thousands of People Gather to Commemorate Srebrenica Massacre', *European Forum for Democracy and Solidarity*, 12 July 2010, http://www.europeanforum.net/news/919/tens_of_thousands_of_people_gather_to_commemorate_srebrenica_massacre.

[5] For example, Serbian President Boris Tadic, Former President of the United States, Bill Clinton, Members of the United States Presidential Delegation of Barack Obama, Croatian President Ivo Josipovic, the European Union High Representative in BH Valentin Inzko, Deputy Turkish Prime Minister Bulent Arinc, and the President of the Republic of Slovenia, Danilo Turk.

2002). Dutch peacekeepers have been blamed by survivors of Srebrenica for witnessing and standing by as Serb soldiers took men and boys to execution sites while deporting their wives, mothers and daughters from Srebrenica.[6] Thirteen years after the genocide, in June 2008, the non-governmental organization 'Mothers of Srebrenica' began a court battle before the International Court of Justice (ICJ) against the UN and the Dutch Government for failing to prevent the massacre. In this landmark case, the plaintiffs claimed that both entities needed to bear responsibility for not preventing the crimes that occurred in the summer of 1995. Lawyers brought the lawsuit in the name of 6,000 survivors of genocide and argued that compensation should be paid to the victims for the loss of loved ones despite the UN claim of immunity (Charter 2008). Alex Hagedorn, acting on behalf of the victims, reportedly stated:

Functional immunity does not mean that international organizations are above the law. Boundless immunity of the UN is both unacceptable and undesirable for the proper functioning and credibility of the UN. (Charter 2008)

In July 2008, the District Court of the Hague ruled that it had no jurisdiction to hear the civil case brought against the UN, given the general immunity of the UN from prosecutions pursuant to international conventions. Such immunity is regarded, according to the Court, as 'absolute immunity' (Appeal Court in The Hague 2010). In explaining its decision, the court stated:

> That the Mothers of Srebrenica seek redress in the court of law for this is wholly understandable… As has been considered before, a substantial general interest is served if the United Nations is not forced to appear before a national court of law. In this field of tensions pros and cons must be balanced … (Judgment, 30 March 2010, para 5.9)

The court recommended that the plaintiffs should take the actual perpetrators of the genocide to court. In May 2010, the Court of Appeal upheld a decision on the immunity of the UN and rejected the Mothers appeal (Radio Netherlands Worldwide 2010). This decision, while not entirely surprising, resulted in disappointment and bitterness among survivors, but also a desire to find a way to shame the UN for failing to act.[7]

This chapter documents a campaign led by two non-governmental organizations, the German Center for Political Beauty (Center for Political Beauty) from Berlin and the Bosnian Society for Threatened People (Bosnian Society) from Sarajevo, which launched a joint project in 2010 called the 'Pillar of Shame' This project aimed to erect a permanent monument in the shape of two UN letters that would be filled with 16,744 shoes symbolically representing 8,372 victims killed in genocide.[8] The monument was

[6] See Hasan Nuhanovic, *Under The UN Flag; The International Community and the Srebrenica Genocide* (DES Sarajevo, 2007), Emir Suljagic, *Postcards from the Grave* (Saqi Books, 2005)

[7] On April 13, 2012, the Mothers of Srebrenica lost their case against the UN before the Supreme Court in Holland. The final ruling was the last legal option in the Netherlands for the Mothers. An appeal would need to be filed at the European Court of Human Rights within 6 months. See Anthony Deutsch, 'UN Can't be Tried for Srebrenica Massacre in Netherlands', *Reuters*, 13 April 2012.

[8] The exact number of people killed in Srebrenica is not known and there are different estimates. The number cited in this paper is taken from the Center for Political Beauty leaflets and materials about campaign.

intended to serve as a permanent reminder of the UN's responsibility for not preventing the Srebrenica genocide. In the first part of the chapter, after outlining the methodology, I situate this inquiry within the theoretical framework of transitional justice, in particular symbolic reparation. The chapter then analyses the campaign and evaluates its impact in BH and Germany. It argues that campaigns of symbolic reparation such as 'Pillar of Shame' need to be documented, brought to light and analysed. Such 'bottom top' campaigns are largely under-analysed, yet are of the utmost importance as civil society and local justice mechanisms, which promote participation and sense of local ownership of justice (Olsen et al. 2010).

This research draws from media, documentaries and interviews with the two founders and leaders of campaign and one of its volunteers from BH. With permission and consent of the interviewees, the analysis draws from interview and email correspondence with Phillip Ruch, the executive director of Center for Political Beauty, Belma Zulcic, the co-ordinator of the project 'Pillar of Shame' with the Bosnian Society, and Aldijana Mustafic from Bihac, a BH volunteer who gave us permission to reproduce a number of photos about the campaign. The interviews were carried out during fieldwork in BH in November and December 2011. With consent and courtesy of the Center, the chapter draws also on internal office documents sent to the author via email and official web page produced for the purpose of campaign. Finally, it draws on informal conversations about this project with representatives of the 'Mothers of Enclave of Srebrenica and Zepa', based in Sarajevo.

Mobilizing Civil Society Across National Borders

The past two decades have seen new social movements, organizations and networks emerging at the transnational level (Batliwala 2002, p.393). In transitional states, civil society actors engage in practices, politics and contests over strategies for dealing with the past. Many organizations operate solely at the local level while others draw on national, regional, or even transnational organizational structures (Boesenecker and Vinjamuri 2011, p.347). In the absence of international justice mechanisms, the creative energy for transition is coming 'from below' (Campbell and Ni Aolainn 2003, p.871). In settings where national and international justice systems are ineffective, victim and survivor groups, community and civil society organizations have been the engines of change (McEvoy and McGregor 2008, p.3). For the purpose of this chapter, the term 'civil society' refers to actors who associate voluntarily to advance common interests, and who are officially separate from the formal state apparatus, which often translates into an assumption that civil society actors are grounded in a common set of beliefs, values, interests and networks (Boesenecker and Vinjamuri 2011, p.348; Anheier 2004, p.22).

There are various definitions of transitional justice but the one most commonly cited is proposed by the International Centre for Transitional Justice as 'a response to systematic or widespread violations of human rights that seeks recognition for victims and to promote possibilities for peace, reconciliation and democracy' (ICTJ 2011).

In order to confront legacies of abuse, a variety of transitional justice mechanisms have been developed that fall within two broad categories: judicial and non-judicial. The former focuses on civil or criminal trials at local, national, or international levels, while the latter encompasses activities such as truth and reconciliation commissions, amnesty, vetting, lustration processes, reparation, memorialization, reconciliation, institutional reform, security sector reform and demobilization, disarmament and reintegration. The concept of transitional justice has evolved and is no longer focused on the role of states but has become more complex, integrating various levels of analysis (international, national and local) and different actors, such as states, international organizations, non-governmental organizations, epistemic communities and global policy networks (Crossley-Frolick 2011, p.33).

An important mechanism of transitional justice processes is symbolic reparation, including public acknowledgment, responsibility and remorse for crimes committed. Reparations are an increasingly frequent feature of post-conflict processes and there is a growing conviction that doing justice in these contexts requires not only doing something against the perpetrators but also doing something for victims (Rubio-Martin 2009; Pablo de Greiff 2006). In BH, the Srebrenica genocide has mobilized survivors and served as a catalyst for the formation of transnational activism that seeks to re-invent and broaden the meaning of accountability and justice. Accountability and justice is a more expansive concept than that designated by the international community's emphasis on prosecutions. Such an emphasis not only limits but also may undermine the search for comprehensive and contextually relevant justice (Hovil and Okello 2011, p.343).

Justice should not be confined solely to the legal processes and outcomes carried out by state entities, or even those of customary law or local courts, but should also extend to processes carried out by members of civil society (Simic and Daly 2011, p.480). Some studies have shown that local communities may understand 'justice' very differently from the way in which it is defined by official institutions (Stover 2005). In the former Yugoslavia, transitional justice strategies have been reduced to the prosecution of war crimes and full cooperation with the International Criminal Tribunal for the former Yugoslavia (ICTY). As such, the ICTY is the main actor of transitional justice, while other mechanisms have been marginalized and largely ignored (Armakolas and Vossou 2008).

However, sustainable peace requires more than trials and the prosecution and punishment of a relatively few high-placed officials. It requires more than state-sponsored or administered forms of truth-telling and reparations. Transitional justice is a complex and long-term process that needs to involve all layers of a society and 'anything that a society devises to deal with a legacy of conflict and/or widespread human rights violations' (Roht-Arriaza 2006, p.2).

Civil society organizations play a critical role in deconstructing authoritarianism, rebuilding the state and establishing a firm foundation for strong democracies. Through their various activities they have the potential to ensure that any transition is genuinely transformational by promoting justice mechanisms that adequately engage with and include citizens (Hovil and Okello 2011, p.333). It is important to consider the ways in which individuals and communities 'within particular localities' appropriate, negotiate with, understand and seek to transform transitional justice discourse and practice (Hinton 2010, p.1). As Kent argues, far less attention has been paid to these *unofficial* local practices of remembering, commemorating and responding to the violence of the past in transitional justice body of literature (Kent 2011, p.435).

In this chapter, however, the dynamic explored is not between state and civil society but between the international community embodied in the UN and civil society. The 'Pillar of Shame' targets a global community and decision makers and asks not only for their remorse but also for legal and moral accountability since they failed to prevent genocide. Some scholars and commentators argue that although the UN Secretary-General at the time, Boutrous Boutrous-Ghali, understood the situation on the ground, in BH, he did not make the war priority, nor did he think it deserved to be (Nettelfield 2010, p.65; Ajami 1996). Although the UN General Assembly has stated that, 'we in the United Nations share in that responsibility [for genocide in Srebrenica]', the UN has done no more than express sorrow and partial responsibility for not preventing murder (UN General Assembly 1999, para. 6). Disappointed with the UN's conduct, survivors have decided to raise the issue of shame and blame that resides within the UN. As Wachman argues, there is an assumption embedded in such campaigns that shaming will cause a desirable reaction in the targeted object and that, once subjected to international scrutiny and denunciation, an object of shame will necessarily take steps to remedy and improve a situation (2001, p.260). Similarly, Nathan states that 'by publicizing abuses, human rights NGOs seek to mobilize psychological and political pressure against violators' (1999, p.139). In the Pillar campaign, embedded in naming and shaming has been the expectation from organizers that the UN would react to the campaign and offer support. Although it is debatable whether shaming can promote human rights, it is still a compelling alternative to remaining silent in the face of injustice, when one is not able to exert more tangible pressure on the regime one reviles (Wachman 2001, p.265).

The Pillar of Shame campaign is an example of a particular type of civil society justice mechanism commonly called 'symbolic reparations' that may (depending on the context and purpose) bring forward the complexities and complicities of often overlapping and shifting categories of perpetrators, victims and bystanders; seek reparations or to rebury and remember the dead; and combine and blend reparations, reflection, truth-seeking, multiple truths, reburial and remembrance (Arriaza and Roht-Arriaza 2010).

Walker calls attention to the 'expressive' dimension of all types of reparations, which is the 'communicative act of expressing acknowledgment, responsibility, and intent to do justice' (2010, p.529). This expressive dimension takes a variety of forms including (but not limited to) art installations, literature projects, media programs, theatre, symbolic courts or popular tribunals, material reparations, remembrance and reflection activities. Such activities may 'partly fill the vacuum when leaderships are reluctant to initiate' them; and they may 'tap into the agency of survivors … , foster the integration of cultural practices, promote participation and a sense of ownership,' although they are not without limits or problems (Dudai and Cohen 2010, p.234). One limitation of the 'Pillar of Shame' campaign was a lack of engagement with the Serb population in the Republika Srpska (RS) and Serbia.

In BH, it is still necessary to reconcile competing versions of individual, public and collective memory. The 'Pillar of Shame' campaign attempts to establish a transnational 'collective memory' and develop a culture of remembrance beyond national borders. Yet this campaign failed to cross ethnic boundaries. One country, BH, is still deeply divided, with an 'enduring enmity between Bosnia's three peoples' who live in ethnic ghettos created after the war (Parish, 9 April 2012). The situation in BH could be described as post-conflict, yet with a strong potential to return to violent actions in order to address the traumatic past (Kurspahić 2003; Erjavec and Volčič 2007). The war in BH, and also in the former Yugoslavia more broadly, was the result of organized and planned nationalistic politics, especially Serbian policies from the mid-1980s onwards (Woodward 1995; MacDonald 2002; Ramet 2005). As a result, the fact that there was a mutual lack of interest from both sides in working together (organisers of the campaign and local civil society organizations in the RS) is of no surprise. Likewise, the reluctance to engage with a similar campaign run at the same time by 'Women in Black' (WiB) from Belgrade, Serbia, is of no surprise. Nonetheless, this does not necessarily decrease the significance of the Pillar campaign, in which local and German actors sought to jointly re-invent justice and demand survivors' rights to truth.

'Pillar of Shame' Across Europe: From Berlin to Srebrenica

Srebrenica was a collapse of our humanistic claim (Phillip Ruch, 2011).

Phillip Ruch, the executive director of the Center for Political Beauty, a non-profit organization from Berlin, is the founder of the campaign tiled 'Pillar of Shame'. Together with the Bosnian Society, he decided to act on the ICJ's failure to bring justice to the survivors of Srebrenica genocide. The focus of this campaign was on the 'UN betrayal' in BH, and its goal was to erect a permanent sculpture in the Srebrenica cemetery. The installation of the monument was originally planned for the 2011 commemoration of the genocide; but due to a lack of financial means, the installation was delayed, and subsequently suspended.

It was planned that the monument would take the form of a sculpture comprising two huge letters ('U' and 'N') filled with 16,744 shoes, representing 8,372 victims. The erecting of some sort of permanent monument that would 'shame and blame' the UN is also a long-term wish of the survivors of Srebrenica. As Zulcic reported, 'The survivors of genocide have always wanted to erect a monument in Srebrenica on which the names of all people who could, but did not, prevent genocide should be written down'.[9] Still, according to Zulcic, the original idea to connect the

[9] The author was told that the names most often mentioned as intended to be written on the monument are: General Philippe Morillon of France, Commander of the United Nations Protection Force (UNPROFOR) based in Srebrenica, Yasushi Akashi, the chief UN envoy in the former Yugoslavia, Boutros Ghali, former UN Secretary-General, and Thomas Karremans, the commander of Dutchbat troops in Srebrenica at the time of Srebrenica genocide.

monument to shoes was Ruch's, who was inspired by the journalist who threw his shoe at former US president George Bush in Iraq.[10] The shoe represents to Zulcic, 'a scream, a protest and a legacy of crime committed' and to Ruch, 'empty shoes on the street are example of loneliness, of not having a shelter, being abounded … and this is the feeling I wanted to catch that people had in July 1995'.[11]

Ruch's personal and emotional involvement with the Srebrenica genocide began during his studies when he was introduced to the BH war as part of his academic course. Around that time, he befriended a classmate who had left a besieged Sarajevo when he was thirteen. A strong sense of injustice haunted Ruch and propelled a desire to assist survivors of the genocide in some way and to initiate broader discussion about how genocide can be prevented. As an artist, Ruch has been determined to make the injustice of Srebrenica visible and public. In July 2009, he was instrumental in establishing a street performance in Berlin in front of the Bundaberg gate. The street performance was a reconstruction of the UN crisis meeting that had unfolded a day before Srebrenica failed. The performance was based on original testimonies of high-ranking UN officials who met that day and was played in the street for four days. Thousands of Berliners watched this play and Ruch and his colleagues made a video of the performance that they showed in schools, universities and other public places. He said that:

> The play gave a close detail of this day, how genocide progressed and how Western decisions are being done. In the background of this play, original footage was played so people get to see not only how decisions were made, but consequences of these decisions.

Although this performance was, according to Ruch, 'well done, structured and extremely well followed by audience', none of the German media reported on it, despite Ruch's invitation to German media houses. Building on this performance, Ruch initiated the idea to collect shoes that will serve as a permanent symbol of shame to be erected in Srebrenica/Potocari. Ruch's overall message with this project has been that:

> The people have not forgotten 1995 and the decision makers involved in the genocide, and we are reminding them that whatever they do, whatever actions they were involved in, this is not something they can get away with. Since it is not possible to sue the UN, because they are legally immune, the only vulnerable place we can hit them is to question the good reputation they enjoy.

---

[10] In December 2008, an Iraqi journalist threw shoes at the then US president, George W. Bush, during a press conference held in Iraq.

[11] The idea of organising a public display of shoes as a memorial to victims of genocide is not new. For example, the Auschwitz Museum in Poland houses a permanent display of about 25,000 pair of shoes titled 'The Wall of Shoes Under Glass'. At the Majdanek Memorial Site, also in Poland, thousands of shoes of Jews killed in the camp were on display in three warehouse buildings until damaged by fire in 2010. Several thousands of shoes from Majdanek were loaned to the United States Holocaust Memorial Museum and are now on display in the Museum's Permanent Exhibition. Similarly, in Budapest, a memorial titled 'The Shoes on the Danube Promenade' was created to honour Jews who were shot at the edge of the river such that their bodies fell into the water, leaving empty shoes behind on the bank.

Fig. 11.2 The artwork made from donated shoes by volunteers in Bihac. (Photo is courtesy of Aldijana Mustafic)

The main aim of the monument is thus threefold: to reveal UN responsibility for not preventing the genocide; to name, shame and point to the representatives of the international community who are guilty of complicity in the genocide; and to symbolize the broken promises of protection, the treason of Bosnia and the shattered dream of western will that the genocide would not take place (*Pressekit*, 2010).

In May 2010, Ruch, along with German and Bosnian–German colleagues, asked the Bosnian Society to become a partner in this project. Together they called on survivors and citizens of BH, and also of Austria, Germany, and other Western European countries, to donate shoes by either mailing them to Berlin or placing them in one of the five open collection centres. There were a number of activists and volunteers who worked on the collection of shoes across the Federation of BH. Both organizations asked for support from the association of survivors in Srebrenica–Potocari and also from the BH public. However, support mainly came from citizens in the Federation of BH and to lesser extent from the RS entity (Fig. 11.2).

The campaign was supposed to have three phases. In the first phase of the project, a pile of shoes symbolically representing missing lives was to be temporarily erected in the front of Bundaberg Gate in Berlin on July 11, 2010 to mark the 15th anniversary of genocide. The second phase was a travelling exhibition of the shoes that would visit the Hague and the ICTY, London, Paris and other cities. This second phase did not materialize. The third phase was building a monument, the

Fig. 11.3 Shoes piled in one of the collection centres. (Photo is courtesy of Aldijana Mustafic)

'Pillar of shame', in Potocari/Srebrenica. This phase also failed to materialize due to lack of financial means.

The first phase of this project was successfully realized. The idea was to publicly condemn the failure of the UN and to remember all those killed and missing in genocide. In less than 2 months, with the help of a number of volunteers, the two organizations mobilized thousands of Bosnians in BH as well as abroad. The cities in which the centres for collecting shoes were established were all in the Federation of BH, the entity with a predominantly Bosniak population. Apart from these established centres, Zulcic reported that many people decided to collect shoes and become involved in campaign on their own initiative. Aldijana Mustafic, one of the volunteers, called Ruch and asked if she could organize the collection of shoes in the city of Bihac. Once Ruch accepted this offer, Aldijana mobilized her community through posters, flyers and media and invited them to donate shoes for campaign. The owner of a coffee shop in Bihac offered his shop as a collection point where people could leave their shoes from 8 am till 11 pm. Aldijana also involved schools and children in the campaign. The children and adults were asked to write messages in the shoes they donated.

Shoes were also collected in Austria, Germany, Italy and other countries, mainly from Bosnian diaspora. A few packages arrived from the US too. However, in RS there were no collection centres, which was, as Zulcic argues, 'expected'. Although Zulcic sent a flyer to press centres in the RS about the campaign, 'this information had not been published'. A few organizations that were of mixed ethnic origin gave their support to campaign but the 'only Serb' organizations did not (Fig. 11.3).

Fig. 11.4 Piling shoes in Berlin, 11 July 2010. (Photo is a courtesy of Phillip Ruch)

In only 40 days, the Center for Political Beauty had collected 16,744 pair of shoes and placed them in front of the Brandenburg Gate in Berlin on July 10, 2010. Berlin was supposed to be only the first in a number of locations for the exhibition but due to lack of financial means it has remained the only one. The largest number of shoes was collected in BH and dispatched to Berlin for the exhibition. Shoes from other countries were directly sent to Berlin by private or organized transportation (Fig. 11.4).

At the same time as the 'Pillar of Shame' campaign was initiated, a similar campaign was organized by WiB, Belgrade. The author interviewed WiB about their campaign 'One pair of shoes, one life' and published an earlier paper on this campaign (Simic and Daly 2011). At the time of that interview, the author was aware that the German Center and Bosnian Society had run a similar campaign in Germany, so the author asked the WiB whether they were also aware of the 'Pillar of Shame' campaign in Berlin. They said that they were, but had never made contact with either the Center for Political Beauty or Bosnian Society. In interviews conducted with Ruch and Zulcic, the author was told that they were both aware of the WiB campaign as well. I found it striking that these two similar campaigns which were organized at the same time (May, June 2010) with the same aim (to erect a permanent monument made of shoes), used the same methodology (invited people to donate shoes and messages to specifically organized collective centres in Belgrade) and were aware of each other's campaigns, did not make contact with each other or explore working together. The two campaigns were simultaneous but with no communication between the organizations that led them.

It is not quite clear why there was no cooperation between these organizations working on the same issue. Ruch said that, 'due to internal conflicts that arose during the campaign amongst the organizations involved, the idea of involving WiB was rejected'. One of the organizations explicitly rejected the notion of cooperating with the WiB. Ruch said he regrets that he did not contact the WiB because, according to him, 'joint action with the WiB, Belgrade would bring much stronger massage, politically and morally, of accountability and reconciliation between Bosnians

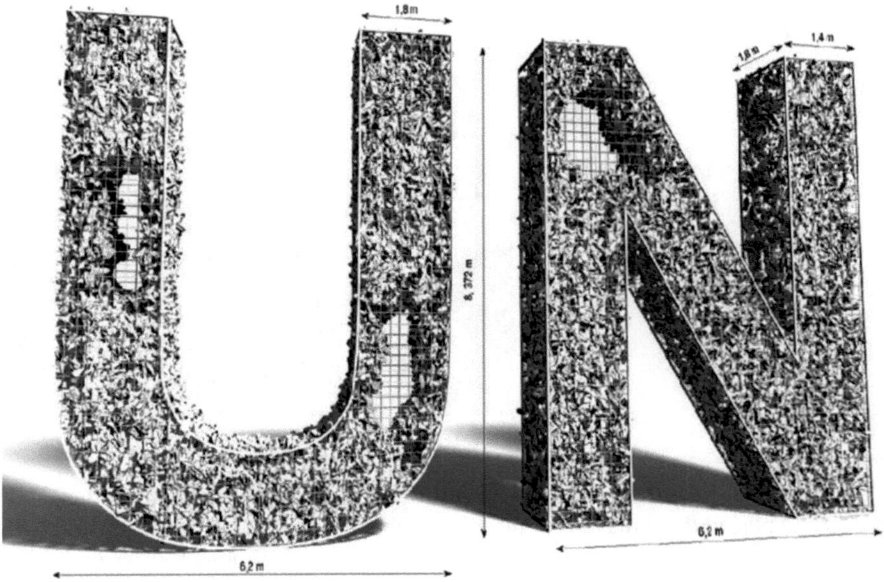

Fig. 11.5 A draft of the monument. The holes will be filled with real shoes found in mass graves. Photo is a courtesy of Phillip Ruch

and Serbians'. On the other hand, Zulcic reported that she believes that they would work together if the campaign entered into its second phase and the exhibition travelled to Belgrade. She also stressed that the key messages of the two campaigns were different. 'Pillar of Shame' was meant to send a message to the international community, while messages from the WiB campaign were aimed at Serbian citizens and asked them to express remorse and guilt for the Serbian government's actions in relation to Srebrenica (Simic and Daly 2011) (Fig. 11.5).

I suggest that another reason for the lack of cooperation may lie in the difficulty of mixing shoes donated by Serbian citizens with those donated primarily by Bosniaks. What would this symbolically mean? Given that no 'Pillar of Shame' collection centres were established in RS, and no potential partners for this project were approached in that part of BH or Serbia, there seems to have been a reluctance to engage with the Serb population, which is often perceived as the ethnic group that should bear collective responsibility for genocide (Franovic 2008). The atrocities committed by the Serbs include mass killings, torture, annihilation, property destruction, execution, elimination, and destruction of cultural, religious and other societal symbols (Wachtel and Bennett 2009). As a consequence, it may be too much to expect survivors to accept shoes from Serbs, even if the shoes were willingly donated for good cause. Although not surprising, this campaign and its methodology demonstrate the complex and difficult environment in which the BH society functions, an environment in which both a lack of trust, resentment and even fear remain pervasive (Subotic 2009; Franovic 2008).

Although in theory anyone could participate in the 'Pillar of Shame' campaign, the organisers emphasized that the campaign should not be used for political gain and, particularly, could not be hijacked by certain political parties. In BH, the Bosnian Society engaged with several primary schools that were based largely in remote areas of Sarajevo. With the consent of directors and teaching staff in these schools, the Bosnian Society entered classrooms to present the campaign to the kids. The aim was primarily educational: to raise awareness about genocide among kids and to invite them to respond by drawing pictures, writing messages, or possibly donating a pair of shoes. The children were encouraged to leave a message in the shoes for the 'international community, to those who had power to do something, to make a difference, to prevent genocide but they decided not to do so'. This was, according to Zulcic, significant for the children. Even the children who could not write, contributed by drawing pictures of war and telling stories about their memories of war, and often about parents they had lost in war.

The majority of the children were refugees and Internally Displaced Persons (IDPs), living below the poverty line. It was intended that some of the shoes that were in good condition would be given to children who lived in extreme poverty and could not afford clothes and shoes. Although the organisers asked for old shoes, some people donated new shoes. Zulcic reported that a few politicians bought new pairs of shoes and brought them to the campaign straight from the shop to show 'how much they care'. As she argues, that was 'not the point at all of this campaign but that by merely giving a pair of shoes one demonstrates the protest ... so the value of shoes or condition of the donated shoes were irrelevant'.

Adults were also encouraged to write messages to 'the UN, international community and European Community' since, as Zulcic argues, 'they are all responsible for not preventing genocide'. Some people donated shoes that had special meaning to them. A man from Capljina donated shoes that he had worn on the day he was forced to enter the concentration camp.[12] He had kept these shoes since that time and did not wear them, and when he heard about the campaign, he decided to donate them. Another woman from Sarajevo donated shoes from her child who was killed during the siege in Sarajevo, and a mother from Srebrenica donated shoes of her child killed in the genocide. All of the messages from the shoes are in the Center for Political Beauty in Berlin, while the shoes are stored in Tuzla.

The first phase of the 'Pillar of Shame' campaign was finalized but with financial difficulty. The two other phases have not materialized because of lack of financial support. It is uncertain whether they will ever eventuate.

---

[12] During the war in BH, concentration camps were set up mostly for Bosnian Muslims by Serb and Croatian authorities. Dretelj-Capljina prison camp for Serbs civilians and Muslims was run by the Croatian Defense Forces (HOS) and later by the Croatian Defense Council (HVO). See *Prosecutor v Jadranko Prlic et al.*, Indictment, 2 March 2004.

Fig. 11.6 Photo of donated shoes with a message. (Photo is a courtesy of Aldijana Mustafic)

(Lack of) Support for the Campaign

> I believe they are hoping that we—the eyewitnesses of the crime committed—will die one day, so that there will no longer be anyone reminding them of the global shame they caused. (Fazila Efendi, returnee to Srebrenica) (Fig. 11.6).

A few employees within UN agencies in BH called the Bosnian Society to give support to the campaign and expressed a willingness to have their photo taken with the shoes for the media. However, as Zulcic reported, on the day when the photographing was scheduled, literally hours before, they all cancelled their appointments for taking the photo and said that:

> They received the threats from their employers that if they do this, they would be fired from their jobs. They did not get any written correspondence about it, but orally they had been told to withdraw from any support for this project. They apologized and said they cannot participate since they were threatened by a term of notice.

There was no public support from any UN agencies for the campaign. None of the international organizations, either formally or informally, gave support to the campaign nor did any call the organisers to inform themselves about the campaign. Zulcic reported that she heard from insiders that 'the UN agencies do not support the campaign, do not like it and will try to prevent it if possible'.

With respect to local politicians, a number of Bosniaks in the Parliament gave support as did the Presidency of BH. Also, Tanja Fajon, a Slovene delegate to the European Parliament, took a photo with the shoes. However, no one from the RS political elite gave support for the campaign. By contrast, in Germany, there has been lot of discussion about 'iconography of shoes'. As Ruch reported, the heated debated has not centred around Srebrenica and the project itself but around the shoes and what the meanings that have been ascribed to them in German history:

> There is no copyright of shoes, concluded one paper published in daily paper in Germany. The discussion was about Auschwitz and that shoes belong to Auschwitz and that you could not use them to represent the genocide in Srebrenica and that by using such artistic language you relativize the uniqueness of Holocaust.

As reported in *Der Tagesspiegel*, Cilly Kugelmann, the Deputy Director of the Jewish Museum in Berlin, insisted that the metaphor of a pile of shoes in a public space 'had already been occupied'. She has reportedly stated that shoes had been used historically in this way as a symbolic demonstration of the Nazi politics of mass murder and accused Ruch of trying to 'transfer emotion between Srebrenica and Auschwitz' (Schoenberner 2010). Rosenbaum contends that a presumption of the uniqueness of Holocaust may be entirely warranted provided that it does not diminish the moral authority that must be accorded to other groups whose members have also been forced to endure death and suffering in their history (2009, p. 3). However, Hancock argues that there is a disturbing trend emerging in Holocaust studies to deny recognition of other peoples, apart from Jews, as victims (2001, p. 123). As a consequence, it is not surprising that any perceived attempt to endanger the Jewish 'unique' position and the use of shoes as a symbol of Holocaust might be received unfavourably by certain commentators. For example, expressed aspirations and requests from Roma people to be recognized as victims have been harshly criticized as an 'insult of memory of Holocaust victims' (Anonymous 1986).

According to some scholars, other groups, such as Roma, 'do not qualify for inclusion in the Holocaust' (Lewy 2000; McFee 2009). In light of this discourse, it comes as no surprise that members of the German public found offensive the perceived attempt of the Pillar of Shame campaign organizers to draw a comparison with the Holocaust deeply offensive. Simply, within Germany, the Holocaust is not comparable with other genocides, including the Srebrenica genocide, to the extent that the victims of other genocides are denied the 'right' to be recognized as genocide victims too. By using Holocaust images and imagery, the Pillar of Shame organizers were thus open to accusations of relativization and of downplaying the Holocaust.

Despite a lack of support from the German media and public and from Serbs in BH, Zulcic reported that the campaign was a success. The fact that many people donated shoes and had a chance through this campaign to express their opinions and positions in relation to genocide was deemed as a measure of success in itself. Yet, as Zulcic argued, 'we were not happy that reaction from UN is lacking'. The campaign organisers also reported a disappointing lack of reaction from other international players. In June and July 2010, a world championship in football took place in South Africa. The Bosnian Society together with the Mothers of Srebrenica sent a statement to the organisers and asked them to pay one minute of silence at the final football game, which was scheduled for July 11. They requested the organisers not to sell the tickets for 8,373 people and to leave those seats empty. They hoped that this action would send a symbolic message that 8,373 people would have been able to be there if they had not been killed at Srebrenica. However, they received a negative response from the organisers, who told them that although they express their empathy, they decided not to involve 'such things in sport'. The organisers also refused the request to mark the anniversary of Mandela's arrest,[13] which is on the same date. Zulcic said this was 'very disappointing and unjustified with the false argumentation … This was an example of a world organization that has no understanding and empathy for victim request'.

Although Zulcic and Ruch remain positive and enthusiastic that the project will be realized one day, Mustafic, a volunteer from Bihac, reported that she was disappointed by the end result of campaign. She feels an enormous responsibility towards the people who donated shoes and worries how she will explain that all their effort was in vain and why the project failed. Mustafic feels betrayed because the campaign did not achieve its end goal of building a monument. There are no indications whether and when the project will be finished. The shoes are stored in a locker in an abandoned factory in Tuzla, BH, awaiting additional campaign funding. The main problem is the lack of financial means to finish the project. In the meantime, the shoes decompose in Tuzla, waiting for a better time.

Conclusion

This case study underscores the complexities of local and transnational understandings and negotiations with justice. As Van der Merwe argues, the notion of justice is inspiring and powerful, but also illusory and elusive, especially for victims of mass atrocities (2009, p.115). The two organizations, the Center for Political Beauty and the Bosnian Society, acted on the failure of international community to be held accountable for failing to prevent genocide. A feeling of disappointment with the international community, and of justice denied and truth untold, was overwhelming among Bosnian survivors and German activists (Subotic 2009). This feeling galvanized a joint collective campaign that crossed the national borders and mobilized people's activism and a sense of ownership and justice for those who had been betrayed by formal justice mechanisms. The impact of many civil society activities, including this one, 'can only be seen years, if not decades, later', thus the impact of this particular social movement on the international community is yet to be seen (Hovil and Okello 2011, p.339).

This chapter has argued that although civil society justice mechanisms can play an important role in societies in transition, they are not without limits. One of the concerns raised in relation to the Pillar of Shame campaign has been the exclusion of Republika Srpska and its citizens and the reluctance to involve WiB from Serbia, which had a similar campaign underway at the same time. This supports the thesis that Bosnian society is still beset with interethnic mistrust, denial and fear (Fischer 2006, p.442). As Franovic argues, the situation in the former Yugoslav region cannot be defined as war since 'we have stopped shooting at each other', but neither is it peace (2008, p.17). Issues surrounding the lack of engagement and participation

[13] Nelson Mandela, former President of South Africa, was arrested on August 5 not July 11.

in the Pillar campaign raise questions about the symbolic nature of the campaign and about the implications for transitional justice if Bosnian people from other ethnic groups, the WiB and the UN, were not interested or not invited or did not want to participate. Rather than confronting existing ethnic tensions, it could be argued that the Pillar campaign exemplified these tensions and brought them into focus.

The adoption of shoes by the Pillar campaign as a symbol of genocide provoked concern amongst the German public and media, which focused solely on the nature of the shoes used in the campaign rather than the campaign itself. The campaign produced heated debate about who possesses copyright to the symbolic use of shoes and whether the metaphor of pile of shoes in a public space has already been 'occupied'. It is unfortunate that the campaign received no other media recognition apart from being accused of appropriating shoes as a sacred symbol of Holocaust.

Despite these shortcomings, campaigns of symbolic reparation such as 'Pillar of Shame' enrich the transitional justice discourse, which has recently acknowledged a 'clash' between 'local' and 'universal justice' norms (Shaw and Waldorf 2010, p. 5). This particular campaign took place in the absence of 'universal justice' and served as a 'local justice' call for the legal accountability of the UN and the international community. Since formal justice mechanisms have been too weak to challenge UN immunity, a transnational network of human rights activists was mobilized to put pressure on the UN by, as Ruch stated, targeting its most vulnerable place: the good reputation it enjoys. In that sense, this campaign is unique since it is externally driven and oriented by German artists, whose aim is to address the UN and bring it to *the moral court* in which survivors and their followers will judge its (non)actions. However, the lack of reaction from the UN and international community to the campaign poses questions as to whether the campaign was able to meet its aim to shame and put pressure on the UN to any noticeable degree.

Campaigns of symbolic reparation lack research and evaluation. As Olsen et al. argue, this is not because they are unimportant, but because of the difficulty of finding systematic accounting for the wide range of processes involved, such as creating museums and memorials (2010, p.37). It is likely that many more civil society justice activities like this have occurred in conflict and post-conflict zones, but they have seldom been researched or theorized. In part, this may be because researchers may not have access to sites, local language, or contacts with local people to document the activities fully and accurately. Likewise, civil society and local justice mechanisms have only recently received attention in the transitional justice literature, thus civil society mechanisms are often overlooked in researchers' analyses of reparation and reconciliation (Simic and Daly 2011, p.491).

This case study fills a gap in the current literature by documenting a campaign of symbolic reparation that crosses the borders of the local context. By taking a deeper, critical look at such campaigns, and exploring the conflicting interests and different responses of the actors involved, the chapter aims to broaden discussion and understanding of justice in specific places and times. Without acknowledging and understanding the contested and often conflicted meanings of justice, attempts at long-term and sustainable peace shall be deemed to fail.

References

Anheier, H. 2004. *Civil society: Measurement, evaluation, policy*. London: Routledge.

Anonymous. 1986. Tragedy of the gypsies. *Information Bulletin* No. 26. Vienna: Dokumentationszentrum des Bundes Jüdische Verfolgte des Naziregimes.

Ajami, F. 1996. The mark of Bosnia: Boutrous-Ghali's Reign of Indifference. *Foreign Affairs* May/June 1996.

Arriaza, L., and N. Roht-Arriaza. 2010. Weaving a braid of histories: Local post-armed conflict initiatives in Guatemala. In *Localizing transitional justice: Interventions and priorities after mass violence*, ed. R. Shaw and L. Waldorf. Stanford: Stanford University Press.

Appeal Court in The Hague, Judgment, 30 March 2010, Case Number District Court: 07–2973.

Armakolas, I., and Vossou, E. 2008. Transitional justice in practice: The International Criminal Tribunal for the former Yugoslavia and beyond. *UNISCI Discussion Papers*, 21–58.

Batliwala, S. 2002. Grassroots movements as transnational actors: Implications for global civil society. *International Journal of Voluntary and Nonprofit Organizations* 13(4): 393–410.

Boesenecker, A., and L. Vinjamuri. 2011. Lost in translation? Civil society, faith-based organizations and the negotiation of international norms. *International Journal of Transitional Justice* 5(3): 345–365.

BBC News. 2002. Dutch Government quits over Srebrenica, 16 April 2002. http://news.bbc.co.uk/2/hi/europe/1933144.stm. Accessed 13 December 2011.

Charter, D. 2008. Victims' relatives battle to Sue UN over Srebrenica massacre. *The Times*, June 19, 2008.

Crossley-Frolick, K. 2011. The European Union and transitional justice: Human rights and post-conflict reconciliation in Europe and beyond. *Contemporary Readings in Law and Social Justice* 3(1): 33–57.

Campbell, C., and F. Ni Aolainn. 2003. Local meets global: Transitional justice in Northern Ireland. *Fordham International Law Journal* 26(4): 871–892.

de Greiff, P. 2006. *The handbook of reparations*. Oxford: Oxford University Press.

Duijzings, G. 2007. Commemorating Srebrenica: Histories of violence and the politics of memory in Eastern Bosnia. In *The new Bosnian mosaic: Identities, memories and moral claims in a post-war society*, ed. X. Bougarel, G. Duijzings, and E. Helms, 141–166. Aldershot: Ashgate.

Dudai, R., and H. Cohen. 2010. Dealing with the past when the conflict is still present: Civil society truth-seeking initiatives in the Israeli-Palestinian conflict (228–252). In *Localizing transitional justice*, ed. Rosalind Shaw and Lars Waldorf. Stanford: Stanford University Press.

Efendi, F. Returnee to Srebrenica. *Stub srama—Pillar of shame*. http://stubsrama.ba/index.php?option=com_content&view=article&id=1&Itemid=3&lang=en. Accessed March 2, 2012

Erjavec, K., and Z. Volčič. 2007. War on terrorism as a discursive battleground: Serbian recontextualization of G. W. Bush's Discourse. *Discourse & Society* 18(1): 123–138.

Fischer, M. ed. 2006. *Peacebuilding and civil society in Bosnia and Herzegovina: Ten years after Dayton*. Berghof Foundation for Conflict Studies, New York.

Franovic, I. 2008. Dealing with the past in the context of ethnonationalism. The case of Bosnia–Herzegovina, Croatia and Serbia. Berghof Occasional Paper no. 29.

Hinton, A.L., ed. 2010. Introduction: Towards an anthropology of transitional justice. In *Transitional justice: Global mechanisms and local realities after mass violence*, pp. 1–22. Rutgers: Rutgers University Press.

Hancock, I. 2001. Review essay: The Nazi persecution of Gypses. *Journal of Genocide Research* 3(1): 120–127.

Hovil, L., and M.S. Okello. 2011. Editorial Note. *International Journal of Transitional Justice* 5(3): 333–344.

ICTJ website, Retrieved from: http://ictj.org/our-work/transitional-justice-issues. Accessed 2 March 2011.

Kent, L. 2011. Local memory practices in east Timor: Disrupting transitional justice narratives. *International Journal of Transitional Justice* 5(3): 434–455.

Kurspahić, K. 2003. *Prime time crime: Balkan media in war and peace*. Washington: United States Institute of Peace Press.

Lewy, G. 2000. *The Nazi persecution of Gypsies*. Oxford: Oxford University Press.

McEvoy, K., and L. McGregor (eds.). 2008. *Transitional justice from below: Grassroots activism and the struggle for change*. Oxford: Hart.

MacDonald, D. 2002. *Balkan Holocausts?* Manchester: Manchester University Press.

McFee, G. 2009. Are the Jews central to the Holocaust? *The Holocaust History Project*, http://www.holocaust-history.org/jews-central/.

Nettelfield, L. 2010. *Courting democracy in Bosnia-Herzegovina: The Hague tribunal's impact in a postwar state*. New York: Cambridge University Press.

Nathan, A. 1999. China and the international human right regime. In *China joins the world: Progress and prospects*, ed. E. Economy and M. Oksenberg. New York: Council on Foreign Relations.

Radio Netherlands Worldwide. Srebrenica Relatives' Case Against UN Rejected, 30 March 2010. http://www.rnw.nl/english/article/srebrenica-relatives-case-against-un-rejected. Accessed 13 March 2012.

Olsen, T., L. Payne, and A. Reiter (eds.). 2010. *Transitional justice in balance: Comparing processes, weighing efficacy*. Washington, DC: US Institute of Peace Press.

Parish, M. (2012). 'Two decades on, Bosnia's divisions are self-imposed. *Balkan Insight*, 9 April.

Pressekit. 16.744 Shoes: The United Nations in Court, May 2010. On file with the author.

Ramet, S. 2005. *Thinking about Yugoslavia*. New York: Cambridge University Press.

Rubio-Martin, R. (ed.). 2009. *The gender reparations: Unsettling sexual hierarchies while redressing human rights violations*. Cambridge: Cambridge University Press.

Roht-Arriaza, N. 2006. The new landscape of transitional justice. In *Transitional justice in the justice-first century: Beyond truth and justice*, ed. N. Roht-Arriaza and J. Mariezcurrena. Cambridge: Cambridge University Press.

Rosenubaum, A. (ed.). 2009. *Is the Holocaust unique?: Perspectives on comparative genocide*. Boulder: Westview Press.

Subotic, J. 2009. *Hijacked justice: Dealing with the past in the Balkans*. Ithaca: Cornell University Press.

Schoenberner, G. 2010. Ein Copyright auf Schuhe?. *Denkmalsdebaten*. Gedenkstattenrundbrief 158/12. On file with author.

Stover, E. 2005. *The witnesses: War crimes and the promise of justice in The Hague*. Philadelphia: University of Pennsylvania Press.

Simic, O., and K. Daly. 2011. One pair of shoes, one life: Steps towards accountability for genocide in Srebrenica. *International Journal of Transitional Justice* 5(3): 477–491.

Simic, O. 2009. Remembering, visiting and placing the dead: Law, authority and genocide in Srebrenica. *Law Text Culture* 13: 273–311.

Shaw, R., and L. Waldorf (eds.). 2010. *Localizing transitional justice: Interventions and priorities after mass violence*. Stanford: Stanford University Press.

UN General Assembly. 1999. *Report of the Secretary-General pursuant to General Assembly resolution 53/35: The fall of Srebrenica*, A/54/549, 15 November 1999.

Urban Walker, M. 2010. Truth telling as reparations. *Metaphilosophy* 41(4): 525–545.

Van der Merwe, H. 2009. Delivering justice during transition: Research challenges. In *Assessing the impact of transitional justice: Challenges for empirical research*, ed. H. Van der Merwe, V. Baxter, and A. Chapman. United States Institute for Peace: Washington, DC.

Wachman, A. 2001. Does the diplomacy of shame promote human rights in China? *Third World Quarterly* 22(2): 257–281.

Wachtel, A., and C. Bennett. 2009. The dissolution of Yugoslavia. In *Confronting the Yugoslav controversies*, ed. C. Ingrao and T.A. Emmert, 13–47. Purdue: Purdue University Press.

Woodward, S.L. 1995. *Balkan tragedy. Chaos and dissolution after the cold war*. Washington DC: The Brookings Institution.

Chapter 12
Afraid to Cry Wolf: Human Rights Activists' Struggle of Transnational Accountability Efforts in the Balkans

Arnaud Kurze* and Iva Vukusic

If he has a conscience he will suffer for his mistake. That will be punishment-as well as the prison

(Dostoyevsky 2000, p.226).

Throughout the 1990s the breakup of the former Yugoslavia led to horrendous conflict among the newly proclaimed independent states. Since, dealing with past war crimes and accounting for mass atrocities has constituted a very intricate and contentious process, mainly led by state-centric international retributive justice initiatives. In fact, the International Criminal Tribunal for the former Yugoslavia (ICTY) was created in 1993 amidst flaring conflict with the aim to push for accountability for war crimes and human rights violations. Over the years, the ad hoc United Nations (UN) Tribunal's work has led to a spillover effect of international criminal

*This chapter is based on Arnaud Kurze's dissertation chapter 5 "Afraid to Cry Wolf: The Struggle of Transnational Accountability Efforts" (Kurze 2012). The authors also wish to thank Jo-Marie Burt, Earle Reybold, Olga Martin-Ortega, Peter Mandaville, Harvey Weinstein, and Nahla Valji for reading earlier drafts of this chapter and providing valuable comments to help clarify certain aspects and are indebted to Vjeran Pavlaković, Christopher Lamont, and other colleagues and friends for the many conversations that helped improve this chapter.

A. Kurze
Center for Global Studies, George Mason University,
3401 Fairfax Drive, MS 1B9, Arlington, VA 22201, USA
e-mail: akurze@gmu.edu

I. Vukusic (✉)
Sense News Agency, ICTY Building, Churchillplein 1 office 080,
2517 JW, The Hague, The Netherlands
e-mail: vukusic.iva@gmail.com

O. Simić and Z. Volčič (eds.), *Transitional Justice and Civil Society in the Balkans*, Springer Series in Transitional Justice, DOI 10.1007/978-1-4614-5422-9_12,

law into domestic judicial systems in the region. As Ruti Teitel, a law professor at New York Law School, points out:

> In the post-Cold War context, we can see that there is an evident transformation in the significance of an expanded role for international criminal justice. Clearly, international criminal justice aims and contributions are complex and in transformed political circumstances inevitably imply diverse understandings of rule-of-law values. Consider, for instance, the extent to which international law at The Hague affords the rule-of-law values of fairness and neutrality, often perceived as fragile or non-existent in domestic processes (2010, 9).[1]

Initially, civil society actors, in particular human rights organizations, supported these retributive justice efforts by participating in different programs, including witness protection and trial monitoring.[2] More recently civil society actors expanded their influence and impact—reaching beyond the initial judicial support, in which NGOs have been invited to participate by state actors—and created a deliberative space to increase victims' voices in society, so-called 'invented spaces' (Jeffrey 2011). Although several scholars have questioned the progress and effectiveness of Bosnian civil society in post-conflict transitions (Belloni 2008; Fischer 2005; Chandler 1998), in a case study Alexander Jeffery underlines efforts by the Bosnia and Herzegovina (BiH) Court to integrate NGOs in transitional justice processes. He describes a court support network, which was established by the BiH Court's Public Information and Outreach Service in 2005 and which aims at integrating the mission of the Court into Bosnian society. To this end, the Court approached various human rights NGOs located in different Bosnian towns, such as Sarajevo, Prijedor, and Mostar, among others. As Jeffrey writes, "[t]he idea was to reach a wide geographical spread and through these organizations establish a sustainable network that would spread information about the Court, and in particular the [War Crimes Chamber of the BiH Court]" (2011, p.352). Such scholarly insight is important, as several past restorative justice attempts across different countries in the region resulted in limited success (Jouhanneau 2010; Grodsky 2009; Pejic 2001). Notwithstanding, social activists and civil society organizations have incrementally increased their role and reach in transitional justice processes.[3] In this chapter, we discuss the ongoing transnational fact-seeking initiative, called the Coalition for a Regional Truth Commission (RECOM), to elucidate the sociopolitical struggle of coalition members to advocate for alternative models to cope with mass atrocity in the former Yugoslavia.

---

[1] See also (Martin-Ortega and Herman 2010).

[2] Other authors, such as (Subotić 2009), and annual activity reports of NGOs, such as Documenta Center for Dealing with the Past and the Humanitarian Law Center, among others, have criticized the effectiveness of this justice cascade model by (Lutz and Sikkink 2001). Ellen Lutz and Kathryn Sikkink argue that through transnational advocacy network of lawyers and activists have fueled domestic judicial processes across Latin America.

[3] Several authors have addressed the question of human rights activism in transitional justice processes and in particular highlighting the important impact of local NGOs in different regions (Roht-Arriaza 2005; Collins 2006; Burt 2009).

Drawing on concepts of sociology of spaces—based on the study of establishing spaces through action and the interdependence of action on spatial structures (Urry 2000)—we illustrate how activists move between different spaces constituted by narratives of justice and truth. More precisely, confronting mass atrocities in post-conflict societies can take place in many different ways, with human rights trials and truth commissions as two very important mechanisms.[4] Our study is particularly interested in the complementary use of truth commissions and trials. To this end, we have carried out over two-dozen semi-structured interviews with key actors, such as human rights activists, representatives of domestic and international judicial institutions, and international organizations, among others.[5] Early on, human rights organizations in the region acted primarily within the legal space. They helped for instance improve domestic war crimes prosecutions by providing victim and witness support. In fact, some of the witnesses and/or victims were initially exposed to intimidation and death threats due to the absence of anonymous testimonies in the courtroom during hearings and the lack of media responsibility.[6] Subsequent projects of human rights activists, such as the RECOM initiative, were therefore an attempt to further develop victim-oriented transitional justice strategies. Additionally, it was an attempt to expand their domestic justice-oriented activities to regional fact-finding efforts. In this study we discuss and illustrate the persisting obstacles that hamper the successful implementation of the RECOM truth commission and human rights activists' attempts of claiming and institutionalizing extra-judicial space thus far.

Indeed, the expansion of so-called truth spaces poses myriad challenges. The attempt of establishing a fact-finding body for the former Yugoslavia has faced different types of opposition, ranging from external critique and politicization across the region to internal disapproval and politics by certain of its members. With the aim of creating a broad regional fact-finding initiative, the RECOM campaign also grapples with different types of victims (including families of victims, prisoners, and veterans, among others) who have, sometimes, conflicting needs and expectations. Different narratives of the past, however, also result in elusive consensus building attempts. Yet another reason was that the initiative's members were part of a regional network from different states that were initially united under of one federal nation-state before the conflict. Now, however, state borders separate this originally common space and, in spite of commonalities among participating actors across the region, several obstacles still impede a successful mobilization thus far. These conditions are unique to the Balkan region, as past truth commissions in South Africa, Peru and East Timor, among others, were established in more homogenous geographical spaces where atrocities were generally committed within the territorial boundaries.

This chapter is organized in two sections. First, we describe the continuous struggle of human rights activists to create a transnational extra-legal space—in

[4] Other forms include reparations, lustration and vetting, among others.

[5] Additionally, the study draws from various reports, policy briefs and strategy papers, among other documentation.

[6] See for instance (Kurze 2012, Chap. 4).

particular a fact-finding commission—to deal with past atrocities across the former Yugoslavia. We focus on internal and external obstacles the movement faces. Second, we discuss issues of multiple narratives of victimhood, as a consequence of the transnational character of restorative justice efforts in the post-conflict Balkans. In order to do so, we draw on several different cases. With our research we show that despite the increasing importance of NGO actors in transitional justice processes in the region, several factors still hamper the successful creation of extra-judicial space and the implementation of alternative accountability mechanisms, such as the regional truth commission RECOM.

The RECOM Initiative: Struggling to Create an Extra-Judicial Space

Several authors have explored the sociopolitical role of NGOs in society using a sociology-of-space perspective in order to illustrate their active involvement in shaping policy processes (Miraftab and Wills 2005).[7] Drawing on Faranak Miraftab and Shana Wills' notion of invited spaces—more precisely, spaces in which state institutions provide opportunities for civil society to participate actively in certain problem areas—Alex Jeffrey recently analyzed the creation of space (invented space) by human rights organizations in BiH to allow for deliberate conceptions of justice that go beyond legal institutions and processes (2011). His study defies a legalist approach, illustrating how activists who initially cooperated with the judiciaries have established alternative ways to implement transitional justice in post-conflict settings. While we employ these concepts to investigate regional transitional justice activities of a number of NGOs across the former Yugoslavia in this chapter, we concentrate on a series of factors that have thus far constrained the effective implementation of regional restorative justice efforts and truth spaces by human rights activists. Initially, however, we will address the problems of international criminal justice efforts promoted by the ICTY, which fueled these alternative accountability efforts.

The recent attempts to institutionalize an interstate fact-finding body—to account for past human rights violations and war crimes in the former Yugoslavia—emerged as a response to the rising critique of international and domestic war crimes prosecutions in the region.[8] In fact, retributive justice mechanisms to cope with the past, such as the ICTY, despite the great global impact on and model character for international humanitarian and criminal law, has only partially helped war-torn and post-conflict societies in the region to transition (Subotić 2009). Some of the issues include: the geographical distance of the court between the Netherlands and the crime scene sites—which has often been criticized by victims/witnesses; the trial of

[7] For an extensive discussion on time and space, see (Urry 2000).

[8] See for instance (Kurze 2012, Chap. 3).

selective cases only (both on the international as well as domestic level); and the politicization of cooperation processes between countries of the former Yugoslavia and the UN tribunal in The Hague (Teitel 2005). Victims thus felt alienated by international and domestic accountability efforts. Increasing critique from victims' associations and human rights organizations were therefore crucial in helping launch an alternative process to improve the relationship between law and society. The objective was to create a deliberate space to account for mass atrocities that was outside the judicial space. Below we briefly trace the genesis and discuss the ontology of the movement in recent years.

The idea behind the RECOM campaign was that progress did not lie in more personnel, better strategies, and on-site presence of the judiciary system, but in the way that those who suffered most during the conflicts were integrated into attempts to cope with the past. In fact, the activities of several nonprofit organizations in this context—many of which often started working at the outbreak of violence in the early 1990s[9] or shortly after—demonstrated the increasing efforts to raise victims' voices in transitional justice processes in the former Yugoslavia. In fall 2005, three established nonprofit organizations in the region—the Humanitarian Law Center in Serbia, Documenta Center for Dealing with the past (Documenta) in Croatia, and the Research and Documentation Center in BiH,[10] discussed the prospects of an independent regional commission to investigate and disclose facts about war crimes and other serious human rights violations in the territory of the former Yugoslavia.[11] By May 2008, these organizations had gained enough momentum and launched the RECOM initiative in Priština, Kosovo, with over 100 NGOs from the region (Coalition for RECOM 2009). Due to the still highly politicized landscape of war-crimes-related issues in the region, the founders of the initiative stressed the importance of establishing a platform offering victims an opportunity to express themselves and to counter the relativization of any crimes against humanity by local and national authorities or justification of crimes committed against opposing sides in the conflict (Coalition for RECOM 2010).

[9] The activities of the Humanitarian Law Center in Belgrade, Serbia, are a good example of documenting war crimes in the former Yugoslavia. The center also promotes victims rights, based on various initiatives, at http://www.hlc-rdc.org/stranice/Linkovi-modula/About-us.en.html. Accessed December 5, 2009.

[10] These various organizations have as their core mission to document and disclose facts about the human rights violations and war crimes committed during the 1990s to educate society and create a voice for victims. Various forms of implementing this mission exist. Documenta, for instance, among other things, engages in commemorative culture, history teaching, and dealing with the past initiatives, thus emphasizing the interactive dialogue with society. The Research and Documentation Center, concentrates its work on documenting missing persons, and has published a comprehensive account of all the war victims in BiH, *The Bosnian book of the dead* (2009), as well as an interactive Google map that shows location, nature of the crime and number of victims. The Humanitarian Law Center, despite its involvement in commemorative culture, is known for its strong legal activities, providing support for victims in court and vis-à-vis state institutions.

[11] The International Center for Transnational Justice (ICTJ) and other prominent NGOs in the region also participated in this discussion.

Interestingly, despite coalition members of the RECOM campaign praising the work of the ICTY at conferences and other public events, their strategy clearly aimed at challenging the existing modus operandi of the UN ad hoc Tribunal. As the prevailing international transitional justice strategy in the former Yugoslavia—particularly advocated by the European Union (EU)—required state cooperation with the ICTY and improving domestic war crimes prosecution, official support of international organizations, such as the United Nations Development Program (UNDP) or the EU, among others, to promote the RECOM initiative was very limited.[12] Despite this drawback, the initiators considered the movement as a response to these retributive transitional justice mechanisms, which in their eyes proved only to be partially successful to cope with the violent past in the region.[13]

What were the tools RECOM coalition members had in mind to implement their goals? According to article 13 of RECOM's final draft statute the commission has six primary objectives. The main goal is to establish facts about war crimes and other grave human rights abuses that occurred during the conflicts in the former Yugoslavia from January 1, 1991 to December 31, 2001.[14] Moreover, it will also look at the socio-political circumstances that led to these crimes and their consequences. The commission also aims at acknowledging "injustices inflicted upon victims in order to help create a culture of compassion and solidarity with victims."[15] Other goals consist of promoting victims' rights, clarify the fate of missing persons and help prevent future human rights violations and atrocities. The RECOM commission's specific functions are listed in article 14 of the final draft statute. Its main tasks consist of collecting information on war crimes and other gross human rights violations as well as information on missing persons. The collected information is then stored in a regional database. Public hearings of victims and other persons about war crimes and human rights violations will also be part of the commission's functions. This function overlaps with the work of domestic judiciaries and has been problematized during several RECOM consultation meetings with lawyers, judges, and public figures (Kurze 2012, chap. 6). Finally, according to the draft statute, the commissioners[16] will provide policy recommendations to help prevent further atrocities and human rights abuses, which will be published in a final report.

[12] Interviews held with several UNDP and EU officials in BiH, Croatia and Serbia during fieldwork between September 2009 and May 2011.

[13] See (Kurze 2012, chap. 3 and 4) and reports published by human rights organizations, including Documenta, the Humanitarian Law Center, Human Rights Watch, Amnesty International, among others.

[14] See Chap. 6 for a discussion on how the RECOM initiative members decided to define this specific timeframe.

[15] See final draft RECOM draft statute of March 26, 2011, http://www.zarekom.org. Accessed on March 1, 2012.

[16] According to the draft statute alleged war criminals and individuals with a political affiliation are ineligible for a position as a commissioner.

As a result, the RECOM movement aims at creating a space for victims to be heard in society, fueling sympathy and understanding. RECOM advocates hope to provide a regional mechanism that takes into account the context of past conflicts. Indeed, as several countries were involved in the breakup of the former Yugoslavia, dealing with past war crimes issues does not stop at national borders, but goes beyond the sovereign territory of the current states. The transnational context to create a regional truth commission, however, creates also several obstacles and challenges that we describe below. With regard to establishing a collective memory, RECOM coalition members also plan on creating a comprehensive database of victims to end the perpetual politicization of the number of victims in the region. In addition, RECOM advocates claim the data collection and the archives could help war crimes prosecutors with evidentiary material, witness handling and searching for the missing. Yet, as we will explain below, these ambitious goals are tainted by internal disagreements of different coalition members.

Since the official constitutional meeting of the RECOM coalition in Priština in 2008 the initiative has faced internal politicking and difficulties.[17] The driving coalition partners of RECOM, such as Documenta and the Humanitarian Law Center, in particular, have grappled with mobilizing coalition partners from BiH, after the head of the Bosnian Research and Documentation Center, for different reasons, refused to give his official support to the coalition at one of the meetings in winter 2008.[18] Void of an essential Bosnian member—BiH constitutes a symbolic member country due to its weighty history during the 1992–1995 conflict—Humanitarian Law Center director, Nataša Kandić, managed to fill the gap created by the loss of the influential Research and Documentation Center by partnering with the Association of BiH Journalists.[19] Yet, the fact that this organization does not essentially concentrate on war crimes reporting has affected its legitimacy within the coalition, according to a prominent member of the initiative.[20] Critique has also come from participating organizations that have deplored the lack of transparency in RECOM's decision-making process (B92 2011). Moreover, the uncertain outcome of whether the commission will be created and the long process in rallying financial and political support—both of which have been fluctuating and vague—has also led to a RECOM fatigue with each of the main partner organizations focusing their energy and resources on domestic and local programs in their respective home countries.[21] In addition to internal obstacles, the initiative's institutionalization

[17] The movement was organized in three phases. The first phase assessed the needs and expectations of victims to create an extra-judicial to deal with past mass atrocities across the region. In phase two participants were incited to provide suggestions and recommendations for creating a regional commission. This took place in local, national and regional consultations and meetings. The last phase aimed at discussing and crafting a draft statute for the commission. It started in May 2010 and lasted until the adoption of a draft statute in on March 26, 2011.

[18] See interview with Mirsad Tokača, director of the Research and Documentation Center in June 2011.

[19] See interview with Nataša Kandić, director of the Humanitarian Law Center, in May 2011.

[20] See interview with official member of RECOM in Zagreb in February 2011.

[21] See supra note 21. See also programs by Documenta, http://www.documenta.hr or the Humanitarian Law Center http://www.hlc-rdc.org. Accessed November 23, 2010.

process of a regional truth space has faced difficulties fueled by other political and international actors in the post-conflict Balkans.

Although the political and institutional structures in the former Yugoslavia have become more favorable for the RECOM initiative in recent years, numerous obstacles still impede the creation of a fact-finding body.[22] In the following we describe the fragile political progress across the region and outline some of the inherent problems. The first important political wave of change in the former Yugoslavia occurred in the early 2000s. The death of Croatia's right-wing leader Franjo Tuđman in 1999 allowed the conservative nationalist era to end in which the narrative of the glorious homeland war to defend the young nation didn't leave any room for discussion of war crimes and human rights violations. Although this nation-building discourse is still promoted by large parts of the conservative political elite in Croatia, the socialist party has gained more political ground in recent years, which contributed to a political shift in Croatia's domestic memory politics. In December 2011 the *Hrvatska Demokratska Zdranka*[23] (HDZ) lost several seats in the legislative elections, ceding its majority rule to a coalition government, further contributing to the political change. As for Serbia, the notorious leader Slobodan Milošević was booted out of power after his 2000 electoral defeat amid rising protests from the streets after he attempted to unilaterally remain in power (Ramet 2010, Chaps. 12 and 13). During the past decade, several reforms and political initiatives demonstrated new governments' willingness to recognize and address war crimes and human rights violations. These efforts, however, are only the tip of the iceberg in a long process that is still ongoing.

Indeed, current political leaders in both countries, Ivo Josipović the president of the Republic of Croatia (who began his first term in February 2010), and Boris Tadić the president of the Republic of Serbia (in his second term, which started in February 2008), have both made important strides to foster a climate of rapprochement in the region. They represent a new political generation that has not been personally involved (be it directly or indirectly) in war crimes or the human rights violations of the 1990s conflicts.[24] In 2007, for instance, Tadić released a statement on Croatian national TV on the eve of the 16th anniversary of Croatia's independence, June 24, 2007, apologizing for crimes committed against the Croatians by

[22] Particularly during electoral campaigns, history is manipulated and old nationalist sentiments exploited by certain political parties or social groups.

[23] Croatian for: Croatian Democratic Union.

[24] In the 1980s, Josipović was a member of the League of Communists of Croatia, playing a key role in the democratic transformation of this party as the author of the first statute of the Social Democratic Party of Croatia (SDP) after Croatia's independence. He left politics in the mid-1990s, pursuing his academic career as a law professor at the University of Zagreb and only reentered the political realm in 2003, when Ivica Račan, then acting Prime Minister, invited him to join the government. Serbia's president, Boris Tadić, a trained psychologist, was part of the Democratic Opposition of Serbia, which was key in overthrowing Milosević in 2000. Politically part of the Democratic Party, he has made multiple symbolic reconciliatory public statements that are a sign of collaboration and understanding of both countries.

members of the Serbian population.[25] While then-president of Croatia, Stjepan Mesić welcomed the remarks, nationalist hardliners at home, such as Aleksandar Vučić, secretary general of the Serbian Radical Party, ferociously criticized him as a national traitor (Popovic 2007). Already in 2004, the Serbian president has made similar remarks while on a visit to Sarajevo, BiH.[26] More recently, on November 4, 2010, Tadić apologized for the massacre that took place 19 years ago in Vukovar, a town in Northeastern Croatia.[27] He said that "[b]y acknowledging the crime, by apologizing and regretting, we are opening the way for forgiveness and reconciliation"; yet not everyone received him with wide open arms. Several mothers of those killed in Vukovar, for instance, turned their backs while he gave his speech (Associated Press 2010). Both of these examples illustrate how the political landscape equals a minefield, as not only right-wing nationalist veterans feel betrayed, but also victims express their discontent with political symbolism that does not go far enough in their eyes. Tadić's Croatian counterpart, Josipović, reciprocated these symbolic steps, and during the November 4, 2010 ceremony in Vukovar, he laid down a wreath in commemoration of over a dozen Serbs that had been killed in a nearby village (Associated Press 2010).[28] In addition, both leaders expressed their political backing of the RECOM initiative, in fall 2010 when RECOM members publicly asked for their support (Andrić and Hadžović 2011).

However, in spite of the symbolic gestures and discourses by heads of state in both of these countries (and across the region) institutional drawbacks remain—ranging from the lack of investigations of war crimes involving high-profile Croatian politicians, such as the former speaker of the parliament, Vladimir Šeks (Amnesty International 2010), to the appointment by the current Serbian government of Zoran Stanković as head of the Serbian Ministry of Health, despite his close ties with indicted war criminal Bosnian Serb General Ratko Mladić (Radio Free Europe/Radio Liberty 2011a).

[25] The rising wave of apologies in the region of the former Yugoslavia is not limited to Serbia and Croatia. In November 2010, Bakir Izetbegovic, Bosniak member of Bosnia-Herzegovina's tripartite presidency, apologized for deaths caused by his ethnic group among other ethnicities. This trend started in 2000 with Montenegrin President Milan Djukanovic, when he apologized for the 1991 shelling of the Croatian coastal city of Dubrovnik in which his country was involved. Since, the Serbian and Croat heads of state have apologized in 2003, and Tadić apologized to Bosnians in Sarajevo in 2004 for Serbian atrocities committed there. Additionally, Josipović has apologized at Jasenovac, a memorial site of a World War II concentration camp, where tens of thousands of people were killed. Public apologies are not the only trend, as there have been political and judicial conciliation as well. The Parliament of Serbia, for instance, voted on a resolution on the 1995 Srebrenica massacre, while Croatia is assisting Serbia in its bid to join the EU (Pejic 2010).

[26] See "Tadić se izvinio građanima Hrvatske," *B92*, 24 June 2007.

[27] Vukovar is situated close to the Serbian border and a war site where Serbian forces took over 200 hospitalized Croats to a nearby pig farm in Ovčara and massacred them in November 1991.

[28] Already in spring 2010, when giving a talk in front of the Bosnian Parliament, Josipović apologized for crimes committed against Bosnians by the Croatian people. He also visited the site of the Ahmići massacre with Bosnian Catholic archbishop cardinal Vinko Puljić and the head of the Islamic Community reis Mustafa Cerić. See "Josipović apologizes for Croatia's role in war in Bosnia," *Croatian Times,* 15 April 2010.

Interestingly, support from international organizations to create RECOM's institutional framework also remains limited and further complicates human rights activists' efforts to account for war crimes. While the Political Affairs Committee of the Parliamentary Assembly of the Council of Europe (CoE) has released a report expressing its support for regional reconciliatory justice mechanisms among states of the former Yugoslavia, such as the regional fact-finding initiative RECOM (Marcenaro 2011), other organizations, including the UNDP and the EU, among others, avoid public statements that engage in direct political or financial support of RECOM.[29] Politicking among Coalition members and the lack of external support, however, were not the only challenges that kept the RECOM initiative from getting off the ground. During the creation of this regional truth space, a debate about the meaning of victimhood has emerged, producing different opposing narratives that we discuss below.

Multiple, Conflicting Narratives of Victimhood

Recent scholarship has grappled with the question of victimhood in post-authoritarian regimes. Drawing on interviews with war criminals and reports of the confessions of perpetrators in post-conflict settings in Argentina, Brazil, Chile, and South Africa, for instance, Leigh Payne analyzes the behavior of perpetrators (in terms of remorse, heroism, denial, or sadism) and the reaction of victim groups (Payne 2008). In the case of the former Yugoslavia, denial still remains an important phenomenon in society. Partly, as we will demonstrate in the following, because state institutions have sustained certain political discourses—such as the foundational myth of the Croatian Homeland War 1991–1995. In this context, veterans have generally enjoyed financial support in form of pensions provided by the state.[30] On the contrary, state institutions across the region have often ignored the fate of civilian war victims and their families. During the RECOM consultation process participating victims association have therefore stressed the need to define the meaning and status of a victim, illustrating RECOM initiators' conundrum of integrating different narratives of the region's looming past.

As a member of a local victims' association from Zvornik, a town in northeastern Bosnia from which nearly all Muslims were expelled during the 1992–1995 war, underlined:

> Persecution of the civilian population can't be compared to the persecution of those who bore rifles and were members of a military formation. (…) This means that a civilian is a civilian, a soldier should not be mentioned because after all he was a member of the army,

[29] See interview with United Nations Development Program and European Union officials in Sarajevo in May 2011.

[30] Veterans in BiH, for instance, have also benefitted from financial and political support by their respective governments. See Popić and Panjeta (2010).

> those are separate issues. However, here I exclusively speak about civilians, people who were taken and killed at their doorsteps or a bit further depending on where one was killed (Cited in Coalition for RECOM 2009, p.8).

This narrative, however, stands in opposition to the RECOM members' goal to establish facts about human rights violations and war crimes of *all* victims.[31] And indeed, in some cases, the meaning of victim includes social groups that do not match the Zvornikan's above definition but include former members of the armed forces. Although the RECOM initiative counts only six veterans associations versus well over one hundred victims associations, this situation demonstrates the inherent predicament of RECOM's leading members to draw bridges among different local and regional civil society organizations during their consultation meetings.

In local and regional consultation meetings, such as in Vukovar in summer 2010 and in Skopje in winter 2010, for instance, members of different branches of the Association of Underage Volunteers of the Homeland War also participated in the discussion.[32] These organizations have been created for persons, who at the time of the war were not considered adults, yet fought in the 1991–1995 Croatian war. As underage participants in the hostilities, however, they are not entitled to any veteran pensions from the Croatian state.[33] Hence, the concerns of one of their representatives with regard to RECOM's task of registering human losses stands in contrast with the statement given by the member of the Zvornika victims association:

> I am in favor of a register of all losses, which would include both civilians and military men, and that list must inevitably include foreign nationals who participated in those conflicts. How are we going to register them? We should include them in the same register, together with the members of international forces. And a separate register should be created for victims, primarily victims of war crimes.[34]

RECOM's policy strategy therefore does not necessarily receive the approval from its main target, support groups of war victims. As a result of its holistic approach, some victims' organizations, such as the Mothers of Srebrenica, do not participate officially as a member of the RECOM coalition because their members insisted on the exclusive definition of noncombatants (Coalition for RECOM 2009, p.8).[35] Narratives by political actors and state institutions also complicate RECOM advocates' mission to establish a regional commission and expand their extra-judicial spaces to deal with the past.

Nationalist discourses generated by (particularly populist) political actors across the region still pose an impediment to the successful creation of truth spaces by human rights activists. As a case in point, after the arrest of Ratko Mladić on

[31] The final draft of the statute was adopted on the fourth Coalition for RECOM Assembly Meeting on 26 March 2011 in Belgrade. The draft is available at http://www.zarekom.org/documents/Proposed-RECOM-Statute.en.html. Accessed on May 2, 2011.

[32] See reports of the consultation meetings at http://www.zarekom.org/Consultations.en.html. Accessed on February 11, 2011.

[33] See informal interview with Documenta director Vesna Teršelić in Vukovar on 14 July 2010.

[34] Supra note 33.

[35] See also Supra note 21.

26 May 2011, the Serbian Radical Party organized a rally consisting of about 10,000 nationalist protestors who gathered—with a small amount of participants rioting—in front of the Serbian parliament in Belgrade to demonstrate against Mladić's extradition to The Hague (Erlanger 2011). A few months earlier, Croatia faced a similar situation with nationalists and veterans mobilizing large parts of Croatian society across the region in order to protest against the ICTY first instance verdict in the General Ante Gotovina case (Radio Free Europe/Radio Liberty 2011b). The verdict was handed down two days before the initial start date of the RECOM signature campaign in Croatia. Given the very tense political climate in the country, human rights activists postponed the launch of the signature campaign to a later date in order to prevent violence among their campaign volunteers and demonstrators.[36] Drawing on the latter case, we will explore the question of state victimhood, which acts as an institutional hurdle to the constitution of truth spaces in society—in addition to the differing narratives of victims mentioned above—and which further exacerbates the restorative justice efforts of NGO activists in the field.

Political symbolism has a very strong effect on the community if it is grounded in lived experience. Benedict Anderson explores the sociopolitical consequences on employing symbols for political means in a longitudinal studies stretching over centuries (1991), Alex Bellamy analyzes the question temporally and spatially more concise, focusing on Croatia (2003, pp.1–6). Not surprisingly, the foundational myth of former Croatian army commander, Gotovina—who has risen to an emblematic war hero figure in Croatian society, and who represents the ontological core of the nation's nascent identity incarnated in a fight of good (Croatia) against evil (Serbia)—has sparked ferocious criticism at the intersection between international and national politics. Despite the Croatian government's international cooperation which led to his arrest and transfer to the ICTY in December 2005, the normative shift in favor of international humanitarian law in the endlessly dragging—and politically highly explosive—extradition issue of Gotovina was incomplete. In fact, during the entire period, politicians strategically politicized various debates on Croatia's national foundational myths (Pavlaković 2008). After the verdict, the Croatian government even took the necessary steps to initiate an appeals process and provide questionable amounts of legal and financial assistance to Gotovina's defense team in The Hague (Croatian Times 2011). These contradictory narratives of victimhood have thus far impeded the successful implementation of a regional truth space and the creation of a transnational truth commission.

Conclusion

This chapter has examined the ongoing struggle of a social movement to establish a restorative justice mechanism in the former Yugoslavia, the RECOM initiative. After the violent disintegration of Yugoslavia in the 1990s, the region has witnessed

[36] See interview with Signature Campaign officials of the RECOM initiative in May 2011.

several retributive and restorative mechanisms to cope with the past, including war crimes tribunals—on the international and (to a lesser, but still noticeable extent) national level—and fact-finding initiatives. During these processes human rights activists have occupied an important intermediary function, communicating and interacting between spaces created by varying justice and truth narratives. The aim of this chapter was to analyze these different, intersecting spaces and the role of civil society within these spaces to help understand why recent efforts to establish a transnational fact-finding mechanism have been unsuccessful thus far. The first part of this chapter addressed the ongoing internal and external struggle of human rights activists to establish an extra-legal space to deal with the past across the former Yugoslavia. In this context, we then analyzed the conflicting impact of different victims groups' narratives that accompanied the institutionalization process. The road to sustainable transition in the region is still long and full of challenges, and future developments will show whether human rights activists will be able to sustain their grass-roots projects and establish a dialog between different victim groups across the former Yugoslavia or whether the described obstacles persist and cannot be overcome to create a transnational truth commission for the Balkan region.

References

Amnesty International. 2010. *Behind a wall of silence: Prosecution of war crimes in Croatia*. Brussels: Amnesty International.

Anderson, Benedict. 1991. *Imagined communities: Reflections on the origin and spread of nationalism*. London: Verso.

Andrić, Gordana, and Eldin Hadžović. 2011. REKOM initiative collects signatures across Balkans. *BalkanInsight.com*, April 26. http://www.balkaninsight.com/en/article/million-signatures-for-the-victims-of-balkan-wars.

Associated Press. 2010. Serbian president apologizes for wartime massacre, November 4. http://www.wtop.com/?sid=2105540&nid=105.

B92. 2011. NGOs fall out over donations. *Život*. http://www.b92.net/eng/news/society-article.php?yyyy=2011&mm=06&dd=30&nav_id=75197&version=print.

Bellamy, Alex J. 2003. *The formation of Croatian national identity: A centuries-old dream*. Manchester: Manchester University Press.

Belloni, Roberto. 2008. Civil society in war-to-democracy transitions. In *From war to democracy: Dilemmas of peace building*, ed. Anna Jarstad, 182–211. Cambridge: Cambridge University Press Cambridge.

Burt, Jo-Marie. 2009. Guilty as charged: The trial of former Peruvian president Alberto Fujimori for human rights violations. *International Journal of Transitional Justice* 3(3): 384.

Chandler, David. 1998. Democratization in Bosnia: The limits of civil society building strategies. *Democratization* 5(4): 78–102.

Coalition for RECOM. 2009. *Report about the consultative process on instruments of truth-seeking about war crimes and other serious violations of human rights in post-Yugoslav Countries*. http://www.korekom.org/webpage/1.

Coalition for RECOM. 2010. Why coalition for RECOM? http://www.zarekom.org/index.php?/webpage/1.

Collins, Cath. 2006. Grounding global justice: International networks and domestic human rights accountability in Chile and El Salvador. *Journal of Latin American Studies* 38: 711–38.

Croatian Times. 2011. Croatian government forms team to help Gotovina, Markac. *Croatiantimes.com*.

Dostoyevsky, Fyodor. 2000. *Crime and punishment*. Hertfordshire: Wordsworth Editions (Trans. Keith Carabine).

Erlanger, Steven. 2011. Pro-Mladic demonstrators in Serbia rally against his extradition—NYTimes.com. *New York Times*, May 29.

Fischer, Martina. 2005. *Peacebuilding and Civil Society in Bosnia and Herzegovina: Ten Years After Dayton*. Berlin: Lit Verlag.

Grodsky, Brian. 2009. International prosecutions and domestic politics: The use of truth commissions as compromise justice in Serbia and Croatia. *International Studies Review* 11(4): 687–706.

Jeffrey, Alex. 2011. The political geographies of transitional Justice. *Transactions of The Institute of British Geographers* 36(3): 344–359.

Jouhanneau, Cécile. 2010. Les Mésaventures Des Projets De Commission Vérité Et Réconciliation Pour La Bosnie-Herzégovine (1997–2006): Une Étude De La Circulation Des Modèles Internationaux De Résolution Des Conflits Mémoriels. In *Le Passé Au Présent: Gisements Mémoriels Et Politiques Publiques En Europe Centrale Et Orientale*, ed. G. Mink and P. Bonnard. Paris: Michel Houdiard Editeur.

Kurze, Arnaud. 2012. *Justice beyond borders? The politics to democratize human rights in the post-conflict Balkans*. Ph.D. Dissertation, George Mason University. Ann Arbor: ProQuest/UMI. Publication Number: 3506226.

Lutz, Ellen, and Kathryn Sikkink. 2001. The justice cascade: the evolution and impact of foreign human rights trials in Latin America. *Chicago Journal of International Law* 2: 1.

Marcenaro, Pietro. 2011. *Reconciliation and political dialogue between the countries of the former Yugoslavia*. Parliamentary Assembly Rapporteur Report. Council of Europe.

Martin-Ortega, Olga, and Johanna Herman. 2010. *Hybrid Tribunals & the Rule of Law Notes from Bosnia & Herzegovina & Cambodia*. Working Paper. Department of Political Science, Lund University, Lund.

Miraftab, Faranak, and Shana Wills. 2005. Insurgency and spaces of active citizenship. *Journal of Planning Education and Research* 25(2): 200.

Pavlaković, Vjeran. 2008. Better the grave than a slave: Croatia and the International Criminal Tribunal for the former Yugoslavia. In *Croatia since independence. War politics, society and foreign relations*, ed. Sabrina P. Ramet, Konrad Clewing, and Reneo Lukić, 446–477. München: Oldenburg.

Payne, Leigh. 2008. *Unsettling accounts: Neither truth nor reconciliation in confessions of state violence*. London: Duke University Press.

Pejic, Jelena. 2001. The Yugoslav truth and reconciliation commission: a shaky start. *Fordham International Law Journal* 25: 1.

Pejic, Nenad. 2010. The weight of wreaths and words. *Radio Free Europe / Radio Liberty*, November 4. http://www.rferl.org/content/The_Weight_Of_Wreaths_And_Words/2211082.html.

Popić, Linda, and Belma Panjeta. 2010. *Compensation, transitional justice and conditional international credit in Bosnia and Herzegovina*. Sarajevo: Royal Norwegian Embassy in Bosnia and Herzegovina and Embassy of Switzerland in Bosnia and Herzegovina.

Popovic, Lidija. 2007. Tadic apology to croats divides Serbia. *Balkan Investigative Reporting Network*, June 26. http://birn.eu.com/en/89/10/3417/?tpl=30.

Radio Free Europe/Radio Liberty. 2011a. Mladic friend approved as Serbian health minister. *News*. http://www.rferl.org/content/mladic_friend_serbian_health_minister/2339279.html.

Radio Free Europe/Radio Liberty. 2011b. Hague tribunal sentences croatian generals to long prison terms. News. http://www.rferl.org/content/un_judgments_due_in_croatian_war_crimes_case/3557888.html.

Ramet, Sabrina. 2010. Politics in Croatia Since 1990. In *Central and Southeast European Politics Since 1989*, ed. Sabrina Ramet, 258–285. Cambridge, United Kingdom: Cambridge University Press.

Roht-Arriaza, Naomi. 2005. *The Pinochet effect: Transnational justice in the age of human rights*. Philadelphia: University of Pennsylvania Press.

Subotić, Jelena. 2009. *Hijacked justice: Dealing with the past in the Balkans*. Ithaca: Cornell University Press.

Teitel, Ruti G. 2005. The law and politics of contemporary transitional justice. *Cornell International Law Journal* 38: 837–862.

Teitel, R.G. 2010. Global transitional justice. *Center for Global Studies Working Paper Series on Human Rights, Global Justice & Democracy* (8).

Urry, John. 2000. Sociology of Time and Space. In *The Blackwell Companion to Social Theory*, ed. Bryan S. Turner, 416–444. Oxford, United Kingdom: Blackwell Publishers.

Chapter 13
From International Courts to Grassroots Organizing: Obstacles to Transitional Justice in the Balkans

Jill A. Irvine and Patrice C. McMahon

Introduction

Despite the May 2011 arrest of Ratko Mladic, there is some consensus on the shortcomings of the International Criminal Tribunal for the Former Yugoslavia (ICTY) (McMahon and Western 2011). Although the court was seen as a judicial organ that would have both a legal and political impact, its effectiveness in both arenas has come under considerable scrutiny. Its primary intent was to prosecute individuals responsible for war crimes, crimes against humanity, and genocide in the Yugoslav wars, though it was also created to contribute to the restoration and maintenance of peace in the region (Hampson 1998). It was clear how the former would happen; it was never articulated how the latter would take shape. Most research suggested or assumed that these broader societal goals would transpire once domestic actors became more engaged in transitional justice and addressing the past (Peskin 2009; Subotic 2009; McMahon and Forsythe 2008; Meernik 2005; Kerr 2004). Yet, only recently, with the establishment of the Regional Truth Commission for the Former Yugoslavia (RECOM) have local organizations come together to organize around this issue. What are the origins of RECOM? What issues and strategies does it advocate? And, finally, does RECOM's development suggest that a regional justice movement is underway in the Balkans?

In the nearly two decades since the establishment of the ICTY in 1993, a substantial body of literature has emerged on transitional justice, its mechanisms, and practices (Forsythe 2011; McEvoy and McGregor 2008; Nettelfield 2010 Olsen et al.

J.A. Irvine (✉)
Women's and Gender Studies Program, University of Oklahoma,
101 Robertson Hall, Norman, OK, 73019, USA
e-mail: Jill.Irvine@ou.edu

P.C. McMahon
Department of Political Science, University of Nebraska-Lincoln,
511 Oldfather Hall, Lincoln, NE 68588, USA

O. Simić and Z. Volčič (eds.), *Transitional Justice and Civil Society in the Balkans*,
Springer Series in Transitional Justice, DOI 10.1007/978-1-4614-5422-9_13,

2010; van der Merwe et al. 2009). The early focus on international law and international institutions (Drumbl 2007; Paust et al. 2006; Akhavan 1998) has given way to more empirical studies that attempt to evaluate mechanisms, highlighting their relative weaknesses, and inability to deliver the desired outcomes (Miller and McMahon 2012; Lamont 2010; Subotic 2009; Grodsky 2007). Most research on transitional justice in the Balkans focuses on the problems with international and national transitional justice mechanisms, underscoring elites' instrumental behavior when it comes to cooperating with the ICTY (Nettelfield 2010; Fletcher et al. 2009; McMahon and Forsythe 2008; Kerr 2004). Little attention has been given to domestic activities because, until recently, there has been relatively little domestic participation and organizing around the topic of transitional justice. Our study seeks to fill this gap, examining RECOM's efforts to establish a regional truth commission, which began officially in 2006. Drawing upon insights from social movement theory and the literature on transnational advocacy networks, we analyze the conditions necessary for a regional justice movement to succeed in the Balkans. We begin by examining the origins of transitional justice in the Balkans, identifying *who* first started to push for accountability and justice. We then consider *what* RECOM is engaged in, explaining its origins and the motivations of local actors who united to form this coalition. We next turn to an analysis of *how* RECOM is doing, identifying the ongoing obstacles to its development, as well as the role and impact of international actors.

Given the recent development of RECOM and dearth of published material on this topic, we relied largely on primary sources, participant observation, and interviews for this research as well as secondary sources. Semi-structured interviews with elected officials, RECOM leaders, victims' and veterans' associations, (both RECOM and non-RECOM members), women's organizations, youth organizations, members of the media, and other civil society organizations were conducted during the summers of 2008 and 2011 in Bosnia-Herzegovina (hereafter Bosnia), Croatia, Kosovo and Serbia, and they constitute an important part of this qualitative research. We initially selected individuals in leadership positions in organizations affiliated with RECOM, but using a snowballing sampling technique we also interviewed individuals who were known to be actively involved in other transitional justice activities. In order to observe and interview a broad array of organizations operating in both urban and semi-urban settings, we conducted a total of 38 interviews in the capital cities and other towns of these countries—both entities of Bosnia and northern Kosovo were included in our visits and interviews. In these semi-structured and open-ended interviews, we asked participants specifically about the formation, goals and impact of RECOM. An important additional source of information was the International Forum on Transition Justice in the Post-Yugoslav Countries which one of us attended in June of 2011. Survey research, when available, as well as secondary material, together allowed us to develop an interpretive analysis of RECOM's intentions and ability to build a grassroots justice movement and to situate our analysis contextually and comparatively.

The arguments we present in this chapter are twofold. First, we contend that although justice and peace are moving forward in the Balkans, ongoing dilemmas underscore important lessons about transitional justice; specifically, grassroots efforts to promote transitional justice must overcome significant obstacles in framing issues, creating coalitions and engaging the state. Different conceptions of justice and priorities among coalition partners, as well as struggles over coalition leadership and political partners have challenged RECOM's ability to create a viable regional movement. Second, we argue that international actors have fundamentally shaped the justice environment in the Balkans by affecting the supply and the demand for transitional justice in both the short-term and the long-run. In the short-term, the international community's decision to establish an international court directly affected the supply of domestic efforts to address the past, unintentionally rendering domestic mechanisms both unnecessary and undesirable. In the years since, the international community's narrow focus on cooperation with the ICTY has meant that it failed to elicit or sustain domestic participation in justice activities. Thus, although the region's transformation has indeed moved forward we contend that domestic progress in this area has been slow in part because of the behavior and policies of international actors. To continue going forward, greater international support for grassroots efforts like RECOM may be necessary to affect real and enduring social change.

From International Prosecution to Domestic Organizing

Although there is still some debate over why the ICTY was established by the United Nations, there is little disagreement over its relative weaknesses particularly its inability to deter violence and stop the conflict. Supporters of the ICTY contended that the court reflected the world's commitment to human rights and justice and it would help the region rebuild their judicial systems and foster peace (Hagan 2003). Skeptics of international criminal justice have, instead, contended that the establishment of the ICTY represented a "fig leaf for inaction" and a way for the international community to "do something" while avoiding a costly war (Hagan 2003, p. 60). Some critics even argued that the ICTY would do more harm than good, provoking a negative backlash in the region (Snyder and Vinjamuri 2003/2004).

Regardless of why the ICTY was created and the theoretical merits of international criminal justice, its utility and viability started to wane by the late 1990s, and ICTY judges already started to think about the court's exit strategy and how it would transfer its responsibilities to local authorities. Meanwhile, research demonstrated clearly how uncooperative Balkan elites were and how hostile or uninformed the average person was to the ICTY (Peskin and Boduszynski 2003). The court, nonetheless, continued to work on its primary goal of prosecution while its other intended objectives, such as fostering progressive liberal change and societal reconciliation, were only rarely and sporadically evident in the ICTY's activities. Obvious contradictions in terms of what leaders say and do related to addressing the past and imparting justice continue to exist, damaging the overall case for transitional justice in the Balkans. In Serbia, in particular, elite decisions that followed ICTY dictates, such as handing Slobodan Milosevic over to the ICTY, happened alongside intense criticism of the ICTY and international justice (Orentlicher 2008). Thus, despite the ICTY's good work, which has culminated in 161 indictments and numerous arrests and prosecutions, even of "big fish" like Milosevic, Karadzic, and Mladic, the international court has seemingly had a minimal impact on political behavior and political culture in the Balkans which remains defined by ethnic nationalism.[1]

Developing an exit strategy and shifting from international to domestic responsibility has proved complicated and domestic responses of any kind are, by definition, politically sensitive and financially costly. Domestic efforts to arrest, detain and try war criminals were considered arbitrary and politically motivated (Donlon 2008), and domestic trials were regarded as occasions for dispensing ethnic justice or exacting revenge (Bohlander 2003). In 2003, ICTY officials devised a Completion Strategy for its requiring cooperation and coordination with local justice systems in the Balkans. Although the Completion Strategy has moved forward, the ICTY has had to revise its timeline for closure because of unfinished and ongoing cases. During this same period, Bosnia, Croatia, and Serbia all established specialized War Crimes Courts to try cases that were transferred from the ICTY. In Kosovo and later in Bosnia, hybrid courts, containing foreign and domestic judges, were established because of the states' fragile institutional structures (Dickinson 2003).

Attempts to establish national truth commissions were uniformly not successful. For example, under pressure from the international community, Serbian leader Vojislav Kostunica, who came to power after the 2000 elections, established the National Truth Commission or "Kostunica Commission." Its stated purpose was to establish "the truth" and gather the facts about victims of violence during the wars of 1991–1995 and 1999. Generally seen as half-hearted and ineffectual, the commission was allowed to die a quiet death in 2003, having produced nothing of value. A similar effort was undertaken in Croatia shortly after Ivica Racan's government came to office in 2000 also producing little. A National Truth Commission was established in Bosnia in 1997, producing only periodic roundtables and discussions (Heil 2000). Perhaps the most successful of these efforts in Bosnia was the Srebrenica Commission, which produced a report in 2004 concluding that "grave crimes" had been committed in the region in 1994 in "violation of international humanitarian law" (Dragovic-Soso and Gordy 2010, p. 204). Nevertheless, while international actors lauded the commission's proceedings and conclusions, the local population viewed the commission with considerably more skepticism (S. Djuderija, personal interview, June 22, 2011).

Brian Grodsky (2007) argues that in the absence of strong domestic pressure and with only moderate international pressure, these forms of "compromised justice" have contributed little to transitional justice in the region. Thus, for almost

[1] See the ICTY's website for more information on indictments and arrests: http://www.icty.org/action/cases/4. Accessed December 12 2011.

two decades, there has been a dearth of domestic engagement and few grassroots organizations that focus squarely on addressing the past and justice. Public opinion, moreover, has reflected considerable apathy, if not overt hostility toward the ICTY and international criminal justice. Surveys from Bosnia are telling; according to a 2002 survey, among international organizations, the ICTY was the least trusted, with 51% of Bosnians in the Federation, but only 4% in the RS indicating that they had trust in the court (International Institute for Democracy and Electoral Assistance 2002). Serbian views illustrate that the impetus for cooperation with the ICTY clearly does not spring from domestic support. Only 11% of Serbs in 2003 approved of cooperation with the ICTY; by January 2006, 16% felt this way, with 46% viewing cooperation with the ICTY as a "necessarily evil" to further relations with Western European countries and the US (International Republican Institute 2006). Many Serbs feel that the ICTY has targeted Serbs unfairly while many Croats object to the idea that any Croats should be tried for war crimes during what they consider a defensive "Homeland" war (G. Bosanac, personal interview, June 17, 2011). More general dissatisfaction with the ICTY has had to do with its lack of transparency, the belief that victims were not well served by the process, and by a general frustration with international involvement in the region.

This level of domestic apathy toward transitional justice is somewhat surprising and unusual because elsewhere in the world, domestic organizations and victims of abuses have generally led efforts to confront the past. As Eric Brahm (2007) observes, "civil society tends to be closely intertwined with the fate of transitional justice," (p. 63) and if governments do not act quickly or appropriately, civil society has weighed in, even initiating its own investigations. Throughout Latin America, in Africa and Asia, there are countless examples of civil society organizations documenting human rights violations and pressuring governments to publicize and respond to these allegations, often at much risk to themselves (Hayner 2002; Bickford 2000). In most cases, domestic groups play a leading role in shaping the accountability mechanisms that have been adopted, bringing their legal expertise to bear and ensuring that proposed policies conform to the country's legal structure. Frequently, domestic groups have had a hand in the construction of truth commissions and they regularly monitor the government's behavior, provide assistance to victims, and, more generally, ensure that reconciliation efforts move forward.

Civil society has many potential and positive roles to play in seeking and promoting justice, but there are also dangers, including leaving the state outside the process, narrowing its scope too much on certain institutions or being co-opted by the state. Although the importance of civic activism and grassroots support cannot be overrated or ignored, the Balkans presents a very different model of transitional justice and the role of society in it because efforts to address the past have been dictated by external actors and imposed from above. Many analyses of the region not surprisingly conclude with the need for genuine local participation or with the expectation that civil society will eventually assume its natural role. Unfortunately, there is neither a blueprint for how to inspire such actions nor a formula to ensure their success (Treves et al. 2005).

Establishing the RECOM Coalition

After more than a decade of internationally dictated transitional justice, a grassroots effort emerged in late 2005, with the goal of establishing a regional commission on human rights violations. According to McEvoy and McGregor (2008), efforts from below arise in a situation where the state has failed to put into place sufficiently robust transitional justice mechanisms. Given the shortcomings of international efforts and states' unwillingness to develop state-led mechanisms, organizations in society responded. Three human rights organizations spearheaded the effort to promote transitional justice at the grassroots: the Humanitarian Law Center in Serbia, Dokumenta in Croatia, and the Research and Documentation Center in Bosnia-Herzegovina. At an initial meeting in September 2005, along with several organizations, they launched a regional process of consultation and discussion on the best mechanisms of truth seeking and truth telling. The organizers emphasized two points: first, that while they supported the formal judicial mechanisms for trying and punishing war criminals, these mechanisms could not fully satisfy the victims' need for information and truth telling; and, second, that any efforts to establish a comprehensive understanding of human rights abuses during the wars would necessitate a regional approach.

Over the ensuing months, consultations were held throughout the former Yugoslavia, involving small, local groups such as human rights organizations, women's groups, victims' organizations, veterans' organizations, and youth organizations.[2] In addition, consultations were held in larger cities with groups of intellectuals, journalists, and artists. Initial discussions focused on the general needs of victims and whether a regional truth commission should be established; later discussions focused on "objectives, tasks, responsibilities and role of RECOM in criminal proceedings, public testimony, and other elements of a Regional Commission" (RECOM 2011c, p. 4). Periodic regional forums, attended by hundreds of representatives of local groups as well as international advisors, considered thematic questions concerning truth seeking and truth telling mechanisms and occasionally heard the personal testimonies of war victims. Practically every aspect of a possible regional truth commission was the subject of heated debate, from the definition of victims to the post-Yugoslav countries to be included in the commission.

A more formal organizational structure for RECOM was established at the Fourth Regional Forum for Transition Justice held in Pristina in 2008. At this meeting, a decision was made to establish a coalition to promote the establishment of a regional

---

[2] For a full list of the RECOM members and information on RECOM's history, goals and strategies, see http://www.zarekom.org. Accessed 21 October 2011.

commission and to solicit membership. Membership in the Coalition for RECOM grew steadily during the next several years. By 2011, RECOM membership has grown from approximately 100 organizations in 2008 to 342 organizations and 1,357 individuals. RECOM members claimed that consultations had reached at least 8,700 individuals (RECOM 2011c, p. 1). Nevertheless, the level of involvement among member organizations varies considerably and many organizations understand their membership to be more an expression of support for the idea than a commitment to participate in the process (S. Sarnavka, personal interview, June 16, 2011; G. Obradović, personal interview, June 17, 2011). Additionally, at least some member organizations have already withdrawn from the coalition even though their names remain on the membership roster, making a real assessment of organizational membership difficult.

In the spring 2010, experts drawn from the RECOM Coalition presented a draft statute for the establishment of a regional truth commission for consideration by RECOM Coalition members (RECOM 2011c, p. 4). The draft statute, adopted by the Assembly of the RECOM Coalition in March 2011, proposed a regional commission of 20 commissioners with a 3-year mandate, appointed by their national governments. Located in Sarajevo, the commission's objective would be to "establish facts about war crimes and other gross violations of human rights committed on the territory of the former SFRJ," as well as "the political and societal circumstances that led to the commission of these acts, and the consequences of these violations" (RECOM 2011a, p. 468). With the statute approved, and armed with public opinion polling that suggested high levels of approval for transitional justice initiatives in all ex-Yugoslav countries except Croatia and the Republic of Serbia in Bosnia, the RECOM Coalition launched a media campaign to familiarize the public with the RECOM process and mobilize popular support. This media effort included television and radio clips, newspaper ads, newspaper supplements, and on-line initiatives (RECOM 2011c). It culminated in a "One Million Signatures" campaign from April 26 to June 30, 2011 to gather signatures on a petition calling for the establishment of a regional truth commission to be presented to national governments. Youth groups, in particular, were enlisted and an effort was made to draw attention to the campaign in the press (M. Mažić, personal interview, June 16, 2011). When the signature drive was completed in June 2011, a total of 543,870 signatures had been collected, with the largest number, 122,540, coming from Bosnia (RECOM 2011c). The campaign fell considerably short of its goals in Croatia where only 19,674 out of the anticipated 220,000 citizens signed and where, as we discuss below, popular resistance to a regional approach to transitional justice remained high. Nevertheless, according to some participants, the effort was effective in drawing public attention to the proposed regional commission (E. Jaković, personal interview, June 18, 2011). In any case, at the Eighth Regional Forum on Transitional Justice in Sarajevo in July 2011, the Coalition decided to move forward with its next phase of activity in which the draft statute will be presented to the national parliaments of the seven ex-Yugoslav states for adoption.

As encouraging and necessary as these grassroots developments are, the coalition is still evolving and its future is unknown. In June 2011 Nataša Kandić, the Director of the Humanitarian Law Centre and one of the founders and leaders of the coalition, resigned as a member of RECOM's Coordinating Council. Although Kandić indicated that she was leaving because of lack of time, interviews in the region suggest that Ms. Kandić is a controversial figure and that some organizations did not join the coalition because of her leadership. A representative of Catholic Relief Services explained that Bosnian organizations were not very active in RECOM, in part, because of Kandić's leadership and because of the perception that the idea for RECOM was coming from Belgrade (G. Bubalo, personal interview, May 16, 2011).As RECOM moves forward, it is evident that the coalition of member organizations faces numerous and perhaps paralyzing obstacles.

A Movement in the Making?

Although RECOM's efforts are relatively recent, theories on transnational social movements provide insight into the domestic conditions essential to the success of a regional justice movement (Tarrow 2005; Keck and Sikkink 1998). Drawing upon cultural theories of social movements that stress identity formation and issue framing, scholars such as Keck and Sikkink demonstrate the importance of framing issues in such a way that they can provide cross border unity while maintaining the flexibility for local applications. Resource mobilization approaches stress the importance of resources—organizational, financial, and leadership—necessary to building an effective transnational movement (McCarthy and Zald 1987). Finally, political process theories of social movements stress the importance of political allies in the formal political sphere (Meyer and Minkoff 2004). The following section discusses how the RECOM coalition tackled the tasks of framing issues, building coalitions, and engaging the state.

Framing Issues

Most important to the success of a transnational social movement is framing; without a clear articulation of the aims of a justice movement and the best means for achieving them, collective action is unlikely to ensue. In the case of RECOM, efforts to frame domestic action on behalf of transitional justice have revolved around the mandate of the proposed truth commission. According to founding member Natasa Kandic, at the minimum the RECOM process aims for "opening public dialogue on facing the past."[3] Nevertheless, from the outset, fierce disagreement erupted about

[3] International Forum on Transitional Justice in Post-Yugoslav Countries, 27 June 2011, Sarajevo, Jill Irvine's notes.

these aims as well as the means of achieving them. Disagreement has revolved around three major questions: justice for whom?; what kind of justice?; and how should justice be pursued? These were all issues that international actors either decided on—on behalf of victims and citizens of the Balkans—or failed to weigh in on, because of the international community's desire to end the conflict quickly without significant loses. Without condemning or praising these decisions, we merely suggest that debates and outcomes today are path-dependent and the initial choices and behavior of the international community has constrained and shaped RECOM's debates and the dilemmas it faces.[4]

Justice for whom is the most crucial and contentious issue because the RECOM process emerged from attempts to address the needs of victims which had not been fully met through formal judicial mechanisms. Are victims only those who had suffered directly or should it include those who had lost a family member? Should the notion of collective victimhood be considered? Were some nations, like Bosnian Muslims, by virtue of the sheer numbers of losses they suffered, be entitled to a special status (M. Tokaca, personal interview, June 20, 2011)? Should this definition of victimhood be based on the particular human rights abuses they or their loved ones had suffered? For some this is self-evident; while others worry that without a specific list of abuses, important classes of war crimes, especially gender-based crimes, might be overlooked (M. Mažić, personal interview, June 16, 2011). Relatedly, members have debated the question of whether perpetrators should be publicly identified. While everyone agrees that the commission should compile comprehensive documentation concerning victims of abuses, there is less agreement about whether and how to compile documentation of the perpetrators of these crimes or who, for that matter, should be considered a perpetrator? Should journalists, for example, whose "message of hate" contributed to the outbreak of war, be considered perpetrators?[5]

What kind of justice, or the kind of justice pursued, is also a subject of debate. Much of the literature on how civil society contributes to transitional justice assumes that while the courts meet out retributive justice, civil society concerns itself with restorative justice, or efforts to promote reconciliation and social peace (Kurze 2010). Judging from the RECOM consultations, this assumption must be reexamined. During the consultation process at local and regional levels, tensions emerged between the truth commission's goal of responding to victims' needs and its goal of promoting reconciliation. Participants in the consultation process across, particularly victims associations, overwhelmingly focused on the need to identify and punish perpetrators. When asked to describe what victims needed from the regional truth commission, the leader of the Movement of Mothers from the Enclave of Srebrenica and Žepa, Munira Subašić, asserted that the "best thing would be if judicial bodies

[4] On this point and the role of the international community in statebuilding, see Patrice C. McMahon and Jon Western, "Introduction: the supply side of statebuilding," in McMahon and Western, eds, *The International Community and Statebuilding* (London: Routledge Press 2012), 1–24.

[5] There was strong sentiment at the most recent regional forum that journalists should be held accountable for their reporting before and during the war. International Forum for Transitional Justice in Post-Yugoslavia Countries, 27 June 2011, Sarajevo, Jill Irvine's notes.

could bring perpetrators to justice."[6] Similarly, a representative from the Serbian Victim's Association in Brčko answered in much the same way, emphasizing the need for local authorities to identify perpetrators and bring them to justice (Association of Missing Serb Fighters and Civilians of Brčko, personal interview, May 22, 2011). Victims continually expressed frustration about the courts' inability to process the high number of potential war crimes cases they had before them in places like Jajce, where according to one RECOM member, "we won't be able to complete all the cases in 100 years."[7] In Brčko and in Sarajevo, victims' association members made similar comments. Given that retributive justice is a main goal of many RECOM Coalition members, it was not clear exactly how or if RECOM will fulfill their expectations. Indeed, relatively few participants in the consultative process expressed a desire to engage in one of the main mechanisms of restorative justice—telling their individual stories (RECOM 2009, 2011a). While the truth commission could gather evidence of crimes and allow victims the opportunity to share their experiences, it was clearly not a judicial body with the power to punish.

Finally, "justice how?" is a question that has continued to arise, even after RECOM's draft statute was completed. Is the Commission only to be tasked with establishing facts about crimes committed during the war, or is it also necessary for the Commission to identify the causes of the war? This goes to the heart of transitional justice for many, who feel that mere documentation of abuses is only a small part of the larger, most important role of a truth commission. That is to say, true justice demands a deeper understanding of causes of, and by extension, of guilt for, the wars and human rights abuses they brought. It was precisely this issue of the "relativization of guilt" that some objected to, while others questioned whether collective (national) guilt and apologies ought to be the business of the commission at all.[8] In any case, it is clear that for those whose main purpose is to create an "accurate" or at least more common narrative of the past, documenting abuses without examining causes is bound to end in failure.[9] Such a new narrative, which should form the basis of education about this period of history, is seen by proponents of examining causes as an indispensable part of the regional commission's aims.

The RECOM statute adopted by the Coalition attempted to address these concerns and competing visions. While the statute emphasized the primary goal of

[6] Speech given at the International Forum for Transitional Justice in Post-Yugoslavia Countries, 27 June 2011, Jill Irvine's notes.

[7] International Forum for Transitional Justice in Post-Yugoslavia Countries, 27 June, 2011, Jill Irvine's notes.

[8] Ivan Pšenica, Savez udruga obitelj zatočenih i nestalih harvatskih branitelja, Hrvatska, speech given at International Forum for Transitional Justice in Post-Yugoslavia Countries, 27 June, 2011, Jill Irvine's notes. A representative of the Government Commission on Missing Person's in Kosovo argued that Serbia must be held responsible for its role in the wars and its obligation to pay reparations. (Prenk Gjetaj speech given at International Forum for Transitional Justice in Post-Yugoslavia Countries, 27 June, 2011, Jill Irvine's notes).

[9] Several panelists from the panel, "Tranziciona pravda iz ugla epistemološke zajednice," reflected this view. International Forum for Transitional Justice in Post-Yugoslavia Countries, 27 June, 2011, Jill Irvine's notes.

establishing a record of human rights abuses, it also endorsed an investigation into the causes of the wars. A precise list was included of the human rights abuses the commission would consider, as well as a disavowal that any particular national group would be accorded greater victim status. Finally, the commission rejected any role in granting amnesty or compensating victims. Nevertheless, various aspects of its proposed mandate continue to be contested by Coalition members as well as the public.

Failure to successfully frame the justice issues around a regional truth commission has made it difficult to garner popular support. Mobilizing widespread public support for a regional truth commission looked to some RECOM activists like "mission impossible" (G. Bosanac, personal interview, June 17, 2011). On the one hand, there was considerable opposition and disappointment with the ICTY throughout the region, which provided an opening for a grassroots strategy. Yet, opposition to a regional, grassroots approach was also high because of differing accounts and perceptions of the war (Stover and Weinstein 2004). This lack of public support for a regional body became increasingly evident during the one million signatures campaign, particularly in Croatia.

Creating Coalitions

The framing of the regional truth commission directly affects the coalition's ability to create a viable regional network of civil society actors and organizations. The literature on social movements draws our attention to the fact that transnational movements must address crucial issues related to membership, leadership, and linkages. These issues have proved difficult for RECOM to resolve thus far, particularly its ability to secure key constituencies and problems with its leadership. Perhaps the biggest challenge involves the cross border character of the RECOM effort, which requires overcoming significant resistance to establishing any regional bodies.

A serious organizational failure that emerged, at least as far as many Coalition members were concerned, was the overly centralized nature of the RECOM structure. As it turns out, RECOM did not emerge spontaneously or organically from many active organizations that finally came together but, instead, the structure is quite centralized, with the preponderance of power and authority resting in the Humanitarian Law Center in Belgrade (RECOM Coalition Members, personal interviews, June 13–July 4, 2011). The problems emanating from this appear to be twofold. First, international donor funding for the RECOM Coalition was funneled through the Humanitarian Law Center. While there may indeed have been sound accounting reasons for wanting the money to originate in one place, it had the result of alienating other organizational partners, particularly the two founding organizations, Dokumenta and the Documentation and Research Center (E. Jakovčič, personal interview, June 18, 2011). Second, the perception of over centralization was reinforced by Kandić's leadership style, which many found "dictatorial" and "imperial."[10] As one leader of a veterans' group IZMIR put it, "this non-violent group is using very hierarchical methods, eliminating groups of undesirables along the way" (G. Bodog, personal interview, June 15, 2011). Another member of the Muslim Victim's Association in Brčko explained that "RECOM is not of interest to us; it doesn't represent us" (confidential interview, May 21, 2011). These remarks illustrate how some organizations in the Balkans feel about the RECOM process. Other NGO leaders complained that agendas for the RECOM regional meetings were routinely ignored or changed at the last minute, and that representatives of organizations deemed "recalcitrant" were deliberately driven from RECOM or silenced in other ways.

The networks and alliances RECOM supporters hoped to build were further weakened by the failure of key constituents to endorse the regional commission. First among them were victims' groups. While several important victims' groups did indeed participate in the Coalition, including the Movement of Mothers of the Enclave of Žepa and Srebrenica, many others did not including the victims associations in the Brčko region. This reluctance to support the RECOM Coalition appears to stem from a variety of reasons, not least of which was the strong association of the RECOM effort with Belgrade and a sense that concerns of local groups were not being heard (M. Tokača, personal interview, June 20, 2011). This was a primary reason why the founding organization from Bosnia, the Research and Documentation Center in Sarajevo, withdrew from RECOM in 2008 (M. Tokača, personal interview, June 20, 2011).

A second key constituency that did not support RECOM in significant numbers was veterans' associations. This is because the role of veterans in the RECOM process was contested from the outset and never resolved. Local consultations were held with numerous veterans' organizations and those, such as Izmir in Croatia, which had been concerned with peace-building, were invited to RECOM meetings and to join the Coalition. Nevertheless, while some individuals joined during the next several years, the major veterans' organizations withheld their support. Many, like those in Croatia, objected that the commission was engaging in the "relativization of guilt" and that any truth commissions should be formed within the confines of their particular states (Z. Pusić, personal communication, June 2011). Others suggested privately that they would like to join but worried that the government would withhold veterans' pensions in retaliation (G. Bodog, personal interview, June 15, 2011). Whatever the reasons, and they are many, this key constituency did not throw its support behind the RECOM effort.

Finally, women's organizations have expressed dissatisfaction with RECOM's approach to gender and transitional justice. Many women's organizations were initially receptive to the Coalition and were among the first members.[11] Over time,

[10] One member described Kandić as brave but criticized her tendency to "see herself as a monument." (RECOM Coalition member, interview with Jill Irvine, 16 June, 2011).

[11] For a discussion of how transitional justice mechanisms in Bosnia-HHerzegovina have failed women see Amnesty International, *'Whose justice?': The Women of Bosnia and Herzegovina are Still Waiting* (London: Amnesty International, 2009); http://www.amnesty.org/en/library/asset/EUR63/006/2009/en/8af5ed43-5094-48c9-bfab-1277b5132faf/eur630062009eng.pdf. Accessed 14 November 2011.

however, this enthusiasm waned. Some leaders of these women's organizations, particularly in Bosnia, felt that RECOM leaders were unwilling to pay serious attention to gender-based crimes and that requests to insert discussion or training related to gender and transitional justice in RECOM meetings were rejected without explanation by RECOM leaders (L. Somun, personal interview, June 20, 2011; N. Zvizdić, personal interview, June 23, 2011). Moreover, the story telling and witnessing central to feminist approaches to transitional justice appeared to be of secondary concern to RECOM organizers. As a result, women's organizations decided to pursue their own transitional justice initiatives and to form a Women's Tribunal modeled after similar tribunals previously held in other regions. Plans to hold such a tribunal are currently underway (N. Zvizdić, personal interview, June 23, 2011).

An additional structural impediment to creating a robust cross regional alliance has been the problem that not all states in the region recognize the independence of Kosovo. To date, Serbia and Bosnia have not recognized Kosovo, calling into question the very basis on which the commission is organized. Organizations unwilling to join RECOM have pointed to this as evidence that such a commission is premature and unworkable in the current circumstances (A. Masović, personal interview, June 22, 2011; S. Madacki, personal interview, June 20, 2011). While RECOM has managed to include participants from all ex-Yugoslav states in the consultation process, many question whether this can be the case once the discussion moves into the formal political arena.

Engaging the State

Literature on social movements also emphasizes the importance of engaging key allies within the state to build an effective social movement. Thus far, RECOM attempts to do this have proved only partially successful. The way in which political elites have instrumentalized transitional justice has been the subject of a great deal of recent attention. Scholars have pointed to ways in which political elites have "hijacked justice" to serve their own political ends, often creating forms of "compromised justice" that undermine the very processes they officially support (Grodsky 2009; Lamont 2010; Peskin 2008; Rangelov 2004; Subotic 2009). In this two-level game, elected officials weigh the necessity of international support, which rests on compliance with the ICTY, against the demands of a domestic electorate, which is perceived as hostile to the aims of the international court. Civil society organizations promoting transitional justice are relatively neglected in this analysis, though they have been the focus of other comparative approaches. Backer (2003), for example, attempts to account for transitional justice outcomes by examining this relationship between civil society and political elites, arguing that government support for transitional justice and the strength of civil society explain different outcomes. What he labels a "hands off the wheel scenario" where an activist civil society faces a lack of political will on the part of government structures seems an accurate description of the situation in the ex-Yugoslav countries we examine here. Nevertheless, there are signs of engagement with the RECOM process at some levels of government.

Political support for the idea of a regional truth commission has been sporadic at best, and thus far subject to the political calculations of particular leaders. The Serbian leadership has given it lukewarm support, although President Tadić has avoided publicly endorsing the commission. On the one hand, supporting the commission allows him to display his willingness to engage in transitional justice efforts to those in the European Union who are watching to see whether he is serious about meeting the conditions necessary for forming a closer relationship with the EU. On the other hand, the Europeans have been almost exclusively focused on Serbia's need to comply with ICTY demands when it comes to transitional justice, and President Tadić therefore has little incentive to take domestic political risks in order to appease EU officials. The Croatian government has been equally cautious, though for different reasons (Amnesty International 2010; Peskin and Boduszynski 2003). While President Josipović publicly endorsed the RECOM process in 2010, leaders of the two major political parties, the SPD and the HDZ, have been reluctant to lend their support.[12] Political support for RECOM in Bosnia runs strictly along ethnic lines; while Serbian leaders in the RS have opposed any transitional justice mechanisms, through judicial or extra-judicial processes, Bosniak leaders in the Federation are generally more receptive. Croat leaders in Bosnia have generally followed the lead of their counterparts in Croatia.

Although political leaders in Kosovo and Montenegro have recently endorsed the regional truth commission, support for RECOM from political leaders elsewhere has been lukewarm. Nevertheless, cooperation has taken place at the level of ministries and governmental departments. Indeed, representatives from the Ministry of Justice and the Prosecutors Office were present at the most recent RECOM meeting, where they expressed strong support for regional cooperation concerning transitional justice.[13] While such participation in the RECOM process cannot compensate for the lack of political will from elected representatives, it nevertheless suggests their willingness to support these processes behind the scenes; such cooperation may form a stronger basis for a regional truth commission in the future. However, the reluctance of elected officials to engage in this process does not bode well for the next step of procuring parliamentary endorsement of the regional commission. There is evidence that international actors, particularly from the EU, may exert more pressure on Balkan political leaders to pay attention to RECOM. Meanwhile in the absence of international pressure and popular support, domestic political elites may not see RECOM as a useful tool for gaining and sustaining their own political legitimacy.

---

[12] MPs from the more liberal SPD ostensibly pledged their support earlier this year but were instructed by the party leadership to refrain from making such statements.

[13] International Forum on Transitional Justice in Post-Yugoslav Countries, 27 June 2011, Jill Irvine's notes.

A Role for the International Community?

RECOM's challenges provide significant insight into why it has taken so long for grassroots organizations to mobilize. Such problems are not necessarily limited to the Balkans. In other parts of the world, civil society initiatives have also had a mixed record of success, and efforts have similarly been stymied by difficulties in framing issues, creating coalitions, and engaging the state effectively. Yet, efforts to address the past are still quite different in the Balkans than in most other parts of the world, such as Latin America and Africa, where transitional justice mechanisms have been shaped primarily by and for domestic forces instead of, as in the Balkans, by external actors for the region and different states. Given the unique situation in the Balkans, how have international actors responded to this grassroots effort, and what role should they play in the future? Using literature on transnational advocacy networks, which theorizes on when and how external actors support domestic organizations, the following section examines the international actors' outreach to domestic groups and how they have tried to influence and assist grassroots and normative change (Risse et al. 1999; Keck and Sikkink 1998;).

Transitional justice in the Balkans has been referred to as "justice from the outside in" (Olsen et al. 2010, p. 201) because of the predominate role of external actors. Much of the scholarly literature criticizes the international community for failing—or only belatedly—trying to reach out to citizens in the Balkans to educate them on what the ICTY does and to build domestic support for transitional justice. In 2004, for example, some self-described "moderate Serbs" acknowledged that they mistakenly believed that there had not been any indictments by the ICTY for crimes committed against Serbs, even though by that point 16 non-Serb individuals had been indicted (Humanitarian Law Center 2006, p. 29). Polls conducted by the Belgrade Centre for Human Rights in 2004 and 2005 confirmed how little Serbs actually knew about the ICTY, with some 77% of the Serbs polled indicating that they did not even know what the ICTY does (Humanitarian Law Center 2006, p. 29).

This ignorance is surprising given that in 1999 the ICTY established outreach programs throughout the region to inform the population and to offset such problems. Although no systematic research has been done on the effects of the ICTY's outreach, one Balkan commentator maintained that, "To be sure, few of the messages coming from The Hague have resonated with local audiences in Serbia as intended" (Rangelov 2004, p. 332). In the same year another commentator concluded that the ICTY's outreach program had very little visibility among the Serbian population (Mahieu 2004). More recently, a member of the war crimes division of the Bosnian national court indicated that a similar ICTY program in Bosnia seems to have had the same effect (Member of Bosnian National Court, personal interview, June 23, 2011). As members of the organization Track Impunity Always put it, the problem is that members of the international community and the ICTY have not supported victim's associations from the beginning and, thus, people in the Balkans did not know or trust what the international community was doing (L. Mamut and S. Korjenić, personal interview, May 17, 2011).

The behavior of the ICTY, its apparent disinterest in reaching out to the public, and the failure of other like-minded international groups to fill this void is at odds with international relations theory that explains how international actors act to encourage normative changes and support like-minded domestic groups. According to Keck and Sikkink (1998), transnational advocacy networks emerge when international groups work closely with domestic organizations to advance liberal norms in targeted countries, with the former providing the later with financial support, technical expertise, and in-kind assistance. The so-called "boomerang effect" that results from these contacts directly and indirectly strengthens the domestic organizations and pushes the targeted governments to change their behavior. Theoretically at least, transnational ties strengthen and empower domestic groups, raise awareness, and help push for domestic change.

In practice, however, outcomes in the Balkans suggest that sometimes transnational ties do not form and transnational advocacy networks do not develop. According to Christopher Lamont (2010), the failure or misfiring of the boomerang mechanism is due to the fact that much of civil society in the Balkans is still comprised of organizations that are hostile to transitional justice or the liberal accountability norms of the international community. That is to say, there is little to no demand for international support to seek truth or advance liberal norms of accountability. In discussing the relationship between the international community, the ICTY and civil society in Croatia he explains that, "To the extent that civil society did engage in the war crimes debate, major segments of civil society were intensely hostile to ICTY investigations of human rights abuses committed by members of the Croatian army" (Lamont 2010, p. 45). Thus, for more than a decade this transnational mechanism did not work in the Balkans; instead, civil society and local NGOs only served to "harden non-compliance" preferences, with local actors mobilizing against the ICTY and transitional justice (Lamont 2010, p. 42). Others in the region suggest that these transnational networks did not operate as theorized because international actors came to region with a finished program of what they wanted to accomplish and did not engage in consultations with local people (Member of Bosnian Ministry of Justice, personal interview, May 17, 2011).

This reluctance of local civil society organizations to engage international organizations and actors on behalf of transitional justice appears to have changed with the RECOM initiative. International NGOs have also played a significant role in the RECOM process, reaching out to members of civil society to strengthen and empower local actors. The International Center for Transitional Justice (ICTJ), for example, attended the first meeting of what would later turn into the RECOM Coalition, and it has remained an important advisor throughout the process, consulting with local organizations, drafting the RECOM statute and mobilizing popular support. The Institute for Missing Persons (ICMP) in Bosnia has also played a key role, particularly in helping with the collection and documentation of evidence concerning missing persons, though it has not taken an official position on transitional justice or reconciliation (M. Holliday, personal interview, May 18, 2011)

By far, the RECOM process has received its greatest financial and political support from the EU. Indeed, the European Instrument for Democracy and Human Rights (EIDHR) and the Dutch Embassy account for some 80% of RECOM's current budget (RECOM 2011b). American and European officials alike initially worried that the RECOM process might draw support and resources away from the ICTY, but these concerns appear to have abated when the ICTY officially endorsed RECOM in 2010, followed by EU bodies. The EU's Enlargement Strategy and European Commission reports praise the RECOM Coalition's effort to establish a regional truth commission. Meanwhile, American involvement in this process appears to be minimal, with officials initially expressing concern that a regional truth commission was an unrealistic goal in the current situation. USIP spokesperson, Neil Kritz, seemed to reflect this view when he cautioned that "A regional approach [to the truth commission] has no realistic future" (Gaffney and Alic 2008).

In total, international support for the RECOM process has been somewhat hesitant and mixed, and as of yet, no international actor has yet applied serious pressure on national governments in the Balkans to approve the RECOM statute. Perhaps most interesting is the impact that perceived international support and influence has on domestic actors. Although some see a regional truth commission as essential to improving their country's prospects for joining the EU, others view the RECOM process as "driven from abroad,"[14] thereby robbing it of legitimacy.[15] Such perceptions may prove crucial in either encouraging or discouraging domestic support for grassroots efforts to promote viable, homegrown justice mechanisms.

Lessons Learned

Although the international community has played a central role in transitional justice in the Balkans, it is not clear from the evidence that this involvement has been either efficient or even that effective. It is certainly true that after more than a dozen years of activity, the ICTY has accomplished a good deal in terms of its primary task to prosecute individuals responsible for war crimes. The establishment of RECOM in 2006 offers some reason for optimism that the international community's other broader outcomes related to peace and justice in the region are also underway. To some degree, the development of RECOM signals a change in attitudes, as well as behavior toward transitional justice. What can this case and the development of RECOM tell us about similar efforts in other parts of the world? We believe that there are lessons for both international and domestic actors.

---

[14] Several panelists on the panel, "Tranziciona pravda iz ugla medjunarodnih organizacija tranzicione pravde," expressed this view (International Forum on Transitional Justice in Post-Yugoslav Countries, 27 June 2011, Jill Irvine's notes).

[15] Vojkan Simić, Ministry of Justice Serbia, Speech given at International Forum on Transitional Justice in Post-Yugoslav Countries, 27 June 2011, Jill Irvine's notes.

First, it is difficult, though possible, to create crossborder networks related to transitional justice. However doing so may limit the scope and the effectiveness of a regional truth commission. Attempting to accomplish important tasks, such as determining the causes of the war or constructing a unified narrative may be extremely difficult in the context of a regional body. Second, any grassroots effort involves a trade-off between organizational efficacy and inclusivity; put differently, there will inevitably be winners and losers in the process and it will be impossible to please everyone. As McEvoy and McGregor have pointed out, there has been little attention paid to the way in which groups and individuals are marginalized in grassroots efforts to construct new justice mechanisms (2008). The RECOM process alerts us to the ways in which concerns that some groups of victims are privileged over others can reduce support for the justice process. It also reminds that victims and veteran's groups make uneasy coalition partners, who may not choose to associate with one another.

This leads us to a third, broad conclusion about efforts to promote reconciliation and societal peace. Despite the persistent theoretical association of civil society initiatives with restorative justice, an examination of the RECOM case suggests that the motivation of many of its participants may, in fact, be retributive in nature. Victims groups, in particular, are focused more on identifying and punishing perpetrators than any other issue. While these tasks may be partially accomplished by the fact-finding competencies of a truth commission, punishment will remain largely in the purview of formal justice mechanisms. Thus, the aims of the truth commission and the motivations of its supporters may be at odds. Finally, the RECOM experience cautions against confusing civil society with popular support. Although groups may come together for a variety of reasons in support of truth seeking and truth telling mechanisms, such efforts may continue to be opposed by the vast majority of the populace. Strategies like the One Million Signatures Campaign may create better knowledge and support, but they are no guarantee of success.

For the international community, it is obvious that its decisions, as well as its failure to decide on important issues—including who is deemed the winner in conflicts and who should be perceived as the victim—fundamentally shape the incentives and behavior of domestic organizations in a path-dependent way. Consequently, the effects of internationally led efforts, while positive in many ways, create certain dynamics and perverse incentives that negatively impact both the supply and demand for grassroots organizing. This leads us to reexamine the international community's choice for international criminal justice and to question its general effectiveness, weighted against the alternatives and in light of the international community's limited resources. The financial cost of using international criminal justice is significant and, by necessity, affects the resources and will for any form of domestic justice. Robert Hayden (n.d.) argues that the ICTY has not only been ineffective in promoting liberalization in the Balkans, but that it has also undermined reconstruction efforts, given the West's limited resources.[16] The ICTY

[16] Robert M. Hayden, "Justice Presumed and Assistance Denied: The Yugoslav Tribunal as Obstruction to Economic Recovery," unpublished paper, used by permission.

budget in 1993 was about $275,000; the next year, its financial needs jumped to almost $11 million (United Nations 2003). For 2010–2011 alone, the UN estimated that the ICTY required more than $300 million to conduct its work. ICTY, "The Cost of Justice" http://www.icty.org/sid/325. This amount is quite a deal more than the budgets for other international organizations involved in helping to rebuild Balkan countries. Without trying to put a price tag on legal justice, the expense of the ICTY until 2011 is approximately $2 billion. Given this, as well as the opportunity costs lost in the process of focusing on international justice, a better investment of resources would be to direct them, from the beginning, toward domestic as well as international processes.

Whether the regional truth coalition promoted by RECOM will prove to be a successful local justice initiative remains to be seen. It will be difficult indeed for RECOM to succeed in the absence of a consensus on the commission's mandate, genuine support from political leaders, and a strong coalition with effective leadership. Nevertheless, sustained international support for domestic processes will undoubtedly increase the chances of success of this homegrown mechanism of transitional justice.

References

Akhavan, P. 1998. Justice in the Hague, peace in the former Yugoslavia? A commentary on the United Nations War Crimes Tribunal. *Human Rights Quarterly* 20(4): 737–816.

Amnesty International. 2010. *Behind a wall of silence: Prosecution of war crimes in Croatia*. London: Amnesty International.

Backer, D. 2003. Civil society and transitional justice: Possibilities, patterns and prospects. *Journal of Human Rights* 2: 297–313.

Bickford, L. 2000. Human rights archives and research on historical memory: Argentina, Chile, and Uruguay. *Latin American Research Review* 25(2): 160–182.

Bohlander, M. 2003. Last exit Bosnia: Transferring war crimes prosecution from the international tribunal to domestic courts. *Criminal Law Forum* 14: 59–99.

Brahm, E. 2007. *Truth and consequences: The impact of truth commissions in transitional societies*. Ann Arbor: University of Michigan Press.

Dickinson, L.A. 2003. The relationship between hybrid courts and international courts: The case of Kosovo. *New England Law Review* 37(4): 1059–1072.

Donlon, F. 2008. Rule of law: From the International Criminal Tribunal for the Former Yugoslavia to the War Crimes Chamber of Bosnia and Herzegovina. In *Deconstructing the reconstruction: Human rights and the rule of law in postwar Bosnia and Herzegovina*, ed. D. Haynes, 257–284. Burlington, VT: Ashgate.

Dragovic-Soso, J., and E. Gordy. 2010. Coming to terms with the past: Transitional justice and reconciliation in post-Yugoslav lands. In *New perspectives on Yugoslavia: Key issues and controversies*, ed. D. Djokic and J. Ker-Lindsay, 193–212. New York, NY: Routledge.

Drumbl, Mark. 2007. *Atrocity, punishment, and international law*. New York, NY: Cambridge University Press.

Fletcher, L., H. Weinstein, and J. Rowen. 2009. Context, timings and the dynamics of transitional justice: A historical perspective. *Human Rights Quarterly* 31(1): 163–220.

Forsythe, D. 2011. *The politics of prisoner abuse: The United States and enemy prisoners after 9/11*. New York, NY: Cambridge University Press.

Gaffney, C., and Alic, A. 2008. First regional truth commission runs into doubts. *Balkan Insight*, 5 August.
Grodsky, B. 2007. Looking for Solidarność in central Asia: The role of human rights organizations in political change. *Slavic Review* 66(3): 442–462.
Grodsky, B. 2009. International prosecutions and domestic politics: The use of truth commissions as compromise justice in Serbia and Croatia. *International Studies Review* 11(4): 687–706.
Hagan, J. 2003. *Justice in the Balkans: Prosecuting war crimes in the Hague Tribunal*. Chicago, IL: University of Chicago Press.
Hampson, F. 1998. The International Criminal Tribunal for the Former Yugoslavia and the reluctant witness. *International and Comparative Law Quarterly* 47: 50–74.
Hayden, R.M. n.d. Justice presumed and assistance denied: The Yugoslav Tribunal as obstruction to economic recovery. Unpublished Paper.
Hayner, P.B. 2002. *Unspeakable truths: Confronting state terror and atrocity*. New York, NY: Routledge.
Heil, A.L., Jr. 2000. A truth and reconciliation commission for Bosnia and Herzegovina? *Washington Report on Middle East Affairs*, p. 26, June.
Humanitarian Law Center. 2006. *Transitional justice report Serbia, Montenegro and Kosovo, 1999–2005*. Belgrade: Humanitarian Law Center.
International Institute for Democracy and Electoral Assistance. 2004. South East Europe Public Agenda Survey. http://www.idea.int/europe_cis/balkans/see_survey.cfm. Accessed 4 April 2002.
International Republican Institute. 2006. Internal survey.
Keck, M., and K. Sikkink. 1998. *Activists beyond borders: Advocacy networks in international politics*. Ithaca, NY: Cornell University Press.
Kerr, R. 2004. *The International Criminal Tribunal for the Former Yugoslavia: An exercise in law, politics, and diplomacy*. New York, NY: Oxford University Press.
Kurze, A. 2010. Learning how to throw the boomerang: Justice and human rights activism after mass atrocities in the Balkans. Paper presented at the 2010 Annual Convention of the International Study Association, New Orleans, LA, February 20.
Lamont, C. 2010. *International criminal justice and the politics of compliance*. Burlington, VT: Ashgate.
Mahieu, S. 2004. Commanders vs. perpetrators? Post-war justice in the former Yugoslavia—The Ovcara Trials before the ICTY and domestic courts. Paper presented at Europa-University Viadrina, Frankfurt, Germany, August 7.
Meyer, David S., and Debra C. Minkoff. 2004. Conceptualizing Political Opportunity. *Social Forces* 82(4): 1457–92.
McEvoy, K., and L. McGregor. 2008. *Transitional justice from below: Grassroots activism and the struggle for change*. Portland, OR: Hart Publishing.
McCarthy, J.D., and M.N. Zald. 1987. Resource mobilization and social movements: A partial theory. In: Social Movements: Perspectives and Issues. Ed. S.M. Buechler and F.K. Cylke. 1997. Mayfield Publishing Company, Mountain View, California. pp 149–172.
McMahon, P.C., and D. Forsythe. 2008. The ICTY's impact on Serbia: Judicial romanticism meets network politics. *Human Rights Quarterly* 30(2): 412–435.
McMahon, P.C., and Western, J. 2011. The Balkans after Mladic. *Foreign Affairs*. http://www.foreignaffairs.com/articles/67962/patrice-c-mcmahon-and-jon-western/the-balkans-after-mladic. Accessed 23 June 2011.
Meernik, J. 2005. Justice and peace? How the International Criminal Tribunal affects societal peace in Bosnia. *Journal of Peace Research* 42(3): 271–289.
Nettelfield, L. 2010. *Courting democracy in Bosnia and Herzegovina: The Hague Tribunal's impact in a postwar state*. Cambridge: Cambridge University Press.
Olsen, T.D., L.A. Payne, and A.G. Reiter. 2010. *Transitional justice in balance: Comparing processes, weighing efficacy*. Washington, DC: United States Institute of Peace.
Orentlicher, D. 2008. *Shrinking the space of denial: The impact of the ICTY in Serbia*. New York, NY: Open Society Institute.

Paust, J., M.C. Bassiouni, M. Scharf, J. Gurule, L. Sadat, B. Zagaris, and S.A. Williams. 2006. *Human rights module: On crimes against humanity, genocide, other crimes against human rights, and war crimes*. Durham, NC: Carolina Academic Press.
Peskin, V. 2008. *International Justice in Rwanda and the Balkans* (Cambridge 2008).
Peskin, V. 2009. *International justice in Rwanda and the Balkans*. Ithaca, NY: Cornell University Press.
Peskin, V., and M. Boduszynski. 2003. International justice and domestic politics: Post-Tudjman Croatia and the International Criminal Tribunal for the Former Yugoslavia. *Europe-Asia Studies* 55: 1117–1142.
Rangelov, I. 2004. International law and local ideology in Serbia. *Peace Review* 15(3): 331–337.
RECOM. 2009. *Report about the consultative process on the instruments of truth-seeking about war crimes and other serious violations of human rights in post-Yugoslav countries, May 2006-June 2009*. Belgrade: Humanitarian Law Center.
RECOM. 2011a. *The consultation process on the establishment of the facts about war crimes and other gross violations of human rights committed on the territory of the former Yugoslavia*. Belgrade: Humanitarian Law Center.
RECOM. 2011b. Process RECOM financial report for period December 15, 2008-August 31, 2011. http://zarekom.org/uploads/documents/2011/07/i_1596/f_1/f_2754_0.xls.
RECOM. 2011c. RECOM development process—Report, May 2006–August 2011 http://zarekom.org/uploads/documents/2011/08/i_1631/f_1/f_2746_en.doc.
Risse, T., S. Ropp, and K. Sikkink. 1999. *The power of human rights: International norms and domestic change*. Cambridge: Cambridge University Press.
Snyder, J., and L. Vinjamuri. 2003/2004. Trials and errors: Principle and pragmatism in strategies of international justice. *International Security* 28(3): 5–44.
Stover, E., and H. Weinstein. 2004. *My neighbor, my enemy: Justice and community in the aftermath of mass atrocity*. New York, NY: Cambridge University Press.
Subotic, J. 2009. *Hijacked justice: Dealing with the past in the Balkans*. Ithaca, NY: Cornell University Press.
Tarrow, S. 2005. *The new transnational activism*. New York, NY: Cambridge University Press.
Treves, T., M.F. di Rattalma, A. Tanzi, A. Fodella, C. Pitea, and C. Ragni. 2005. *Civil society, international courts and compliance bodies*. New York, NY: Cambridge University Press.
United Nations. 2003. Fifth committee takes up 2004–2005 budgets for Rwanda, former Yugoslavia tribunals (Press Release). http://www.un.org/News/Press/docs/2003/gaab3594.doc.htm.
van der Merwe, H., V. Baxter, and A.R. Chapman. 2009. *Assessing the impact of transitional justice: Challenges for empirical research*. Washington, DC: United States Institute of Peace.

Index

O. Simić and Z. Volčič (eds.), *Transitional Justice and Civil Society in the Balkans*, Springer Series in Transitional Justice, DOI 10.1007/978-1-4614-5422-9,

MIX
Papier aus verantwortungsvollen Quellen
Paper from responsible sources
FSC® C105338

If you have any concerns about our products,
you can contact us on
ProductSafety@springernature.com

In case Publisher is established outside the EU,
the EU authorized representative is:
Springer Nature Customer Service Center GmbH
Europaplatz 3, 69115 Heidelberg, Germany

Printed by Libri Plureos GmbH
in Hamburg, Germany